terra australis 24

Terra Australis reports the results of archaeological and related research within the south and east of Asia, though mainly Australia, New Guinea and island Melanesia — lands that remained *terra australis incognita* to generations of prehistorians. Its subject is the settlement of the diverse environments in this isolated quarter of the globe by peoples who have maintained their discrete and traditional ways of life into the recent recorded or remembered past and at times into the observable present.

Since the beginning of the series, the basic colour on the spine and cover has distinguished the regional distribution of topics as follows: ochre for Australia, green for New Guinea, red for South-East Asia and blue for the Pacific Islands. From 2001, issues with a gold spine will include conference proceedings, edited papers and monographs which in topic or desired format do not fit easily within the original arrangements. All volumes are numbered within the same series.

List of volumes in *Terra Australis*

Volume 1: Burrill Lake and Currarong: Coastal Sites in Southern New South Wales. R.J. Lampert (1971)

Volume 2: Ol Tumbuna: Archaeological Excavations in the Eastern Central Highlands, Papua New Guinea. J.P. White (1972)

Volume 3: New Guinea Stone Age Trade: The Geography and Ecology of Traffic in the Interior. I. Hughes (1977)

Volume 4: Recent Prehistory in Southeast Papua. B. Egloff (1979)

Volume 5: The Great Kartan Mystery. R. Lampert (1981)

Volume 6: Early Man in North Queensland: Art and Archaeology in the Laura Area. A. Rosenfeld, D. Horton and J. Winter (1981)

Volume 7: The Alligator Rivers: Prehistory and Ecology in Western Arnhem Land. C. Schrire (1982)

Volume 8: Hunter Hill, Hunter Island: Archaeological Investigations of a Prehistoric Tasmanian Site. S. Bowdler (1984)

Volume 9: Coastal South-West Tasmania: The Prehistory of Louisa Bay and Maatsuyker Island. R. Vanderwal and D. Horton (1984)

Volume 10: The Emergence of Mailu. G. Irwin (1985)

Volume 11: Archaeology in Eastern Timor, 1966–67. I. Glover (1986)

Volume 12: Early Tongan Prehistory: The Lapita Period on Tongatapu and its Relationships. J. Poulsen (1987)

Volume 13: Coobool Creek. P. Brown (1989)

Volume 14: 30,000 Years of Aboriginal Occupation: Kimberley, North-West Australia. S. O'Connor (1999)

Volume 15: Lapita Interaction. G. Summerhayes (2000)

Volume 16: The Prehistory of Buka: A Stepping Stone Island in the Northern Solomons. S. Wickler (2001)

Volume 17: The Archaeology of Lapita Dispersal in Oceania. G.R. Clark, A.J. Anderson and T. Vunidilo (2001)

Volume 18: An Archaeology of West Polynesian Prehistory. A. Smith (2002)

Volume 19: Phytolith and Starch Research in the Australian-Pacific-Asian Regions: The State of the Art. D. Hart and L. Wallis (2003)

Volume 20: The Sea People: Late-Holocene Maritime Specialisation in the Whitsunday Islands, Central Queensland. B. Barker (2004)

Volume 21: What's Changing: Population Size or Land-Use Patterns? The Archaeology of Upper Mangrove Creek, Sydney Basin. V. Attenbrow (2004)

Volume 22: The Archaeology of the Aru Islands, Eastern Indonesia. S. O'Connor, M. Spriggs and P. Veth (2005)

Volume 23: Pieces of the Vanuatu Puzzle: Archaeology of the North, South and Centre. S. Bedford (2006)

Volume 24: Coastal Themes: An Archaeology of the Southern Curtis Coast, Quuensland. S. Ulm (2006)

terra australis 24

Coastal Themes:
An Archaeology of the
Southern Curtis Coast, Queensland

Sean Ulm

ANU
THE AUSTRALIAN NATIONAL UNIVERSITY

E PRESS

ANU
E PRESS

© 2006 ANU E Press

Published by ANU E Press
The Australian National University
Canberra ACT 0200 Australia
Email: anuepress@anu.edu.au
Web: http://epress.anu.edu.au

National Library of Australia Cataloguing-in-Publication entry

Ulm, Sean.

Coastal themes : an archaeology of the Southern Curtis Coast, Queensland.

Bibliography.
ISBN 1 920942 93 9 (pbk).
ISBN 1 920942 96 3 (web).

1. Archaeology — Queensland — Port Curtis Region. 2. Aboriginal Australians — Antiquities. I. Title. (Series : Terra Australis ; 24).

994.01

Series Editor: Sue O'Connor

Typesetting and design: Emily Brissenden

Cover: Bustard Bay from Bustard Head (Photograph: S. Ulm).
Back cover map: *Hollandia Nova*. Thevenot 1663 by courtesy of the National Library of Australia.
Reprinted with permission of the National Library of Australia.

Foreword

THE RESEARCH DOCUMENTED here represents the first systematic archaeological work in this area of the southeast Queensland coast and was undertaken as a major part of a larger, multi-component project concerning archaeology and cultural heritage in the traditional country of Gooreng Gooreng speaking people. Sean's task was to build upon the results of exploratory site survey and excavation to address two key concerns. The first was the relationship of patterns of cultural change in his study area to those described elsewhere in southeast Queensland. The second was to ensure that any such comparisons were taphonomically well-founded, particularly with regard to the analytical integrity of the shell middens upon which he and other coastal researchers in Australia rely so heavily.

Sean took to this task with a vengeance, closely surveying a large area of landscape and excavating an array of site types to provide himself with a solid sample of the archaeological variation thus revealed. Though most were not archaeologically rich, these sites provided substantial grist for Sean's taphonomic mill, prompting him to adapt conjoining techniques to work on bivalve shellfish — a simple but clever innovation — as well as to undertake much more sophisticated work on local variation in correction factors for the radiocarbon dating of marine shell. In the end, he was able to distil the three-phase cultural sequence he describes in this volume, 'hygienically' dated on the basis of reliable correction coefficients and demonstrably high degrees of stratigraphic integrity in his middens. The patterns he identified accord with those from other parts of coastal southeast Queensland, suggesting widespread major restructuring of coastal occupation strategies in the archaeologically very-recent past, and particularly the last 1,000 years. Sean was also able to demonstrate that Aboriginal people were still using the area, and often the same sites that had been used for substantial periods in pre-European times, well after they disappeared from the documentary historical record, thus emphasising the importance of archaeology as an independent record of Aboriginal life in the historical period.

Ian Lilley

Acknowledgements

MANY, MANY PEOPLE contributed to the completion of the research reported in this monograph. My greatest debt in producing this work is to my University of Queensland supervisors Jay Hall (School of Social Science) and Ian Lilley (Aboriginal and Torres Strait Islander Studies Unit) and my de facto supervisor Michael Williams (Aboriginal and Torres Strait Islander Studies Unit) who have been with me every step of the journey.

I thank members of the Gooreng Gooreng community who collaborated on this project and gave me the opportunity to work in their country. In particular I thank Colin Johnson, Hilton (Charlie) Johnson, Mervyn Johnson, Ron Johnson (Sr), Ron Johnson (Jr), Vicki Johnson (and the rest of the Johnson Family), Cedric Williams, James Williams and Michael Williams (and the rest of the Williams Family), Connie Walker and Michael Hill.

I thank the Gurang Land Council Aboriginal Corporation and the Queensland Environmental Protection Agency for their continuing assistance and participation in the project.

The advice and assistance of the Agnes Water-Town of Seventeen Seventy community has been critical to the success of this project. In particular, I would like to acknowledge the support of Denis Dray, Neil Teague and Kris Hall (Queensland Parks and Wildlife Service), Georgia and Tony Isaacs, Jan and David McKauge (The Beach Shacks), Neil, Des, Betty and Katherine Mergard and Glen Finlay (1770 Environmental Tours).

I thank the many specialists who freely (well, mostly) gave up their time to answer many questions and undertake collaborative research on various aspects of the project: Errol Beutel (Queensland Museum) identified bottle glass; Stuart and Shirley Buchanan (Coral Coast Publications) discussed historical information; Maria Cotter (University of New England) provided palaeoenvironmental data; Stephen Cotter (University of Canberra) identified stone raw materials and undertook LA-ICP-MS analyses; Tracy Frank (University of Queensland) provided advice on stable isotopes and obtained stable O and C data; Luke Godwin (Central Queensland Cultural Heritage Management) provided access to reports; Tom Higham (Oxford Radiocarbon Accelerator Unit, University of Oxford) advised on marine reservoir issues; Vojtech Hlinka (University of Queensland) identified fish remains; Ian Loch (Australian Museum) provided live-collected shell samples; Alan Hogg, Fiona Petchey and Matthew Phelan (University of Waikato Radiocarbon Dating Laboratory) undertook the majority of dating and provided advice on sample selection and marine reservoir issues; Paula Reimer (Center for Accelerator Mass Spectrometry, Lawrence Livermore National Laboratory) generously provided advice and datasets crucial to the completion of the marine reservoir study; John Stanisic, Darryl Potter and Thora Whitehead (Queensland Museum) helped identify shellfish and provided live-collected shell specimens; Errol Stock (Griffith University) helped with geoarchaeological sampling and provided advice on geomorphology; John Turunen (Keilar Fox and McGhie Pty Ltd) undertook mapping and digital terrain modelling; Deborah Vale (University of New England) examined and identified fish remains; Steve Van Dyck (Queensland Museum) undertook preliminary identification of marine mammal and marine reptile remains.

Fieldwork and radiocarbon dating was funded by: the Australian Heritage Commission National Estate Grants Program (Chief Investigator: Ian Lilley); Australian Institute of Aboriginal and Torres Strait Islander Studies (G97/6067 and G98/6113) (Chief Investigator: Sean Ulm); Australian Research Council Large (A10027107) (Chief Investigator: Ian Lilley) and Small Grant (00/ARCSO15) (Chief Investigator: Ian Lilley); Australian Institute of Nuclear Science and Engineering (AINSE) (Grant #98/048) (Chief Investigator: Bill Boyd); School of Social Science, University of Queensland (Chief Investigator: Sean Ulm); and, Dr Joan Allsop Australian Studies Fund Award (Chief Investigator: Sean Ulm). The Aboriginal and Torres Strait Islander Studies Unit and School of Social Science at the University of Queensland provided laboratory and storage facilities.

Thanks to Paul Aurisch (Figs 9.9–9.10, 9.12), Ian Lilley (Figs 2.12, 7.3) and the Queensland Environmental Protection Agency (Fig. 2.13) for permission to reproduce photographs. All unacknowledged photographs are my own work. Thanks to Jean Andrews for permission to reproduce Figure 3.3 from *Sea Shells of the Texas Coast* (1971). The Queensland Department of Natural Resources, Mines and Water gave permission to reproduce the aerial photographs in Figures 9.2 and 13.1.

Many people have also assisted in the field and laboratory. Thanks to Paul Aurisch, Desley Badrick, Estelle Baker, Damien (Brutus) Bauer, Delyna Baxter, Joann Bowman, Kylie Bruce, Melissa Carter, Val Chapman, Araluen Cotter, Maria Cotter, Stephen Cotter, Tony Eales, Kyleigh Engeman, Joe Firinu, Victoria Francis, Renee Gardiner, Greg Gilles, Jay Hall, Colleen Hanahan, Vojtech Hlinka, Angela Holden, Alex Hunt, Charlie (Hilton) Johnson, Amanda Kearney, Jenna Lamb, Michelle Langley, Daniel Leo, Sarah L'Estrange, Ian Lilley, Mark Limb, Patricia Livsey, Geraldine Mate, Emma Miles, Lucile Myers, Carly Naughton, Lad Nejman, Alison Neuendorf, Stephen Nichols, Antje Noll, Sue Nugent, Cyndi Osborne, Bev Rankine, Coen Rankine, Ian Rankine, Chris Reid, David Reid, Pene Reid, Ralph Reid, John Richter, Evelyn Riddle, Richard Robins, Yolanda Saurez, Stephen Skelton, Linda Tebble, Joy Thompson, Sophie Thompson, Thomas Twigg, Phyllis Ulm, Deborah Vale, Helen Vale, Kim Vernon, Kristen Ward, Bill Wedge, Shaun Wiggins, Cedric Williams, Dany Williams, Michael Williams, Alex Wisniowiecka, Amy Wood and Paul Wood. In particular I would like to single out Chris Chicoteau, Jill Reid, Karen McFadden, Gail Robertson, Catherine Westcott and Nathan Woolford for being regular (and sometimes long-term) field participants. Tony Eales, Victoria Francis, Kim Vernon, Gail Robertson and Jenna Lamb examined various artefacts for use-wear and residues.

For discussing aspects of the research I thank Tony Barham, Bryce Barker, Bill Boyd, Maria Cotter, Stephen Cotter, Daniel Cummins, Bruno David, Iain Davidson, Tony Eales, Luke Godwin, Alice Gorman, Jay Hall, Michael Haslam, Fiona Hook, Jenna Lamb, Sarah L'Estrange, Ian Lilley, Harry Lourandos, Tom Loy, Ian McNiven, Rob Neal, Jon Prangnell, Paul Rainbird, Jill Reid, Richard Robins, Annie Ross, Mike Rowland, Errol Stock, Tam Smith, Bruce Veitch, Peter Veth, Louis Warren, Catherine Westcott and Nathan Woolford.

For assistance in preparing this manuscript for publication I thank Sue O'Connor, Marjorie Sullivan, David Frankel, Bill Boyd, Ian Lilley, Catherine Westcott, Jill Reid, Antje Noll and the staff at Pandanus Books, especially Emily Brissenden.

My extended family has once again kept me relatively sane during the completion of this research. Thanks to the Rankine Family, Reid Family, Ulm Family, Williams Family and my Unit Family. They have shown me that there is a world outside my office walls, although at times I found it hard to believe. Finally, a special thanks to Jill Reid for endless support and too many years without weekends.

Contents

List of Figures

List of Tables

calculated by deducting the mid-point of the equivalent marine model age of the charcoal determination from the ^{14}C age of the paired marine shell sample. $\Delta R\sigma = \sqrt{(\sigma\text{marine model age}^2 + \sigma\text{marine shell age}^2)}$ (Gillespie 1982). This method is illustrated for pair NZA-12117/Wk-8326 in Fig. 4.3.

1

Introduction: investigating the archaeology of the southern Curtis Coast

Introduction

Southeast Queensland is one of the most intensively studied archaeological provinces in Australia, incorporating some 73 dated Aboriginal sites (Ulm and Reid 2000, 2004). Over the last 15 years, general syntheses of regional archaeological patterns have emerged which emphasise significant increases in site numbers and use since the mid-Holocene, especially on the coast (e.g. Hall 2000; Hall and Hiscock 1988; McNiven 1999; Morwood 1987; Walters 1989; Ulm and Hall 1996). Many of these studies emphasise the primary role of marine resources in the elaboration of social complexity in the region.

The research presented in this monograph assembles a regional archaeology for the southern Curtis Coast, at the northern extremity of southeast Queensland, based on theoretical and methodological issues emerging out of studies in southeast Queensland and throughout Australia. The structure of late Holocene Aboriginal settlement and subsistence systems is investigated through documentation of the distribution, composition and chronology of sites in the region. These findings constitute the first new large-scale regional archaeological investigation conducted in southeast Queensland since McNiven's (1984, 1985, 1988, 1989, 1990a, 1990b) study of the Cooloola region in the 1980s. It provides detailed understanding of the archaeological resources of the area to explain patterns of regional interaction. It thus helps to integrate the archaeology of the study area with that of the wider southeast Queensland region and other adjacent regions and ultimately contributes to a broad understanding of continent-wide patterns.

The continental narrative

> Taken together ... [the] evidence does indeed seem to suggest that "something happened" in Australia about 2000 BC. But when each piece of evidence is examined individually, the apparent pattern becomes blurred (White 1994:225).

Ethnohistoric accounts and oral histories represent recent Aboriginal societies as a complex mosaic of localised land-using groups, actualised in the landscape by prescribed physical and social boundaries, mediated by complex patterns of intergroup alliance, cosmology, marriage and exchange. It is estimated that this pattern emerged in the mid-to-late Holocene and is associated with fundamental structural changes in the archaeological record, including increases in the rates of site establishment and use, evidenced by increases in discard of cultural materials, particularly stone artefacts; changes in stone artefact technologies, rock art styles and fishing technologies; the increased use of some marginal landscapes, such as offshore islands and the arid zone; changes in resource use, including the intensive utilisation of new foods such as cycads, cereals and some marine resources; evidence for long-distance exchange networks; an increase in the establishment of bounded cemeteries; and increased external contact, evidenced by the introduction of the dingo, fishhooks and some forms of watercraft (e.g. Beaton 1982; Bowdler 1981; David 2002; Flood 1980, 1999; Flood et al. 1987; Godwin 1997; Hiscock 1994; Lourandos 1980a, 1983, 1985, 1988, 1993, 1997; Lourandos and Ross 1994; Mulvaney 1969; Smith 1986; White and O'Connell 1982).

These changes have been variously explained by one or a combination of five main arguments:

— as an artefact of site preservation factors, including differential destruction and visibility (e.g. Bird 1992; Fanning and Holdaway 2001; Godfrey 1989; Head 1983; O'Connor and Sullivan 1994a, 1994b; Rowland 1983, 1989);

— as a product of environmental factors, particularly resource productivity and availability (e.g. Bailey 1983; Beaton 1985; Morwood 1987; Rowland 1983, 1989, 1999; Walters 1989);

— as a consequence of population growth and changes in demographic structure (e.g. Beaton 1985, 1990; Hall and Hiscock 1988; Hughes and Lampert 1982; Lampert and Hughes 1974);

— as related to the introduction and/or development of new technologies (e.g. Beaton 1985; Sullivan 1987; Vanderwal 1978); and

— as associated with changes in social structure, especially trends towards socio-economic intensification (e.g. Barker 1996, 2004; David 2002; Lourandos 1997).

To put these arguments into perspective, it must be remembered that an Australian 'deep past' has only been widely accepted since the 1960s, on the basis of pioneering excavations conducted by Mulvaney in the Central Queensland Highlands (Mulvaney and Joyce 1965). In fact, much of the enduring framework of Australian archaeology was mapped out by the early 1970s (e.g. Mulvaney 1969; Mulvaney and Golson 1971). A significant archaeological database accumulated during the 1970s, but it was the early 1980s that signalled a major turning point in the development of Australian archaeology. Heated debate during this period, labelled by Lourandos and Ross (1994) as the 'Intensification Debate', focussed critical attention on issues of change, particularly during the late Holocene. Although major recent change was recognised from the earliest excavations undertaken in Australia (e.g. Hale and Tindale 1930), serious consideration of the late Holocene archaeological record is often treated as synonymous with the work of Lourandos (e.g. 1980a, 1980b, 1983, 1985, 1988, 1997). Issues of change in Aboriginal societies were discussed by others (e.g. Hallam 1977; Hughes and Lampert 1982; Jones 1977; Lampert and Hughes 1974; Mulvaney 1969; White 1971), but it was Lourandos' major works that had the most significant impact on the acknowledgement of dynamism and were seminal in stimulating new research directions. Employing a socially-oriented approach, Lourandos pointed to a wide range of changes in the archaeological record of the last 5,000 years and linked these to continent-wide processes of socio-economic intensification which resulted in trajectories towards more intensive production and productivity in Aboriginal societies (see also Barker 2004; David 2002; Williams 1988).

Lourandos' contribution was valuable in exploring alternative explanatory paradigms and helping to focus attention on the mid-to-late Holocene, and away from a preoccupation with initial colonisation and early sites. Since the early 1980s, socio-economic intensification arguments, both explicit and implicit, have increasingly been incorporated into normative accounts of Australian archaeology (Lourandos and Ross 1994). However, by emphasising supraregional trajectories of change, some advocates of intensification devalued the importance of local and subregional trajectories as a primary locus of change, instead amalgamating diverse sequences from widely separated regions to define overarching patterns. Lourandos' (especially 1997) schema in effect homogenised distinct local and regional trajectories into generic patterns which fail to adequately contextualise sequences within local frameworks. Instead, trajectories are modelled on gross characteristics of the archaeological record, which tend to homogenise significant regional variability, emphasising widespread cultural continuities within changing structures. Lourandos clearly sees 'distinct regional and local signatures' as firmly embedded in 'patterns on a general continental scale' (1997:306). The problem is that such higher-order archaeological syntheses are necessarily selective and emphasise similarities rather than differences, resulting in 'the distortion of an archaeological record of variation in adaptive responses into a record of homogeneous response' (Claassen 1991:249). As Frankel (1995:654) has argued, individual 'sites and regions each have a particular, and not necessarily related, history of environmentally and historically contingent developments'. Regional cultural trajectories need to be disarticulated from the 'continental narrative' (Frankel 1993:31) to enable independent characterisation of local behavioural variability. In effect, the limits of archaeological variability, or at least their interpretation, have been predetermined by expectations deriving from the continental narrative. A corollary of this is that regional sample inadequacy has often been transcended by invoking the continental narrative: the region is not seen as separate from the whole and so can be explained in terms of the whole rather than the part.

At the heart of this problem is our inability to recognise and account for the diversity of cultural and historical trajectories evident in the archaeological record. For example, in synthesising evidence for distinctly different Holocene trajectories for nine major regions across northern Australia, Lourandos (1997:166) concluded that:

> Evidence for more significant socio-cultural and demographic changes appears during the late Holocene, from around 4,000 or so years ago, increasing after about 3,000 years, and particularly in the last 1,500 years or so.

It is difficult to reconcile this abstract statement with the detailed information presented for each of the regions in previous pages (Lourandos 1997:126–65). For example, the site of Mickey Springs 34 in the upper Flinders River region, demonstrates increases in the rate of stone artefact deposition after 8,000 BP and most intensive occupation after 3,400 BP (Lourandos 1997:132). On the Keppel Islands off the central Queensland coast, seven of the eight excavated sites were only occupied after 1,200 BP, with greatest intensity of occupation after 700 BP (Lourandos 1997:138–9). For the Alligator Rivers region of Arnhem Land, Lourandos goes to some lengths to argue that 'fairly steady socio-cultural and demographic changes (including possible increases in population) took place from around 6,000 BP and continued more or less up to the historical period' (1997:152). In the Central Queensland Highlands Lourandos (1997:169) concedes 'a notable alteration in pattern' where 'late Holocene cultural changes are followed by apparent reversals in site use after about 2,000 years ago'. In such generalised chronologies asynchroneity of several millennia in the timing of regional changes become insignificant. By (re)moving to this very abstract synthetic level, unique regional patterns and trajectories are made insignificant by the continental narrative.

Recent detailed studies across Australia have emphasised the variability of behavioural responses in terms of magnitude, chronology and nature at the regional level, for example, in southeast Cape York Peninsula (David and Chant 1995; Morwood and Hobbs 1995), Torres Strait (Carter 2002; McNiven 2003), the islands off the central Queensland coast (Barker 2004; Rowland 1996), the arid zone (Smith 1993; Veth 1993), the Kimberley region (O'Connor 1999; Veitch 1996), southwest Tasmania (Allen 1996; Cosgrove 1995) and southeast Queensland (McNiven 1999; Ulm and Hall 1996) (Fig. 1.1). New understandings of regional patterns have also emerged through the critical synthesis and evaluation of existing archaeological datasets, particularly gross patterns of radiocarbon dates and rates of site establishment and use (e.g. Bird and Frankel 1991a, 1991b; David and Lourandos 1997; Holdaway and Porch 1995). These studies demonstrate a more complex view of the human past than had been allowed by the conventional continental narrative — different sites, places, landscapes and regions were found to have been used differentially through time.

Fundamental elements of our understanding of the mid-to-late Holocene have been challenged in recent years as well. The eel trapping systems and swamp management facilities in southwest Victoria, central to Lourandos' original arguments (1980b, 1983) may date to the early Holocene rather than the late Holocene, as was originally thought (Heather Builth, Monash University, pers. comm., 2003). Backed stone implements, once thought to be exclusively a post-4,500 BP innovation across the southern two-thirds of the continent (e.g. Bowdler and O'Connor 1991; White and O'Connell 1982) have been identified in early Holocene deposits in the upper Mangrove Creek region (Hiscock and Attenbrow 1998). Grindstones implicated in widespread (re)occupation of the arid zone during the late Holocene have been recovered from deposits dating to the late Pleistocene (Fullagar and Field 1997; cf. David 2002; Smith 1986). Complex food processing techniques for toxic plants, once exclusively associated with the late Holocene (Beaton 1982), have been discovered in terminal Pleistocene deposits in southeast Western Australia (Smith 1996). In central Australia the Pleistocene occupation of desert areas has challenged the status of these environments as marginal landscapes, only occupied during the late Holocene (Smith 1993; Veth 1993). Central Australian engravings, once thought to relate to early low intensity open alliance networks, now appear to overlap, both stylistically and chronologically, with highly regionalised recent painted art, challenging the conventional temporal distinction of these types and their associated behaviours (Ross 2002; see also David 2002:181–5). Simple linear increases posited for marine fish deposition rates in southeast Queensland (Walters 1989) are now considered to be unsustainable (Ulm 2002a). Patterns of decreasing stone artefact discard at some sites after c.1,000 BP have also been identified in the upper Mangrove Creek region (Attenbrow 1987, 2003, 2004) and in the Hunter Valley to the north of Mangrove Creek (Hiscock 1986a). The recognition of such regional diversity shifts our focus from simply attempting to place these changes in the mid-to-late Holocene, to establishing an understanding of them.

Regional findings which appear at odds with the continental narrative have prompted archaeologists to (re)turn to detailed local level chronology-building and to examine basic assumptions underlying abstract synthetic studies at the continental level (e.g. Bird and Frankel 1991a; Fanning and Holdaway 2001; Hiscock and Attenbrow 1998; Holdaway et al. 1998; Holdaway et al. 2002; Ulm and Hall 1996). This shift to a conceptual emphasis on local and regional historical archaeological and environmental histories has not, however, been matched by an equivalent shift to the finer-grained methodologies needed to provide resolution of these issues. Three fundamental concerns are particularly important in this regard: chronological control, sampling and taphonomy. Control of time is, of course, critical to the validity of our constructions of the past and, at a very basic level, accurate site chronologies are required to provide the framework within which the study of culture change can begin. A second set of concerns can be subsumed under the heading of sampling. Despite the routine investigation of a wide range of site types in Australia (e.g. shell middens, stone artefact scatters, rock art sites etc), understanding is largely based on narratives developed from a handful of deeply stratified rockshelter deposits. The third issue of taphonomy revolves around processes of differential preservation and post-depositional modification of archaeological materials, particularly with those processes impacting site integrity. Methodological advances to address all three issues are critical to the continuing development of our understanding of Holocene Australia.

Refining mid-to-late Holocene chronologies

Establishing secure regional chronologies remains a fundamental key to building meaningful accounts of intra- and inter-regional sequences in Australia. In the absence of unambiguous time-marking artefacts (e.g. metals, ceramics etc), radiocarbon dates are almost universally employed in Australia as the basis for periodising site components and defining regional chronologies. Previously employed relative chronologies based on assumed changes in stone artefact technologies (e.g. Ross 1981) have been shown to be erroneous or dubious at best (Hiscock and Attenbrow 1998). Large sequences of radiocarbon dates are now routinely used to compare cultural chronologies at the local, regional and continental scales (e.g. Bird and Frankel 1991a; Holdaway and Porch 1995; Lourandos and David 1998; Ulm and Hall 1996). Assessment of the validity of individual dates and suites of dates has therefore become increasingly important as cultural chronologies are progressively refined and higher resolution understandings sought for particular archaeological questions. However, most studies have assumed, rather than demonstrated, direct comparability between individual radiocarbon determinations.

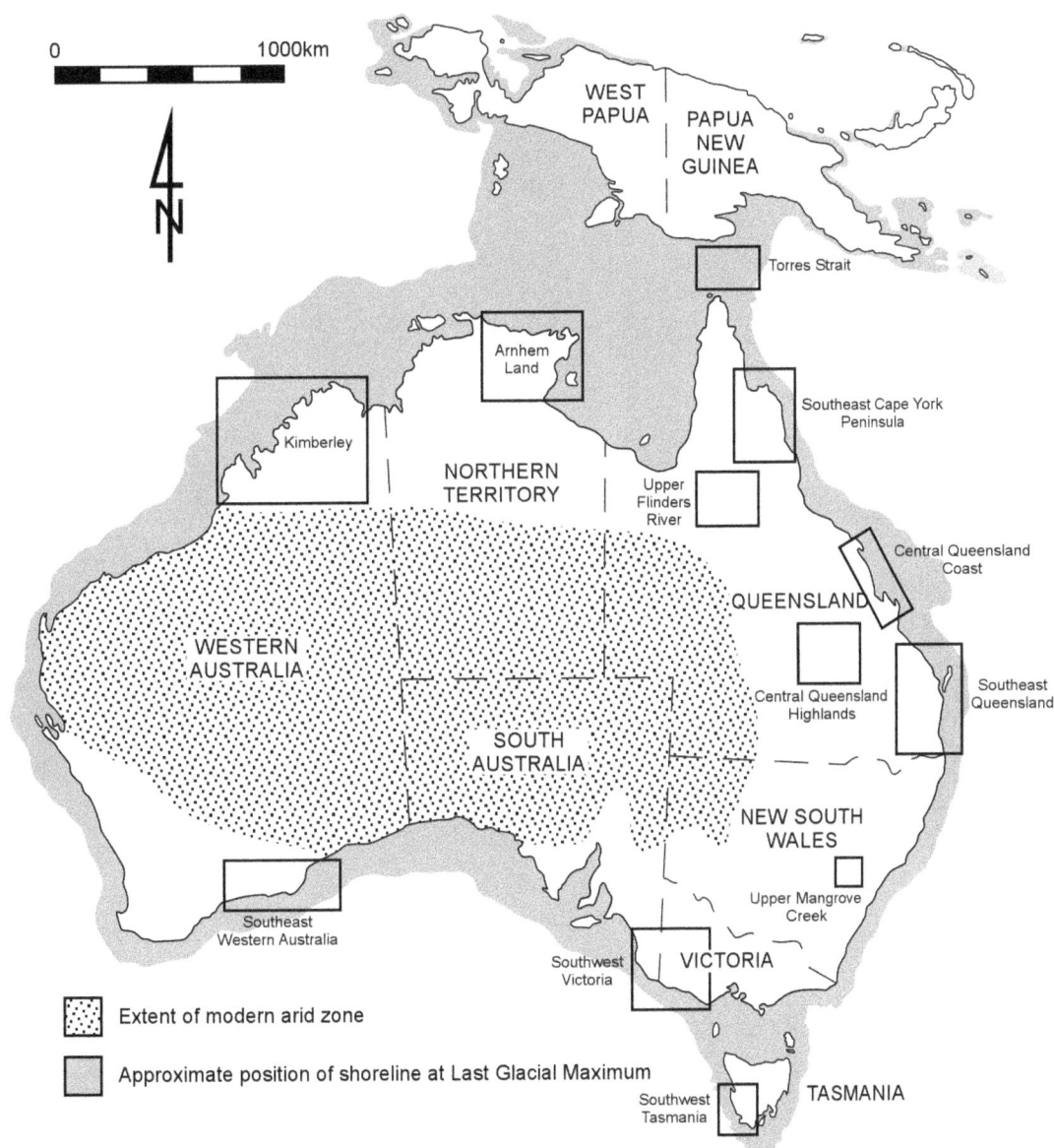

Figure 1.1 The Australian region, showing places mentioned in the text (after Allen and O'Connell 1995:vi).

At least two key problems impact on the use of radiocarbon results in Australian archaeology. First, radiocarbon sampling regimes frequently target only basal deposits, inhibiting our understanding of the chronology of other parts of occupational sequences, especially hiatuses and terminations (David 2002:37; Smith and Sharp 1993; Ulm and Hall 1996). In fact, in the widespread absence of termination dates it is common practice to assume that sites are continuously occupied until truncation by European invasion. The small number of dates available also tends to encourage views of occupational continuity rather than discontinuities in sequences. If sites prove to be of Pleistocene antiquity they tend to be subject to more extensive dating than mid-to-late Holocene sites. There is also a distinct bias in dating rockshelters over open archaeological deposits. These biases are evident in the Queensland radiocarbon dataset, the only large-scale synthesis of dates available for Australia (Ulm and Reid 2000). Overall, dated archaeological sites in Queensland have an average of 3.3 dates/site. Sites dating to the Pleistocene, however, have an average of 12.8 dates/site versus Holocene sites which average 2.4 dates/site. Rockshelters are preferentially dated, with an average of 4.6 dates/site. When these biases are combined rockshelters dating to the Pleistocene have an average of 12.8 dates/site whereas open sites dating to the Holocene have an average of 2.2 dates/site (see Ulm 2004a) (Table 1.1). The basic structure of the Queensland dated site dataset also exhibits a strong geographical bias towards the east coast, particularly the southeast corner (Fig. 1.2). A related problem is that the chronology of many critical sequences remain poorly understood. In particular, few assemblages are dated to the early to mid-Holocene (10,000–5,000 BP). In fact, out of the 834 radiocarbon dates available for Queensland, only 77 (9.2%) dates from 36 (14.1%) sites fall into this interval. If we are to develop an understanding of the changes evident in the mid-to-late Holocene we must first address the early Holocene period as a critical interval to assess the distinctiveness or otherwise of these later Holocene changes.

In some areas and for some time periods this situation is changing, with a small number of researchers recognising the limitations imposed on our understanding of Holocene sequences by low resolution chronologies. David (2002), for example, has attempted to address this situation in southeast Cape York Peninsula by obtaining large sequences of radiocarbon dates, especially accelerator mass spectrometry determinations, based on charcoal samples. He has also implemented a thorough sampling program to date terminal occupation deposits. As a result of this approach he was able to demonstrate regional abandonment of the Ngarrabullgan area around 600 BP (David and Wilson 1999).

The second key problem for chronology-building concerns environmental factors which impact directly on the samples selected for radiocarbon dating. For charcoal samples, important considerations include whether the sample is from a long-lived or short-lived plant and how the sample is related to changes in the production and distribution of ^{14}C in the biosphere. Various studies have shown that ^{14}C activity in the southern hemisphere is consistently lower than that in the northern hemisphere, at least for the recent past (e.g. Hogg et al. 2002; McCormac et al. 1998). Moreover, detailed regional studies in the northern hemisphere have demonstrated distinct regional differences which can impact on the accuracy of radiocarbon determinations measured on charcoal. Marine samples (e.g. shell) are also affected by regional differences in the availability of ^{14}C. Variation in ^{14}C activity in marine environments, although related to changes in atmospheric activity, depend greatly on local and regional factors, such as hinterland geology, tidal flushing and terrestrial water input. Such factors are highly variable and can introduce uncertainties of up to several hundred years into dates obtained on marine samples in some parts of the world (Reimer and Reimer 2001). These issues have received much attention in Pacific archaeology where determinations on marine samples are routinely scrutinised (e.g. Anderson 1991; Spriggs and Anderson 1993) and major resources have been devoted to resolving regional correction factors (e.g. Dye 1994; Phelan 1999). In Australia, however, only very limited investigations have been conducted despite routine dating of marine and estuarine shell material (e.g. Bowman 1985a, 1985b; Bowman and Harvey 1983; Gillespie 1977). For other areas of the world regional offsets of up to 1,000 years have been documented (Reimer and Reimer 2001), highlighting a key problem in a country where marine shell from open coastal sites is routinely dated.

The limitations of Australian chronologies are thrown into sharp relief against those available for the island Pacific. Here passionate argument hinges on periods of less than a century, and it is common for very precise radiocarbon chronologies to be cited and compared (Anderson 1991; Spriggs and Anderson 1993). The continuing investigation of such issues has considerable ramifications for normative models of social and economic change in Aboriginal societies in the late Holocene, where demonstrated contemporaneity between sites and regions is critical to the validity of abstract regional- and continental-scale models. Accurate site chronologies are therefore critical for situating sites within long-term patterns of land-use.

Table 1.1 Average number of dates/site for various subsets of the radiocarbon dates available for archaeological sites in Queensland (Ulm and Reid 2000).

SAMPLE	DATES (n)	SITES (n)	AVERAGE DATES/SITE
All sites	834	255	3.27
Pleistocene sites	269	21	12.81
Holocene sites	565	234	2.41
Rockshelter sites	449	97	4.63
Open sites	385	158	2.44
Pleistocene rockshelter sites	231	18	12.83
Holocene open sites	347	155	2.24

Rockshelters versus open sites: some sampling issues

Archaeology in Australia, as elsewhere, has been dominated by excavations of deeply stratified rockshelter deposits which have provided the chronological framework for our understanding of the continent's human past. Even in coastal Australia, where the recent archaeological record is dominated by shell middens, accounts remain based on rockshelter sequences (e.g. Barker 2004; David 2002; Hall 1999; Hiscock 1988; Lourandos 1997; Mulvaney and Kamminga 1999).

This general focus on limited numbers of intensively investigated sites and a limited range of site types inhibits our ability to develop sophisticated understandings of temporal and spatial variability in past Aboriginal lifeways. Some analyses have completely excluded the consideration of open sites. Lourandos and David (1998:109), for example, discounted open sites from a broad consideration of regional trends across several regions of northern Australia. In a related study, David (2002:117) only used rockshelter data to investigate regional occupation trends in north Queensland, while acknowledging that 'other kinds of sites should also be considered if we wish to address overall regional occupational trends'. Similarly, accounts of the regional archaeology of the Whitsunday Islands on the central Queensland coast, are based almost entirely on evidence from small excavations conducted in rockshelters (Barker 1996, 2004), despite open sites featuring in the ethnohistoric and archaeological records (Rowland 1986). Other examples abound in the Australian archaeological literature.

A major limitation of this approach is that rockshelter deposits are heavily biased in favour of the limited range of behaviours likely to have taken place in rockshelter contexts. Ethnographic and ethnoarchaeological studies have demonstrated the limited functionality of rockshelter sites (e.g. Binford 1978; Gorecki 1991; Nicholson and Cane 1991; Parkington and Mills 1991). Walthall's (1998:226) wide-ranging review concluded that 'habitation of rockshelters by mobile hunter-

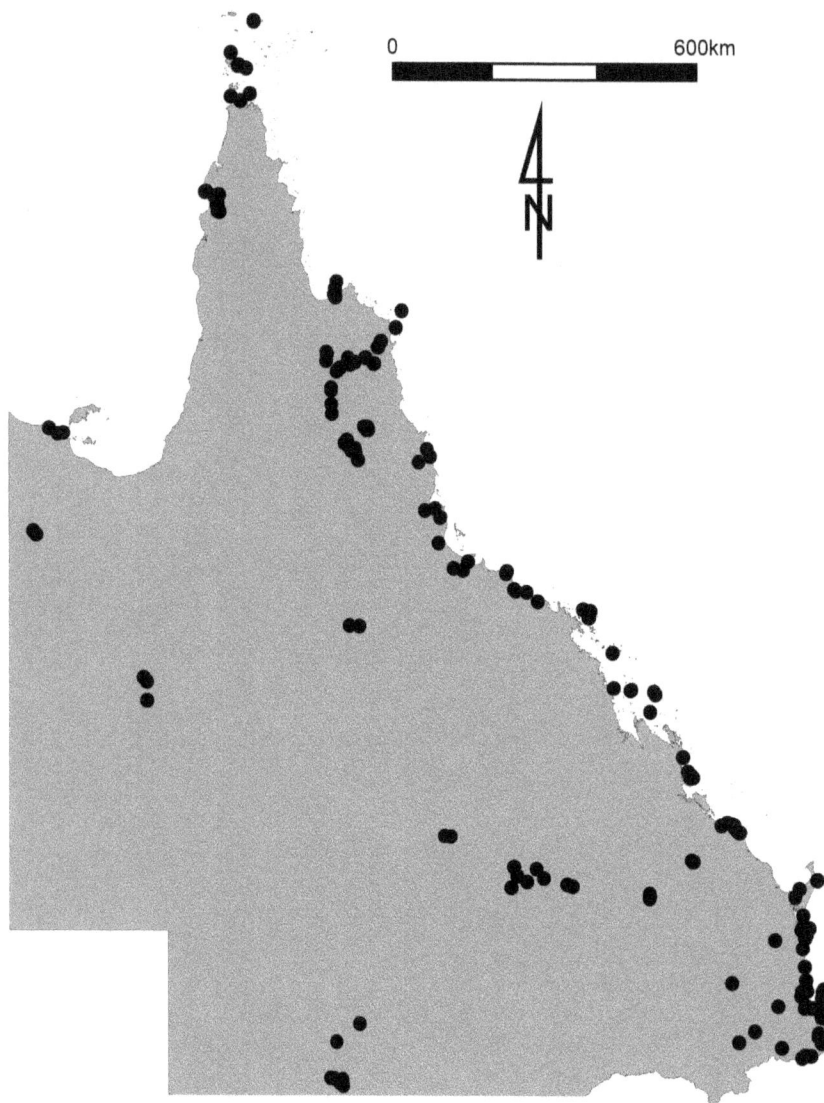

Figure 1.2 Distribution of dated archaeological sites in Queensland with a Holocene component (Ulm and Reid 2000:3).

gatherers was structured and that the activities conducted within them were highly standardized. Use of these shelters was normally confined to periods of inclement weather, either cold and/or wet seasons. The residential use of such sites was generally of brief duration and confined to occupations by either male hunters or family-based foraging groups'. Therefore, rockshelter assemblages are likely to overrepresent activities of males and small groups.

Of course, research is always constrained by resource imperatives, but our confidence in the accuracy and replicability of regional cultural sequences must be reduced when only very limited numbers of sites, or site types, and very limited sampling of archaeological materials, provide the basis for constructions of regional (pre)histories.

Rockshelters versus open sites: some taphonomic issues

The rationale for the bias towards rockshelters in regions dominated by open sites is rarely made explicit. When it is acknowledged at all, anecdotal claims are frequently cited, concerning the lack of integrity perceived to be inherent in open sites. Indeed, Lourandos (1996:18) has argued that rockshelter deposits provide a 'sounder' dataset as they 'are not subject to the same degree of post-depositional modification as open sites'. Frankel (1993:26) has similarly argued that only 'caves or shelters have the potential for showing continuity or discontinuity of use'. Implicit in this argument is the notion that rockshelter deposits are somehow exempt from post-depositional modification, such as that documented for open deposits, particularly those located in coastal landforms (e.g. Lourandos 1996, 1997). As Hofman (1992:5) has argued, however, virtually 'all archaeological collections were once surface deposits, and any argument or assumption that buried assemblages are more suitable or reliable for behavioral analyses must be demonstrated'. Walthall (1998:225) and others (e.g. Collins 1991) have also pointed out that rockshelter deposits may, in fact, be *prone* to post-depositional modification because the 'restricted space within a rockshelter, combined with frequently long and intense periods of use, means that such sites are subject to postdepositional disturbance'.

Several studies have demonstrated that a high degree of post-depositional movement of cultural material between stratigraphic units can occur without damaging the physical appearance of strata or strata boundaries (e.g. Hofman 1986, 1992; Villa 1982). In Australia, conjoin analyses undertaken on stone artefact assemblages from well-stratified sandstone rockshelters in the Central Queensland Highlands by Stern (1980) and Richardson (1992, 1996) found significant vertical and horizontal movement of conjoining artefacts, despite apparently well-defined stratigraphic sequences. These findings call into question basic assumptions about the integrity of the rockshelter deposits which form the basis of our understanding of the archaeology of Australia.

Another continuing problem is that interpretations based on evidence collected from open, especially coastal, sites have been heavily criticised over the last two decades, owing to presumed uncertainties in site preservation related largely to erosional processes (e.g. Bird 1992, 1995; Godfrey 1989; Head 1983, 1986, 1987; O'Connor and Sullivan 1994b; Rowland 1983). Geomorphological processes, especially the possible impact of sea-level change and erosion on the representation of archaeological materials, have featured prominently in discussions of many regions, such as Arnhem Land, southeast Queensland and southwest Victoria. The problem is particularly acute in areas without major rock formations, close to the coast, and with a dominance of sandy sediments. Such factors have obvious implications for the representation of coastal archaeological sites pre-dating the end of major sea-level change in the mid-Holocene: the archaeological record will be truncated and biased towards the last 5,000 years or so. Yet over the last 4,000 years the dominant coastal landscape processes in southern Queensland appear to have been toward progradation of the shoreline rather than recession. Clearly, it is essential in this context to have control of landscape formation processes in any area where open sites are to be discovered and interpreted. Recent landscape approaches, employing a wide range of archaeological and environmental data, have begun to explicitly redress some of these problems, particularly in arid areas. For example, Holdaway and Fanning (Fanning and Holdaway 2001; Holdaway et al. 1998; Holdaway et al. 2002) adopted a conjunctive approach employing detailed archaeological recording with modelling of geomorphic landscape dynamics in arid western New South Wales. They found major discontinuities in the regional sedimentary record indicating erosion and general instability of land surfaces prior to 2,000 years ago (see also Robins 1999). The clear implication is that any archaeological record of Aboriginal occupation of this area prior to this time has been destroyed.

The major task for advancing our knowledge of mid-to-late Holocene Australia therefore remains a basic archaeological one: to construct and compare detailed individual sequences from a range of site types, at the local and regional level, to establish the existence of trends independent of site-specific taphonomic and/or environmental factors. As Frankel (1993:31) suggested a decade ago, '[p]erhaps we should work outward to the broad picture from accumulated data and concepts developed in detailed local sequences … [rather than] [s]hort-term, or broad-scale research projects [that] can only lead to large-scale narrative'. It is hard to escape Holdaway et al.'s (2002:352) conclusion that 'detailed records from a number of well-studied locations will be needed before general explanations can be put forward and tested'. This study begins the process of forming a basic regional understanding of the variability of the archaeological record in one region: the southern Curtis Coast.

Background to the study

The stimulus for this study derives from earlier work (Ulm 1995, 2002a; Ulm and Hall 1996) which synthesised the radiocarbon date and fish bone datasets from southeast Queensland to test propositions about cultural change. This synthesis of site use and discard patterns demonstrated that the apparent trend towards increased site creation and use, identified in the radiocarbon record at 1,000–1,200 BP, is also reflected in structural changes in the regional archaeological record. Most significant is the widespread appearance of shellfish remains, as evidenced by the dramatic increase in coastal shell middens, a site type which only appeared in the mid-Holocene. Several other studies have also documented significant subsistence transformations in southeast Queensland in the late Holocene, reflecting a general broadening of the subsistence base (e.g. Morwood 1986, 1987; Walters 1986).

Despite the volume of research conducted in southeast Queensland, many key issues remain poorly understood. Although the widespread appearance of intensive shellfishing appears to be a late Holocene phenomenon, limitations of the database, particularly a lack of comparability in recovery and analytical techniques, preclude detailed consideration of these developments in the context of other changes, such as representation of fish remains and patterns of stone procurement and use (see Ulm 2002a). Establishing the antiquity and nature of marine fishing in the region also remains a highly contentious issue. Walters (1992a, 1992b, 2001) has forcefully argued that marine fishing was only regularly incorporated into Aboriginal subsistence and settlement regimes in southeast Queensland in the last 2,000 years as part of permanent and intensifying coastal settlement. Others, however, have pointed to limitations imposed by taphonomic factors and recovery techniques and the presence of earlier fish remains to the north and south of the region as the basis for alternative interpretations (McNiven 1991a; Ross and Coghill 2000; Ulm 1995, 2002a). Other scholars regard the evidence from southeast Queensland as fundamentally problematic owing to its heavy reliance on open sites in active sedimentary contexts (David 1994; David and Chant 1995; Lourandos 1996). Ulm and Hall (1996) have also highlighted potential problems in the accuracy of radiocarbon chronologies in the region arising from a heavy reliance on marine and estuarine shell samples.

These contested accounts of cultural change in southeast Queensland represent a microcosm of the wider debates about the nature and causes of late Holocene changes played out across the continent.

The study region

The southern Curtis Coast study area is located at the northern end of the southeast Queensland bioregion, just south of the Tropic of Capricorn. The area extends along the coast from Wreck Rock in the south to Hummock Hill Island in the north and inland to Miriam Vale and Bororen in the west. This region covers a total land area of about 1,200km², with a high water shoreline length of over 500km. The boundaries of the study region are located 70km northwest of Bundaberg and 20km southeast of Gladstone. For the purposes of this study, the southern Curtis Coast is defined as the coastal landscapes between the mouth of Baffle Creek and Rodds Bay. This region is characterised by broad, curved, sandy beaches anchored to rocky headlands backed by high dunes in the south and extensive tidal estuaries and freshwater wetlands bordered by low transgressive dunes in the north. Further inland, the near-coastal ranges formed of local granites contribute to a complex drainage system which maintains freshwater coastal swamps and wetlands and which flows both north into Rodds Harbour and south into Baffle Creek.

Since 1993, archaeological surveys and excavations have been conducted in the study area as part of the Gooreng Gooreng Cultural Heritage Project, an interdisciplinary Aboriginal cultural heritage investigation of the Burnett-Curtis region of southeast Queensland. The project was initiated by the Aboriginal and Torres Strait Islander Studies Unit at the University of Queensland in collaboration with Aboriginal people from the study area. The broader project includes studies in archaeology, history, contemporary social landscapes and linguistics in a region which coincides with the area broadly identified as the country of Gooreng Gooreng speakers at the time of European invasion (see Clarkson et al. n.d.; Jolly 1994; Lilley et al. 1998; Lilley and Ulm 1995; Lilley et al. 1996; Lilley et al. 1997; Ulm 2002b, 2002c; Ulm et al. 1999a, Ulm et al. 1999b).

To date, the archaeological component of the Gooreng Gooreng Cultural Heritage Project has concentrated on the Cania Gorge area as an inland component of the study, where Aboriginal occupation has been dated to the early Holocene (Chapman 1999, 2002; Eales 1998; Eales et al. 1999; Lilley et al. 1998; Westcott 1997; Westcott et al. 1999a; Westcott et al. 1999b), and the southern Curtis Coast, centred on the Town of Seventeen Seventy, as a coastal component and subject of the present study.

Aims of the study

The major task of this research is to map the dynamics of archaeological change and continuity on the southern Curtis Coast through documentation of the distribution, size, composition and chronology of sites in the region. The research assembles a regional archaeology problematised in terms of theoretical and methodological issues emerging out of archaeological studies in southeast Queensland. Sampling targetted a range of sites in order to develop a basic understanding of coastal land-use across the region through time. Excavations and analyses of eight open coastal sites located on six separate estuaries are described and discussed. Differences in site structure, content and chronology are used to establish a framework to describe variability in the regional archaeological record. Patterns in resource use are investigated through systematic comparison of individual site histories, thus permitting a more detailed understanding of spatial and temporal variability in regional land-use. The articulation of this variability with those from neighbouring regions is considered and areas for further investigation identified.

The project has a number of specific objectives:

— to create an independent regional archaeological dataset, based on a consistent data recovery strategy, to establish a first-order understanding of regional land-use histories;

— to establish a reliable and robust regional chronology, based on dating initiation, periodicity and termination of site occupations and evaluation of marine and estuarine reservoir factors;

— to evaluate site integrity in order to enhance the contribution of open coastal shell midden deposits to understandings of late Holocene lifeways; and

— to use findings to investigate key interpretations of the southeast Queensland archaeological record and promote a broader understanding of patterns of regional interaction in the wider southeast Queensland region.

2

The study region: the southern Curtis Coast

Introduction

This chapter contextualises the study with brief outlines of the physical environment, Aboriginal cultural setting and previous archaeological investigations in the region. The first section covers geology, climate, hydrology, flora, fauna and palaeoenvironment to provide a background to regional landscape development, resource availability and dating issues. Local ethnographies, documentary histories and oral histories are reviewed in the second section to create an overview of recent Aboriginal lifeways and historical transformations in the region. The final section summarises previous archaeological work in the broader region before focussing specifically on the results of Gooreng Gooreng Cultural Heritage Project investigations on the southern Curtis Coast. Information from dated but as yet unexcavated sites is also presented.

Physical setting

For the purposes of this study, the southern Curtis Coast is defined as the coastal landscapes between the mouth of Baffle Creek and Rodds Bay. This area is also commonly known as the 'Discovery Coast'. It is located in southeast Queensland, just south of the Tropic of Capricorn. The study area extends along the coast from Wreck Rock in the south to Hummock Hill Island in the north and inland to Miriam Vale and Bororen in the west (Figs 2.1–2.2). This region covers a total land area of about 1,500km², with a high water shoreline length of over 500km. The boundaries of the study region are located 70km northwest of Bundaberg and 20km southeast of Gladstone, between latitudes 24°20' and 23°58' south and longitudes 151°26' and 152°00' east.

The study area falls within the northern end of the well-defined southeast Queensland bioregion and exhibits two major landscape provinces — the Burnett-Curtis Coastal Lowlands in the southeast and the Burnett-Curtis Hills and Ranges in the north and west (Sattler 1999; see also Coaldrake 1961). The former are characterised by fine-grained sediments; alluvium; coastal and estuarine sediments; a broad coastal plain; high rainfall (1,100mm); low elevation (<50m); and eucalypt and melaleuca forests and woodlands. The latter are characterised by acid volcanics; metamorphics; localised basic volcanics; small areas of elevated sediments; hills and ranges; alluvial valleys; high rainfall (900mm); medium elevation (<250m); and eucalypt woodlands and araucarian microphyll rainforest (Young and Dillewaard 1999).

The coast is characterised by extensive estuaries and embayments bordered by saltflats and claypans. Recent (<6,000 year old) Holocene sediments dominate the geology of the area with beach ridges up to 2km wide bordering Bustard Bay. The Baffle Creek catchment dominates the western two-thirds of the study area, although its outlet to the ocean is located 24km south-southeast along the coast from Wreck Rock at the southern-most point of the present study region. The region is characterised by broad curved sandy beaches anchored to rocky headlands backed by high dunes in the south and extensive tidal estuaries and freshwater wetlands in the north (Figs 2.3–2.4). Further inland, the near-coastal granite ranges contribute to a complex drainage system which maintains freshwater coastal swamps and wetlands that flow both north into Rodds Harbour and south into Baffle Creek.

Geology and geomorphology

The southern Curtis Coast comprises a relatively restricted range of rock types and landforms dating from 235–213 million years ago to the last 6,000 years (QDEH 1994:33). The basal geology of all but the extreme west of the study area is dominated by rhyolites and granites assigned to the Agnes Water Volcanics formation of the Toogoolawah Group (Ellis and Whitaker 1976; Stevens 1968). The Agnes Water Volcanics drape an uneven underlying land surface composed of heavily weathered and undifferentiated granites of Permian to Triassic age and adjacent Palaeozoic sediments (Ellis and Whitaker 1976). The thickest accumulations of the formation fill palaeotopographic lows that now persist as headlands along the contemporary coastline (Stephen Cotter, Cooperative Research Centre for Landscape Evolution and Mineral Exploration, University of Canberra, pers. comm., 2001; see also Ulm et al. 2005).

Although the coast is punctuated by rocky headlands, the study area is characterised as a depositional coastline with low north-northwest trending Holocene beach ridges and swales oriented roughly parallel to the modern coastline, trailing northwards from the northern side of almost every estuary of note (Hopley 1985:76–7). Beach ridges and sand masses consist of fine-grained quartz sands and dispersed heavy mineral sands including rutile, ilmenite and zircon (Connah 1961). The coast in the study area is open to full oceanic conditions, which has significantly impacted on coastal sedimentation and erosion regimes. This situation is relatively unusual on the Queensland coast, as to the south the mainland is protected by the marine sand accumulations of Fraser, Moreton, and North and South Stradbroke Islands and Cooloola, and to the north by the Great Barrier Reef.

One of the primary issues in evaluating the regional archaeological record is consideration of palaeoenvironmental factors, particularly the potential effects of sea-level change and erosion on site survival and visibility. Accumulating geomorphological evidence suggests that there may have been minor variations in sea-level along the eastern Australian coast since 6,000 BP. Larcombe et al. (1995) have recently presented a model of episodic post-glacial sea-level rise based on a detailed study of radiocarbon dates from the central Great Barrier Reef shelf (between Hayman Island and Cape Tribulation) for the last c.12,000 years. On this basis, they identify a peak in sea-level at c.8,500 BP at c.–11m, a regression at c.8,200 BP at –17m, followed by a rapid rise to c.–5m at c.7,800 BP.

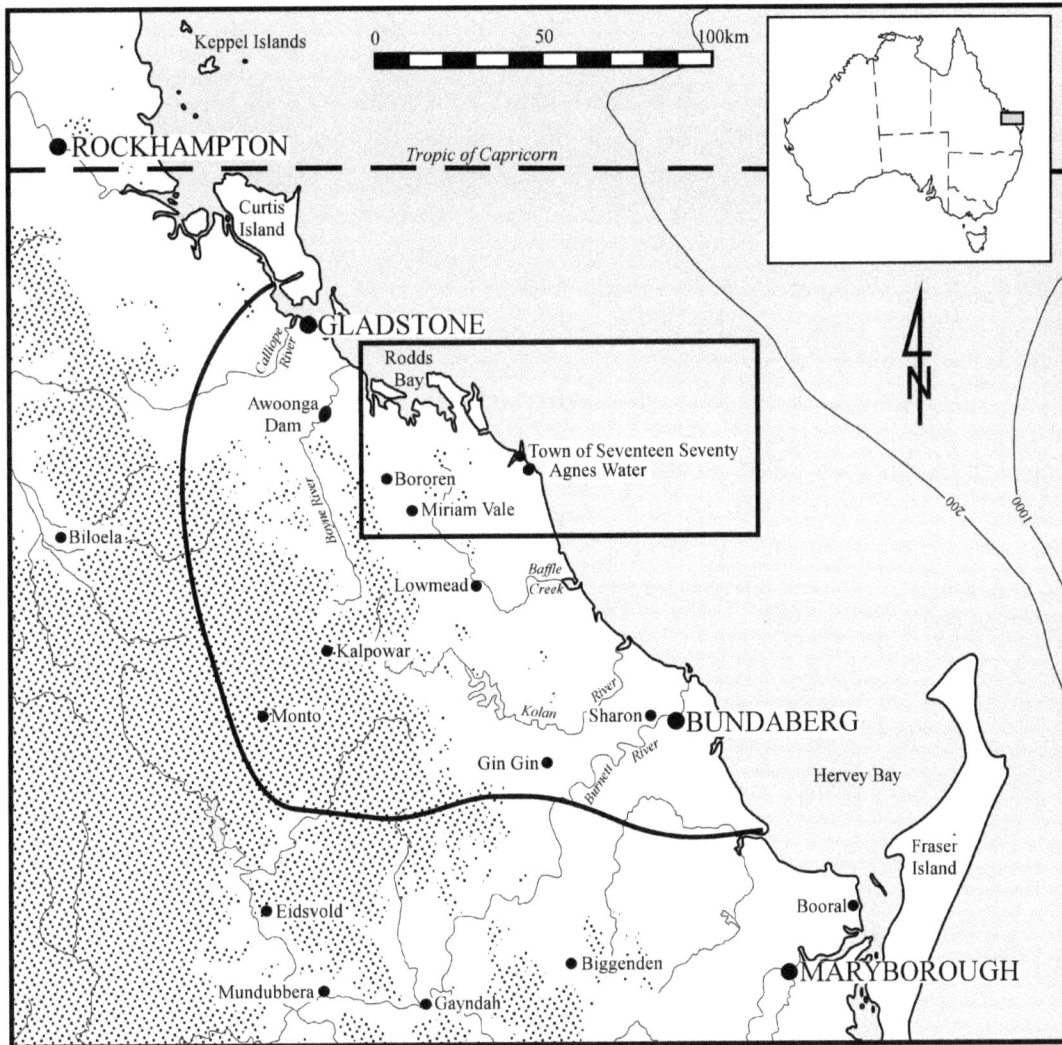

Figure 2.1 The Gooreng Gooreng Cultural Heritage Project study area, showing major towns and the general distribution of Gooreng Gooreng speakers (heavy line) (after Horton 1994; Williams 1981). The southern Curtis Coast study area is shown by the box.

Sea-level remained relatively stable until c.6,800 BP before a rise to a short stillstand at –2m at c.6,000 BP and then to the Holocene stillstand of +1.65m at c.5,500 BP until c.3,700 BP, when sea-levels dropped to approximately modern values. This model contrasts with earlier sea-level curves for northeastern Australia, which have suggested stabilisation at current levels at 6,000±500 BP (e.g. Hopley 1983; Lambeck and Nakada 1990; Thom and Roy 1983).

The current model of sea-level change has significant ramifications for understanding the archaeology of the study area, as much of the land within 2km of the present coastline is of very low elevation, interspersed with large freshwater swamps and wetlands and extensive estuarine systems. Field surveys and examination of aerial photographs revealed a regular system of parallel transgressive beach ridges extending over much of the study area (particularly between Round Hill and Falls Creeks), suggesting major transformations of the coastal landscape over time. The assignment of the majority of these changes to the late Holocene is supported by a preliminary series of four radiocarbon dates from a pollen core taken from freshwater wetlands adjoining the southwest margin of Round Hill Creek (on the inland side of a major series of transgressive beach ridges). A basal date for swamp formation of around 3,000 years ago is consistent with recent arguments for sea-level retreat (Maria Cotter, Centre for Coastal Management, Southern Cross University, pers. comm., 1999).

In the absence of sea-floor studies in the study region, local bathymetric contours are employed to approximate the general position of palaeoshorelines. This approach is particularly problematic on sandy coastlines with broad, low gradient continental shelves owing to sediment accumulation on the sea-floor. Despite these uncertainties, this approach can provide a general impression of the approximate positions of palaeoshorelines relative to the modern coastline (see Barker 1995; McNiven et al. 2002). The –200m bathyometric contour, located 75–95km northeast of the current coastline, provides an approximation for the position of the shoreline at the Last Glacial Maximum c.18,000 BP (Fig. 2.1). The study area is located just south of the point where the continental shelf broadens to some 250km wide. Over 200km^2 of continental shelf immediately adjacent to the study region has therefore been inundated over the last 12,000 years, since sea-levels breeched the steep continental slope and began to inundate the continental shelf. In the early Holocene the coast would have been within 1.5km of the current shoreline throughout most of the study area although, for reasons outlined below, it is unclear how these changes impacted on the form and development of estuaries. Taken together, this evidence suggests a geologically very recent origin for many of the coastal landforms which are the subject of this study, including numerous tidal estuaries, extensive intertidal and subtidal flats, low sandy beach ridges and cheniers (see below).

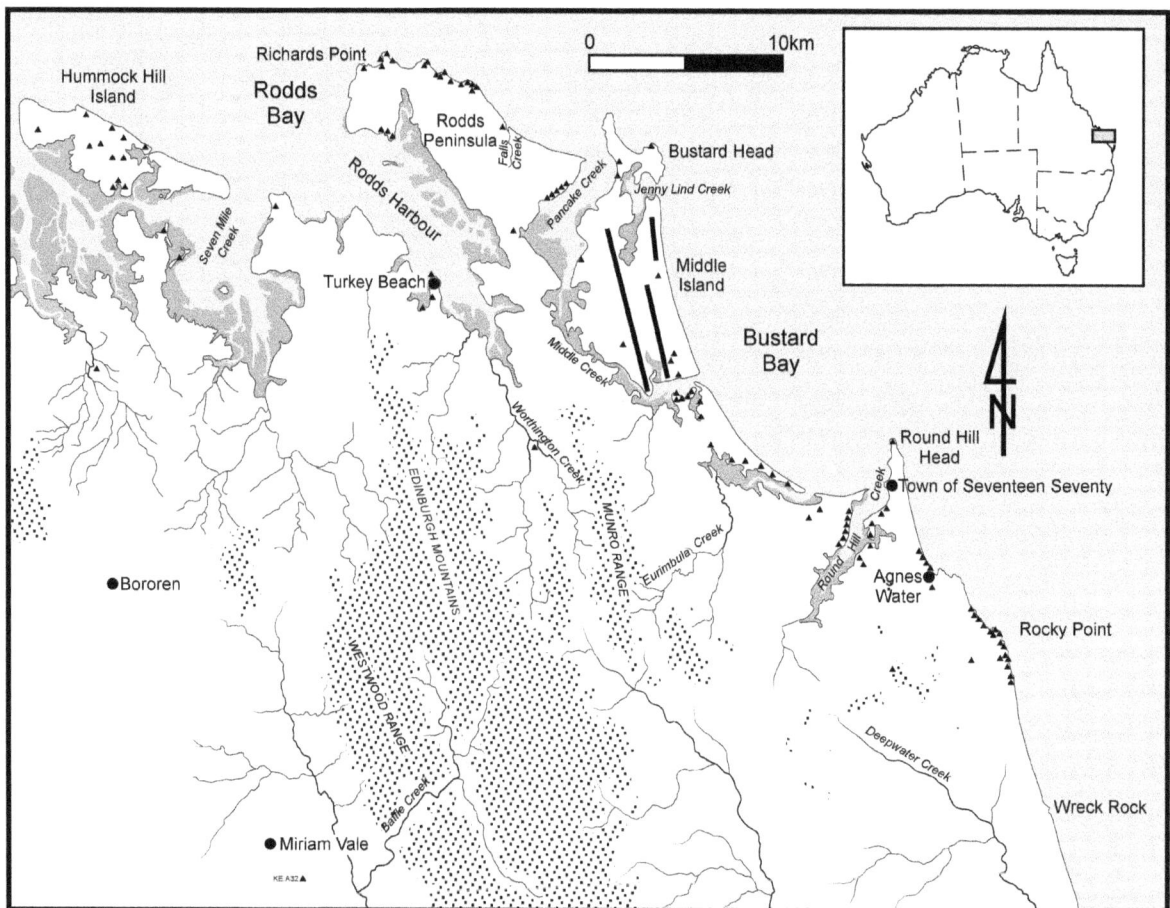

Figure 2.2 The southern Curtis Coast study area, showing all recorded archaeological sites as triangles (after Ulm and Lilley 1999). Site designations are shown for sites which are not illustrated in Figs 2.9–2.11. Heavy black lines on Middle Island indicate the general location of extensive low density shell deposits.

Climate

The Curtis Coast has a subtropical maritime climate characterised by dry mild winters and hot humid summers influenced by the southeast trade winds, topography and the moderating influences of the ocean. The region experiences occasional monsoon influences, although cyclones are more frequent features, as are the major frontal systems common in more southerly latitudes. These varied influences generate marked variability in rainfall, temperature and prevailing wind conditions (QDEH 1994:11). Average minimum and maximum temperatures range from 22.8°C and 28.9°C respectively in the summer to 13.4°C and 20.9°C respectively in the winter (QDEH 1994:13). Rainfall is summer-dominated, with January and February commonly the wettest months and August and September the driest, with mean annual rainfall at the Town of Seventeen Seventy of 1,318mm (QDEH 1994:11). Major factors which influence the distribution of rainfall include topographic influences of mountain ranges, geographic influences such as the orientation of the coastline to the prevailing water-saturated winds, and occasional cyclones causing extreme rain events from November to April.

Hydrology

The southern Curtis Coast is transected by numerous creeks and rivers which form an extensive network of interconnected estuaries. Numerous minor seasonal tributaries drain into estuarine creeks from the low subcoastal ranges in the west. The Munro Range, Edinburgh Mountains and Westwood Range divide the catchments of Baffle, Round Hill, Eurimbula and Middle Creeks to the south and east from Worthington and Seven Mile Creeks to the north (Olsen 1980a:4; see Fig. 2.2). The major influences on water movement within these tributaries are prevailing tides and weather conditions, although freshwater inflow associated with periods of high intensity rainfall can cause heavy runoff. In the south, Round Hill, Eurimbula, Middle, Jenny Lind and Pancake Creeks are generally shallow, mangrove-fringed estuaries characterised by sandy bottoms merging to silt and clay in the upper reaches (Olsen 1980a:3). To the north, Rodds Harbour and Seven Mile Creek exhibit deeper channels near the mouth and extensive flats and zones of silty sand upstream with large areas of mangroves with claypans bordering grassy or layered eucalypt forest (Olsen 1980a:3; see Table 2.1).

Tidal processes of the southern Curtis Coast are influenced by the presence of the southern extremities of the Great Barrier Reef, ocean floor topography and coastal geology, such as inshore islands and headlands (QDEH 1994:17). The tidal effects of estuaries also contribute to the amplification of tidal range, with an average maximum tidal range of 2.43m at Pancake Creek in the approximate centre of the study region (QDOT 1998). Like all coastal regions, the area is subject to both wind- and storm-generated waves which modify the configuration of the shoreline. Unfortunately, only scant research into coastal erosion processes has been undertaken in the area and the effects of erosion on the representation of archaeological materials in open beach contexts is difficult to assess. Anecdotal evidence and field observations suggest that storm-surge activity exacerbates local erosion (Fig. 2.5).

Table 2.1 Creek/estuary characteristics in the study region (Olsen 1980a:17–25).

CREEK	ESTUARY (km2)	MANGROVE (km2)	SAMPHIRE/ CLAYPAN (km2)	SHORELINE LENGTH (km)	WATER AREA (km2)
Round Hill	14.06	4.66	3.21	38.8	6.08
Eurimbula	6.8	2.98	1.06	22.2	1.76
Middle	17.41	5.64	4.61	53.8	7.15
Jenny Lind	6.96	2.5	1.58	40.8	2.88
Pancake	25.6	8.21	4.78	59	12.61
Rodds Harbour	72.54	17.32	6.34	97.5	48.88
Seven Mile	55.57	14.39	9.88	82.5	31.2
Total	198.94	55.7	31.46	394.6	110.56

Flora

The ecological complexity and diversity of the study region reflect its status as part of a transitional zone between tropical and temperate provinces, with a zoogeographical boundary identified at about latitude 25° south (Endean et al. 1956; Knox 1963). This overlap generally translates into high rates of floral and faunal diversity, with representation of both tropical and temperate species. The region supports an extensive range of herblands, grasslands, heaths, scrubs and tall shrublands, and open and closed forests (QDEH 1994:45). Mixed herblands on foredunes include goat's foot convolvulus (*Ipomoea pes-caprae*) and coastal jack bean (*Canavalia rosea*). Extensive wet and dry heathlands occur on poorly-drained sandy-loam soils, comprising a number of species generally less than 2m in height, including banksia (*Banksia* sp.), teatree (*Melaleuca* sp.), and grass trees (*Xanthorrhoea* sp.). Beach ridges support tall, open paperbark forests, dominated by teatrees (*M. leucadendra* and *M. dealbata*) in association with weeping cabbage palm communities (*Livistona australis*) and bracken fern (*Pteridium esculentum*) (Fig. 2.6) (QDEH 1994:48). The distribution of closed forest is limited, with a relatively restricted tall, closed-forest community found bordering Eurimbula Creek, with hoop pine (*Araucaria cunninghamii*) emergent above a notophyll vine forest (QDEH 1994:49). Estuaries exhibit extensive fringing vegetation communities consisting of combinations of some 13 mangrove species dominated by the grey mangrove (*Avicennia marina*), spotted mangrove (*Rhizophora stylosa*) and yellow mangrove (*Ceriops tagal*) (Dowling 1980; Olsen 1980a). Seagrass beds (dominated by *Zostera capricornia*) are typically found in sheltered waters where water clarity allows sufficient light penetration for photosynthesis, including Round Hill Creek, Rodds Harbour, Pancake Creek and Mort Creek (Olsen 1980b; QDEH 1994) (Fig. 2.7). These estuarine habitats provide important habitats for fish, shellfish and crustaceans as well as turtles, dugongs and water fowl. Important local plant resources include cycads (*Cycas megacarpa*), bracken fern, grass tree (*Xanthorrhoea* sp.), bungwall fern (*Blechnum indicum*), weeping cabbage palm and pandanus (*Pandanus* sp.). Some of these resources are associated with often complex specialised processing technologies.

Fauna

The region's terrestrial fauna is diverse and includes seven species of amphibians (all frogs and toads), 60 species of mammals (including bats, echidnas, koalas and kangaroos), 59 species of reptiles (including lizards and snakes) and 288 bird species (including shorebirds, waterbirds, seabirds and birds of prey) (QDEH 1994:59–65). Common macropods include the eastern grey kangaroo (*Macropus giganteus*), whiptail wallaby (*M. parryi*), swamp wallaby (*Wallabia bicolor*) and rufous bettong (*Aepyprymnus rufescens*). Other commonly recorded mammals include the Gould's long-eared bat (*Nyctophilus gouldii*), short-beaked echidna (*Tachyglossus aculeatus*), common ringtail possum (*Pseudocheirus peregrinus*) and yellow-footed marsupial mouse (*Antechinus flavipes*) (Woodall 1991). Occasional dingoes (*Canis lupus dingo*) have also been sighted in remote parts of the study region. Emus (*Dromaius novaehollandiae*) and the Australian brush-turkey (*Alectura lathami*) are locally common terrestrial birds. Other common bird species include the abundant Pacific black duck (*Anas superciliosa*), common noddy (*Anous stolidus*), beach thick-knee curlew (*Esacus magnirostris*) and eastern curlew (*Numenius madagascariensis*). The terrestrial environment has been impacted by localised intensive logging and long-term cattle grazing. At least one local extinction is apparent, that of the eastern bustard or plain turkey (*Choriotis australis*) (Growcott and Taylor 1996:25).

Common marine fauna found along the Curtis Coast include a number of whale species (including the humpback *Megaptera novaeangliae*), four species of dolphins (including the common bottle-nose *Tursiops truncatus*), dugongs (*Dugong dugon*) and turtles (including loggerhead *Carreta caretta* and green *Chelonia mydas*). Rodds Harbour supports the largest dugong population along the Curtis Coast (QDEH 1994:66). As a transition zone, the area is also a wintering destination for some whales and migratory waterbirds.

A total of 148 species of fishes from 69 families is recorded for the Curtis Coast (QDEH 1994:68). Useful data are provided by Lupton and Heidenreich's (1996) detailed study of Baffle Creek just to the south of the study area. The lower estuarine component of this fisheries resource assessment covered habitats similar to the coastal estuaries in the study area. Despite significantly depressed regional rainfall levels (25% under the annual average) before and during the survey period, 55 fish and nine crustaceans were recorded. The larger fish species were (in order of abundance) flat-tail mullet (*Liza dussumieri*), sand mullet (*Myxus elongatus*), whiting (*Sillago ciliata, S. maculata* and *S. sihama*), yellowfin bream (*Acanthopagrus australis*), blue-tail mullet (*Valamugil seheli*), sea mullet (*Mugil cephalus*) and garfish (including *Arrhamphus sclerolepis* and *Hyporhamphus ardelio*) (Grant 1993; Lupton and Heidenreich 1996). Commercial finfish catches for the region reflect this pattern, with mullet, whiting and bream accounting for 64.1% of commercial catches (Olsen 1980a:11). Early historic accounts suggest that fish may have been more abundant in the past with both Banks and Parkinson impressed by the quantities of fish in the bay and estuary (Banks in Beaglehole 1963; Parkinson 1773). Mud crabs (*Scylla serrata*) and sand crabs (*Portunus pelagicus*) are also common.

Bivalves are common local marine mollusc fauna, attributed by Woodall et al. (1991, 1993) to the prevalence of sandy and muddy substrates and the paucity of rocky platforms. Unfortunately, the only marine mollusc studies available are based on observations of dead beach-washed specimens recovered from exposed beaches. Such inventories can only be taken as a general guide to living mollusc populations in the region. The contemporary estuarine mollusc fauna along the southern Curtis Coast is dominated by rock oyster (*Saccostrea glomerata*), found in mangrove and rocky habitats, and gastropods such as the hercules club shell (*Pyrazus ebeninus*), and members of the family Potamididae, including the mud creepers *Telescopium telescopium* (Fig. 2.8) and *Terebralia sulcata* (Roughley 1928; Shanco and Timmins 1975). Midden deposits are

Figure 2.3 General view of estuary and near-coastal ranges across Jenny Lind Creek. Such estuaries are common in the region and support diverse populations of marine life. Facing southwest.

Figure 2.4 Freshwater wetlands adjoining the upper reaches of Round Hill Creek. This swamp is part of a vast network of interconnected wetland areas located in the swales between ridge systems. Facing north.

Figure 2.5 Example of recent bank recession of the north bank of Middle Creek, showing concrete stairs once used to access the beach from the top of the erosion bank some 15m away from the contemporary section. Facing northwest.

Figure 2.6 Beach ridge vegetation in Eurimbula National Park, including weeping cabbage palm communities (*L. australis*) and bracken (*P. esculentum*).

Figure 2.7 Mud ark (*A. trapezia*) in a dense paddock of seagrass (*Z. capricornia*) in the middle of Round Hill Creek. The presence of mud ark valves in sites dated to 4,000 BP indicate that seagrass beds are a long-term feature of the local environment. Facing west.

Figure 2.8 Cluster of telescope mud whelk (*T. telescopium*) on the mangrove fringe of Eurimbula Creek. These whelks are abundant throughout the region, although virtually absent from archaeological deposits. Facing west.

dominated by rock oyster (*S. glomerata*) and the mud ark (*Anadara trapezia*) (Fig. 2.7), an estuarine bivalve. This last species is uncommon in the study area today and generally sparse in the coastal waters of Queensland (Chappell and Grindrod 1984:222) and is not included in estuarine inventories for the region (Shanco and Timmins 1975). Contemporary open coast mollusc populations are relatively depauperate, with scattered populations of milky oyster (*S. amassa*) on exposed rocky headlands (Olsen 1980a:11, 13) and pipi (*Donax deltoides*) on stretches of high energy sandy coast. Freshwater bivalves have been collected from both Eurimbula Creek and Deepwater Creek and include *Alathyria pertexta* and *Velesunio ambiguus* (Woodall et al. 1991, 1993).

Palaeoenvironment and environmental change

Shoreline progradation, erosion and sea-level change are the major factors responsible for coastal landscape changes in the region since the early Holocene. Minor changes in sea-level can have a dramatic impact on the configuration of low gradient coastlines such as those of southeast Queensland. None of the estuaries in the study area has a significant or permanent freshwater outflow and therefore none is likely to have an incised channel, suggesting that local estuary formation is related to the final stages of sea-level rise. Episodes of mid-to-late Holocene dune-building towards the seaward margins are evidenced in the alignment of transgressive dune systems parallel to the modern shoreline. Of particular interest is the fact that the overall sedimentation regime in place since the mid-Holocene is positive, like elsewhere in southeast Queensland, with prograding rather than receding landforms the dominant landscape features (see Cotter 1996). Contemporary patterns of erosion appear to be relatively recent, impacting primarily on the exposed coastline and estuary mouths. The absence of archae-ological materials on transgressive dunes bordering the coastline supports the notion that the coast was still actively prograding until the last few hundred years. As Lourandos (1997:225) noted for southwestern Victoria, the regional coastal landscape is a prograding or depositional form and therefore recent sites are, in fact, more susceptible to erosion than older deposits (i.e. more recent archaeological deposits are in closer proximity to the modern shoreline than older ones). Periods of chenier formation also changed landscapes in specific areas since the mid-Holocene. Radiocarbon analyses of chenier deposits at Mort Creek suggest several episodes of formation, the earliest ceasing around 3,300–3,100 BP and a later episode occurring between 2,400 and 2,100 BP (Chapter 7). These dates broadly coincide with major periods of chenier formation identified at 3,550 BP and 2,500 BP at Broad Sound, 250km to the north, where they are linked to phases of decreased sediment supply (Cook and Polach 1973).

As part of the Gooreng Gooreng Cultural Heritage Project, a large suite of cores was taken from swamp deposits, archaeological sites and deposits adjacent to archaeological sites in an attempt to integrate local environmental and archaeological records. The most significant cores have been obtained from a permanently water-logged area of the Deepwater Creek drainage basin at the northern end of the Deepwater Section of Eurimbula National Park. Several cores up to 6m in length and dating up to 13,150±120 BP (OZD-757) were obtained from this deposit. It is anticipated that analysis of these cores will provide a detailed long-term vegetation history of the region. Preliminary study of another core dated to 3,240±60 BP (OZD-755) establishes the presence of mangrove pollen in present swamp areas draining into Round Hill Creek. Unfortunately, detailed analysis of these cores, which would provide an independent model of local palaeo-environmental and sedimentary development, remained unavailable at the time of writing (Maria Cotter, School of Human and Environmental Studies, University of New England, pers. comm., 2003). A limited coring program has demonstrated that the low sand ridges abutting the rhyolitic core of the peninsula formed on the northern bank of the confluence of Tom's Creek and Round Hill Creek and adjacent sand ridge residuals on the intertidal flats to the west are underlain by bluish-grey muds (associated with a date of 1956±57 BP (NZA-13385)), suggesting deposition in an estuarine environment (Chapter 13). Preliminary examination indicates that these sediments are rich in organic material and dominated by mangrove pollen and are therefore interpreted as mangrove facies.

In the absence of further data in the immediate region, the closest palaeoenvironmental studies available come from Fraser Island to the south and the Whitsunday Islands to the north. Longmore (1997a, 1997b; Longmore and Heijnis 1999) found a record of a mid-Holocene dry period in perched lakes on Fraser Island and of a major shift in vegetation after the Last Glacial Maximum from Araucariaceae to Casuarinaceae. A more detailed Holocene palaeoenvironmental record is available from Whitehaven Swamp on Whitsunday Island (Genever et al. 2003). Analysis of a 5.77m deep core dated to 6,957±58 BP (Wk-9389) at 5m suggests relative environmental stability since the mid-Holocene, with an emphasis on open and coastal woodland communities, and a reduction in rainforest over the last few thousand years. In contrast to the Fraser Island data, the distribution of rainforest taxa are associated with a moister mid-Holocene climate.

Shell midden composition can also be a useful broad palaeoenvironmental indicator on the assumption that midden contents will generally reflect local resource availability. It should be noted that shellfish communities oscillate in response to a wide range of often very localised conditions such as changes in substrate, temperature and salinity as well as predation pressure (Claassen 1998). Although middens in the region contain large quantities of mud ark (*A. trapezia*), estuary surveys revealed very low extant populations, suggesting that *A. trapezia* was more abundant in the past. Researchers in northern and northwestern Australia frequently cite the closely related *A. granosa* as 'not found in the mangrove-lined coast' (Hiscock 1997:445) and preferring 'the silty-sandy substrates of open beaches' (Hiscock 1997:447 citing Broom 1985). Although few studies have been conducted on *A. trapezia* ecology, it is known to be very similar to *A. granosa*, with the possible exception of a frequent (perhaps symbiotic) association with seagrass, which requires shallow water with relatively low turbidity (Inglis 1992). Shell assemblages dating to the last 2,000 years show a consistent decline in *A. trapezia* abundance with a concomitant increase in the representation of oyster. The presence of *A. trapezia* in local assemblages by 4,000 BP suggests that estuarine conditions were already established in the region by this time, but that conditions less favourable to this taxa have prevailed in the recent past.

Oyster (*S. glomerata*) representation also sheds light on past environmental conditions in the area. Oysters require a hard substrate for attachment and occur on sheltered rocky shores and mangroves in the mid-intertidal zone. In the absence of rocky shores in most sheltered estuaries in the region, oysters often attach to the aerial prop roots of mangroves. At the Seven Mile Creek Mound, the high proportion of non-artefactual stone in the deposit and the presence of oyster bases still attached to gravels suggest that oysters were collected from the extensive shell/rock debris on the intertidal flats adjacent to the site (Chapter 6). However, such debris beds do not occur at other major estuaries in the region, indicating that oysters may have been primarily collected from mangrove substrates. Examination of oyster bases from sites such as Eurimbula Site 1 (Chapter 12) show a high proportion of valves exhibiting distinct concavities suggesting a growth position around a cylindrical substrate. The presence of oyster bases with these features may therefore be used as a proxy for the presence of mangrove communities near to the site in question. The increasing representation of oyster in excavated assemblages dated to the last 2,000 years in estuaries where rock is absent (or minimal) is consistent with an expansion of mangroves.

Another taxon, the telescope mud whelk (*T. telescopium*), is almost entirely absent from middens on the southern Curtis Coast but is the most common shellfish species in many estuaries today, regularly occurring in densities up to $100/m^2$ (Fig. 2.8). Shanco and Timmins (1975:153) note this taxon in pools of water amongst the rearward mangroves of Eurimbula Creek. The almost complete absence of *Telescopium* from middens in the area is suggestive of a recent (<200 years) invasion by this species. A single *Telescopium* base has been observed on the surface of the Eurimbula Creek 2 site (see Chapter 11) which dates to within the last 200 years, supporting a recent origin for the proliferation of this taxa in the region. *Telescopium* is frequently recovered from shell middens pre-dating 1,000 BP in northern Australia (e.g. Beaton 1985; Hiscock 1997; Robins et al. 1998:119), where ethnographic data from western Cape York Peninsula indicate that it is a preferred food source (Cribb 1996; Isaacs 1987:176). It therefore seems unlikely that this taxon would not be collected had it been available in the local environment.

Overall, geomorphological data and patterns of mollusc availability are consistent with the establishment of estuarine conditions by 4,000 BP, with small areas of mangroves being progressively replaced by the dense mangrove forests of today. All data suggest that the modern coastal configuration is relatively recent and that, at least since the mid-Holocene, the coastal landscape appears to have consisted at all times of a broad mosaic of estuaries, sedgelands and wetlands.

Cultural setting

As elsewhere in Australia, introduced diseases and frontier violence caused major reductions in Aboriginal populations in the study area (e.g. Butlin 1983; Campbell 2002). Population reduction in many areas contributed to demographic restructuring to accommodate groups displaced by the advancing European frontier, with remnants of groups aggregating along traditional social lines. Thus by the time European observers started to enter areas such as the southern Curtis Coast, it is likely that significant changes in demographic structures and land-use strategies had already occurred. The ethnohistoric record for the southern Curtis Coast is generally very sparse, with considerable conflict both among documentary sources and between documentary and Aboriginal oral histories (see Clarkson et al. n.d. for a detailed discussion). In one of the earliest discussions of

Aboriginal lifeways in this area, Curr (1887:122, 126) used the term 'Maroonee' to describe the people occupying the coastal areas of Rodds and Bustard Bays and inland to the Many Peaks Range. Mathew (1914) also placed 'Meerooni' in this area. Brasch (1975) suggested that a dialect of Gooreng Gooreng labelled 'Guweng' occurred in this same geographical area. Tindale (1974) termed this group 'Goeng' (also listing Meerooni, Gurang Gurang and Yungkong), and suggested it covered an area from Miriam Vale south to the mouth of Baffle Creek. Popularist accounts (e.g. Buchanan 1994, 1999; Growcott and Taylor 1996) represent the Bustard Bay area as occupied by the Meeroni. Buchanan (1994) recounts a conversation with Elsie and Bertha Bowden in the 1970s who were resident on Middle Island between 1910 and 1974 who stated that the local 'tribe' was called 'Meeroni', although it is unclear whether they derived this term from actual interaction with local Aboriginal people or from second-hand sources. Williams (1981:62) was not certain about including the area fringing Bustard Bay within Gooreng Gooreng country, despite extensive interviews with Aboriginal people from the study area. Clarkson et al. (n.d.) suggested that as the various terms refer to country today identified with Gooreng Gooreng speakers (see also Curr's 1887 language list for the Meerooni of Baffle Creek where 'no'= 'Gooraong'), they are unlikely to refer to the name of languages *per se*, but may be those of dialect groups or subgroups of Gooreng Gooreng (Burke 1993:8).

While Aboriginal people in the study area undoubtedly had links to the north, most of their cultural ties at the time of European contact were to the south. The Gooreng Gooreng language, for example, is more closely related to Gubbi Gubbi and Waka Waka rather than to languages in the Rockhampton area (Jolly 1994). Ethnohistoric accounts document a clear schism between Gooreng Gooreng and Bayali peoples and their Darambul neighbours to the north, which manifested itself in warfare and separate alliance networks (Horton 1994:112).

The very few documentary sources which relate specifically to the southern Curtis Coast begin with Banks' observation on 23 May 1770 of two Aboriginal men walking along the beach just south of Bustard Bay (Beaglehole 1963:65). The following day, a party from the *Endeavour* went ashore at Bustard Bay to inspect the country, noting 'innumerable Oysters, Hammer oysters and many more sorts' and a recently-vacated occupation site:

> Those who stayd on board the ship saw about 20 of the natives, who came down abreast of the ship and stood upon the beach for some time looking at her, after which they went into the woods; we on shore saw none. Many large fires were made at a distance from us where probably the people were. One small one was in our neighbourhood, to this we went; it was burning when we came to it, but the people were gone; near it was left several vessels of bark which we conceivd were intended for water buckets, several shells and fish bones, the remainder I suppose of their last meal. Near the fires, for there were 6 or 7 small ones, were as many peices of soft bark of about the length and breadth of a man: these we supposd to be their beds: on the windward side of the fires was a small shade about a foot high made of bark likewise. The whole was in a thicket of close trees, defended by them from the wind; whether it was realy or not the place of their abode we can only guess. We saw no signs of a house or any thing like the ruins of an old one, and from the ground being much trod we concluded that they had for some time remain in that place (Banks in Beaglehole 1963:67).

Other members of the landing party also reported the tail of a land animal at the camp to those that remained on the ship (Pickersgill in Bladen 1892:218). Setting sail and journeying north the next day Banks noted fires at some distance 'tho not many' (Banks in Beaglehole 1963). Banks also commented on the rich coastal resources of the area:

> On the shoals and sand banks near the shore of the bay were many large birds ... On the shore were many birds, one species of Bustard, of which we shot a single bird as large as a good Turkey. The sea seemd to abound in fish ... on the mud banks under the mangrove trees were

innumerable Oysters, Hammer oysters and many more sorts among which were a large proportion of small Pearl oysters (Banks in Beaglehole 1963:66).

Subsequent sources (mainly from ships and exploratory vessels) make passing references to sightings of Aboriginal people, material culture or smoke from campfires in the general region (e.g. Flinders 1814; Oxley 1825; Perry and Simpson 1962). In August 1802, Flinders (1814:15–6) noted bark canoes, turtle remains and scoop nets at the southern end of Curtis Island while the ship's artist made sketches of dwellings. In 1846, MacGillivray (1852:57) made the following observations while visiting Port Curtis:

> During our stay at Port Curtis, we had no inter-course whatever with the natives, although anxious to establish friendly communication. With the aid of the spyglass, we could occasionally make out a few, chiefly women, collecting shell-fish on the mud flats of the main land, and their fires were daily seen in every direction.

Although there have been suggestions of patterns of coastal transhumance related to water shortages (e.g. Oxley 1825), early historical sources document the presence of Aboriginal populations on the coast throughout the year (Burke 1993; Clarkson et al. n.d.) and water stored in coastal sandmasses appears to have been perennially available (Buchanan 1999).

The most recent ethnohistoric account for the study area dates from October 1846, when Colonel George Barney on board the *Cornubia en route* to Gladstone encountered Aborigines close to their camp while searching the southern entrance of Bustard Bay for freshwater with which to fill the ship's casks. Barney was shown a small freshwater soak in dense scrub about 100m from the base of Round Hill Head (McDonald 1988:10).

From around 1850 onwards Europeans made increasing inroads into the study region. The Wide Bay and Burnett Pastoral Districts were declared by 1848 and Port Curtis by the 1856 census. The Moreton Bay Penal Settlement had been broken up in 1840 and declared open for free settlement in 1842 but by this time squatters were already at the 50 mile settlement limit (Taylor 1967:40). Taylor (1967:62) notes that in the Burnett and Wide Bay Pastoral Districts, large coastal areas remained unused by Europeans in the early period of settlement and retained resident Aboriginal populations into the late 1850s and early 1860s. Aboriginal oral histories attest to continuing use of the area into the early twentieth century. A senior local Aboriginal community member, Connie Walker (pers. comm., 1999), recalls that her family used to visit the Round Hill Creek area from Greenvale Station, near Lowmead, to fish and to make boomerangs, shields and 'nulla nullas'.

Colonial impact, notably in the form of frontier violence and introduced diseases, precipitated the demographic collapse of local Aboriginal social groups and virtual abandonment of the near-coastal landscape by the late nineteenth century. Of the Tooloola of the Gladstone area to the immediate north, Curr (1887) estimated that by 1882 a pre-European population of 700 had been reduced to 43. During the 1850s, the Native Mounted Police were active in the region and several massacres are known to have occurred in the Miriam Vale area (Clarkson et al. n.d.). In the main, by the late nineteenth century Aboriginal populations in the region had coalesced into fringe camps at major European townships such as Miriam Vale in the west and Gladstone in the north (Roth 1898). Cox (1888) described wax figures taken from an Aboriginal camp at Miriam Vale in 1864 and Roth (1898, 1909) describes nets, clubs and shields made by people living in Miriam Vale at the time of his visit in 1898. Although Aboriginal people may have occasionally visited the area after the 1920s from local Aboriginal population centres such as Berajondo and Gladstone, the entire region was effectively depopulated by the removal of Aboriginal people to reserves and missions (particularly Barambah, Woorabinda and Bogimbah) under the provisions of the *Aborigines Protection and Restriction of the Sale of Opium Act 1897* (Blake 1991; Evans 1991).

Today Gooreng Gooreng people maintain active connections to country on the southern Curtis Coast through a range of activities, including residence, camping, fishing, cultural festivals and conducting cultural heritage impact assessments. Contemporary Gooreng Gooreng understandings of landscape are informed by a narrative tradition that articulates particular or general features of the natural and cultural world to specific mythological figures, historical figures or events, or to each other (Clarkson et al. n.d.; Williams 1981).

Previous archaeological research

Knowledge of the archaeological record of the study region was extremely limited prior to the 1990s. In fact, prior to 1993 only eight sites were recorded on the Queensland Environmental Protection Agency's (EPA) Indigenous Sites Database for the study area. This lack is particularly evident in comparison to the extensive investigations conducted in the Central Queensland Highlands to the west (e.g. Beaton 1977; Johnson 1979; Morwood 1979, 1981, 1984; Mulvaney and Joyce 1965; Richardson 1992, 1996; Stern 1980; Walsh 1984), the Keppel and Whitsunday Islands and adjacent coast to the north (e.g. Barker 1989, 1991, 1995, 1996; Rowland 1981, 1982, 1985, 1992, 2002) and the Great Sandy Region and Moreton Bay to the south (e.g. Frankland 1990; Hall 1982; Lauer 1977, 1979; McNiven 1990a, 1991a, 1991b, 1993, 1998; McNiven et al. 2002; Ulm and Hall 1996).

Despite its paucity, archaeological research in the Burnett-Curtis region in fact extends back to the 1890s, when interest arose in the extensive Burnett River Engravings site located in the bed of the Burnett River near Bundaberg. This site was included in early discussions of Aboriginal rock art by Mathews (1897, 1910) and Elkin (1949), as well as in more recent works by Quinnell (1976) and Maynard (1976, 1979). Sutcliffe (1972, 1974), who supervised a salvage operation at the site in the early 1970s, wrote a draft Master of Arts thesis on the salvage process, but apparently never submitted the manuscript for examination. Rola-Wojciechowski (1983) undertook a quantitative analysis of a sample of the engravings at the site as part of her Honours thesis at the University of New England. She suggested that the predominantly non-figurative art displayed affinities with other sites in southeast Queensland, but was distinct from assemblages recorded for the Central Queensland Highlands. Similar conclusions were reached in Chapman's (1999, 2002) recent study of the regional pigment art assemblage.

In the latter part of the nineteenth century and throughout the twentieth century, various Aboriginal artefacts from the general region were donated to or otherwise acquired by private collectors, local museums, the Queensland Museum and the Anthropology Museum at the University of Queensland. Unfortunately, many of these artefacts are not well-documented and it is difficult to determine whether they truly originated in the study area.

In 1915, Hamlyn-Harris conducted a surface collection of 330 stone artefacts and samples of shell from a 'feeding-ground' at Bargara, east of Bundaberg, after the site was brought to his attention by Lionel C. Ball of the Queensland Geological Survey. Ball had donated a small collection of 24 stone artefacts from the site to the Queensland Museum earlier that year. Ball (1915) noted that the 'chips are scattered in some profusion along the summits of the sand ridges' which 'rise immediately behind the beach'. This pattern of site location on beach ridges in the study area has been confirmed by recent research (see Lilley 1994a; Lilley and Ulm 1995; Ulm and Lilley 1999).

Hamlyn-Harris (1915:104) noted 'secondary chipping' or retouch on some artefacts in the collection. The surface collection consisted of 'shells, together with one blank (unfinished) axe of silicified sandstone, two primitive stone tools also made of the same material, a large quantity of flakes, chips, scrapers, drills and gouges made of silicified sandstone, jasperoid, petrified wood, quartz, etc., and a basalt hammer' (Hamlyn-Harris 1915:104). Hamlyn-Harris (1915:104) concluded that many of the raw materials represented in the collection derived from non-local sources and 'had evidently been brought some distance by them for this purpose'. In 1979, this stone artefact collection was analysed by students from the University of Queensland (Findlay 1979; Horsfall

1979; Horsfall and Findlay 1979). The presence of numerous stone artefacts, including retouched flakes, and abundant shellfish remains described for the site led Horsfall and Findlay (1979:6) to suggest a pattern of 'intensive and/or prolonged use of the site'.

In 1970, Marks reported the existence of earthen circles ('bora rings') at Rosedale Station and at Koonawalla and Woodside, both near Bororen. In a 1976 synthesis of archaeological sites in the Wide Bay-Burnett regions prepared by the Queensland Museum, only a single archaeological site is recorded for the study region (a stone arrangement on Hummock Hill Island). From the late 1970s, several archaeological sites in the area were recorded by State Government staff, including the Crevasses Art Site (JE:A25) at Cania Gorge, which was recorded in 1982, and a quarry (KE:A05) and axe grinding grooves (KE:A06) on the coast just south of Agnes Water, recorded in 1978. Smith (1980) reported local advice that a stone arrangement had been destroyed during construction of the road through the Town of Seventeen Seventy. It was not until the mid-to-late 1980s, however, that the number of reported Aboriginal archaeological sites in the study region increased markedly with the growing numbers of surveys carried out by independent archaeological consultants as part of cultural heritage impact studies (see Alfredson 1987, 1989, 1990, 1991, 1992, 1993; Barker 1993; Davies 1994; Duncum 1991; Gorecki 1995; Hall 1980a, 1981, 1985; Hatte 1992; Hill 1978; Lilley 1980, 1994a, 1994b, 1995a, 1995b, 1995c, 1995d, 1995e; Neal 1986; Reid et al. 2000; Spencer 1995; Ulm 2000a, 2001). With the exception of a few broad overviews (e.g. Gorecki 1995), the majority of these consultancy reports have focussed on geographically-limited development impact areas and adopted discovery-oriented research designs rather than predictive site location models. Further, most of these studies have been undertaken in coastal or near-coastal areas, creating a sampling bias against subcoastal and inland parts of the region.

Many of the foregoing surveys report little or no archaeological material (e.g. Alfredson 1989, 1990, 1991, 1992; Hall 1980a, 1981; Hill 1978; Lilley 1994b, 1995a, 1995b, 1995c, 1995d, 1995e; Spencer 1995), although some sites were surface-collected and several subjected to limited analysis (see Alfredson 1987). This apparent scarcity of archaeological material in the region led Alfredson (1990:10) to comment that:

> The present survey results are in keeping with the results of the other archaeological surveys conducted in the region. The relatively sparse archaeological record does not indicate a large Aboriginal population inhabiting this part of the coast. It seems that the well-watered country behind the coastal ranges rather the [sic] coast may have been the main focus for the regional Aboriginal population.

On the basis of the small number of artefacts observed east of Monto, Spencer (1995:21) suggested that 'in spite of a richly diverse resource base … few people lived and foraged here', and concluded that 'more studies in this region will shed further light on this matter, but until then we must assume that the lithic resources had little importance to the Aboriginal occupants of the area' (Spencer 1995:22).

During an impact assessment survey for Awoonga Dam, southwest of Gladstone, Hall (reported by Lilley 1980) identified four sites on the banks of the Boyne River. Only one exhibited significant integrity, the others having been severely disturbed by European activities. Hiscock (1982) conducted the first detailed archaeological investigations and excavations in the study region while following up Hall's (Lilley 1980) original study. Hiscock excavated a total of $1m^2$ in addition to recording surface artefact scatters and the attributes of samples of the artefacts they contained. The undisturbed site covered approximately $8,700m^2$ and exhibited artefact densities up to $29/m^2$. The assemblage was dominated by flakes but also included cores, retouched flakes and flaked pieces, with artefacts restricted to the top 6cm of the deposit. No organic material suitable for radiocarbon dating was recovered. Hiscock (1982:20) argued that the site was less than 1,000 years old on the grounds that the deposit was shallow, the river bank itself was probably very recent and the artefacts did not include backed blades or points suggestive of an earlier age. He (1982:19) concluded that the 'archaeological evidence does therefore suggest that Aboriginal activities were concentrated close to the Boyne River'.

In contrast with the foregoing picture, several studies conducted over the last decade have revealed abundant Aboriginal cultural materials in the region, particularly on the coast (Figs 2.2, 2.9–2.11). Long-term local residents in the area were aware of many archaeological sites in the region, concluding that:

> Axe flakings and remains of shell fish on Middle Island, Eurimbula, Round Hill Creek and
> Agnes Water [provide] the evidence that these were the main camping grounds of the Meerooni
> in the Bustard Bay area (Growcott and Taylor 1996:5).

In 1986 two shell middens were recorded by Neal (1986) on the west bank of Seven Mile Creek as part of an environmental impact assessment. In the same year, Rowland (1987) conducted a 'broad cursory investigation' of the study area as part of general surveys of the coast between Elliott Heads and Turkey Beach. Prior to his survey, only nine sites were recorded on the Queensland State Site Register for the Bundaberg 1:250,000 map sheet. Rowland located a further seven sites, although only one, KE:A11 just to the south of the Town of Seventeen Seventy, was reported to be of any significant size. Rowland (1987:17) noted that 'substantial middens are rare', sites are located 'either atop rocky headlands or in sheltered estuaries', and that 'smaller scatters of shells are located along open beaches'. He (1987:17) concluded that:

> Whether this is a true reflection of Aboriginal settlement patterns in the area or an expression of
> geomorphological factors affecting preservation and visibility is a problem still to be resolved.
> Certainly the extent of erosion along the open coastal dune systems of the area would suggest
> that the loss of sites may be an important factor affecting the above pattern.

Two other localised surveys have been conducted in the area, one on Facing Island (Ringland 1978), where seven extensive shell midden sites were located, and one taking in part of Eurimbula National Park, where an extensive stratified midden deposit on the western bank of Round Hill Creek was identified (later registered as KE:A49–KE:A54) (Godwin 1990).

In 1993, Burke (1993) conducted selective surveys of the coast between Raglan Creek in the north and Agnes Water in the south and up to 1km inland from the mainland coastline as part of the Curtis Coast Scan project (see QDEH 1994). Although it extended south to overlap with Rowland's survey at Round Hill Creek, Burke's study was primarily concerned with the area to the north of Rodds Peninsula. In the southern Curtis Coast study area, Burke (1993) documented 93 sites, including shell middens (n=77), stone artefact scatters (n=12), quarries (n=2), stone-walled fishtraps (n=1) and scarred trees (n=1). Overall, Burke (1993) found that sites were most commonly located on level or gently inclined dune surfaces in low energy estuarine environments.

Gooreng Gooreng Cultural Heritage Project

Since 1993, archaeological surveys and excavations have been undertaken on the southern Curtis Coast as the coastal component of the Gooreng Gooreng Cultural Heritage Project, augmented by a number of cultural heritage impact studies (Lilley 1994a, 1995e). Together these investigations were designed to expand the scope of the earlier, more limited or project-specific surveys discussed above. Of particular interest were questions concerning the antiquity of human occupation in the coastal region and whether the concentration of sites in estuaries and near absence of material on ocean beaches noted by Rowland (1987) reflected past Aboriginal behaviour, recent geological processes or patterns of archaeological research.

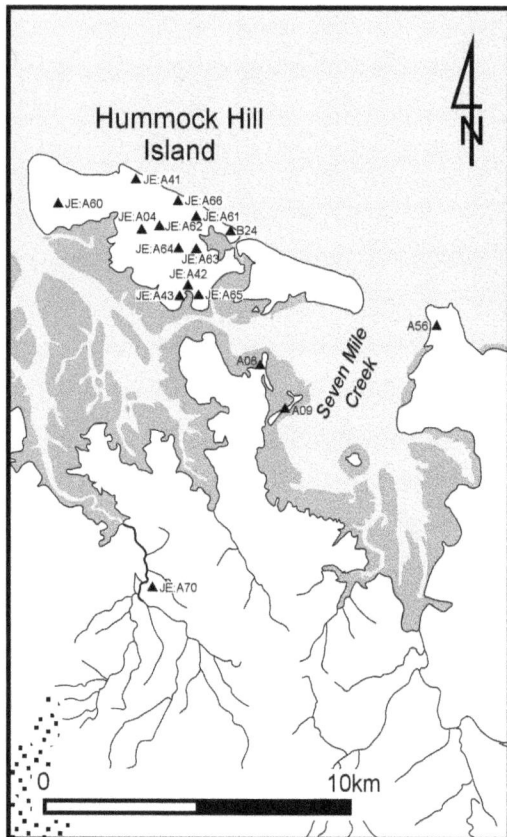

Figure 2.9 Northern segment of the southern Curtis Coast study area showing recorded archaeological sites as triangles in the Hummock Hill Island and Seven Mile Creek areas. EPA site numbers are shown. Sites without a 'JE:' prefix have a 'KE:' prefix which has been omitted owing to space limitations.

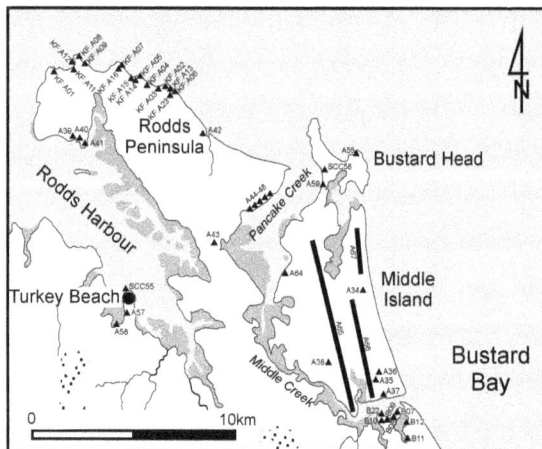

Figure 2.10 Central segment of the southern Curtis Coast study area showing recorded archaeological sites as triangles in the Turkey Beach, Rodds Peninsula and Middle Island areas. EPA site numbers are shown where available. Sites without a 'JF:' prefix have a 'KE:' prefix which has been omitted owing to space limitations. Sites with a 'SCC' prefix are currently unregistered sites.

On the coast, systematic site surveys undertaken as transects were conducted in all major environmental zones including open beaches and rocky headlands, marine estuary, swamp and wetland margins, and the coastal ranges. To date, however, the majority of investigations have focussed on near-coastal landscapes. The entire open coastline between Wreck Rock and Richards Point has been systematically surveyed, as have the lower estuarine margins of Round Hill, Eurimbula, Middle, Jenny Lind, Pancake, Falls and Mort Creeks (Figs 2.1, 2.9–2.11). Inland areas have proven more difficult to access owing to restricted access to freehold land, a lack of visibility and an absence of access tracks in many areas. The whole of Middle Island was intensively surveyed by Lilley (1994a) using a grid of drilling-lines graded across the island for mineral sand exploration. Some small transects have also been undertaken on the northeast margin of a large swamp which dominates the western half of the Deepwater Section of Eurimbula National Park and parts of Round Hill National Park (Lilley et al. 1997). Poor visibility away from coast and estuary margins was found to be a major impediment to site detection (see also Burke 1993:23, 32–3).

Survey crews were deployed to sample the different micro-environments that were encountered. The general strategy involved walking through the subject area in line-abreast at one visual distance separation between walkers and up to 50m in total width, focussing on areas of high ground surface visibility. In some areas, the crew was split into two teams to examine different zones, such as foredunes and backdunes, simultaneously. While field crews took advantage of every opportunity to examine soil profiles in road cuttings, creek banks and the like, no excavation, augering or other subsurface testing was undertaken during the survey phase of the study (cf. Burke 1993:23–4). Site locations were established with the use of topographic maps, aerial photographs and Global Positioning System (GPS) readings.

A major complication in effectively conducting and evaluating archaeological survey data from the region was a lack of access to the earlier site recordings made by Burke (1993) and held by the Queensland Environmental Protection Agency. Although abbreviated survey results are available in Burke's (1993) report, an absence of detailed site descriptions and inconsistencies in location data prevented accurate ground-truthing of previously recorded sites. These details have recently been obtained under the provisions of the *Freedom of Information Act 1992*. As a result of these problems, some of the sites reported by Lilley et al. (1997:Table 2) had already been recorded by Burke (1993). Further confusion arose as some of the pre-allocated EPA site numbers assigned to Burke had already been allocated to other sites by the time they came to be registered (Melissa Carter, Queensland Environmental Protection Agency, pers. comm., 1999) and many of Burke's

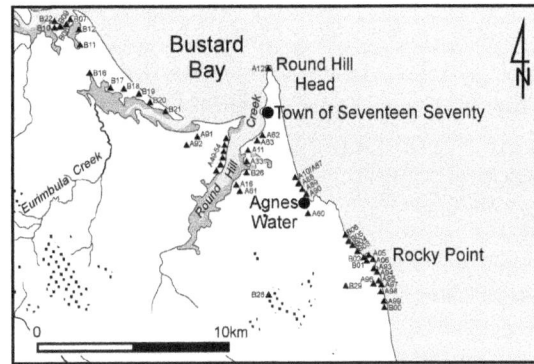

Figure 2.11 Southern segment of the southern Curtis Coast study area showing recorded archaeological sites as triangles in the Agnes Water, Round Hill Creek, Eurimbula Creek and Middle Creek areas. EPA site numbers are shown. All sites have a 'KE:' prefix which has been omitted owing to space limitations.

sites were subsequently conflated on the basis of proximity when entered onto the database (Burke's original 93 sites from this area were registered as only 51 sites). Additionally, Burke (1993:Appendix 5) assigned the same pre-allocated site numbers to more than one site on a number of occasions (e.g. KE:A37–KE:A40, KE:A44–KE:A46), which was not addressed during the registration process. It should be noted that most of the site numbers listed in QDEH (1994:Appendix XVI) were among Burke's (1993) pre-allocated site numbers, which were superceded when registered by the Queensland Environmental Protection Agency. To simplify the multiple site designations, Appendix 3 presents a key which links actually-registered sites to Burke's (1993) field numbers and pre-allocated site designations, as well as those assigned by other researchers, including those employed on the Gooreng Gooreng Cultural Heritage Project.

Appendix 2 is an attempt to synthesise the survey data for the area included in Rowland (1987), Burke (1993), Lilley et al. (1997) and the Queensland Environmental Protection Agency's (EPA) Indigenous Sites Database (obtained under the *Freedom of Information Act 1992*). This task was complicated by variations in site definitions and recording strategies combined with problems in site provenance data related primarily to inadequate location data. For example, the site Lilley et al. (1997) and Ulm et al. (1999a) refer to as Eurimbula Site 1 (SCC43) is registered as six separate sites on the Queensland Environmental Protection Agency site database (KE:A49–KE:A54), which is a secondary conflation of Burke's (1993) original 20 separate sites.

Clearly, these variations would render spurious any absolute quantitative comparisons between the sites presented in Appendix 2. The general descriptions of the nature of observed archaeological materials, however, provide a basis for broadly characterising the archaeological record of the study region.

A total of 79 Aboriginal cultural places is listed in Appendix 2, including recording details, basic location data and a brief site description. In most cases this description includes all information available for the site. Figures 2.9–2.11 show site locations. Site types include stone quarries, axe grinding grooves, stone-walled tidal fishtraps (Fig. 2.12), scarred trees (Fig. 2.13), shell middens (Figs 2.14–2.16), stone artefact scatters and contact period sites. Low density shell middens are clearly the most common archaeological expression of Aboriginal behaviour in the area, dominated, without exception, by mud ark (*A. trapezia*) and/or rock oyster (*S. glomerata*) (see also Burke 1993:40). Stone artefacts are commonly associated with larger shell middens (Fig. 2.17) and/or rocky headlands. Lithic raw materials are dominated by locally available rhyolitic tuff and quartz with occasional non-local silcrete and banded rhyolite artefacts. Flaked glass artefacts also occur on the surface of two large shell midden complexes. Several glass artefacts exhibit clear evidence for use in plant processing activities (Ulm et al. 1999b).

These extensive surveys have revealed a consistent pattern of site location. Extensive, stratified shell midden deposits with evidence for multiple occupations and diverse activities are limited exclusively to tidal estuary margins in close proximity to their mouths. Such sites have been located on the northern and southern banks of Round Hill, Middle and Pancake Creeks. All are exposed in erosion faces extending over several kilometres. Relatively small, low density surface scatters of marine shell and stone artefacts made on local materials were noted on exposed headlands. Low density single-taxa scatters of shellfish remains on exposed coastal headlands are likely to be accumulated by birds rather than humans (especially Pacific gulls) (see Claassen 1998:72). Although occasional artefacts are encountered on these headlands, they have not been noted to co-occur with shell scatters. The extremely exposed situation of these shell deposits means that they are unlikely to survive long periods of time, with the absence of sediments preventing the possibility of burial. Midden deposits on the open beaches themselves were generally very limited in extent and composition and often deflated. Beach ridges and transgressive dunes located adjacent to the modern coastline contain abundant, extensive but shallow site complexes (Lilley 1994a; Lilley and Ulm 1995).

Prior to the commencement of the research reported in this monograph, two of the largest complexes of shell midden deposits in the area were test excavated: the Mort Creek Site Complex (KE:A41) on the west bank of Mort Creek at the northern end of Rodds Peninsula (see Chapter 7 for details) and Eurimbula Site 1 (KE:A49–KE:A54) on the west bank of Round Hill Creek (see Chapter 12 for details). The sites were radiocarbon dated to 2,300 cal BP and 3,200 cal BP respectively.

Other dated archaeological sites in the region

In addition to the eight excavated and radiocarbon dated sites discussed at length in this monograph, four other sites were subject to basic recording and radiocarbon dating. Results are briefly described below and are incorporated into the regional synthesis in Chapter 14.

Table 2.2 Radiocarbon dates from unexcavated sites in the study region (see Appendix 1 for full radiometric data for each determination). ABM=Agnes Beach Midden; MISS=Middle Island Sandblow Site; RHCM=Round Hill Creek Mound; WCM=Worthington Creek Midden. Dates on shell were calibrated using a ΔR value of +10±7, except Wk-10090 where ΔR= –155±55 (see Chapter 4 for further details).

SITE	SQUARE	XU	DEPTH (cm)	LAB. NO.	SAMPLE	$\delta^{13}C$ (‰)	^{14}C AGE	CALIBRATED AGE/S
ABM	–	–	–	Wk-10969	charcoal	-27.1±0.2	266±87	485(289)0*
ABM	–	–	–	Wk-11280	*D. deltoides*	0.8±0.2	674±47	420(294)247
MISS	A	1	0	Wk-7679	*D. deltoides*	1.1±0.2	980±50	642(545)494
MISS	B	1	0	Wk-10091	*D. deltoides*	0.9±0.2	730±39	452(362)282
MISS	C	1	0	Wk-10092	*D. deltoides*	1.2±0.2	958±40	620(534)492
MISS	D	1	0	Wk-10093	*D. deltoides*	0.9±0.2	559±42	279(234)0*
RHCM	–	–	–	Wk-10090	*A. trapezia*	-0.3±0.2	1910±42	1810(1623)1482
WCM	–	–	5	Wk-10089	*S. glomerata*	-3.4±0.2	349±60	modern

Agnes Beach Midden (ABM)

Recent storm-surge activity exposed a small discrete *in situ* layer of shell dominated by pipi (*D. deltoides*) and stone artefacts in a truncated frontal dune on the high energy beach adjacent to the town of Agnes Water (Latitude: 24°12′28″S; Longitude: 151°54′12″E) (Figs 2.11, 2.14). This layer of shell is part of a shell midden/artefact scatter originally recorded by Rowland (1987) and re-recorded by the author (Lilley et al. 1997; Ulm and Lilley 1999). The site comprises several small, low density surface scatters of shell and occasional stone artefacts. The site is registered on the Queensland Environmental Protection Agency's (EPA) Indigenous Sites Database as KE:A10 and KE:A87 and Queensland Museum Scientific Collection Number S233 (Fig. 2.11). Despite extensive surveys this is one of the few stratified deposits containing organic material located on the open coast in the study region. After cleaning back the erosion section, a small area (10cm × 10cm × 10cm) of the *in situ* layer was removed *in toto* for laboratory analysis. Two radiocarbon dates were obtained from the recovered material to investigate local open water marine reservoir factors (see Chapter 4 for further details). The shell sample comprises valves of pipi while the charcoal sample is based on large blocky fragments of charcoal removed from the exposed lens in direct association with the shell fraction. The shell/charcoal pair (Wk-11280/Wk-10969) indicate that this site was occupied around 300 years ago (Table 2.2).

Middle Island Sandblow Site (MISS)

The Middle Island Sandblow Site is an extensive shell and stone scatter exposed on the surface of the sandblow located on the northeast section of Middle Island, immediately south of Bustard Head. The exposure is c.2.8km long and covers a minimum area of 140,000m². The approximate centre of the site is situated 15.5km northwest of Round Hill Head (Latitude: 24°03′44″S; Longitude: 151°45′56″E). The site is registered on the EPA's Indigenous Sites Database as KE:A67 and Queensland Museum Scientific Collection Number S234 (Fig. 2.10). It is located on the eastern, active face of the sandblow, parallel to the ocean beach to the east and bordered by Jenny Lind Creek to the north and west. The vast majority of shell and stone material is exposed immediately upslope of extensive pumice rafts located on a low area of trailing dunes, partially vegetated by dense stands of immature black she-oak (*Allocasuarina littoralis*) separating the active sandblow from

Figure 2.12 Stone-walled tidal fishtrap at Richards Point on Rodds Peninsula. This is one of two such features recorded in the study region. The second fishtrap is located in Mort Creek 4km to the south (see Chapter 8). Facing north (Photograph: Ian Lilley).

Figure 2.13 Scarred tree on eucalypt at Agnes Water (KE:A60) (Photograph: Environmental Protection Agency).

Figure 2.14 Agnes Beach Midden. Facing southwest.

Figure 2.15 Worthington Creek Midden. Facing northwest.

Figure 2.16 Middle Island Sandblow Site, showing pipi (*D. deltoides*) scatter on the surface of the sandblow. A range of cultural and non-cultural material is exposed on the eastern margin of the sandblow. Facing north.

Figure 2.17 Middle Island Sandblow Site, showing microgranite grinding implement located on the surface of the sandblow. Obviously modified and apparently unmodified pieces of stone are common along the eastern margin of the sandblow. Facing south.

the frontal dunes bordering the beach. Cultural deposits of shell and stone have been artificially concentrated by active dune-building processes resulting in deflation and conflation of archaeological materials onto a surface which continues to be differentially exposed by recent sand movement. Shell densities range from $<1/m^2$ to $150/m^2$ across the site (Fig. 2.16). In all areas, pipi (*D. deltoides*) dominates the shellfish assemblage, comprising at least 99% of all visible shellfish remains. Other shellfish taxa include lined nerite (*Nerita lineata*), hercules club whelk (*P. ebininus*), rock oyster (*S. glomerata*) and mud ark (*A. trapezia*). In addition, occasional toothless clam (*Anodontia edentula*) and pearly nautilus (*Nautilus pompilius*) shells were also noted. The pearly nautilus is a deep water species which is sometimes washed up on eastern Queensland beaches. While not documented as a food source, historical records document the use of nautilus shell to manufacture oval-shaped pendants at Cooloola and Fraser Island to the south (McNiven 1998). Occasional stone artefacts are distributed across the site (Fig. 2.17). The remains of at least six macropods, four birds and five cattle were also observed, all thought to be of a very recent origin.

Burke visited the site area during surveys conducted in 1993. While no mention of it is made in either of the two reports arising from the Curtis Coast Scan project (Burke 1993; QDEH 1994), notations made on the mounts of Burke's 35mm slides of this area held by the EPA include the words 'not cultural'. The cultural status of at least parts of this deposit was recognised by Lilley and Ulm in 1994 (see Lilley 1994a). Although all cultural material appears to be restricted to the surface, the site is significant as the only large open beach archaeological site known between Fraser Island to the south and Facing Island to the north. During June 2001, basic attributes of all stone material visible on the surface of the sandblow was recorded. Five test excavations (Squares A–E) were also conducted, spaced across the length of the exposure to assess the subsurface extent of the site and to obtain samples for radiocarbon dating.

In total, 342 pieces of stone weighing over 300kg were recorded across the eastern margin of the sandblow. The assemblage included backed blades, flakes, broken flakes, flaked pieces, cores, hammerstones and grinding implements manufactured on a range of raw materials including rhyolitic tuff, microgranite, quartz, quartzite and silcrete. Most of the stone assemblage, however, is not obviously modified and its cultural status remains equivocal. Excavation revealed that all cultural material is restricted to surface deposits. The excavated material was dominated by pipi (*D. deltoides*) with minute quantities of bone, non-artefactual stone and pumice (Table 2.3). The shell valves were very friable, probably reflecting long periods of sun-bleaching and physical erosion by wind-blown sand particles. Given the high mobility of the sandmass and the completely unconsolidated sediments, none of the exposed cultural material is thought to be in primary depositional context. The four radiocarbon dates available for the site were obtained on samples of pipi (*D. deltoides*) valves (Table 2.2). The dates indicate that much of the material currently exposed on the surface dates to the last 500 years.

Table 2.3 Middle Island Sandblow Site, Squares A–E: summary excavation data and dominant materials.

SQUARE	SIZE (m²)	DEPTH (cm)	VOLUME (l)	pH	PIPI (g)	OTHER SHELL (g)	BONE (g)	ROCK (g)	CHARCOAL (g)	ORGANIC (g)
A	0.25	10	40	8.5	124.4	–	1.1	1.9	–	2.2
B	1	5	c.100	7	197.4	–	<0.1	–	0.1	0.8
C	1	5	c.100	NA	152.9	22	<0.1	3.6	<0.1	2.1
D	1	5	c.100	NA	734.4	–	0.1	0.3	0.2	35
E	1	5	c.100	NA	–	115	–	1.8	–	<0.1
Total	4.25	–	c.440	–	1209.1	137	1.2	7.6	0.3	40.1

Round Hill Creek Mound (RHCM)

The Round Hill Creek Mound is a large shell mound located in open eucalypt woodland on a low rocky terrace close to the eastern margin of Round Hill Creek (Latitude: 24°12′09″S; Longitude: 151°51′56″E). Dense shell material is spread over an area of at least 16m x 10m with a minimum depth of 50cm visible in an exposed section created by bulldozer activity. The visible assemblage is dominated by mud ark (*A. trapezia*) with oyster (*S. glomerata*), stone artefacts, bone and charcoal also noted. The site was initially recorded by Rowland (1987) and subsequently visited by Godwin (1990), Burke (1993) and the author (Lilley et al. 1997; Ulm and Lilley 1999). The site is registered on the EPA's Indigenous Sites Database as KE:A16 and Queensland Museum Scientific Collection Number S235 (Fig. 2.11). A radiocarbon date was obtained on a sample of *A. trapezia* valves removed from an apparently undisturbed area half-way down the exposed section (Table 2.2). The date indicates that the site was in use around 1,600 cal BP.

Worthington Creek Midden (WCM)

The Worthington Creek Midden is an extensive, shallow, linear shell midden exposed in section on the west bank of Worthington Creek (Latitude: 24°07′25″S; Longitude: 151°40′59″E). The midden material is located along the top margin of a high (c.4m) creek erosion bank (Fig. 2.15). Sandstone is exposed at the base of the section, overlain by a thick layer of light brown clays and a thin veneer of eroding top soils containing the shell material. Shell is visible along a segment of bank 350m in length and up to 5cm deep. The shell assemblage is dominated by oyster (*S. glomerata*) with scallop (*Pinctada albina sugillata*) the only other taxon observed. This site has not yet been assigned an EPA site number and has the Gooreng Gooreng Cultural Heritage Project identification number of SCC64. It is registered as Queensland Museum Scientific Collection Number S236. A radiocarbon date was obtained on a small sample of oyster (*S. glomerata*) valves removed from an apparently undisturbed area containing the deepest material visible in the erosion section. It should be noted that oyster was the only sample material available and it is acknowledged that oyster is not ideal for radiocarbon dating purposes owing to uncertainties introduced by the growth structure of the shell (Wang and van Strydonck 1997). The 'modern' result dates occupation of the site to the recent past, although the subsurface context and absence of European materials confirms an Aboriginal origin for the deposit (Table 2.2).

Discussion

At this stage it is unclear how processes of erosion and progradation on the exposed coast have affected the differential preservation and visibility of sites, as has been observed elsewhere (e.g. Bird 1992; Godfrey 1989; Head 1987; Rowland 1989). Intensive surveys of the open coastline suggest an almost complete absence of cultural material on the low frontal dunes bordering the ocean beach and a preponderance of material on the lower margins of major estuaries. That this pattern is an accurate reflection of past Aboriginal behaviours rather than an artificial pattern created by differential preservation of cultural materials by erosion of the exposed coast is supported not only by the absence of material on current coasts, but also by the absence of cultural material on particular coastal landforms. Although scatters of midden shell and stone artefacts have been recorded on numerous headlands on the exposed coast, they are very sparse compared to the abundance of material on estuary margins, suggesting a qualitative pattern of settlement preference in these areas. This pattern may in part reflect patterns of resource distribution, with lower creek margin sites often situated adjacent to a range of potential resource zones. Significantly, despite the proximity of large midden deposits on the lower estuaries to open beach habitats (in some cases <100m), only very occasional pipi (*D. deltoides*) valves have been observed or recovered from excavated deposits. Only one site in the region, the Middle Island Sandblow Site (KE:A67), contains significant quantities of this species. *D. deltoides* is the diagnostic signature of Aboriginal use of open coasts in southeast Queensland (e.g. Hall 1980b; McNiven 1990a, 1998). The pattern on the southern Curtis Coast suggests that either this species was consumed and discarded at locations other than those identified, and which have been obscured or destroyed by erosional processes, or that the open coast was not a primary focus of resource extraction in this region.

McNiven (1985, 1989) identified a similar pattern at sites at the mouth of the Maroochy River and on the Inskip Point peninsula at Cooloola. He suggested that in southeast Queensland 'major ocean beach shellfish (i.e. pipi) exploitation only occurred in contexts far removed from estuarine environments' (McNiven 1989:47), arguing that estuarine environments were preferentially exploited by people over the open beach owing to a greater productivity and diversity of resources in estuaries.

Results of previous archaeological work confirm Aboriginal occupation of the coast in this region from at least 3,000 BP, and conform with other dates obtained for the Queensland coast (Ulm et al. 1995; cf. Nicholson and Cane 1994). In particular, the dates are similar to the earliest dates obtained at the site of Booral in the Great Sandy Straits to the south (Bowen 1998; Frankland 1990), as well as those from the Keppel Islands just to the north (Rowland 1985, 1992). The general structure of the archaeological record of the region is qualitatively similar to adjacent areas to the north (Burke 1993; Border 1994) and south (McNiven 1990a, 1998) which have featured in discussions of late Holocene change in Aboriginal societies on the Queensland coast (e.g. McNiven 1999; Ulm and Hall 1996). The investigation of the archaeological resources of the southern Curtis Coast therefore has the potential to contribute to discussions of these wider issues.

Summary

The overview presented in this chapter is a baseline synthesis of the physical environment and known archaeological resources in the study area. High rates of floral and faunal diversity in the study region are related to its position as a transitional zone between temperate and tropical provinces. Although the present configuration of the coastal environment was largely created by Holocene sea-level rise, the major landscape features appear to have been in place shortly after the last major shift in sea-levels around 4,000 BP, with shoreline progradation the major active landscape process since that time. Preliminary investigations demonstrate that the southern Curtis Coast exhibits a rich and diverse archaeological record dating to the late Holocene which has the potential to contribute to discussions of wider issues in southeast Queensland archaeology.

3

Methods of investigation

Introduction

This chapter outlines the methods used to address data recovery and analytical issues. Excavation strategies and methods are outlined and sampling issues discussed. Analytical procedures adopted for the major classes of recovered materials are also presented. Detailed justification of the use of these methods is presented in Ulm (2004a:55–74).

Excavation strategy

Excavations targetted a range of sites located throughout the study area in order to develop a basic understanding of coastal land-use and construct a regional chronology upon which to examine continuities and disjunctions within and between individual site sequences (Table 3.1). Access to sites was constrained by land tenure and the remoteness of some parts of the study region. Sites were selected for excavation to represent both the most common site types encountered in the study region (large linear shell middens and smaller middens), as well as rarer site types (such as mounded shell middens and stone quarries). Excavated sites are located in a range of environmental contexts representing the diversity of landscapes in the study region. Only sites which exhibited material suitable for ^{14}C dating were selected in order to contribute to the regional chronology.

Detailed intra-site pedestrian transect surveys were used to characterise the surface distribution of cultural materials. Visibility was frequently limited to eroded sections on creek banks or crab burrow spoil and otherwise disturbed areas on the surface, owing to vegetation and leaf litter coverage and the subsurface location of many *in situ* deposits. Owing to resource limitations, excavations were generally located on or adjacent to the densest and/or deepest exposures of cultural material observed, or in the case of large sites, at several of these locations. A key problem with this strategy is the assumption that visible concentrations of cultural material are an accurate indicator of intra-site diversity and not simply a product of differential visibility. The linear form of many sites investigated and large-scale exposure along eroding creek banks mitigated this problem to some degree at most sites, although it was shown in some excavations that apparent concentrations of material in the eroding section were not encountered in adjacent excavated deposits (see discussion of Squares A–B, Pancake Creek Site Complex, Chapter 8). Detailed excavation and analysis of small (often adjacent) 50cm × 50cm sampling units maximised data recovery within resource constraints while minimising damage to the archaeological record.

Table 3.1 Summary of excavated sites (arranged north to south).

SITE	SITE TYPE	YEAR EXCAVATED	SITE AREA (m²)	AREA EXCAVATED (m²)	SIEVE SIZE (mm)	EARLIEST DATE (cal BP)
Seven Mile Creek Mound	mounded midden	2000	200	1	3	3904
Mort Creek Site Complex	midden	1998	2500	1.75	3	3310
Pancake Creek Site Complex	linear midden	1998	22500	2	3	667
Ironbark Site Complex	quarry/midden	1998	150000	4.75	3	1519
Eurimbula Creek 1	midden	1999	10	1	3	171
Eurimbula Creek 2	midden	1999	10	0.25	3	modern
Eurimbula Site 1	linear midden	1999	100000	4.25	3	3205
Tom's Creek Site Complex	midden	1999	50000	1.5	3	966

Excavation methods

A key concern of this study was to employ a consistent data recovery strategy to establish comparable datasets from a range of stratified sites in the region. At all sites excavation was limited to small test pits, with excavations conducted as single or adjoining 50cm × 50cm squares. Where several adjacent 50cm x 50cm grid squares were excavated, squares were labelled alphabetically (A, B, C etc) in a clockwise direction from the northeast square. The southeast corner of each 50cm x 50cm grid square was used as the reference point for horizontal (*x* and *y*) coordinates of plotted objects. All squares were excavated by trowel in shallow arbitrary excavation units (XUs or spits) within stratigraphic units. Excavation procedures generally followed the model outlined by Johnson (1979). All sites were excavated until cultural material ceased to be recovered. Five elevations (four corners and centre-point) were recorded at the beginning and end of each excavation unit, using a local datum, autoset level and stadia rod. Plan view photographs were taken at the completion of each excavation unit. Section drawings and photographs were made of all sides of every pit after cleaning at the end of the excavation. Specific artefacts, ecofacts and features within excavation units were also photographed *in situ*. In addition, photographs were regularly taken to record the excavations in progress and the general site context. Artefacts and ecofacts encountered *in situ* during excavation, generally above 30mm in maximum dimension, were plotted three-dimensionally at their central points and removed as individual finds. This strategy was applied to stone artefacts, charcoal concentrations, large bone elements, articulated bivalves and other cultural material encountered towards the base of cultural deposits to enhance resolution for spatial analysis. The volume of all sediment removed was measured using a graduated bucket and weighed on a tared spring balance to the nearest 0.1kg.

Excavated sediments were gently dry-sieved onsite through 3mm mesh to reduce sample mass while minimising damage to fragile elements such as fish bone. Sieving stations were set-up on tarpaulins covering the ground surface to prevent contamination of underlying sediments. All sieve residues were retained and bagged in the field. Grain-size analysis, pH readings and Munsell

Soil Color® Chart tests were each completed in the field for all XUs. Sediment samples (c.200g) were taken for each XU from the material that passed through the 3mm sieve. Each excavation crew kept a daily general log as well as detailed notes of each XU. All excavated materials were transported to the laboratory facilities at the Aboriginal and Torres Strait Islander Studies Unit, University of Queensland, Brisbane. This generic set of excavation methods was modified and/or augmented according to specific site conditions. Deviations are noted in specific site chapters.

Radiocarbon dating and calibration

This study establishes an independent absolute chronology based on a large series of radiocarbon dates. At the commencement of the project, 12 radiocarbon determinations from two sites had been obtained from the study area. At least four of these date the formation of a natural chenier deposit on Rodds Peninsula. The dating program aimed to determine the initiation, periodicity and termination of site occupations. A further 68 radiocarbon dates were obtained for this project, comprising 53 dates on archaeological deposits, 10 dates on environmental samples and five dates on live-collected shell specimens. Since marine shell is the primary form of cultural material represented in the archaeological record of coastal landscapes of the area it provided an ideal medium for dating.

All conventional radiocarbon determinations and sample preparation for accelerator mass spectrometry (AMS) determinations (including CO_2 production) were undertaken by the University of Waikato Radiocarbon Dating Laboratory to reduce the effects of interlaboratory variation in sample preparation and counting procedures. AMS dating was conducted by the Rafter Radiocarbon Laboratory of the New Zealand Institute of Geological and Nuclear Sciences (IGNS). Charcoal samples were washed in hot 10% HCl to remove possible contaminants. Shell samples were cleaned and washed in an ultrasonic bath before acid-etching (2M HCl) for 100 seconds to minimise the possibility of contamination through isotopic exchange between the sample and its environment. All aragonite shell samples (i.e. *A. trapezia*) were either subjected to XRD analysis (up to Wk-8328) or feigl staining (from Wk-8553 onwards) (Friedman 1959) to establish the absence of recrystallised $CaCO_3$ (calcite) in the shell structures. All conventional samples were converted to benzene through hydrolysis and ^{14}C activity measured by liquid scintillation counting (LSC) (Higham and Hogg 1997). For AMS samples, CO_2 was converted to graphite before introduction to the mass spectrometer. Radiocarbon ages are reported as conventional radiocarbon ages (Stuiver and Polach 1977). That is, they are corrected for isotopic fractionation but not corrected for marine reservoir effect or any other factor (cf. Kelly 1982; Thom et al. 1981). The conventional radiocarbon ages include a laboratory error multiplier of 1.22 (Higham and Hogg 1997).

Marine and estuarine reservoir effects

Global studies have demonstrated major variations in radiocarbon dates obtained from contemporaneous samples living in marine and estuarine environments (Reimer and Reimer 2000). These differences arise primarily from variations in the carbon reservoirs in which these organisms grow. Marine and estuarine reservoir differences are a major issue in the investigation and dating of coastal archaeological deposits where failure to take these factors into account can result in dating errors of up to several hundred years. Although general correction values have been proposed for various regions (e.g. Reimer and Reimer 2000), in areas where no local studies have been undertaken, reliance on generic regional correction values can reduce confidence in the accuracy of individual radiocarbon dates obtained on marine and estuarine samples. As part of this study, five marine shells live-collected between AD 1904 and AD 1929 and 12 shell/charcoal paired samples from archaeological contexts were radiocarbon dated to determine local marine and estuarine reservoir offset values. This research found considerable variation in estuarine reservoir effects between individual estuaries. Estuary-specific correction factors are proposed for each estuary to account for local factors. Full details are presented in Chapter 4.

Radiocarbon age calibration

Radiocarbon ages are usually reported in ^{14}C years rather than calendar years. The primary difference between the two time-scales is caused by variability in the proportion of radioactive carbon in the biosphere through time and space. Conversion (or calibration) of radiocarbon dates from ^{14}C years to calendar years is possible by reference to records of known variability in radioactive carbon activity derived from dated tree-ring and coral-varve sequences. Such sequences, calculated from growth bands, enable direct comparisons to be made between radiocarbon dates and true calendar ages to derive a correction for a specific time period and region to convert ^{14}C years to calendar years.

Conventional radiocarbon ages were converted to calendar years using the CALIB (v4.3) computer program developed by the Quaternary Isotope Laboratory, University of Washington (Stuiver and Reimer 1993). Determinations based on charcoal were calibrated using the atmospheric decadal dataset of Stuiver et al. (1998a) with no laboratory error multiplier. Charcoal ages were reduced by 41±14 years to correct for ^{14}C variation between northern and southern hemispheres (McCormac et al. 2002). Note that the incorporation of the southern hemisphere offset error assumes that each atmospheric conventional radiocarbon age derives from an independent secondary carbon reservoir (see Jones and Nicholls 2001 for discussion). Dates on marine samples (e.g. marine and estuarine shell) were calibrated using the marine calibration model dataset of Stuiver et al. (1998b) with a variable ΔR correction value defined for each estuary (see Chapter 4). The calibrated ages reported span the 2σ calibrated age-range. Samples too young for use of the calibration curves are reported as 'modern'. Where a calibrated intercept or age-range is reported as '0*', a 'negative' or 'modern' age BP is indicated owing to uncertainties introduced by nuclear testing. The dates presented on either side of the bracketed dates represent the 2σ calibrated age-range of the radiocarbon date using the calibration procedure outlined above. The date/s in the brackets represent the intercept/s of the radiocarbon age with the calibration curve. In parts of the calibration curve exhibiting short-term variation in atmospheric radioactive carbon activity, or where radiocarbon ages have large standard errors, it is common to have multiple intercepts (i.e. multiple calibrated ages for any given radiocarbon date) which are equally probable. Note that a new version of the CALIB (v4.4) calibration program incorporating an option to use a southern hemisphere calibration curve for the last c.1,000 calendar years was released in late 2003. Radiocarbon dates presented in this study have not been recalibrated using this new dataset, although the new recommended southern hemisphere offset of 41±14 years is employed. The small difference in calibrated ages does not significantly alter any data presented in this study.

Dating terminology

Following convention, calibrated radiocarbon dates are referred to in this study as years cal BP; for example, 2,340 cal BP. Specific uncalibrated (i.e. conventional) radiocarbon ages are reported with their accompanying error estimate and laboratory number; for example, 2,340±55 BP (Wk-5353). Where a precise date is unnecessary, a generalised date is used; for example, c.2,300 years ago (referring to the calendrical, or calibrated time-scale). Occasionally very generalised dates are cited to refer the reader to a general time interval; for example, c.3,000 BP (very roughly around 3,000 years ago; calibration is irrelevant in such cases).

Laboratory analyses

In the laboratory, all 3mm sieve residues were gently wet-sieved with freshwater and air-dried prior to analysis. Small finds were also wet-sieved, with the exception of charcoal and other fragile materials, which were cleaned using tweezers to remove adhering sediments. Excavated material was sorted into the following generic categories using tweezers: organic material (i.e. roots, leaf litter, seeds, scats, insect remains etc), shell (separated into marine bivalve, freshwater bivalve, marine gastropod and terrestrial gastropod), crustacean carapace, bone (separated into fish bone and non-fish bone), charcoal, stone artefacts, non-artefactual stone, pumice, pigment and coral. Material in each category was weighed to the nearest 0.1g on an A&D EK 1200 electronic balance for samples above 10g, while those below 10g were weighed on a Shimadzu AW120 electronic balance to the nearest 0.0001g. Detailed laboratory procedures for the major categories of recovered materials are presented below. This generic set of analytical methods was modified and/or augmented according to specific assemblage requirements. Deviations are noted in relevant chapters detailing the results from specific sites.

Invertebrate remains

This category includes both marine and terrestrial shellfish remains as well as crustacean carapace. Shell was identified to the lowest taxonomic level possible using diagnostic features and a comprehensive shell reference collection specifically constructed for this project with the assistance of the Malacology Section, Queensland Museum (Appendix 5) and standard reference works (Coleman 1981; Lamprell and Healy 1998; Lamprell and Whitehead 1992; Wilson and Gillett 1979). Minimum number of individuals (MNI), number of identified specimens (NISP) and shell weight per taxon were used to characterise shellfish abundance. Intra-specific size measurements were undertaken on whole shells of principal taxa represented in excavated assemblages throughout the region — *A. trapezia* and *S. glomerata*. Fragmentation rates were calculated for *A. trapezia* (NISP/100g) to investigate site-specific differences in post-depositional processes. The diversity of recovered shellfish assemblages was examined using Simpson's index (see below). Finally, excavated *A. trapezia* assemblages from each site were subject to bivalve conjoin analyses to examine aspects of site integrity and depositional processes. Details of all of these analyses are presented below. Figures 3.1–3.2 show bivalve terminology. Figure 3.3 shows gastropod terminology.

Minimum number of individuals (MNI)

The MNI measure is the minimum number of individual shellfish necessary to account for the number of diagnostic elements identified in an assemblage. For symmetrical bivalves (e.g. *A. trapezia, D. deltoides*), the highest number of hinges of one side is taken as the MNI for that sampling unit. For asymmetical bivalves (e.g. *S. glomerata*) shells were separated into upper (lids) and lower (bases) valves and the greater number taken as the MNI for that sampling unit. Spires of gastropods were used to calculate MNI. Although MNI calculations are often adopted to avoid the impact of fragmentation on NISP calculations (e.g. Mowat 1995:83), in highly fragmented assemblages, few diagnostic attributes amenable to MNI calculation may be represented, potentially rendering the MNI a severe underestimate of abundance.

Number of identified specimens (NISP)

The NISP measure is the number of shell fragments identified to a particular taxon. The major limitation of this method for application to shellfish is the level of identifiability of fragmented shell. As laboratory analysis proceeded it was found that virtually all shell in the 3mm sieve residue was identifiable to species as analyst experience increased and the comparative reference collection became more comprehensive. Although NISP has been criticised for over-representing the abundance of taxa with distinctive sculpture attributes (e.g. Mowat 1995), it is useful for intra- and inter-site comparison of individual taxa within and between shell assemblages and for examining shell fragmentation rates (see below).

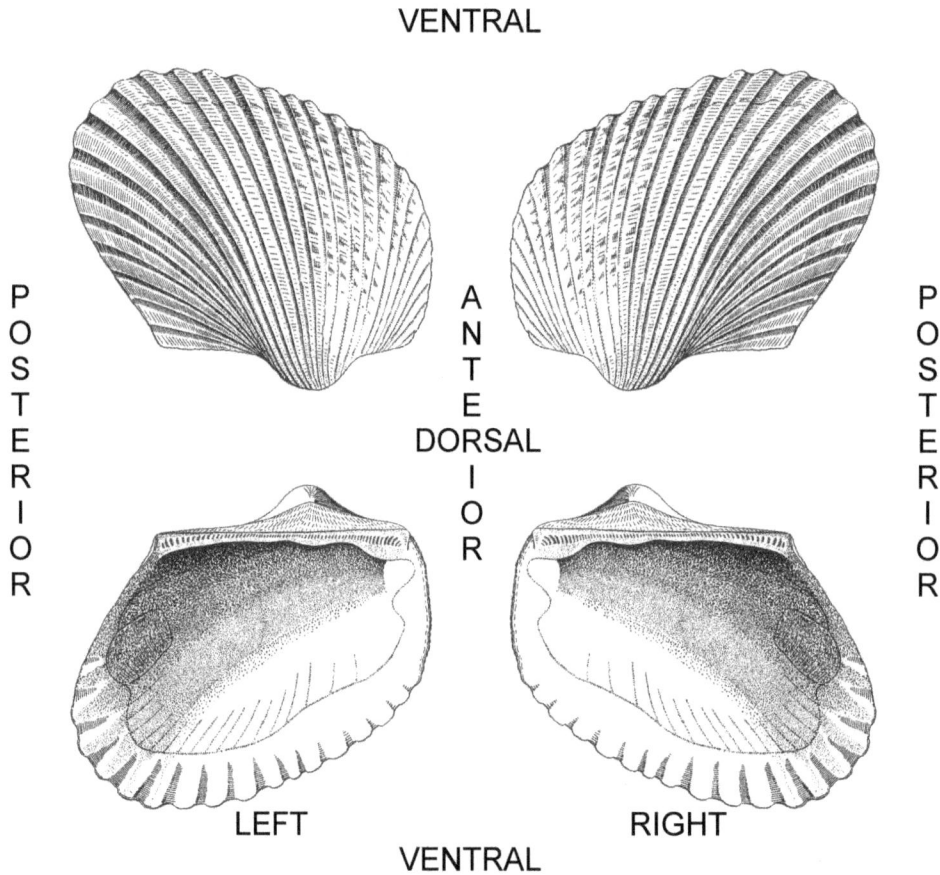

Figure 3.1 Bivalve terminology (after Claassen 1998:21; Hedley 1904).

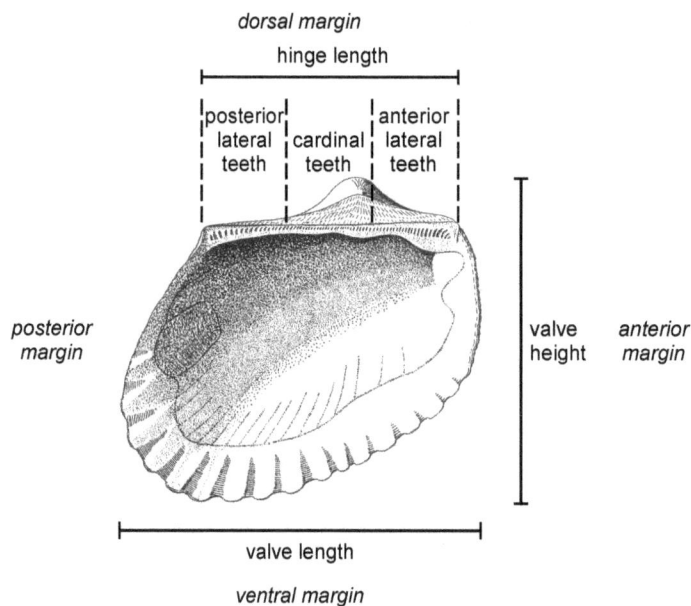

Figure 3.2 Bivalve terminology and measured attributes (after Claassen 1998:21; Hedley 1904).

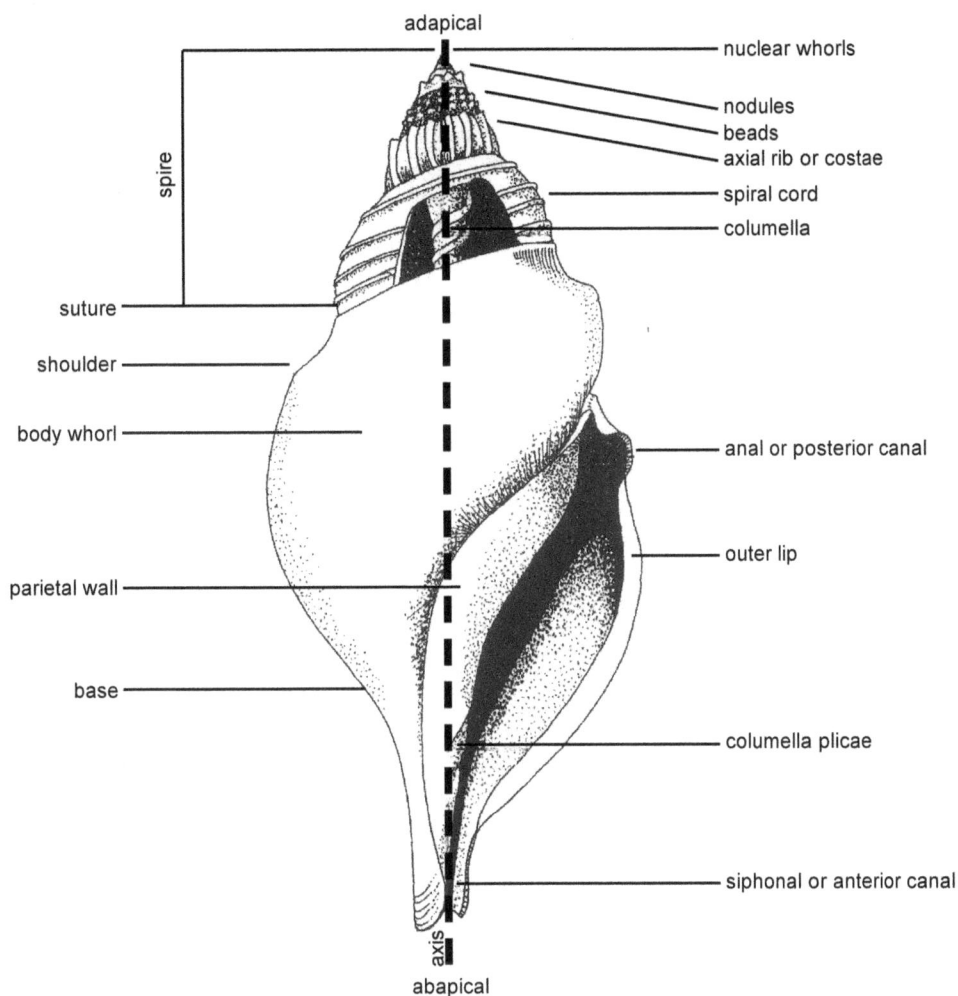

Figure 3.3 Gastropod terminology (after Andrews 1971:42; Claassen 1998:19).

Species diversity and similarity measures

In this study, the diversity of recovered shellfish assemblages was examined using two methods: the Shannon-Weaver Function (H′) and Simpson's Index of Diversity (1–D) (Krebs 1989; Reitz and Wing 1999; Simpson 1949).

Shannon-Weaver Function:

$$H' = \sum_{i=1}^{s}(p_i)(Log\ p_i)$$

[3.1]

where:

H′ = information content of the sample

p_i = the relative abundance of the ith taxon within the sample

Log p_i = the logarithm of p$_i$ (to base e in this study)

s = the number of taxonomic categories

High Shannon-Weaver Function values are associated with higher numbers of taxa exhibiting higher degrees of evenness in abundance across taxa. Disproportionate representation of individuals in only a few of the same taxonomic categories will lead to lower values (Reitz and Wing 1999:105).

Simpson's Index of Diversity:

$$1\text{–}D = 1\text{–}\sum_{i=1}^{s}(p_i)^2 \qquad\qquad [3.2]$$

where:
D = Simpson's Index
p_i = the proportion of individuals of species i in the assemblage
s = the total number of species

This index expresses the probability that two individuals selected at random from a community will belong to different species.

Similarity coefficients between mollusc assemblages within and between sites and from different time periods were calculated using the percentage similarity measure (Reitz and Wing 1999:107–9; Krebs 1989:304):

$$PS = \sum \text{minimum}(p_{1i}, p_{2i}) \qquad\qquad [3.3]$$

where:
P = percentage similarity between samples 1 and 2
p_{1i} = percentage of species i in community 1
p_{2i} = percentage of species i in community 2

Bivalve conjoin analyses

An innovative method of identifying conjoins in assemblages of bivalves was developed to examine aspects of site integrity and depositional processes in the region. Articulated live-collected and excavated specimens of *A. trapezia* were used to derive reliable criteria for identifying unarticulated valve-pairs. These criteria were used as the basis of a systematic bivalve conjoin analysis method which successfully identified valve-pairs among the unarticulated *A. trapezia* valves excavated from the Seven Mile Creek Mound. A subsequent blind test confirmed the reliability of the method. Full details are presented in Chapter 5. Valve-pairing methods were applied to all excavated assemblages in the region containing whole *A. trapezia* valves.

Vertebrate remains

All bone recovered from analysed squares was initially examined by Deborah Vale (2002, 2004). Vale separated fish remains from the bone assemblage and undertook basic characterisation studies of the fish bone assemblage including: body part representation, vertebral sizing, identification rate, number of fragments, NISP and MNI. MNI was calculated for each excavation unit, which may overestimate the MNI for the site as a whole. Fish NISP is calculated on the basis of the number of specimens identified to family or species only (Vale and Gargett 2002). Fish taxa were identified using a comparative reference collection assembled for northern New South Wales and southern Queensland and adapted for the central Queensland coast (for example, the tiger flathead, *Platycephalus indicus*, uncommon in New South Wales, was included). Owing to the limitations of the reference collection, however, Vale made taxonomic identifications to the family level only, with the exception of the bream, *Acanthopagrus australis*, which was identified on the basis of distinctive diagnostic cranial elements. Preliminary identification of non-fish remains was undertaken by Stephen Van Dyck (Mammals and Birds Section, Queensland Museum). No further analysis of this non-fish material was conducted.

Stone artefacts
Artefactual stone exhibits a range of technological and descriptive diagnostic attributes consistent with deliberate modification, transport and/or use. Technological attributes include the presence of a bulb of percussion, platform and/or point of force application. In the absence of unambiguous signs of modification or use, stone objects were classified as artefacts if they were of exotic petrological origin. Raw materials were identified by Stephen Cotter (Cooperative Research Centre for Landscape Evolution and Mineral Exploration, University of Canberra), utilising a comprehensive reference collection compiled for the region. Artefactual stone was categorised as either core, flake, broken flake, flaked piece, manuport or other. The size of each artefact was characterised to the nearest 0.1mm by measurements of the maximum length (longest longitudinal axis), width (longest lateral axis) and thickness (maximum height between ventral and dorsal surfaces) using vernier calipers. Limited use-wear and residue studies have been conducted on parts of the stone artefact assemblage. Methods employed in these studies are outlined in the relevant chapters.

Non-artefactual stone
Non-artefactual stone is lithic material native to the sedimentary context of the site with no obvious signs of modification, transport and/or use. The non-artefactual stone was weighed, but no further analysis of this material was conducted.

Charcoal
Charcoal is the carbonised remains of organic material. Samples of charcoal from most sites were subject to radiocarbon isotope analysis. The charcoal was weighed, but no further analysis of this material was conducted.

Organic material
Organic material refers to plant remains other than charcoal including roots, leaves, seeds, twigs, wood, scats, insect casings. The organic material was weighed, but no further analysis of this material was conducted.

Age-depth curves and analytical units

Analytical units for synthesising results are based on defining chronostratigraphic units for each excavated 50cm × 50cm square. Individual excavation units were assigned to 500 year intervals between 0–5,500 BP. For several sites this process is straightforward with all deposits assigned to a single 500 year interval (Seven Mile Creek Mound, Eurimbula Creek 1, Eurimbula Creek 2). For other sites, age-depth models were established based on linear regressions through available data points. Where dates are not available from a square radiocarbon determinations are extrapolated from adjacent squares where there are no obvious discontinuities in depth, sediments and stratigraphy between the units.

Summary

The generic field and laboratory methods used to investigate the archaeological record of the southern Curtis Coast are outlined in this chapter. In addition to employing well-established standard archaeological techniques, this study also employs detailed analyses of local marine and estuarine reservoir effects on radiocarbon dates to improve the accuracy of radiocarbon determinations and develops a procedure for identifying bivalve conjoins in excavated shellfish assemblages in order to investigate aspects of site integrity. These technical studies are discussed in detail in the following chapters.

4

Marine and estuarine reservoir effects in central Queensland: determination of ΔR values

Introduction

This chapter forms the first of two chapters addressing technical issues critical to the rest of the study. Uncertainty in radiocarbon ages introduced by variation in marine and estuarine reservoir effects through space and time are a major issue in the investigation and dating of coastal archaeological deposits. In areas where no local studies have been undertaken reliance on generic regional correction values can reduce confidence in the accuracy of individual radiocarbon dates obtained on marine and estuarine samples (e.g. mollusc shell, fish bone etc) and the chronologies constructed on the basis of these dates. As a technical component of this study a series of five marine shell specimens live-collected between AD 1904 and AD 1929 and 12 shell/charcoal paired samples from archaeological contexts were radiocarbon dated to determine local ΔR values. The object of the study was to assess the potential influence of localised variation in marine reservoir effect in accurately determining the age of marine and estuarine shell from archaeological deposits in the area.

The marine reservoir effect (also known as the oceanic reservoir correction factor, marine shell correction factor or surface ocean water reservoir effect) of 450±35 ^{14}C years calculated for Australia by Gillespie (1975, 1977, 1991; see also Gillespie and Polach 1979; Gillespie and Temple 1977) is based on samples from Torres Strait and southern Australia. Several Australian studies have suggested the possibility of significant deviations in regional marine reservoir signature from this generalised value (e.g. Hughes and Djohadze 1980; Murray-Wallace 1996; Spennemann and Head 1996; Woodroffe et al. 1986:75, 77; Woodroffe and Mulrennan 1993). Additionally, Spennemann and Head (1996:35) note that the present 'use of universal ocean reservoir factors not only potentially masks chronological variation, but potentially invalidates some observations *in toto*'. However, despite routine dating of marine and estuarine shell from archaeological deposits in Queensland (see Ulm and Hall 1996; Ulm and Reid 2000, 2004), no systematic evaluation of the applicability of this generalised marine reservoir effect has been undertaken in the region.

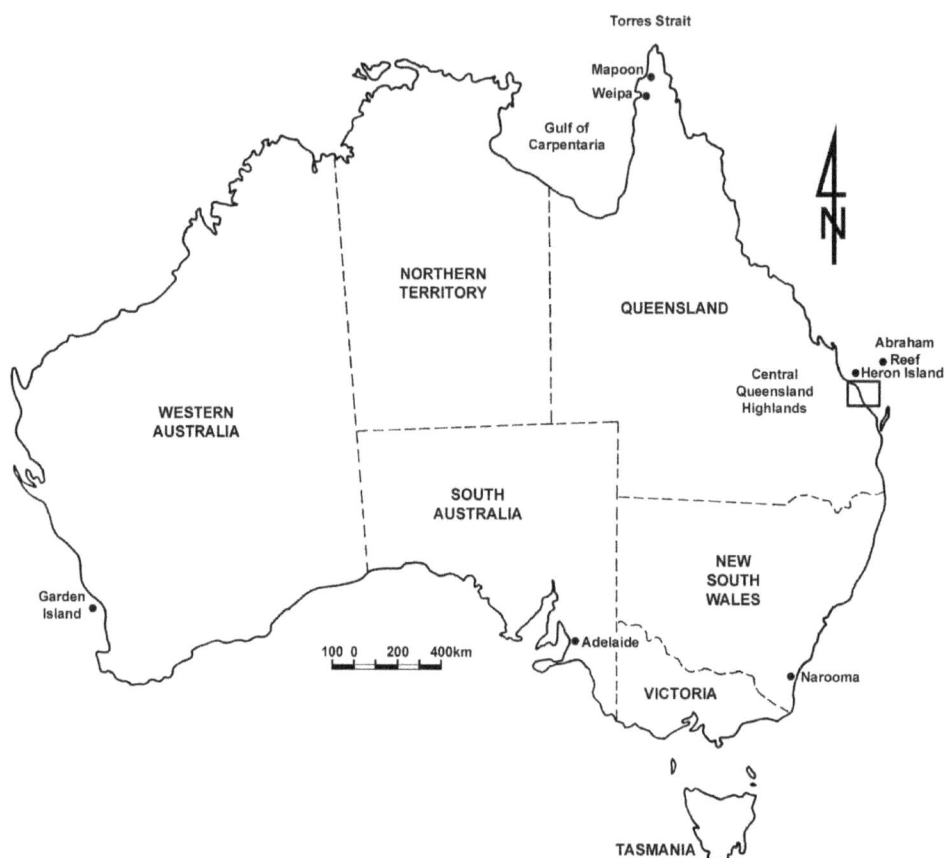

Figure 4.1 Map of Australia, showing places mentioned in the text. The boxed area indicates the southern Curtis Coast study area.

Since 1993, 20 archaeological sites have been test excavated in central Queensland under the auspices of the Gooreng Gooreng Cultural Heritage Project (see Lilley and Ulm 1995, 1999). Investigations have focussed on a coastal area centred on the Town of Seventeen Seventy and an inland area centred on Cania Gorge, approximately 100km from the coast (Figs 4.1–4.2). Because this project constitutes the first detailed archaeological investigation in the region, a basic objective was to develop a cultural chronology for the two areas to enable systematic comparison within and between inland and coastal sequences as well as to relate findings to results from southeast Queensland (e.g. Ulm and Hall 1996) and the Central Queensland Highlands (e.g. Morwood 1984). To date, radiocarbon determinations from Cania Gorge have been based exclusively on wood charcoal samples (n=38), while those from sites on the coast include wood charcoal (n=27), marine shell (n=5) and estuarine shell (n=32) (see Ulm and Reid 2000, 2004 for details). To examine the potential influence of local marine reservoir effect on the comparability of radiocarbon ages obtained on various sample materials, a limited program of dating live-collected marine shell specimens of known historical age and archaeological shell/charcoal paired samples was undertaken to improve confidence in calibration of ages from the study area and to ensure comparability between shell ages and between shell and terrestrial (charcoal) samples.

In addition to the significance of this study for interpretations of cultural chronologies in Australian coastal archaeology, the results have broader implications for studies of coastal geomorphology in eastern Australia. In recent studies there is a heavy reliance on marine radiocarbon ages to establish chronologies for sea-level change (e.g. Lambeck and Nakada 1990; Larcombe et al. 1995), reef and coral cay development (e.g. Chivas et al. 1986), coastal dune sequences (e.g. Pye and Rhodes 1985), and storm event frequency (e.g. Hayne and Chappell 2001; Nott and Hayne 2001).

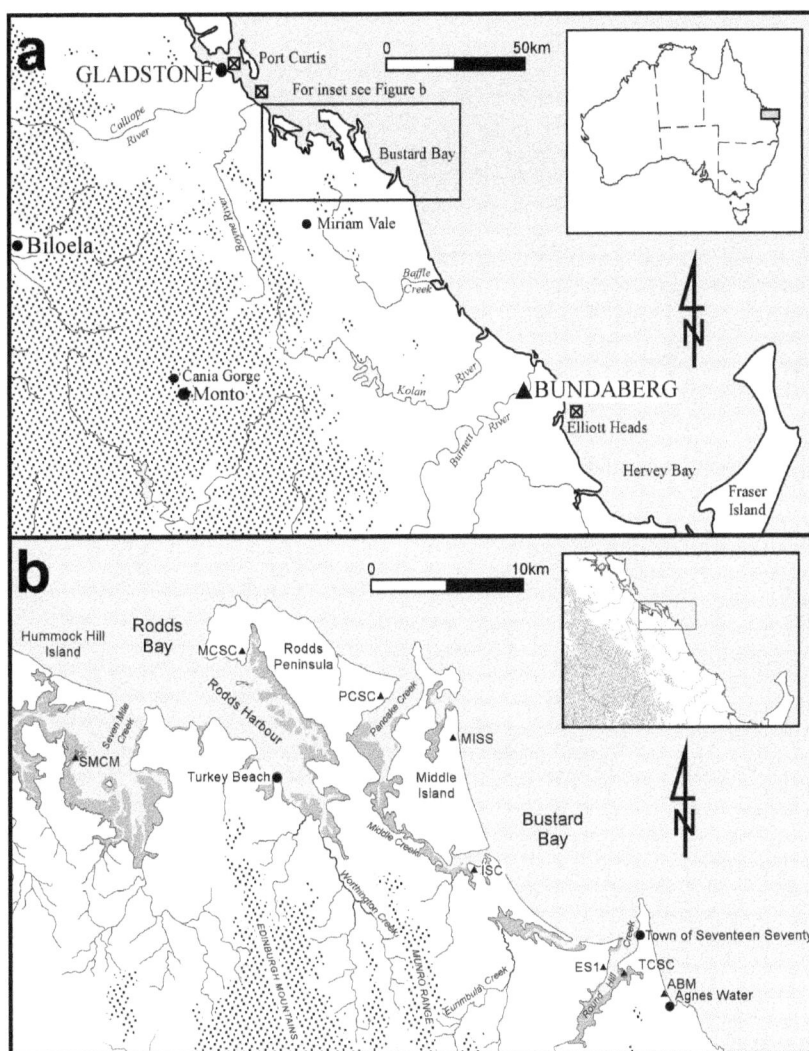

Figure 4.2 (a) Part of the central Queensland coast showing the location of 'pre-bomb' live-collected shell specimens (⊠) and (b) estuaries and archaeological sites (▲) discussed in the text. SMCM=Seven Mile Creek Mound; MCSC=Mort Creek Site Complex; PCSC=Pancake Creek Site Complex; ISC=Ironbark Site Complex; ES1=Eurimbula Site 1; TCSC=Tom's Creek Site Complex; ABM=Agnes Beach Midden.

Background

A basic assumption of the radiocarbon dating method is that the concentration of ^{14}C in the biosphere is uniform through space and time. Early in the development of the ^{14}C method, however, it was recognised that certain environments exhibited much slower rates of mixing than the atmosphere, indicating significant variation within and between some ^{14}C reservoirs. In particular, marine shells (which derive carbon principally from dissolved inorganic carbon in ocean waters) exhibited a systematic age difference from contemporary terrestrial samples on a regional basis which allowed calculation of a regionally-specific age offset. These differences have been documented primarily through the dating of marine shell specimens of known historical age (see Stuiver et al. 1986:Table 1; Reimer and Reimer 2000).

Global variation in marine reservoir effects evident in marine shell carbonates are mainly caused by incomplete mixing of upwelling water of 'old' inorganic carbonates from the deep ocean, where long residence times (>1,000 years) cause depletion of ^{14}C activity through radioactive decay, resulting in very old apparent ^{14}C ages (Mangerud 1972). Estuarine reservoirs are even more complex with the interaction and incomplete mixing of ^{14}C from both terrestrial reservoirs and marine reservoirs from tidal action (Little 1993; Robinson and Thompson 1981:50; Stuiver and Braziunas 1993:155). This is generally not the case in open coast contexts, where the effect of terrestrial runoff on near-shore ^{14}C activity is attenuated by unrestricted circulation that ensures rapid mixing and distribution of atmospherically derived ^{14}C. In estuarine environments, on the other hand, shellfish (and other estuarine organisms) can obtain a significant proportion of their carbon from CO_2 dissolved in terrestrial rainwater runoff (i.e. enriched in ^{14}C activity relative to depleted marine waters). The more atmospheric CO_2 gained by marine and estuarine shells, the closer the ^{14}C age value should be to the expected coeval terrestrial sample age. Fluctuations in ^{14}C through time are amplified in estuarine contexts because of significant regional variability in rainfall magnitude and periodicity combined with the effects of relative sea-level changes on local circulation patterns, such as changes in sedimentation that limit the interaction of estuarine water bodies with the open ocean (e.g. entrance bars). Additional factors are also important in specific estuaries; for example, hinterland geology (e.g. Dye 1994; Ingram 1998; Spennemann and Head 1998) and intra-estuary variability in rainwater input and circulation patterns (Little 1993). Given this dependent relationship, we would expect significant variation in ^{14}C activity between individual estuaries, regardless of their proximity (Spennemann and Head 1996). The combined effect of these factors can create an estuarine reservoir age offset of up to several hundred years from the conventionally applied regional open surface water figure. For example, Kirch (2001) has documented strongly negative ΔR values for shallow tropical lagoons in the Pacific with limited exchange with the open ocean. Paired samples from two sites in the Mussau Islands in the Bismarck Archipelago dated by Kirch (2001) indicate ΔR values of up to –320, with a further pair from Ofu Island, American Samoa, yielding –230.

The potential magnitude of such factors in the southern Curtis Coast study area can be gauged by examining salinity profiles and intra-estuary circulation models. Extended periods of depressed salinity coinciding with periods of distinctly seasonal annual rainfall have been documented in many estuaries on the central Queensland coast. Although the major influences on water movement within these tributaries are prevailing tidal and weather conditions, freshwater inflow associated with periods of high intensity rainfall can cause heavy runoff, which produces short-to-medium-term fluctuations in estuary salinity and turbidity (Olsen 1980a:5). Olsen (1980a) notes that tidal flushing of estuaries is generally high, except for a period of depressed salinity between January and March, representing significant terrestrial freshwater rainfall input (see also Lupton and Heidenreich 1996 for similar data for catchments immediately south of the study area).

Regional differences in marine reservoir effect are generally determined through one or a combination of three major methods:

1. direct radiocarbon dating of pre-AD 1955 ('pre-bomb') live-collected shells of known historical age;
2. radiocarbon dating shell/charcoal paired samples from high integrity archaeological contexts that are assumed to be contemporaneous; and,
3. radiocarbon dating and/or paired radiocarbon and uranium-thorium (^{230}Th/^{234}U) dating of live corals or long-lived live shells with clear annual growth bands.

In the first method, marine shell specimens of known historical age must have been live-collected prior to AD 1955 ('pre-bomb') and the date and location of collection known with confidence. After AD 1955 ('post-bomb') natural levels of ^{14}C activity in marine environments were enriched as a result of detonation of nuclear and thermonuclear weapons in the atmosphere (e.g. Druffel and Griffin 1993, 1999; Peck and Brey 1996). Even samples collected prior to AD 1955 may require correction for fossil fuel depletion resulting from large-scale fossil fuel combustion beginning in the late nineteenth century (Aitken 1990; Taylor 1997:69). Dating shell/charcoal paired samples is also potentially problematic because it must be assumed that the samples selected are contemporaneous and that association is not simply the result of post-depositional processes or excavation procedures (see Ingram 1998, for example, who assumes contemporaneity between shell/charcoal paired samples collected from 305mm thick excavation units). Results of paired samples can therefore be difficult to interpret because variation could result from (1) temporal variation in ocean properties through time; (2) a lack of true association (contemporaneity) between the samples (e.g. old wood or taphonomic factors); and/or (3) localised reservoir signatures. Recent studies have demonstrated that examination and radiometric dating of certain coral species with well-defined annual growth structures can provide the most accurate determination of marine reservoir effects (Reimer and Reimer 2000). Unfortunately, such studies are limited largely to tropical regions with long-term coral records.

In recent years, regional marine reservoir effect has commonly been expressed as ΔR (e.g. Erlandson and Moss 1999; Higham and Hogg 1995; Ingram 1998; Ingram and Southon 1996; Kennett et al. 1997; Phelan 1999; Reimer and Reimer 2000). Stuiver et al. (1986; Stuiver and Braziunas 1993) modelled global marine ^{14}C activity using a simple box diffusion global carbon cycle model of marine reservoir responses to variation in atmospheric ^{14}C activity. Regional deviations from the modelled marine calibration curve (ΔR) were calculated using radiocarbon ages on live-collected marine shell samples of known historical age (Stuiver et al. 1986:Table 1). ΔR is the difference between the conventional radiocarbon age of a sample of known age from a specific locality (P) and the equivalent age predicted by the modelled global marine calibration curve (Q); therefore ΔR= P–Q (Stuiver et al. 1986:982).

Australian marine reservoir studies: a review

The original study

Gillespie (1977) established the conventionally employed marine reservoir effect for marine shells grown in Australian waters of 450±35 years using radiocarbon ages of six shell specimens live-collected between AD 1875 and AD 1950 from four locations around the Australian coast. Three samples were from the central Torres Strait and one each from Narooma in New South Wales, Adelaide in South Australia and Garden Island in Western Australia (Table 4.1, Fig. 4.1). The six shells returned a weighted apparent mean age of 475±35 (D^{14}C= –57.7±4‰), which was reduced to 450±35 (D^{14}C= –55±4‰) after correction for the fossil fuel dilution effect (Gillespie 1991; Gillespie and Polach 1979; Gillespie and Temple 1977; see Mangerud and Gulliksen 1975 for details of fossil fuel correction procedure). It is important to note that these values are not based on the conventional radiocarbon ages of the samples but rather calculated from D^{14}C values that had been 'age corrected for ^{14}C decay from year of live collection and 1950' (Gillespie and Polach 1979:410) (i.e. the reservoir age of 450±35 is the error-weighted mean of the conventional radiocarbon age minus the age of the samples in AD 1950 plus fossil fuel correction, see Robinson and Thompson 1981:47) (see Gillespie 1977; Gillespie and Polach 1979; Gillespie and Temple 1977).

Stuiver et al. (1986:Table 1) used these results (uncorrected for fossil fuel dilution) to calculate the ΔR correction for Australian waters of –5±35 for application to marine calibration curves (e.g. Stuiver and Braziunas 1993:Fig. 17A–N). In a footnote, Stuiver et al. (1986:1021–Note e) state that the Australian sample ages reported in their paper as the basis of the ΔR= –5±35 value had been calculated from the D^{14}C reported by Gillespie and Polach (1979:411) after removing an age-correction (see Stuiver et al. 1986:1021). This statement is incorrect, however, because the age-corrected D^{14}C results reported by Gillespie and Polach (1979) produce the same radiocarbon ages as those reported by Stuiver et al. (1986:Table 1), which are presented as conventional radiocarbon ages (i.e. age-correction removed). The ΔR= –5±35 value reported by Stuiver et al. (1986:Table 1,

Fig. 10B; Stuiver and Braziunas 1993:Fig. 14) is thus not the weighted mean of the difference between the age-corrected ^{14}C ages (rather than the conventional ^{14}C ages) and the global marine model based on the known historical ages of the specimens. In Table 4.1 the reported radiocarbon ages were calculated from the reported D^{14}C values after removing the age-correction. Note that there are also slight discrepancies in these ages as reported by Stuiver et al. (1986:1020), Gillespie (1991:15), and Bowman (1985a:Table 1) owing to rounding factors. This error was also repeated in the original version of Reimer and Reimer's (2000) internet *Marine Reservoir Correction Database* but has since been corrected at my suggestion.

Table 4.1 Original 1970s series of radiocarbon dates obtained on live-collected marine shell samples from Australian waters presented by Gillespie (1977; Gillespie and Temple 1977). δ^{13}C is an assumed value of 0±2 (Gillespie and Polach 1979:410). Historical ages of shell samples were converted to equivalent global marine model ages using data from Stuiver et al. (1998a). ΔR was calculated by deducting the equivalent marine model age of the historical age of the shell sample from the ^{14}C age of the shell sample (after Stuiver et al. 1986:1020). ΔRσ= $\sqrt{(\sigma}$historical age^2+σmarine model age^2+σ^{14}C age^2) (Gillespie 1982). The uncertainty in the marine model age includes estimated error in the calibration dataset (derived from Stuiver et al. 1998a). Error-weighted means are calculated using formulae in Ward and Wilson (1978). Samples: Mo=*Mactra obesa*; Pb=*Pinna bicolor*; Pm=*Pinctada margaritifera*; Pl=*Proxichione laqueata*; Dd=*Donax deltoides*; Kr=*Katelysia rhytiphora*.

SITE	LAB. NO.	SAMPLE	HISTORICAL AGE (YEAR AD)	MARINE MODEL AGE	D^{14}C[a] (‰)	^{14}C AGE (YEARS BP)[b]	ΔR (^{14}C YEARS)
Torres Strait	SUA-354/1	Mo	1875±3	476±6	−58±8	553±68	77±68
Torres Strait	SUA-354/2	Pb	1875±3	476±6	−56±10	536±85	60±85
Torres Strait	SUA-357	Pm	1909	451±7	−49±10	443±84	−6±85
Garden Is.	SUA-355	Pl [c]	1930	458±10	−55±10	474±85	16±85
Adelaide	SUA-393	Dd [de]	1937±2	463±8	−70±10	596±86	132±87
Narooma	SUA-356	Kr	1950	473±13	−58±10	480±85	7±86
Error-weighted mean:							50±33

a   Value corrected for ^{14}C decay between year of live collection and AD 1950 (Gillespie and Polach 1979:410) (see text).

b   Value calculated from the D^{14}C presented by Gillespie and Temple (1977) and Gillespie and Polach (1979) after removal of the age-correction (see text).

c   Bowman (1985b:Table 1) reports as *Pinna bicolor*.

d   Bowman (1985b:Table 1) reports as *Pinetoda margaritifera*.

e   Gillespie and Temple (1977:29) and Gillespie and Polach (1979:411) report as *Conuber incei*.

Another source of difference in the results is the use of the 1986 calibration curve to calculate the ΔR values presented by Stuiver et al. (1986:Fig. 10B, Table 1) and Stuiver and Braziunas (1993:Fig. 16). Stuiver and Braziunas (1993:155) state that no 'attempt was made to update the regional ΔR determinations, because minor changes in calibration results for the last few centuries would result in corrections about equal to the rounding error (up to 5 yr) of the original data set'. Stuiver et al. (1998b:1135) note that although calibration curves were again updated in 1998 (see Stuiver et al. 1998a), data are virtually identical for the 0–7,000 cal BP interval, and they therefore recommend the use of the 1993 marine calibration curves (Stuiver and Braziunas 1993:Fig. 17A–N). In the recent *Marine Reservoir Correction Database*, however, Reimer and Reimer (2000) have recalculated worldwide ΔR values using the 1998 calibration dataset. In the present study, I have recalculated previous ΔR values for the six samples presented in Table 4.1 using the 1998 calibration curve. Using the 1998 instead of the 1986 model results in a reduction of ~20 years in the marine model ages and a corresponding change in ΔR values.

The standard deviations reported with the model marine ages are derived from Stuiver et al. (1998a) combined with any estimate of error in the historical age of the year of live collection of the sample. Similarly the ΔR standard deviation combines the error estimates of the ^{14}C age and the marine model age. Using these procedures and the methods outlined by Ward and Wilson (1978), I found that the six specimens formed a statistically indistinguishable group with an error-weighted mean of ΔR= +50±33 (T'=1.90; $\chi^2_{5:0.05}$=11.07) (Table 4.1).

Further studies

Various studies have attempted to evaluate the applicability of the Gillespie correction factor to particular Australian regions. These studies have been based on four classes of data: (1) additional dating of 'pre-bomb' live-collected shell specimens (Bowman 1985a, 1985b; Bowman and Harvey 1983; Gill 1983; Rhodes et al. 1980); (2) radiocarbon dating of coral cores with clear growth bands (Reimer and Reimer 2000); (3) shell/charcoal paired samples from archaeological and natural deposits (Gillespie and Temple 1977; Head et al. 1983; Horsfall 1987; Hughes and Djohadze 1980; Luebbers 1978; Ross et al. 2000; Thom et al. 1981); and (4) dating of 'post-bomb' live-collected shells (Gillespie and Polach 1979; Rhodes et al. 1980). The following discussion will briefly review the results of these studies with a particular focus on Queensland and eastern Australian studies.

Dating 'pre-bomb' live-collected shell specimens

Rhodes et al. (1980) tested the validity of the Gillespie correction for the Gulf of Carpentaria by dating two marine shells live-collected in 1903 by Charles Hedley offshore from Mapoon, c.80km north of Weipa (Table 4.2) (Rhodes et al. 1980). Note that the ^{14}C ages reported in Table 4.2 have had the age-correction for difference in time between live collection in AD 1903 and AD 1950 calculated by Rhodes et al. (1980:Table 1) removed. The results are statistically indistinguishable from the six results produced by the original Gillespie study ($T'=4.63$; $\chi^2_{7:0.05}=14.07$). Prior to the present study, these are the only other live-collected specimens dated in Queensland since the original Gillespie (1977) study. Dating of live-collected specimens elsewhere in Australia has also reinforced the general applicability of the Gillespie value, with all results statistically indistinguishable from the original value (see Bowman 1985a, 1985b; Bowman and Harvey 1983; Gill 1983).

Table 4.2 Radiocarbon dates obtained on live-collected marine shell samples of known historical age from the Gulf of Carpentaria (Rhodes et al. 1980). Samples: A=*Anadara* sp.; Tt=*Telescopium telescopium*. See caption for Table 4.1 for details of calculations.

SITE	LAB. NO.	SAMPLE	HISTORICAL AGE (YEAR AD)	MARINE MODEL AGE	^{14}C AGE (YEARS BP)	ΔR (^{14}C YEARS)
Mapoon	ANU-1828	A	1903	454±7	576±60	122±60
Mapoon	ANU-2092	Tt	1903	454±7	436±60	−18±60
Error-weighted mean:						52±42

Radiocarbon dating of coral cores with clear growth bands

Reimer and Reimer (2000) have recently calculated two ΔR values for the central Queensland coast from high precision Δ^{14}C results for annual and biannual coral (*Porites australiensis*) core samples presented by Druffel and Griffin (1993, 1995, 1999). The first core was from the outer edge of Abraham Reef, part of Swains Reef, located c.200km east of Gladstone (22°S, 153°E) and spans 323 years from AD 1635–AD 1957 (Fig. 4.1). The core site is well-flushed by open ocean waters. The radiocarbon record exhibits pronounced variation between AD 1680 and AD 1730 that does not correspond to variation documented for atmospheric radiocarbon from tree-ring studies (Druffel and Griffin 1993). Druffel and Griffin (1993) argue that these variations are attributable to changes in source waters and vertical mixing in the western Coral Sea, related to long-term changes in the nature of El Niño/Southern Oscillation (ENSO) events. The second core site is at Heron Island, c.60km east of Gladstone (22°S, 152°E), spanning 106 years from AD 1849–AD 1955. Reimer and Reimer (2000) calculated ΔR values by averaging biannual Δ^{14}C data from AD 1800–AD 1900 for Abraham Reef and for AD 1849–AD 1899 for Heron Island (Table 4.3). The results are indistinguishable and combine to give a ΔR= +11±7 ($T'=0.13$; $\chi^2_{1:0.05}=3.84$).

Table 4.3 ΔR values for Abraham Reef and Heron Island coral cores (after Reimer and Reimer 2000). Samples: *Porites australiensis.* See caption for Table 4.1 for details of calculations.

SITE	LAB. NO.	SAMPLE	HISTORICAL AGE (YEAR AD)	MARINE MODEL AGE	[14]C AGE (YEARS BP)	ΔR (^{14}C YEARS)
Abraham Reef	WH&AA Series	Coral	1850	487±8	500±6	13±10
Heron Island	WH&AA Series	Coral	1874	476±8	484±6	8±10
Error-weighted mean:						11±7

Dating shell/charcoal paired samples from archaeological sites

In the absence of live-collected specimens and well-dated coral cores, several studies have been made of stratigraphically associated shell/charcoal paired samples. As part of early investigations, Gillespie and Temple (1977:Table 4; see also Gillespie 1977; Gillespie and Polach 1979:Table 6) calculated the marine reservoir age for 37 marine shell samples from nine archaeological sites by comparing them with the ^{14}C age of charcoal samples assumed to be temporally equivalent because of archaeological association. These shell/charcoal pairs came from five sites on the New South Wales coast and one site each in the Northern Territory, Victoria and Tasmania. The age difference between the shell and charcoal samples ranged from –820 to +725, with a mean of 283 years (shells older than charcoal). Variance from the expected 450±35 value discussed above was attributed primarily to the lack of integrity of midden deposits, with the investigators concluding that 'these middens are not ideal sites for the determination of past relationships between terrestrial and marine sample activities' (Gillespie and Polach 1979:418). Hughes and Djohadze (1980) conducted a study of paired samples from the Bass Point midden in south New South Wales, finding a mean age difference of 270 years (shells older than charcoal). They suggested that the similarity of this result to Gillespie and Temple's (1977) mean of 283 years indicated that the marine reservoir effect for this region may be less than the Gillespie value of 450±35 years. However, the apparent similarity of this figure to that derived from paired shell/charcoal samples by Gillespie and Polach (1979) was probably fortuitous because of the contingent nature of local reservoir conditions outlined above. Studies by Head et al. (1983) and Luebbers (1978), on the other hand, have generally shown good agreement with the generalised correction factor. In Queensland, only two limited studies of shell/charcoal paired samples are available. Horsfall (1987:404) presented a single pair from the Bramston Beach Midden 1 just south of Cairns indicating ΔR= –364±69, with an apparent age difference of only 40 ^{14}C years. This result is much lower than expected and most probably resulted from a lack of contemporaneity between the shell and charcoal samples selected. Horsfall (1987:181) noted that this paired sample was part of a wider program of assessment of marine reservoir effect in north Queensland but no further research was conducted. Data presented by Ross et al. (2000) from Peel Island in Moreton Bay, southeast Queensland, indicated potentially significant deviations from the open ocean ΔR value but full results are not available.

Dating 'post-bomb' live-collected shell specimens

Another approach to determining regional variation in marine reservoir effect has been to date post-AD 1950 ('post-bomb') live-collected shells (Table 4.4). Gillespie and Polach (1979:Table 5; see also Gillespie 1977:Table 4) presented results of three such dates on samples collected from Queensland in AD 1973 (two from Moreton Bay and one from near Mackay), while Rhodes et al. (1980:Table 1) report a single determination from a specimen collected from a water depth of 25m offshore from Edward River on the Gulf of Carpentaria in AD 1978. Differences indicated variation in ^{14}C activity of source waters and therefore also local and regional oceanographic processes. Although two of the four 'post-bomb' specimens show good agreement, the results are difficult to interpret because of the lack of detailed regional modelling of post-AD 1950 alteration to carbon reservoirs resulting from nuclear detonations.

Table 4.4 Post-AD 1950 live-collected shell (Gillespie and Polach 1979:Table 5; Rhodes et al. 1980:Table 1). Samples: Mep=*Mytilus edulis planulatus*; Pe=*Pyrazus ebeninus*; V=*Volachlamys* sp.; Ss=*Saccostrea succulata*.

SITE	LAB. NO.	SAMPLE	HISTORICAL AGE (YEAR AD)	% MODERN
Macleay Is.	SUA-218/1	Mep	1973	105.9±0.8
Macleay Is.	SUA-218/2	Pe	1973	104.6±0.8
Charon Point	ANU-1173	Ss	1973	116.0±0.7
Edward River	ANU-2099	V	1978	119.7±0.9

Summary

Attempts to investigate marine reservoir effects on the eastern Australian seaboard by dating shell/charcoal paired samples from archaeological sites and 'post-bomb' live-collected specimens have met with limited success. The general agreement of studies of 'pre-bomb' live-collected shells with the original Gillespie value has encouraged researchers to apply the original value rather than more recent calculations. Many researchers have thus either applied the erroneous ΔR= –5±35 value (e.g. Hiscock and Hughes 2001) or rejected the use of any marine reservoir estimate (e.g. O'Connor 1999; Veitch 1999). The ΔR values from the coral cores at Abraham Reef and Heron Island currently provide the most detailed information for eastern Australian near-shore waters. Reimer and Reimer (2000) combined the seven independent ΔR values available for northeast Australia to recommend a ΔR= +11±5 (revised value after Torres Strait values were corrected for age, and the Mapoon samples were added at my suggestion).

The present study: methods

A search of the Queensland Museum and Australian Museum malacology collections for pre-AD 1955, live-collected shells from the coast between Hervey Bay and Rockhampton resulted in the identification of 13 specimens representing three species from three locations with sufficient documentation for the purposes of this study. Five specimens from three locations were selected for radiocarbon analysis (Fig. 4.2, Table 4.5). A single sample from each location was dated by conventional liquid scintillation counting (LSC) and a duplicate sample from the Gladstone and Port Curtis groups was submitted for accelerator mass spectrometry (AMS) determination. The three live-collected species selected for this study are all filter-feeding bivalves with limited mobility and are consequently good candidates for examining local reservoir conditions. Several studies have indicated that detrital-feeders (such as grazing gastropods) are potentially problematic as ingested organic carbon from diverse sources can become incorporated into shell structures through metabolic action (Hogg et al. 1998; Tanaka et al. 1986). Two of the three species dated (*Donax deltoides* and *Anadara trapezia*) are common in coastal Aboriginal archaeological deposits in the area (Ulm and Lilley 1999).

Table 4.5 Radiocarbon ages obtained on 'pre-bomb' live-collected marine shell samples from central Queensland. Samples: Dd=*Donax deltoides*; At=*Anadara trapezia*; Vs=*Volachlamys singaporina*. See caption for Table 4.1 for details of calculations.

SITE	LAB. NO.	SAMPLE	HISTORICAL AGE (YEAR AD)	MARINE MODEL AGE	δ^{13}C (‰)	^{14}C AGE (YEARS BP)	ΔR (^{14}C YEARS)
Elliott Heads	Wk-6994	Dd	1925±5	455±5	−0.6±0.2	400±60	−55±60
Port Curtis 1	Wk-8457	Vs	1929	458±5	0.3±0.2	460±60	2±60
Port Curtis 2	NZA-12120	Vs	1929	458±5	0.9±0.2	570±60	112±60
Gladstone 1	Wk-8456	At	1904	453±10	0.3±0.2	480±50	27±51
Gladstone 2	NZA-12119	At	1904	453±10	−0.8±0.2	360±60	−93±61
Error-weighted mean:							1±26

No pre-AD 1955 live-collected shells were from fully estuarine contexts in the study area, so the study of possible estuarine marine reservoir effects relied on examination of shell/charcoal paired samples from excavated archaeological sites. In addition, paired samples are useful in examining variation in ΔR through time because the dating of live-collected specimens (see above) is valid strictly for the period of collection (i.e. the recent past). Twelve shell/charcoal pairs from seven archaeological sites representing five separate estuaries and the open beach were submitted for dating (Table 4.6). The samples were either identified during excavation where associations between samples were obvious (e.g. where charcoal was located *in situ* inside shell valves) or selected in the laboratory after consideration of the integrity of the association between samples (e.g. samples derived from densely packed layers of midden material where the possibility of post-depositional movement is reduced). For all estuarine pairs, *A. trapezia* was selected as the shell component of the paired samples to reduce variation introduced by differences in the relationship of particular species to the carbon cycle. *A. trapezia* samples ranged from 10–40mm long with an average of 25mm. Inglis (1992) found that *A. trapezia* attains a size of 20–30mm in length within 12 months with large individuals (>40mm) growing less than c.1mm/year up to 70–80mm. The majority of samples are thus likely to have had relatively short life-spans (<2 years), although sufficient to minimise short-term variation in reservoir conditions. The shell component of the pair from the open beach consisted of *D. deltoides* in the absence of other shell material suitable for dating. The dated sample consisted of valves of mature shellfish averaging 37mm in length, indicating an age of <10 months (Murray-Jones 1999). Only the largest blocky fragments of charcoal available were selected for dating although no attempt was made to identify the species of wood involved (cf. Higham and Hogg 1995). For technical details on sample pretreatment procedures see Chapter 3.

Results

Live-collected known-age samples

A single 19.6g specimen of *D. deltoides* (Lamarck 1818), variously known as pipi or eugarie, was dated. T.L. May collected this specimen, as well as five others, in the 1920s from Elliott Heads (24°04′S, 151°09′E) just south of Bundaberg (Queensland Museum MO63339) (Fig. 4.2, Table 4.5). A more precise record of the time of live collection is not available, and a collection date of AD 1925±5 is assumed in calculating an equivalent model marine age of 455±5 years. *D. deltoides* is a littoral sand dweller on high energy beaches and would therefore be unlikely to be associated with estuarine water circulation patterns (Coleman 1981; Lamprell and Whitehead 1992). The radiocarbon age determination of 400±60 BP (Wk-6994) is equivalent to ΔR= −55±60 (Table 4.5).

Two specimens of *Volachlamys singaporina* (Sowerby 1842), commonly known as the Singapore scallop or Cuming's scallop, were dated. They were from a collection of four small, articulated shells (with the desiccated animal enclosed) collected by M. Ward and W. Boardman during dredging in July 1929 from shallow water (16–22m) at Port Curtis just southeast of Gladstone (Australian Museum C369716) (Fig. 4.2; Table 4.5). Coordinates (23°55′S, 151°23′E) place the collection site offshore of the mouth of the Boyne River. *V. singaporina* prefers shell debris habitats and occurs in shallow water throughout northeast Australia (Lamprell and Whitehead 1992). The total shell weights for each of these specimens is low (8.8g, 1.7g, 1.5g, 0.6g). The largest sample was conventionally dated to 460±60 BP (Wk-8457), equivalent to ΔR= +2±60, while the second largest sample was dated by accelerator mass spectrometry (AMS) to 570±60 BP (NZA-12120), equivalent to ΔR= +112±60. The two dates are indistinguishable at the 95% confidence level and combine to give an error-weighted mean of 515±42 BP (T'=1.68; $\chi^2_{1:0.05}$=3.84), equivalent to ΔR= +57±42 (T'=1.68; $\chi^2_{1:0.05}$=3.84).

Two specimens of *A. trapezia* (Deshayes 1840), variously termed mud ark or Sydney cockle, were dated. These specimens came from a collection of four small valves live-collected pre-1904 from Gladstone (Australian Museum C018788) (Fig. 4.2, Table 4.5). Coordinates (23°51'S, 151°16'E) place the collection site under the present eastern margin of the city of Gladstone. This area consisted of intertidal flats fronting Port Curtis before extensive land reclamation in the second half of the twentieth century. The specimens were presented to the Australian Museum by Charles Hedley (1862–1926) in 1904. The accompanying museum label simply dates the specimens as pre-1904. Hedley stayed briefly in Gladstone *en route* to a field excursion to Masthead Island in October-November 1904 (Hedley 1906). The coincidence of the dates for the fieldtrip and the donation to the museum suggests the samples were collected during the 1904 fieldtrip. Furthermore, in a 1904 paper, Hedley (1904:206) described the distribution of *A. trapezia* (under the name *Arca lischkei*; see Murray-Wallace et al. 2000) as ranging from 'Bass Straits [sic] to Moreton Bay' but in a subsequent paper as reaching 'the tropics' (Hedley 1915:51). For the purposes of this study, I therefore assume that the *A. trapezia* specimens were live-collected in October-November 1904. *A. trapezia* is common on intertidal mudflats in southeast Australia and is strongly associated with the presence of seagrasses (Inglis 1992; Murray-Wallace et al. 2000; Sullivan 1961). The small size of the four individual specimens (11.5g, 4.3g, 4.6g, 4.5g) led to the adoption of the same dating strategy as that for *V. singaporina* (see above). The largest sample was conventionally dated to 480±50 BP (Wk-8456), equivalent to ΔR= +27±51, while the second largest sample was dated by accelerator mass spectrometry (AMS) to 360±60 BP (NZA-12119), equivalent to ΔR= –93±61. The two dates are indistinguishable at the 95% confidence level and combine to give an error-weighted mean of 431±38 BP (T'=2.36; $\chi^2_{1:0.05}$=3.84) equivalent to ΔR= –22±39 (T'=2.28; $\chi^2_{1:0.05}$=3.84).

The five ΔR calculations presented in Table 4.5 are indistinguishable at the 95% confidence level from the two averaged coral ΔR values for central Queensland presented in Table 4.3 (Reimer and Reimer 2000) and combine to give an error-weighted mean of ΔR= +10±7 (T'=7.17; $\chi^2_{6:0.05}$=12.59). This correction is currently the best estimate of variance in local open water marine reservoir effect from the modelled global marine calibration curve for the central Queensland coast. Although this new value is not significantly different from the generalised northeast Australian value of ΔR= +11±5 recommended by Reimer and Reimer (2000), it is based on a larger number of independent local samples with reduced error estimates. Pooling the five new ΔR values (Table 4.5) with the seven used to calculate the northeast Australian value (Tables 4.1–4.3) changes this value to +12±7 (T'=12.16; $\chi^2_{11:0.05}$=19.68).

Archaeological shell/charcoal paired samples

Seven Mile Creek

Four shell/charcoal paired samples were dated from the Seven Mile Creek Mound (SMCM) (see Chapter 6 for site details) (Fig. 4.2, Table 4.6). The eight dates indicated extremely rapid accumulation of the dense shell matrix over a period of less than 350 years between c.3,900–3,600 cal BP. The deposit thus presented an ideal opportunity to examine localised marine reservoir factors over a relatively short time-span at the beginning of the known chronology of human occupation of the region. The paired samples were selected at intervals down the 90cm deep densely packed shell deposit from a single 50cm × 50cm column. Samples consisted of large, articulated *A. trapezia* valves plotted during excavation paired with single fragments of blocky charcoal from the same excavation unit identified during excavation or during laboratory analysis of sieve residues. Single pieces of charcoal were used to avoid combining fragments representing possibly separate events (see Ashmore 1999).

The four ΔR values are significantly different (T'=199.00; $\chi^2_{3:0.05}$=7.82). The charcoal date NZA-12272 is clearly at odds with the results of the other seven determinations and indicates a lack of association with the paired shell sample (Wk-8324). This charcoal sample is derived from close to the top of the deposit (7–10cm) and was probably introduced through percolation down the midden profile, infilling the interstices in the porous shell matrix. If this pair is excluded, the stratigraphically lower three pairs show good agreement with an error-weighted mean of the remaining three values yielding $\Delta R= -155\pm55$ ($T'=1.23$; $\chi^2_{2:0.05}=5.99$). This value is significantly different from the generalised local open ocean value of $\Delta R= +10\pm7$ ($T'=8.65$; $\chi^2_{1:0.05}=3.84$) calculated above. This result based on replicated paired samples provides strong evidence that between c.3,900–3,600 cal BP the Seven Mile Creek estuary exhibited a $\Delta R= -155\pm55$.

Round Hill Creek

Analyses were undertaken on three separate shell/charcoal pairs from archaeological sites bordering Round Hill Creek (Fig. 4.2, Table 4.6). The creek has a large catchment area and is broad and shallow with an entrance bar inhibiting complete tidal flushing. Significantly, Round Hill Creek is bordered by extensive freshwater wetlands on its southwest margins (see Olsen 1980a:17), with freshwater influxes significantly depressing salinity during the wet season. This suggests that there may be complex hydrological factors operating periodically to introduce large volumes of dissolved atmospheric CO_2 into the estuary system. On this basis it was predicted that the estuarine shell dates would reflect uptake of terrestrial carbon mobilised in freshwater runoff, with dates more closely approximating results from the charcoal samples.

A single shell/charcoal pair (Wk-3944/Wk-5215) was obtained from Eurimbula Site 1 (ES1) on the west bank of Round Hill Creek from a discrete lens of *A. trapezia* located 30–40cm below ground surface (see Chapter 12 for site details). The lens was bulk sampled from the section of the test pit and returned to the laboratory for sorting. The charcoal sample comprises small fragments from the 3mm sieve. The paired samples exhibit an apparent difference of 790 ^{14}C years, with a ΔR value of +458±174 (Table 4.6). The shell age is much older than predicted and the most probable explanation for this wide discrepancy is a lack of a close temporal association between the shell and charcoal samples. Although the apparently discrete shell lens from which the samples derive appeared to be a secure stratigraphic context, it is possible that bulk sampling of the lens from the section resulted in contamination by more recent charcoal fragments. Alternatively, some or all of the fragmented charcoal selected for dating may have been intrusive – possibly introduced by the tree root activity or crab burrowing which was noted during excavation (see Specht 1985). Taylor (1987:Table 5.3) notes that contamination of the sample with <20% modern carbon would be sufficient to result in the observed discrepancy. Given the probable lack of association between the shell and charcoal samples this result is not considered as a reliable indicator of local reservoir conditions.

On the opposite side of Round Hill Creek, two separate shell/charcoal paired samples were submitted from the Tom's Creek Site Complex (TCSC), located at the confluence of Tom's Creek and Round Hill Creek (see Chapter 13 for site details) (Fig. 4.2, Table 4.6). The first pair (Wk-7838/Wk-7686) came from a dense layer of shell c.22cm below ground surface. The layer consisted almost exclusively of *A. trapezia* and the rock oyster *Saccostrea glomerata*, with occasional fragments of blocky charcoal that were recovered from the 3mm sieve residue for dating. The paired samples exhibit an apparent difference of 90 ^{14}C years with $\Delta R= -305\pm61$. Unfortunately, no absolute result was obtained from the second pair (Wk-7682/Wk-7681), which consisted of a valve of *A. trapezia* packed with charcoal. This was because the very recent age of the charcoal did not allow a precise assignment of age to be made, although the sample is clearly younger than 200 years (%Modern= 99.5±0.6). The shell component of the sample returned a ^{14}C age of 620±50 BP. If the charcoal sample is assumed to exhibit a ^{14}C age of 0 (=AD 1950), the paired samples exhibit an apparent difference of ≥620 ^{14}C years with $\Delta R= \geq+620$ (Table 4.6). Because assignment of an absolute value for ΔR is not possible this result is excluded from further consideration.

Middle Creek

Although Middle and Pancake Creeks are joined together by a narrow channel (see Fig. 4.2), they exhibit different morphological attributes and very different hydrological patterns (Olsen 1980a) that require them to be treated as potentially separate estuarine carbon reservoirs. A single paired sample (Wk-8558/Wk-8557) was obtained from the Ironbark Site Complex (ISC) on the lower southern margin of Middle Creek (see Chapter 9 for site details), consisting of whole shells and charcoal fragments collected from the 3mm sieve residue. Although the pair returned an apparent age difference of 390 ^{14}C years, a ΔR value could not be calculated because the charcoal age is modern at one sigma (Table 4.6).

Table 4.6 Shell/charcoal paired samples from the southern Curtis Coast. ^{14}C ages obtained on charcoal samples were reduced by 41±14 years to correct for ^{14}C variation between northern and southern hemispheres (McCormac et al. 2002). An estimate of the atmospheric calibration curve error, derived from an average of estimated error in the 1σ span of the age, was also included. Therefore, atmospheric age σ= √(σ^{14}C age^2+σsouthern hemisphere offset^2+average of calibration curve error^2) (Gillespie 1982). Note that the incorporation of southern hemisphere offset error in this formula assumes that each atmospheric conventional radiocarbon age derives from an independent secondary reservoir (see Jones and Nicholls 2001 for discussion). The 1σ range of the ^{14}C value was converted to the equivalent global marine model 1σ range using atmospheric ages interpolated from INTCAL98 to the same calendar year as MARINE98 (Stuiver et al. 1998a). ΔR was calculated by deducting the mid-point of the equivalent marine model age of the charcoal determination from the ^{14}C age of the paired marine shell sample. ΔRσ= √(σmarine model age^2+σmarine shell age^2) (Gillespie 1982). This method is illustrated for pair NZA-12117/Wk-8326 in Fig. 4.3.

SITE	LAB. NO.	SAMPLE	^{14}C AGE (YEARS BP)	EQUIVALENT MARINE MODEL AGE	ΔR (^{14}C YEARS)
Seven Mile Creek					
SMCM	NZA-12272	charcoal	1260±80	1592±113	-
SMCM	Wk-8324	*A. trapezia*	3540±80	3540±80	+1948±139
SMCM	NZA-12117	charcoal	3500±60	3801±71	-
SMCM	Wk-8326	*A. trapezia*	3610±70	3610±70	−191±100
SMCM	NZA-12273	charcoal	3570±60	3854±70	-
SMCM	Wk-8327	*A. trapezia*	3780±60	3780±60	−74±92
SMCM	NZA-12118	charcoal	3660±60	3959±76	-
SMCM	Wk-8328	*A. trapezia*	3750±60	3750±60	−209±97
Round Hill Creek					
ES1	Wk-5215	charcoal	1600±160	1932±164	-
ES1	Wk-3944	*A. trapezia*	2390±60	2390±60	+458±174
TCSC	Wk-7686	charcoal	540±50	935±35	-
TCSC	Wk-7838	*A. trapezia*	630±50	630±50	−305±61
TCSC	Wk-7681	charcoal	modern	modern	-
TCSC	Wk-7682	*A. trapezia*	620±50	620±50	≥+620±50
Middle Creek					
ISC	Wk-8557	charcoal	200±140	627±140	-
ISC	Wk-8558	*A. trapezia*	590±60	590±60	see text
Pancake Creek					
PCSC	Wk-6993	charcoal	700±140	1059±112	-
PCSC	Wk-6992	*A. trapezia*	800±80	800±80	−259±137
Mort Creek					
MCSC	Wk-7458	charcoal	1970±80	2279±65	-
MCSC	Wk-6987	*A. trapezia*	2260±50	2260±50	−18±82
MCSC	Wk-7458	charcoal	1970±80	2279±65	-
MCSC	Wk-7836	*A. trapezia*	2320±50	2320±50	+42±82
Open Coast					
ABM	Wk-10969	charcoal	266±87	609±153	-
ABM	Wk-11280	*D. deltoides*	674±47	674±47	+65±160

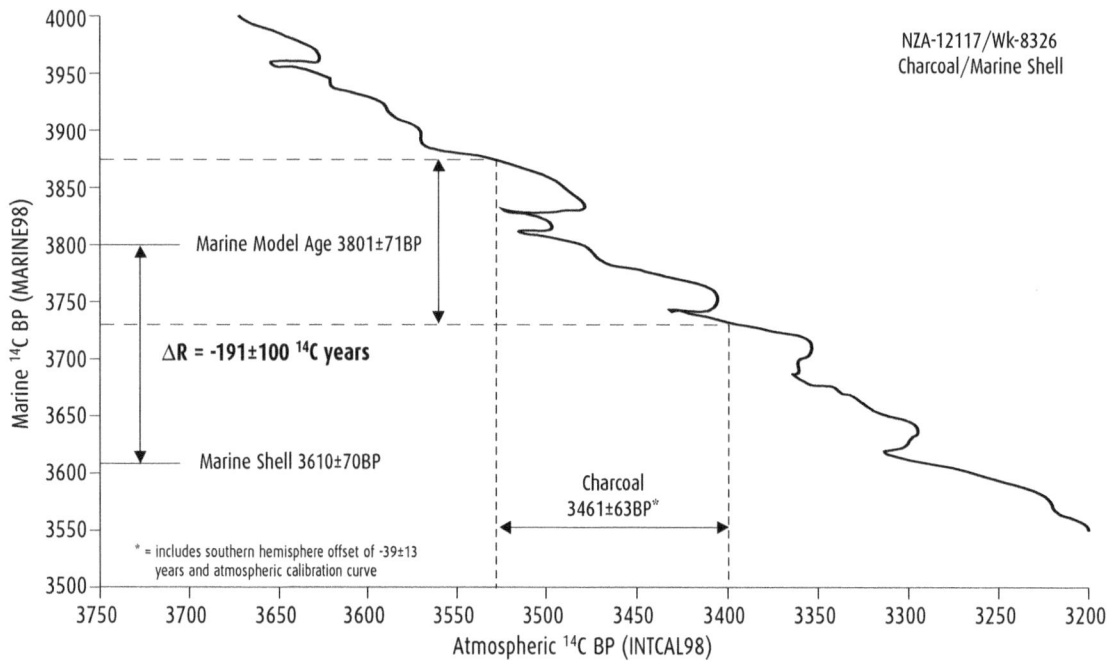

Figure 4.3 Example of ΔR calculation method for pair NZA-12117/Wk-8326 (see caption for Table 4.6).

Pancake Creek

A shell/charcoal pair (Wk-6992/Wk-6993) was obtained from the Pancake Creek Site Complex (PCSC) on the north bank of Pancake Creek, exhibiting an apparent age difference of 100 ^{14}C years with ΔR= –259±137 years (see Chapter 8 for site details) (Fig. 4.2, Table 4.6). The large error estimate associated with this value makes it indistinguishable from the generalised local open ocean value of ΔR= +10±7. The paired sample consisted of a valve of *A. trapezia* tightly packed with blocky charcoal fragments, indicating a secure stratigraphic association. This site is located close to the broad entrance of the creek, which exhibits good tidal flushing and no obvious large terrestrial water input (Olsen 1980a). Terrestrial CO_2 dissolved in water stored in the beach ridges lining the north bank of the creek (possibly with long residence times) may have been introduced into shellfish beds in the intertidal zone.

Mort Creek

At the Mort Creek Site Complex (MCSC), a single blocky charcoal sample (Wk-7458) was paired with two separate shell samples (Wk-6987 and Wk-7836) of *A. trapezia* derived from the middle of a densely packed shell lens located c.20–30cm below ground surface (Table 4.6) (see Chapter 7 for site details). It was predicted that the result would show good agreement with the open ocean because the site is located near the mouth of Mort Creek (also known as Morris Creek), which can be considered part of Rodds Harbour which has a high tidal range and thus good tidal flushing. Also, Mort Creek is a relatively minor estuary with a small catchment and does not have backing swamps, suggesting minimal freshwater input. The samples were derived from a extremely dense layer of shell increasing the possibility of minimal post-depositional movement within the matrix. The two shell samples (Wk-6987 and Wk-7836) show good agreement, with apparent age differences of 290 and 350 ^{14}C years, respectively. The values of ΔR= –18±82 and +42±82 are indistinguishable at the 95% confidence level and combine to give an error-weighted mean of ΔR= +12±58 (T'=0.27; $\chi^2_{1:0.05}$=3.84).

Agnes Beach
A single shell/charcoal pair (Wk-11280/Wk-10969) was obtained from the Agnes Beach Midden (ABM) on the high energy beach adjacent to the town of Agnes Water. It came from a discrete layer of *D. deltoides* and stone artefacts exposed in an eroding frontal dune (Fig. 4.2, Table 4.6). The charcoal sample is based on large blocky fragments of charcoal removed from the exposed lens. As the only stratified deposit containing organic material located on the open coast in the study area (see Ulm and Lilley 1999) the site presented an ideal opportunity to examine local open water marine reservoir factors. The paired samples exhibit an apparent difference of 408 ^{14}C years with a ΔR value of +65±160 (Table 4.6). The large error estimate makes this value indistinguishable from the coral and live-collected shell ΔR values presented in Tables 4.3 and 4.5 and combine with these to give an error-weighted mean of ΔR= +10±7 (T'=7.29; $\chi^2_{7:0.05}$=14.07).

Summary
As expected, the conventional radiocarbon age of estuarine shell samples was consistently older than that of paired charcoal samples (Table 4.6). Major discrepancies exist between the three sets of pairs from Round Hill Creek, with only one result (ΔR= –305±61) considered free of obvious interpretation problems. If the pair from Eurimbula Site 1 (Wk-5215/Wk-3944) and the indeterminate results from the Tom's Creek Site Complex (Wk-7682/Wk-7681) and the Ironbark Site Complex (Wk-8558/Wk-8557) are excluded, the results from the other eight paired samples support the prediction that estuarine shell samples will exhibit a lower marine reservoir correction factor than the local open ocean value of ΔR= +10±7. Direct comparison of the ^{14}C ages shows that shell samples are 90 to 390 ^{14}C years older than their paired charcoal sample. The open beach pair result from the Agnes Beach Midden provides further support for the regional applicability of the ΔR= +10±7 based on coral and live-collected shell samples.

Temporal variability or ΔR(t)
The ΔR values calculated above must be considered as a first approximation only because they do not account for variation in specific reservoir parameters (e.g. reservoir size and exchange rates between the atmosphere and ocean) through time. In the absence of additional information, it is assumed that temporal changes in ΔR for a specific region must coincide with changes in the global mixed surface layer model ocean (Stuiver et al. 1998b:1135). Time-factored ΔR(t) (t=time) for marine and estuarine environments can be calculated through large-scale studies of annual coral records and/or paired shell/charcoal samples from a variety of periods (e.g. Ingram 1998; Kennett et al. 1997). Preliminary examination of ΔR values of the Abraham Reef coral record through time, for example, indicates shifts from up to ΔR= +39±10 in AD 1730 to ΔR= –41±8 in AD 1950 (Paula Reimer, Centre for Accelerator Mass Spectrometry, Lawrence Livermore National Laboratory, pers. comm., 2001; Druffel and Griffin 1993, 1995, 1999).

Discussion

The five dates reported from pre-AD 1955 live-collected shell specimens from between the Elliott River and Gladstone combined with the two previous coral ΔR values indicate, at least for the recent past, that open ocean reservoir influences are in the order of ΔR= +10±7. These results support the general conclusions of previous studies and are statistically indistinguishable from the five dates on live-collected shells from Torres Strait (Gillespie and Temple 1977) and the Gulf of Carpentaria (Rhodes et al. 1980), although the error estimates are reduced. The error-weighted mean of all 12 ΔR values available for 'pre-bomb' live-collected Queensland shells and corals indicate that a generalised ΔR= +12±7 (T'=12.16; $\chi^2_{11:0.05}$=19.68) is appropriate for open marine contexts in northeast Australia and supersedes the erroneously calculated ΔR= –5±35 value recommended by Stuiver et al. (1986:Fig. 10A, Table 1). Relative consistency observed between determinations on live-collected specimens from northeast Australia is probably largely a function of the broad, shallow continental shelf that mitigates the influence of ^{14}C-depleted deep ocean upwelling by ensuring mixing through wave and current action (Taylor 1987:8).

These results stand in marked contrast to results obtained from estuaries on the basis of shell/charcoal paired samples from archaeological contexts. Results indicate estuary-specific values of ΔR= –155±55 for Seven Mile Creek and ΔR= +12±58 for Mort Creek. Less confidence is placed in the ΔR values calculated for other estuaries because of suspected problems of association (e.g. Round Hill Creek) or the existence of only a single paired result (Pancake Creek). However, as a first approximation, I suggest that ΔR= –259±137 for Pancake Creek, and ΔR= –305±61 for Round Hill Creek. Figure 4.4 schematically represents all the ΔR values calculated in this study. The high level of variation observed in regional estuarine ΔR values is best explained as a function of the structures of the carbon reservoirs in which the samples were formed. Many coastal archaeological sites, and all of those from which shell/charcoal paired dates are reported here, are in environments of restricted circulation: they are located in estuaries where significant variation in terrestrial carbon input and exchange with the open ocean conditions might reasonably be assumed to deviate from that of the mixed surface layer of the ocean. Intra-estuary spatial variation in ΔR values is considered to be negligible. Studies by Hogg et al. (1998:184) found that tidally-dominated estuaries with catchments free of calcareous materials (such as those of the study area) 'should exhibit a uniform distribution of ^{14}C throughout the estuary'.

The potential impact of differences in ΔR between estuaries can be illustrated through a case study of the Seven Mile Creek Mound, the oldest dated open site on the central Queensland coast. Table 4.7 and Figure 4.5 present the calibrated results of four conventional radiocarbon dates of shell samples (*A. trapezia*) using four different ΔR values: ΔR= –5±35 (recommended by Stuiver et al. 1986 and Stuiver and Braziunas 1993); ΔR= +50±33 (the corrected version of this value; see above); ΔR= +10±7 (calculated for the open ocean of the study region; see above); and ΔR= –155±55 (actual determined value for the Seven Mile Creek estuary based on shell/charcoal paired samples from the Seven Mile Creek Mound; see above). Conventional radiocarbon ages were calibrated with the CALIB (v4.3) computer program (Stuiver and Reimer 1993) using the marine calibration dataset of Stuiver et al. (1998a). The calibrated ages reported span the 2σ age-range. Table 4.7 shows overall variation in the central tendencies of the calibrated radiocarbon ages between the generalised Australian value of ΔR= +50±33 and the actual estuary-specific value of ΔR= –155±55 is in the order of 300 years. The difference between the regionally determined open ocean value of ΔR= +10±7 and the estuary-specific determined value is ~250 years.

The identification of variability in estuarine ΔR values has significant implications for archaeology in Queensland. Some 25% (n=216) of all radiocarbon determinations from the state are based on marine or estuarine shell samples (Ulm and Reid 2000). In some coastal regions, the reliance on marine and estuarine shell samples for dating is often pronounced. For example, more than 95% (n=21) of dates from the Keppel Islands are on marine shell (Rowland 1981, 1992; Ulm and Reid 2000).

This limited study demonstrates the potential for significant variation in local marine reservoir effect within the study area. Clearly, radiocarbon age determinations based on marine specimens need to be considered in the context of local conditions (such as location, hydrology, geology, sedimentology). In confined areas such as estuaries, deviation of marine reservoir effect from global open ocean values may be pronounced. Possible adjustments of several hundred years to these determinations to take into account local marine reservoir effect has obvious implications for arguments for regional cultural change based on tightly defined radiocarbon chronologies. In estuarine contexts in Queensland, a localised estuary-specific correction factor must be calculated that takes into account alterations in reservoir parameters through time (Spennemann and Head 1996).

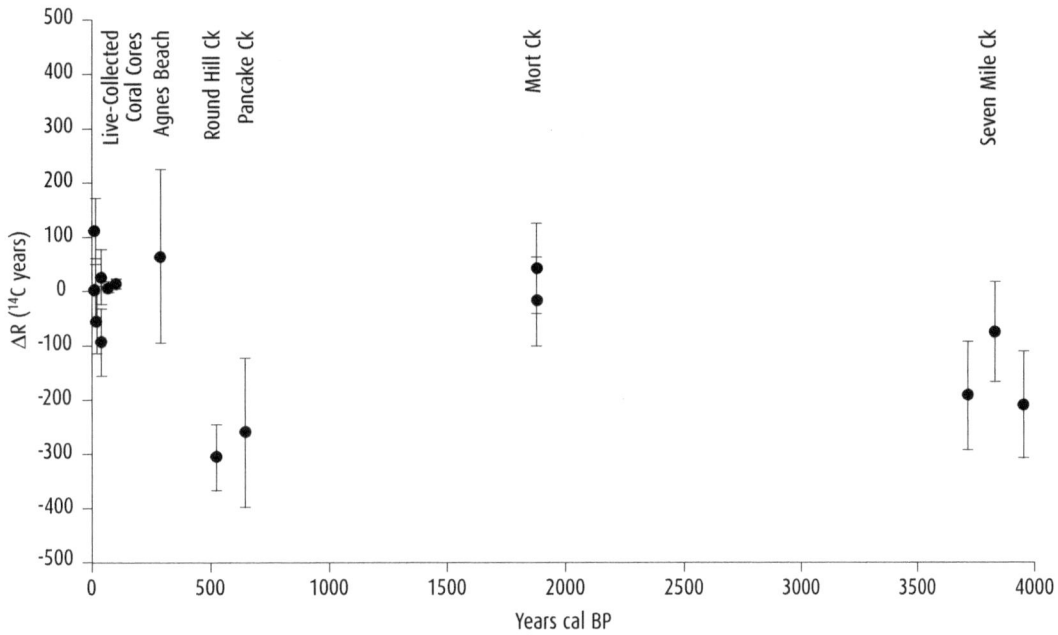

Figure 4.4 ΔR values calculated for the southern Curtis Coast.

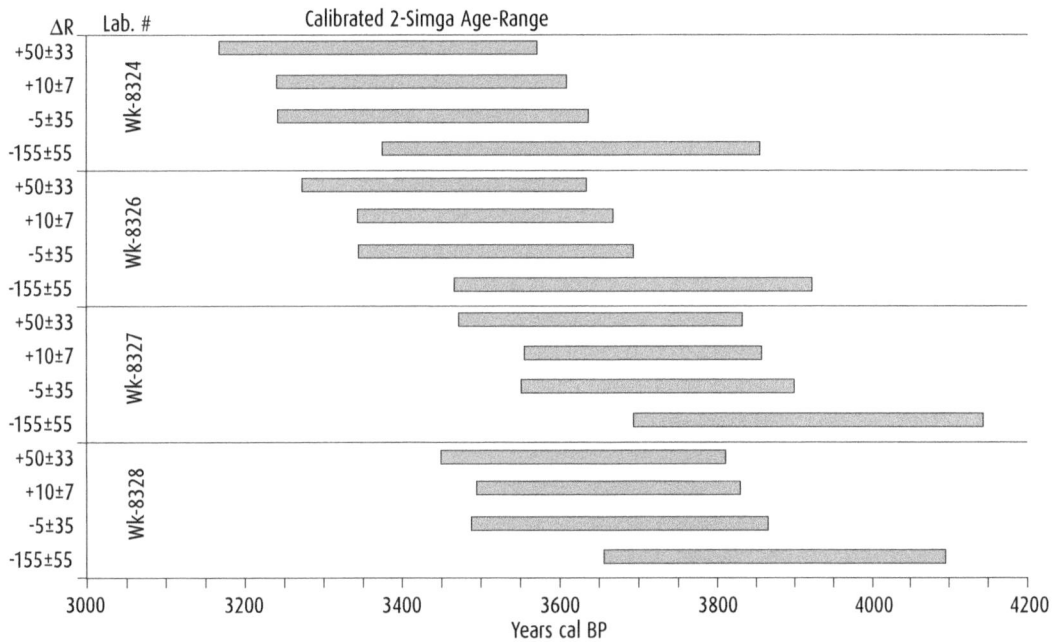

Figure 4.5 Calibrated radiocarbon age-ranges from the Seven Mile Creek Mound, using various ΔR values (see Table 4.7).

Table 4.7 Calibrated radiocarbon ages from the Seven Mile Creek Mound, using various ΔR values.

LAB. NO.	CALIBRATED AGE/S ΔR= +50±33 (CAL BP)	CALIBRATED AGE/S ΔR= +10±7 (CAL BP)	CALIBRATED AGE/S ΔR= −5±35 (CAL BP)	CALIBRATED AGE/S ΔR= −155±55 (CAL BP)
Wk-8324	3567(3362)3164	3606(3399)3235	3633(3429)3237	3850(3608)3372
Wk-8326	3630(3442)3269	3665(3470)3339	3690(3487)3338	3919(3684)3462
Wk-8327	3827(3637)3465	3855(3687)3550	3893(3700)3548	4140(3904)3688
Wk-8328	3808(3618)3444	3824(3651)3491	3859(3677)3485	4089(3867)3652

Summary

A review of the literature on Australian marine reservoir effects indicates that the routinely applied ΔR value of -5 ± 35 for northeast Australia is erroneously calculated. The values determined in this study suggest a minor revision to Reimer and Reimer's (2000) recommended value for northeast Australia from $\Delta R = +11\pm5$ to $+12\pm7$, and specifically for central Queensland to $\Delta R = +10\pm7$, for near-shore open marine environments. In contrast, data obtained from estuarine shell/charcoal pairs demonstrate a general lack of consistency, suggesting estuary-specific patterns of variation in terrestrial carbon input and exchange with the open ocean. Preliminary data indicate that in some estuaries, at some time periods, a ΔR value of more than -155 ± 55 may be appropriate. In estuarine contexts in central Queensland, a localised estuary-specific correction factor is recommended to account for geographical and temporal variation in ^{14}C activity.

For the purposes of this study, the following ΔR values will be applied in the calibration of marine and estuarine radiocarbon dates: local open waters (i.e. non-estuarine) $+10\pm7$; Seven Mile Creek -155 ± 55; Mort Creek $+12\pm58$; Pancake Creek -259 ± 137; and Round Hill Creek -305 ± 61. For estuaries where a specific ΔR could not be determined (e.g. Middle Creek) the local open ocean $\Delta R = +10\pm7$ will be applied in default. In the absence of local high resolution $\Delta R(t)$ records, temporal changes are assumed to coincide with changes in the modelled global ocean (Stuiver et al. 1998b:1135).

5

Bivalve conjoin analyses: assessing site integrity

Introduction

Conjoin (also refitting or cross-mending) analyses of stone artefact, ceramic and faunal assemblages have long been employed to assess the integrity of various archaeological deposits (see, for example, collected papers in Cziesla et al. 1990; Hofman and Enloe 1992). Only two systematic studies have been conducted in Australia, both concerning rockshelters in the Central Queensland Highlands (Richardson 1992, 1996; Stern 1980; see also Leavesley and Allen 1998). No comparable studies are available for open coastal midden sites despite explicit and implicit reference to this site type as stratigraphically problematic (e.g. Gillespie and Polach 1979; Lourandos 1996, 1997; Roberts 1991; Stone 1989, 1992, 1995). In this chapter I present results of experiments which (a) establish methods for effective conjoin analysis of *Anadara trapezia* (mud ark or Sydney cockle) valves recovered from coastal middens; (b) apply the methods to a case study of the *A. trapezia* assemblage excavated from the mid-Holocene Seven Mile Creek Mound as part of this study; and (c) assess the replicability of these methods through a blind test. The chapter also briefly outlines the range of conjoin analyses undertaken for archaeological purposes and discusses some limitations of these studies. This is the first known application of bivalve conjoin analysis to an archaeological deposit in Australia and the first known application anywhere of these methods to *A. trapezia*. The overall aim of the experiments was to evaluate the potential of using bivalve conjoin analyses to assess the integrity of excavated shell midden deposits on the southern Curtis Coast.

Background

Conjoin analyses in archaeology seek to reassemble objects or parts of objects broken or separated in the past in order to reach understandings beyond those attainable by considering the separated items in isolation. Archaeological conjoin analyses are undertaken for three main purposes (in addition to restorative or conservation work). First, most studies have been undertaken as part of technological analyses of stone artefacts to explicate manufacturing and reduction sequences (e.g. Fullagar 1990; Jones 1987; Leach 1984; for Australian examples see Hiscock 1986b, 1993, in press; Knight 1990; Luebbers 1978). Second, conjoin studies of stone artefacts, ceramics and bone have been used in intra-site spatial analyses to investigate the differential use of space (e.g. Singer 1984; Sullivan et al. 1991; Todd and Stanford 1992). Third, the vertical and horizontal separation of conjoining objects has contributed to understandings of depositional and post-depositional processes, especially assessments of stratigraphic integrity (e.g. Bollong 1994; Cahen 1978; Cahen et al. 1979; Cahen and Moyersons 1977; Villa 1982). It is this last application that is of central concern here.

In Australia, conjoin analyses have had limited application in assessments of stratigraphic integrity. Several studies cite incidental or non-systematic stone artefact conjoin identifications as indicators of stratigraphic integrity (e.g. Flood and Horsfall 1986:19; Fullagar et al. 1996:770). In the first detailed study conducted, Stern (1980) conjoined stone artefacts from the top four excavation units of the Native Well 1 rockshelter in the Central Queensland Highlands to examine vertical displacement within sandy shelter deposits. The analysis indicated that the magnitude of movement of stone artefacts in the deposit could not support the very precise chronology of late Holocene stone artefact technological change proposed by Morwood (1979). Richardson (1992, 1996) conducted an extensive conjoin analysis of the stone artefact assemblage from Kenniff Cave in the Central Queensland Highlands, demonstrating significant vertical and horizontal movement of conjoining artefacts despite a well-defined stratigraphic sequence. Like Stern, Richardson (1996:81) argued that the poor chronological resolution available could not support 'a precise chronology for the appearance or disappearance of implement classes' (see also Mulvaney 1969:213). These findings called into question basic assumptions about the integrity of apparently well-stratified sandstone rockshelter deposits which form the basis of our understanding of the archaeology of eastern Australia.

Open shell middens are frequently cited as lacking integrity, especially in comparison to rockshelter deposits. Indeed, Lourandos (1996:18) has argued that rockshelter deposits provide a 'sounder' dataset as they 'are not subject to the same degree of post-depositional modification as open sites'. Several studies have shown that shell deposits can be (and have been) created and modified by a wide range of taphonomic (Roberts 1991) and depositional processes (Stone 1989, 1992, 1995). Over recent years, a number of approaches to assessing the integrity of coastal shell deposits has been developed, including stratigraphic analyses, fragmentation studies, studies of midden composition (Attenbrow 1992), shell size studies (Carter et al. 1999) and studies of foraminifers (Lilley et al. 1999). However, many of these studies have produced equivocal results. Stratigraphy is often difficult to define in sandy coastal deposits and is often related to differential moisture content and soil profile development rather than depositional processes. Rowland (1994) has also critiqued simplistic models of midden composition showing that items such as pumice, coral and minute gastropods, long-thought to be diagnostic of non-cultural origin of shell deposits are in fact commonly used by hunter-gatherers.

Materials commonly used in conjoin analyses include stone, ceramic, bone, wood, glass and metal (Hofman 1992). Although several studies have reported refitting shell artefacts and engraved shell art (Brown 1981; Hofman 1985; Phillips and Brown 1978), a literature review identified only one study that reported analysis of bivalve conjoins (Koike 1979). Koike investigated the depositional history of a deposit through pairing clam shell valves as part of a shell midden study. Koike (1979:67) argued that, assuming that valve-pairs were connected by ligaments when discarded, the separation of valve-pairs within a deposit may contribute to an understanding of site formation processes and post-depositional movement of shells in the deposit.

Koike (1979) attempted to pair 2,089 whole clam (*Meretrix lusoria*) valves (53% of the total clam assemblage) recorded using three-dimensional plotting from a discrete shell layer (180cm × 80cm × 30cm) in an abandoned Kofun Period dwelling pit in the Natsumidai site, Japan. A total of 380 valve-pairs was identified. A deep occlusion and ridges at the apex coupled with the shared colour pattern of paired *Meretrix lusoria* valves were used to identify valve-pairs. Koike selected whole and unfaded valves from the assemblage and sorted them by side (left or right), length and colour pattern. Analysis proceeded 'by comparing successive right valves to a given left valve until a colour match and tight-closing fit were obtained' (Koike 1979:66). It is unclear at what point manual refitting attempts were abandoned. Seventy percent of the identified pairs were separated by 20cm or less, with most movement evident on the horizontal plane parallel to the slope of the deposit.

Claassen (1998:87) cautioned that the separation of left and right valves in an excavated deposit may not be a valid indicator of the degree of disturbance in a deposit as disarticulated valves can separate upon impact with the ground (see Muckle 1985:62 for results of replication experiments). However, although horizontal separation of paired valves during discard events is likely for this reason, significant vertical separation is not.

Bivalves have a distinct advantage over stone artefacts in conjoin studies owing to the fact that they primarily enter archaeological deposits as food refuse and are unlikely to be reused or recycled as raw materials. Therefore, the assumption that two valves of an articulated bivalve were discarded at the same time is more easily sustained than for stone artefacts or ceramics where significant intervals may elapse between individual discard events (e.g. Lindauer 1992). Although bivalve shells may be used or reused as tools (e.g. scraper) or ornaments (e.g. pendants), the incidences of such artefacts in large shell midden deposits in Australia will be low and edge-damage, use-wear and other analytical studies can be used to identify such uses (e.g. Przywolnik 2003). The validity of individual bivalve conjoins can also be independently checked by, for example, thin section comparison of growth structures, stable isotope analyses and/or radiometric dating of paired valves. Another advantage is that the conjoining material can be directly dated and determination/s obtained on one valve in the pair can be automatically extended to the other to determine deposition rates and/or periods of site disturbance between the two objects. Bivalve conjoins also avoid the problem of assuming that multiple conjoins result from a single breakage event (Larson and Ingbar 1992): there are only two parts to a bivalve conjoin.

Conjoined bivalves are occasionally mentioned in Australian midden studies to establish either cooking and preparation methods, demonstrate the non-cultural origin of deposits or illustrate the integrity of deposits. O'Connor and Sullivan (1994a; Sullivan and O'Connor 1993) have cited the presence of articulated bivalves in shell deposits investigated in the Kimberley region as evidence for a natural origin of the deposit. Robins et al. (1998:112) used the presence of 'hinged' *A. granosa* shells and an internally consistent radiocarbon chronology in support of the integrity of the Bayley Point 3 shell mound in the Gulf of Carpentaria. They argued that these attributes are consistent with a cultural origin for the mounded shell deposits rather than scrub fowl (*Megapodius reinwardt*) nesting activity as the latter is characterised by constant reworking of shell deposits (Mitchell 1993). Elsewhere, Peacock (2000) and Claassen (1991) have associated articulated valves with non-subsistence activities such as raw material for artefact manufacture.

The conjoin analyses presented in this chapter were undertaken to achieve a number of basic objectives:
— To establish reliable criteria and methods for effective valve-pair identification of *A. trapezia* in coastal middens on the southern Curtis Coast through an analysis of the morphological attributes of articulated specimens.
— To apply these methods to a case study to identify conjoins in an excavated assemblage of unarticulated *A. trapezia* valves from the Seven Mile Creek Mound.
— To assess the reliability of these methods through a blind test.
— To assess results in terms of site integrity and formation processes.

General methods and approach

A. trapezia was selected for this valve-pairing study for three main reasons. First, it is present in most recorded shell middens in the region, making this method potentially applicable to a range of excavated assemblages (Ulm 2002b; Ulm and Lilley 1999). Second, this mollusc is still present in tidal estuaries in the region, making studies of live-collected specimens possible. Third, individual specimens exhibit considerable morphological variability, making it unlikely that valves will conjoin tightly unless they are from the same individual.

A. trapezia have slightly asymmetrical valves (Figs 3.1–3.2). Valves exhibit an inflated anterior with tapering to the posterior and ventral margins. Tightly interlocking corrugations occur along the anterior, ventral and posterior margins at the terminations of strong radial ribs. The anterior and posterior lateral teeth are more pronounced than the cardinal teeth, with all enclosed in a rectangular hinge structure (c.1–3mm wide) which is straight along its dorsal edge and concave towards the cardinal teeth on its ventral edge. The presence of numerous interlocking teeth of various sizes, patterns and angles on conjoining hinges precludes tight conjoins unless relatively precise alignment is achieved. Lamprell and Healy (1998:54) note that in juveniles of this taxon, the left valve encompasses (i.e. is slightly larger than) the right valve and the dorsal margin is more angulate at its anterior and posterior terminations.

Attributes for identifying probable conjoins were established by studying articulated archaeological and live-collected specimens. A total of 158 articulated *A. trapezia* was plotted *in situ* during excavation of the Seven Mile Creek Mound (see below and Chapter 6 for site details). Five attributes were measured for each matched pair of whole valves: length, width, height, weight and hinge length (Fig. 3.2). Linear regressions of left versus right valve attributes show that both hinge length and weight are strongly correlated in paired valves (Table 5.1). During 2001 fieldwork, a sample of 10 live-collected *A. trapezia* specimens was obtained from Mort Creek and Round Hill Creek in the study area. Linear regressions of left versus right valve attributes showed that both hinge length and weight were strongly correlated in paired valves (Table 5.2), confirming the results obtained from the sample of archaeological articulated specimens from the Seven Mile Creek Mound.

Table 5.1 Attributes of articulated *A. trapezia* specimens recovered from the Seven Mile Creek Mound in rank order of correlation coefficient. Note that specimens were only included if there was no damage inhibiting accurate measurement of each attribute on either valve in a pair.

ATTRIBUTE	n	RANGE DIFFERENCE	MEAN DIFFERENCE	r^2
Hinge length (mm)	146	0–0.5	0	0.9998
Weight (g)	155	0–3.9	0.8	0.9813
Length (mm)	155	0–3.6	0.7	0.9697
Width (mm)	154	0–6.3	1.4	0.8561
Height (mm)	156	0–2.5	0.8	0.8546

Table 5.2 Attributes of live-collected *A. trapezia* specimens in rank order of correlation coefficient.

ATTRIBUTE	n	RANGE DIFFERENCE	MEAN DIFFERENCE	r^2
Weight (g)	10	0.2	0.8	0.9994
Hinge length (mm)	10	0–0.5	0.1	0.9982
Length (mm)	10	0.1–2.1	0.8	0.9928
Height (mm)	10	0–1.4	0.9	0.9854
Width (mm)	10	0.4–6.4	3.3	0.9127

This examination of articulated *A. trapezia* specimens shows that the hinge and associated articulating teeth are the most symmetrical features of valve-pairs, providing a well-defined rectangular-shaped area amenable to measurement. Weight is the next most strongly correlated variable, despite clear variation in other valve-pair attributes indicating considerable valve dimorphism. Width and height exhibit relatively low correlation coefficients probably owing to more pronounced valve dimorphism in these attributes and minor variability introduced by uncertainty in the selection of measuring points. Hinge length and weight of valves are therefore considered to be the most reliable measures for defining a narrow range of probable conjoins in any given assemblage. Manual refitting of left and right valves can then be used for confirmation, based on the observation that the extent and pattern of teeth on hinges prevents tight or closely fitting pairs unless the valves are from the same individual.

Seven Mile Creek Mound bivalve conjoin analysis

Introduction

The Seven Mile Creek Mound is located on the fringe of Seven Mile Creek, a tributary of Rodds Harbour (see Chapter 6). A single 1m² pit divided into four 50cm × 50cm squares (A–D) was excavated to a depth of 117cm. Excavation proceeded in shallow (<3–8cm), arbitrary excavation units within identified stratigraphic units. Analysis of Square A revealed a deposit dominated by oyster (*Saccostrea glomerata*) and mud ark (*A. trapezia*), dating to 3,600–3,900 cal BP. The presence of numerous articulated *A. trapezia* (n=158) and occasional *Tichomya hirsutus* valves throughout the deposit suggested rapid burial and settling of these shells in the matrix before their ligaments deteriorated and the individuals disarticulated. Dead bivalves articulated by ligaments tend to gape as their adductor muscles are no longer able to pull the valves shut (Cadée 2002; Claassen 1998:18). However, the presence of many closed valves suggests that there is a process whereby at least some shells are forced together, indicating that closing of ligament-articulated valves occurred during discard. An alternative explanation is that valves failed to open during cooking and were discarded. Meehan (1982:97) notes that *A. granosa* do not open easily even after heating, while my observation of *A. trapezia* indicates they are similar in this regard. The presence of these articulated shells *in situ* suggested the possibility that further pairs might be identified through refitting separated valves and that this might contribute to an understanding of site integrity and depositional history. An understanding of the despositional history of the site was considered particularly valuable because the 300 year sequence had no discernible stratigraphic features, although the internally consistent radiocarbon chronology indicated that rapid deposition has occurred at the site in a number of sequential events (see Chapter 6).

Unarticulated valves were not plotted in three dimensions during excavation. Provenance of individual valves has a uniform horizontal error of ±25cm while vertical error ranges from ±3–8cm depending on excavation unit. Given these uncertainties, each valve conjoin set has a different level of vertical resolution, constrained by the size of the excavation units containing the valves. Therefore, the separation distances cited for an individual conjoin set are maximums.

Aims

Conjoin analysis at the Seven Mile Creek Mound was employed to address three specific research issues:

— To provide a practical application of the criteria and methods for effective valve-pair identification of *A. trapezia* developed on the basis of the morphological attributes of articulated specimens.

— To establish the degree of site integrity by examining the vertical distribution of identified conjoins through the dated sequence.

— To contribute to an understanding of site use and occupational intensity.

The null hypothesis based on the radiocarbon chronology (see Chapter 6) is that the c.300 year deposit represents a series of occupational events rather than continuous occupation. Mechanical damage and displacement will thus have had the most impact on the terminal deposits from the previous occupation event. These are also the deposits most likely to be impacted by other agents, such as animal burrowing (Roberts 1991). The structure and circumscribed extent of the Seven Mile Creek Mound enhances the potential of the site to reveal evidence for occupational periodicity and the contemporaneity of deposits. The frequency and distribution of bivalve conjoins within the deposit may therefore be used in some instances to define depositional events in the absence of clear stratigraphic information or to test stratigraphic inferences. More spatially-extensive sites would be expected to exhibit a more diffuse series of loci of repeated occupation activity.

Methods

Largely intact unarticulated *A. trapezia* valves from Square A were separated from the excavated assemblage (n=608) and individually bagged with basic provenance details. Length, width, height, weight and hinge length were measured for each valve (Fig. 3.2). Other valves were excluded from the study owing to the presence of attributes which biased one or more valve measurements (e.g. hinge damage, broken margins, attached oysters etc). The 608 whole valves were sorted by hinge length, weight and side (left or right) in descending order and assigned an arbitrary identification number (1–608). Unlike many stone artefact conjoin analyses which look at the closest material first, this method makes no assumptions about provenance of valve-pairs (i.e. valves from the same unit or adjacent unit were not assumed to have a greater probability of exhibiting conjoins than more distant units). Valve-pairs were identified by manually refitting successive right valves to left valves on a trial-and-error basis until a conjoin was identified. Refitting attempts for an individual valve were abandoned after 10 attempted refits forward and 10 backward from the valve's position on the descending size scale. This procedure was adopted to take into account the range of valve-pair dimorphism observed in articulated specimens, measurement uncertainty and the clumping of size-selected valves within limited size-ranges.

For example, in Figure 5.1 right valve 157 would be compared with left valves 156, 154, 153, 152, 151, 150, 149, 142, 138 and 137 in ascending valve size and with 158, 159, 160, 161, 164, 165, 168, 170, 172 and 173 in descending valve size. This procedure would then be repeated for right valve 162 and so on. Even though right valve 157 was found to pair with left valve 154, the full 20-refit-attempt procedure was undertaken to confirm the validity of the identified valve-pair. This method underestimates the number of conjoining valves that actually occur in the deposit for several reasons. First, any pair of valves where there was ambiguity about the match was rejected. Second, only a very limited amount of time was allocated to the assessment of each potential match (see below). The conjoin analysis of the 608 whole valves involved in the order of 6,000 individual refitting attempts.

Results

A total of 56 valve-pairs was identified in the assemblage, distributed between XU2–24 (Table 5.3). Seventy-nine percent of valve-pairs are vertically separated by less than 10cm, 55% less than 5cm and 98% less than 20cm apart. The small distances separating valve-pairs supports the impression gained from the radiocarbon chronology that shell was deposited in extremely rapid episodes (see below for further discussion).

There is no significant relationship between the size of valves in conjoining pairs and distance separating them, as measured by average hinge length versus maximum valve separation distance (correlation coefficient, r^2=0.1539). This discounts size-sorting as a major determinant of valve position within the deposit and reinforces the proposition that the structural properties of the matrix, constructed of closely interlocked shells, prevented significant post-depositional movement.

#	133	134	135	136	137	138	139	140	141	142	143	144	145	146	147
R		0		0			0	0	0		0	0	0	0	0
L	0		0		0	0				0					

#	148	149	150	151	152	153	154	155	156	157	158	159	160	161	162
R	0							0		**0**					0
L		0	0	0	0	0	0		0		0	0	0	0	

#	163	164	165	166	167	168	169	170	171	172	173	174	175	176	177
R	0			0	0		0		0			0	0	0	
L		0	0			0		0		0	0				0

Figure 5.1 Schematic representation of conjoin identification procedure. O=individual *A. trapezia* valve. In this matrix, left and right valves are arranged in descending size order. See text for details.

Bivalve conjoin analysis blind test

Introduction
The conjoin analysis provided encouraging results for further valve-pairing studies. However, the potential subjectivity in the identification of valve-pairs during the manual refitting stage indicated that an independent evaluation of the efficacy of the method should be undertaken. A blind test was designed for this purpose based on replicated attributes of valves used in the archaeological case study.

Aims
The objectives of the blind experiment were:
— To assess the reliability of the valve-pairing method applied to the Seven Mile Creek Mound assemblage.
— To identify sources of error in the valve-pairing method.

Methods
An independent collection of live-collected *A. trapezia* was unavailable for the blind test owing to the small size of museum holdings of this taxon. An experimental archaeological set of 500 valves was created with 50 known valve-pairs and 400 unpaired single valves. The control sample of 100 valves that form 50 known articulated pairs was selected from valve-pairs encountered as articulated specimens during excavation of the Seven Mile Creek Mound, Squares A–D. The original 316 valves (158 valve-pairs) in this assemblage were reduced to a sample of 100. Valves from articulated specimens from Square A (44 valves) were excluded as many had been used or modified for radiocarbon dating and stable isotope studies. The remaining valve-pairs were reduced to a sample of 50 by maintaining the hinge size distribution evident in the original sample of 158 valve-pairs. A further 400 valves thought not to form a pair with any other in the sample were selected from valves used in the Seven Mile Creek Mound, Square A case study above. These were arranged in ascending order by height (the attribute with the least significant correlation coefficient), and then assigned an arbitrary reference identification number (1–500) in ascending order of height. The 500 valves were then resorted in ascending order by hinge length, weight and side (as per the archaeological case study above). Refitting methods proceeded as outlined above. This blind test involved in the order of 5,000 individual refitting attempts.

Results

The overall success rate for correct identification of articulated values in the blind test was 94%. Six valves out of 500 (1.2%) were misidentified as conjoins. On re-examination these were rejected as valid conjoins because they did not fit tightly together at both the articulating hinge teeth and the corrugated margins. One misidentified conjoin set actually contained a valve from a known articulated valve-pair. Five valve-pairs were identified as new conjoins in the single-valve dataset. These valve-pairs were confirmed on re-examination (Table 5.4). These pairs were not identified in the previous study. Twenty-two valves (11 valve-pairs) in the 100-valve control sample were not identified as conjoins during the blind test. On re-examination these were clearly conjoins. The vertical position of these valves in the spreadsheet sorted in ascending order by hinge length, weight and side (left/right) shows that all are within 10 valves up and down of each other on the descending size scale and thus the fact that they were missed during the blind test cannot be ascribed to limitations of the methods employed. Thus a total of 28 individual valves out of the 500 in the blind test dataset was either misidentified as conjoins or missed altogether (5.6%).

Discussion

With the additional five valve-pairs identified from the Seven Mile Creek Mound, Square A during the blind test, a total of 20% of the whole *A. trapezia* valve assemblage was refitted (122/608). Up to 80% of the whole *A. trapezia* valves in the assemblage are therefore pair-less. This latter figure must be considered as a maximum, however, given the probability that some conjoins were missed (see below). Although the overall success rate for correct identification of articulated valves was relatively high (94%), the error rate (4.4%) is largely composed of the 11 articulated valve-pairs in the control sample which were missed in the blind test. The blind test demonstrated that the conjoin identification methods adopted were adequate to identify *all* known conjoins present in the blind test assemblage.

Two major factors are thought to have structured the valve-pair identification rate determined through the blind test. The primary factor is the short amount of time spent attempting the manual refit and evaluation of any particular valve-pair. Relatively precise alignment is required to achieve tight interlocking of conjoining valves owing to the presence of numerous interlocking teeth of various sizes, patterns and angles on conjoining hinges. As Larson and Ingbar (1992:151–2) observed, 'the number of refits or conjoins found is directly proportional to the amount of time expended in seeking them'. A secondary factor in the blind test is thought to be the fact that some of the articulated control samples had minor edge-damage that might have biased identification, with otherwise conjoining valves rejected owing to the inability of valves to interlock at the corrugated margins. Given these limitations, the number of conjoins identified in any given bivalve assemblage should be considered a *minimum* as this number will almost certainly underestimate the true number of conjoins in any assemblage. Although theoretically there can only be a finite number of conjoins in any assemblage, limitations of resources allocated to refitting mean that it is unlikely that any but the most thorough and time-consuming analyses will identify every conjoin. The term 'minimum number of conjoins' is therefore adopted to describe the conjoins identified in this study. Given these constraints, no unequivocal statements can be made about the *absence* of conjoins or proportion of conjoins to non-conjoins in any given assemblage.

Combining the results of the initial conjoin analysis with those of the blind test, a total minimum number of conjoins of 61 was identified from the Seven Mile Creek Mound, Square A. The distribution of conjoins is positively skewed with most pairs (55.74%) separated by 5cm or less and over 80% separated by 10cm or less. Only 3.28% of the identified conjoins are separated by

Table 5.3 Identified *A. trapezia* conjoin sets, Seven Mile Creek Mound, Square A.

CONJOIN SET	XU		MIN. SEPARATION (cm)	MAX. SEPARATION (cm)	MID-POINT (cm)	± (cm)
	L	R				
Set 1	21	21	0	3.6	1.8	1.8
Set 2	21	18	7.1	16.86	11.98	4.88
Set 3	21	21	0	3.6	1.8	1.8
Set 4	20	19	0	7.1	3.55	3.55
Set 5	21	23	3.48	10.4	6.94	3.46
Set 6	21	20	0	7.44	3.72	3.72
Set 7	21	24	6.8	13.36	10.08	3.28
Set 8	9	4	13.62	21.24	17.43	3.81
Set 9	19	21	3.84	10.7	7.27	3.43
Set 10	21	21	0	3.6	1.8	1.8
Set 11	20	20	0	3.84	1.92	1.92
Set 12	3	3	0	3.28	1.64	1.64
Set 13	12	9	7.44	14.94	11.19	3.75
Set 14	20	21	0	7.44	3.72	3.72
Set 15	9	9	0	3.84	1.92	1.92
Set 16	20	20	0	3.84	1.92	1.92
Set 17	20	18	3.26	13.26	8.26	5
Set 18	4	2	3.28	10.04	6.66	3.38
Set 19	20	20	0	3.84	1.92	1.92
Set 20	19	21	3.84	10.7	7.27	3.43
Set 21	18	20	3.26	13.26	8.26	5
Set 22	20	20	0	3.84	1.92	1.92
Set 23	13	13	0	4.68	2.34	2.34
Set 24	20	20	0	3.84	1.92	1.92
Set 25	13	13	0	4.68	2.34	2.34
Set 26	20	22	3.6	10.92	7.26	3.66
Set 27	9	10	0	7.22	3.61	3.61
Set 28	21	21	0	3.6	1.8	1.8
Set 29	3	2	0	6.26	3.13	3.13
Set 30	20	21	0	7.44	3.72	3.72
Set 31	2	2	0	2.98	1.49	1.49
Set 32	4	3	0	7.06	3.53	3.53
Set 33	10	10	0	3.38	1.69	1.69
Set 34	3	4	0	7.06	3.53	3.53
Set 35	11	10	0	7.44	3.72	3.72
Set 36	19	22	7.44	14.18	10.81	3.37
Set 37	16	16	0	4.64	2.32	2.32
Set 38	13	13	0	4.68	2.34	2.34
Set 39	3	2	0	6.26	3.13	3.13
Set 40	20	20	0	3.84	1.92	1.92
Set 41	2	2	0	2.98	1.49	1.49
Set 42	10	10	0	3.38	1.69	1.69
Set 43	19	19	0	3.26	1.63	1.63
Set 44	20	20	0	3.84	1.92	1.92
Set 45	3	3	0	3.28	1.64	1.64
Set 46	6	5	0	6.64	3.32	3.32
Set 47	4	4	0	3.78	1.89	1.89
Set 48	24	24	0	2.96	1.48	1.48
Set 49	3	3	0	3.28	1.64	1.64
Set 50	5	5	0	2.98	1.49	1.49
Set 51	4	4	0	3.78	1.89	1.89
Set 52	4	3	0	7.06	3.53	3.53
Set 53	23	23	0	3.32	1.66	1.66
Set 54	22	23	0	6.8	3.4	3.4
Set 55	4	4	0	3.78	1.89	1.89
Set 56	18	18	0	6.16	3.08	3.08

Table 5.4 Additional *A. trapezia* conjoin sets identified during the blind test, Seven Mile Creek Mound, Square A.

CONJOIN SET	XU L	XU R	MIN. SEPARATION (cm)	MAX. SEPARATION (cm)	MID-POINT (cm)	± (cm)
Set 57	21	21	0	3.6	1.8	1.8
Set 58	21	19	3.84	10.7	7.27	3.43
Set 59	20	20	0	3.84	1.92	1.92
Set 60	21	21	0	3.6	1.8	1.8
Set 61	19	19	0	3.26	1.63	1.63

more than 15cm (Table 5.5, Fig. 5.2). The site, therefore, appears to exhibit a high degree of stratigraphic integrity. This impression is reinforced when it is considered that the lack of more detailed resolution means that these maximum separation measurements probably overestimate the actual separation distance between valves in many instances.

The large number of articulated valve-pairs encountered both *in situ* (n=22) and through conjoin analysis (n=61) in the Seven Mile Creek Mound, Square A suggest that a large proportion of the *A. trapezia* assemblage was still connected by ligaments or otherwise closely associated when discarded. Although horizontal data are not available, I suggest that both vertical and horizontal movement is largely explicable as a function of some valves becoming separated upon discard and moving down the slope formed by the margins of the mounded deposit. The characteristics of the matrix mean that the separation of valves forming valve-pairs is probably largely a function of deposition events rather than post-depositional movement of valves. If this proposition is accepted, the distribution of valve-pairs can contribute to an understanding of the sequence and duration of depositional events and therefore periodicity of occupation.

Other lines of evidence can be used in conjunction with bivalve conjoin analyses to investigate aspects of site integrity. In simulated heating and mechanical destruction experiments, Robins and Stock (1990) found that *A. trapezia* was resilient up to 600°C, equivalent to a small camp fire, but showed marked susceptibility to mechanical destruction (fragmentation) under higher experimental temperatures of 800–1,000°C. Such high temperatures are not thought to be achievable using the fuel resources available to foragers in southeast Queensland (Robins and Stock 1990:86–7). This suggests that in the absence of very high temperatures, fragmentation of *A. trapezia* will be largely the result of mechanical damage to the shell itself. The high proportion of whole *A. trapezia* shells encountered, the virtual absence of charcoal and the lack of evidence for burning on the surface of examined shells does not suggest extended heating during cooking, direct disposal into fires or that shells were later exposed to fires built close to discarded shells (see Robins and Stock 1990:89). There is no strong correlation between the abundance of bivalve conjoins in any area of the deposit (as defined by XU) and differential fragmentation rates (compare Figs 5.3 and 5.5). In fact, the low fragmentation rate of *A. trapezia* suggest that shells in the deposit were not exposed to temperatures above 600°C or significant mechanical damage owing to post-depositional agents acting on the deposit (e.g. trampling etc).

The vertical distribution of overlapping conjoin sets indicates that the site can be divided into at least two separate sequences of accumulation (XU2–12, XU18–24), separated by zones that may also be associated with these events (Fig. 5.3). The internal consistency of the radiocarbon chronology coupled with the apparently internal coherence of multiple

Table 5.5 Summary of maximum distance separating all conjoined *A. trapezia* valve-pairs, Seven Mile Creek Mound, Square A.

DISTANCE	#	%
0–5cm	34	55.74
5–10cm	15	24.59
10–15cm	10	16.39
15–20cm	1	1.64
20–25cm	1	1.64
Total	61	100

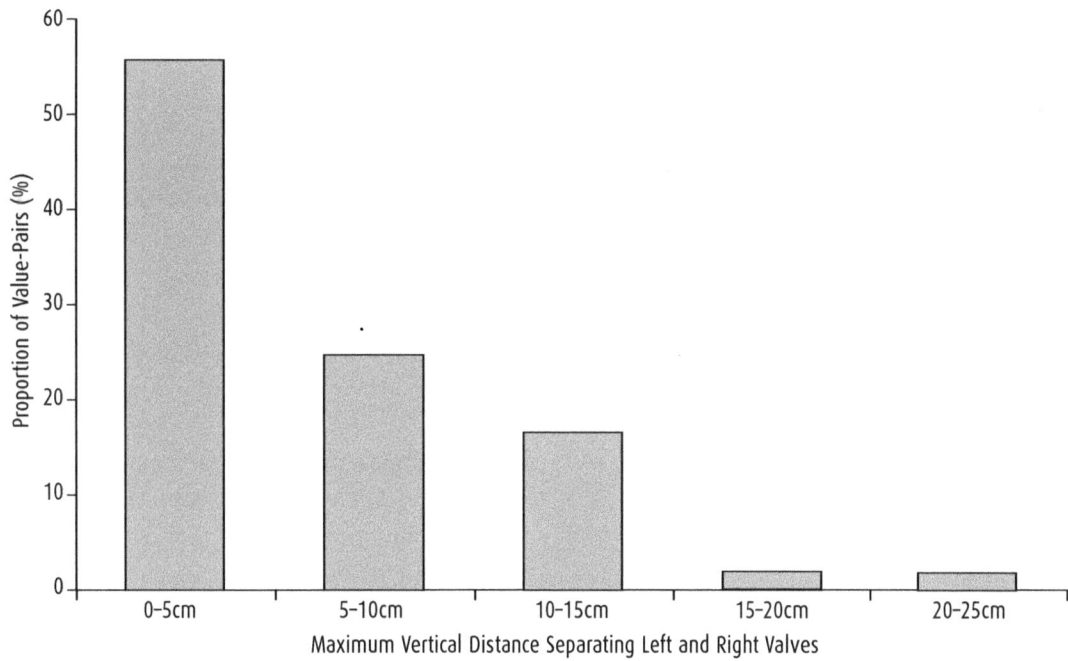

Figure 5.2 Maximum vertical distance separating all conjoined valve-pairs, Seven Mile Creek Mound, Square A.

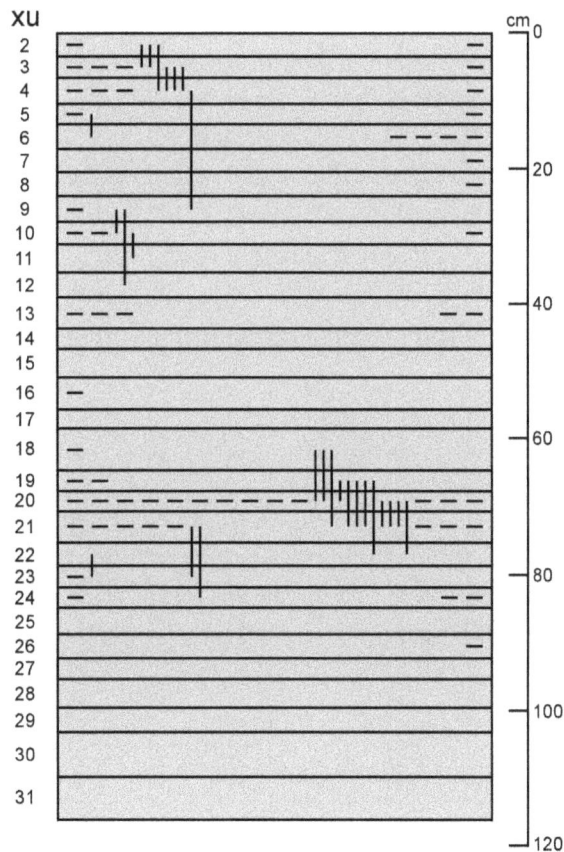

Figure 5.3 Distribution of identified *A. trapezia* valve-pairs (n=61), Seven Mile Creek Mound, Square A. An additional 22 valve-pairs encountered as articulated specimens during excavation are shown as short horizontal lines down the right hand side of the figure (see Chapter 6). Line termination points indicate the vertical mid-points of the excavation units from which conjoining valves were recovered. Short horizontal lines indicate valve-pairs identified within excavation units. Not to scale on the horizontal axis.

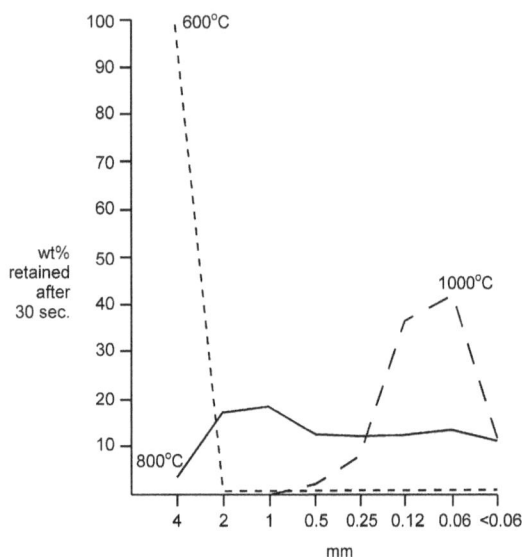

Figure 5.4 Particle size distribution of *A. trapezia* after heating at various temperatures and then mechanical destruction for 30 seconds (after Robins and Stock 1990:98).

conjoined bivalves suggests that the distribution of conjoins may be used as a proxy for layer thickness. As Huchet (1991:46) suggested, the 'presence of a large percentage of conjoinable pieces of bone, stone or other material found within a constricted area can be taken as a sign that deposition occurred within a limited time period or during a single event'.

Conjoin analyses undertaken in this study are restricted to vertical associations in a single 50cm × 50cm test pit. The identification of bivalve conjoins across adjacent excavation squares and beyond would make a major contribution to the understanding of horizontal site integrity, site structure and activity area identification. Also, the establishment of an independent stone artefact conjoin dataset for these sites would contribute to understandings of taphonomic processes acting on different types of material. Unfortunately, stone artefacts are a relatively rare component of shell middens in southeast Queensland, reducing the potential of this line of enquiry in this region. The resolution of the valve-pairing from the Seven Mile Creek Mound would have been dramatically enhanced had all whole valves been plotted in three dimensions during excavation. This would have enabled precise assessment of the horizontal and vertical relationship between conjoining valve-pairs rather than minimum and maximum separations. Additional information may also be useful, such as the orientation, inclination and side-up of each valve (Hofman 1992).

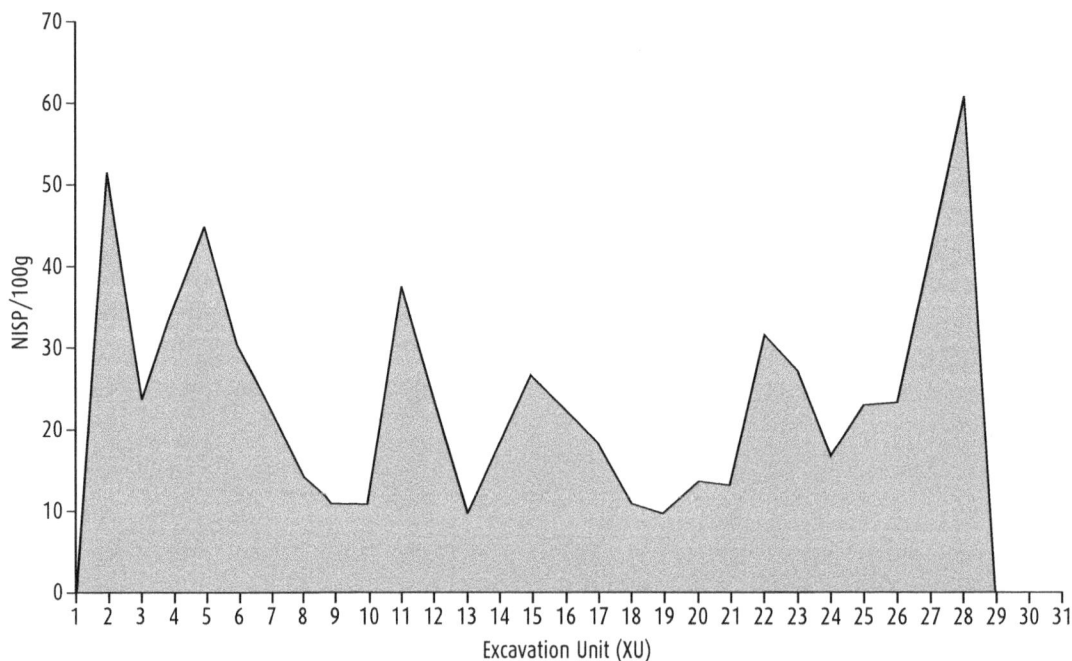

Figure 5.5 Fragmentation of *A. trapezia*, Seven Mile Creek Mound, Square A, expressed as the number of fragments per 100g of shell.

The techniques developed here could be applied to the examination of many excavated assemblages. Investigations of coastal sites have been conducted around the entire length of the Australian coastline (see Hall and McNiven 1999). With the possible exception of southwestern Australia, large shell middens have frequently been targeted for excavation. Bivalves suitable for conjoining are relatively common in these deposits, including *Donax deltoides*, *A. trapezia* and *A. granosa*. Valves of these taxa are robust and are frequently encountered as whole specimens in shell middens dating from the mid-Holocene. In some circumstances more fragile taxa may also be suitable, such as *Trichomya hirsutus* and *Mytilus* sp.

Summary

Articulated live-collected and excavated *A. trapezia* were used to derive reliable criteria for identifying unarticulated valve-pairs. These criteria were used as the basis of a systematic conjoin method which successfully identified valve-pairs among the unarticulated *A. trapezia* valves excavated from the Seven Mile Creek Mound. A subsequent blind test confirmed the reliability of the method but indicated that the amount of time devoted to assessing the conjoinability of individual valves was the single major determinant for maximising the number of valve-pairs identified. In the Seven Mile Creek Mound case study, the small distances which separate the majority of valve-pairs attest to the stratigraphic integrity of the site and may indicate episodic mound accumulation. In certain circumstances, bivalve conjoining may be a useful adjunct to conventional approaches to shell midden analyses, involving very basic characterisation of assemblage composition, with the potential to contribute an independent form of evidence to our understanding of site integrity (and resolution), discard patterns and periodicity of occupation. Shell conjoin analyses therefore provide one avenue to investigate differential deposition, dispersal and disarticulation patterns in bivalve assemblages.

The successful application of bivalve conjoining to the Seven Mile Creek Mound provides the basis for applying valve-pairing methods to other shell-dominated assemblages in the region as an adjunct to traditional approaches to the evaluation of site integrity. The techniques outlined in this chapter were applied to all excavated assemblages where whole *A. trapezia* valves were recovered. Results are presented in individual site chapters (Chapters 6–13).

6

Seven Mile Creek Mound

Introduction

Archaeological excavations of the Seven Mile Creek Mound revealed a dense cultural deposit dominated by marine shell dated to c.3,950 cal BP. This result provides some of the earliest evidence of focussed marine exploitation from an open archaeological site on the Queensland coast. This chapter describes the site and its stratigraphy, chronology and contents, followed by a discussion of the implications of the data for understanding the archaeology of the study region.

Site description and setting

The Seven Mile Creek Mound is a discrete shell mound located on a low, sandy ridge isolated on tidal flats fringing Seven Mile Creek, approximately 35km southeast of Gladstone (Latitude: 24°04'01"S; Longitude: 151°31'11"E) (Fig. 6.1). Hummock Hill Island and Innes Head shelter the Seven Mile Creek estuary, which is fringed by extensive communities of spotted mangroves (*Rhizophora stylosa*) and to a lesser extent grey mangroves (*Avicennia marina*). Claypans and saltflats fringe most of the foreshore, with a small brackish zone in the upper reaches of the estuary maintained by freshwater inflow from Seven Mile Creek (Olsen 1980a:18). Extensive tidal flats adjacent to the site support dense populations of rock oyster (*Saccostrea glomerata*) and hairy mussel (*Trichomya hirsutus*). The base of the mound is less than 1.5m above mean high tide level. The mound is fringed by a number of mature Moreton Bay ash trees (*Eucalyptus tessellaris*) and its surface is covered by introduced grasses. Vegetation surrounding the mound is open woodland with mature trees including the forest red gum (*E. tereticornis*), burdekin plum (*Pleiogynium timorense*) and quinine tree (*Petalostigma pubescens*). The understorey is generally open, with frequent occurrences of introduced weeds including prickly pear (*Opuntia stricta*) and lantana (*Lantana camara*) (Fig. 6.3). A four wheel drive track runs along the saltflat margin to the north of the site. Grazing in the area is evidenced by the presence of cattle faeces on and around the mound site. The site measures 20m east-west × 10m north-south × 0.8m high (Fig. 6.2). Estimated volume is 44m³ based on digital terrain modelling (Fig. 6.2). Scattered low density shell material is visible in disturbed areas away from the mound itself and several stone artefact scatters with rounded quartz nodules and flaked material were found on the tidal flats to the southeast of the mound.

Although shell mounds are a relatively common feature of the coastal archaeological record of northern Australia, these site types are generally restricted in distribution to locations above the Tropic of Capricorn and are almost exclusively late Holocene in age (e.g. Bailey 1999; Beaton 1985; Veitch 1999). The Booral Shell Mound on the western shore of Great Sandy Strait dating to 2,950±60 BP (Beta-32046) is the only mound investigated on the Queensland coast to the south of the study area (Frankland 1990). Godwin and Ulm (2004) have also recently reported a probable shell mound complex at Buxton, just south of Bundaberg. Early reports indicate that other mounds may have existed in southeast Queensland that have been destroyed (Anon. 1877 cited in McNiven 1994a:1). Despite exhaustive surveys, and an otherwise rich Aboriginal archaeological record, only three shell mounds are known for the 500km stretch of shoreline between Hummock Hill Island and Agnes Water (Ulm and Lilley 1999). Shell mounds at Round Hill Creek and Hummock Hill Island have been damaged by development activity, leaving the Seven Mile Creek Mound the only known remaining intact example of this site type in the region.

The site was first described by Neal (1986), then a consultant archaeologist for Telecom Australia, who identified it from a circling helicopter, although Rowland (1987) was unable to locate the site during a field inspection later that year. Local informants suggested to Rowland that it was quartz tailings from quarrying activities. In late 1999 the mound was relocated by the author

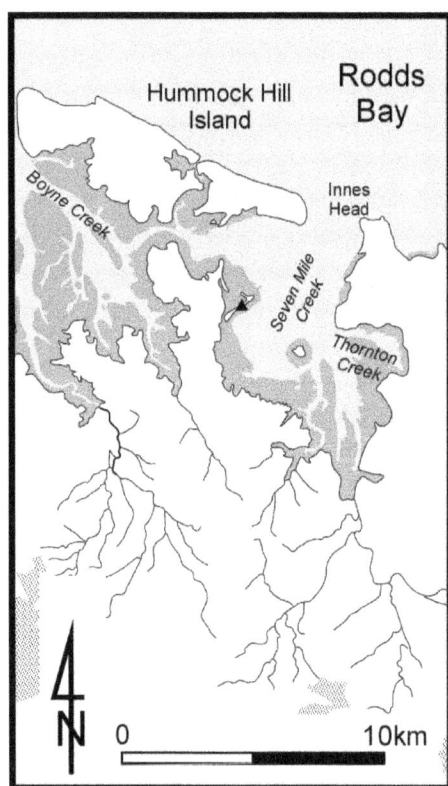

Figure 6.1 The Seven Mile Creek catchment area showing the location of the Seven Mile Creek Mound as a triangle. Dark grey shading indicates the general extent of mangrove, saltflats and claypans.

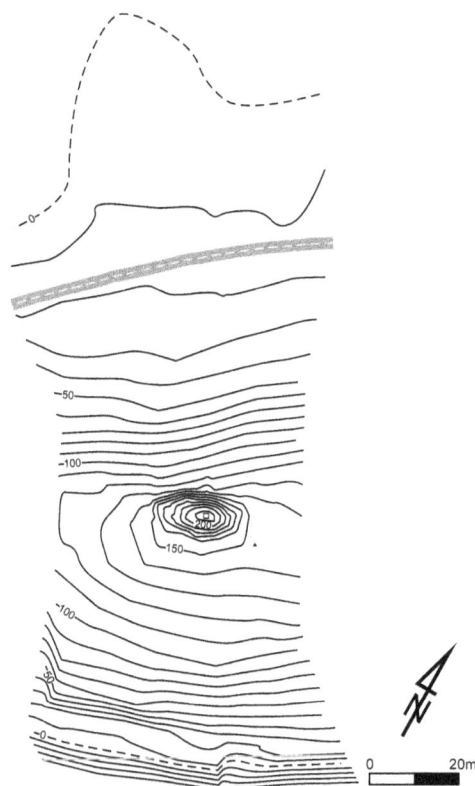

Figure 6.2 Topographic map of the site area, showing the position of the excavation. The site datum is shown as a triangle. Dashed lines indicate the location of the tide strand line during the period of excavation. The wide shaded line denotes a 4WD track on the edge of the saltflats. Contours in 10cm intervals.

during pedestrian transect surveys conducted as part of the Gooreng Gooreng Cultural Heritage Project (Lilley and Ulm 1999) and targetted for excavation. It is registered on the Queensland Environmental Protection Agency's (EPA) Indigenous Sites Database as KE:A09 and Queensland Museum Scientific Collection Number S229. A note on the initial radiocarbon dates and descriptions of the excavation have been presented elsewhere (Ulm 2000b, 2002c).

In addition to the mound, Neal (1986) recorded a sparse shell and stone artefact scatter on the peninsula immediately north of the site (KE:A08). Several sparse shell midden scatters have also been recorded on Hummock Hill Island to the north (Alfredson 1993). A stone arrangement at the summit of Hummock Hill is entered in the EPA database (KE:A04). Alfredson (1993) identified this site 'as a probable surveyor's trig point' although her subsequent research failed to locate records for the establishment of such a point in archival records. A shell mound is the major archaeological feature recorded on the island with a depth of up to 40cm (JE:A65). Burke (1993), however, noted that only a small portion of the deposit had not been damaged by bulldozer activity and water erosion. On the eastern margin of Seven Mile Creek, Burke (1993) located an isolated low density surface shell scatter at Innes Head during surveys between Innes Head and Thornton Creek (KE:A56) (see Fig. 2.9).

Excavation methods

A single 1m^2 pit divided into four 50cm x 50cm quadrants (Squares A–D) was excavated into the highest part of the mound between 26 January and 16 February 2000 (Fig. 6.2). The excavation grid was located so as to maximise the potential depth of deposit and to avoid the sloping margins of the mound. Excavation proceeded in shallow, arbitrary excavation units (XUs) averaging 3.75cm in depth and 12.75kg in weight. Excavation ceased at a maximum depth of 117cm below ground surface when cultural material ceased to be recovered (Figs 6.4–6.5). A total of 130 XUs was removed, distributed as follows: Square A (31 XUs), Square B (34 XUs), Square C (31 XUs), Square D (34 XUs). A total of 1,631kg of

Figure 6.3 General view of the Seven Mile Creek Mound. Note the low fringing mangroves on the intertidal flats through the gap in trees at rear left of frame. Facing south.

Figure 6.4 General view of the completed excavation. Note the dark zone of grass root penetration at the top of the deposit. Facing north.

Figure 6.5 General view of completed excavation (Squares D–A). Note the stratigraphic break at the base of the shell deposit. Facing north.

sediment was excavated. Sediments were gently dry-sieved through 3mm screens onto a plastic tarpaulin 20m south of the excavation to prevent contamination of underlying sediments and to minimise the risk of airborne dust for excavation personnel. *Anadara trapezia* (n=158), charcoal (n=74), stone (n=59) and bone (n=3) specimens encountered *in situ* during excavation were plotted three-dimensionally. Articulated *A. trapezia* specimens encountered *in situ* were plotted and removed whole to allow examination of contents under controlled laboratory conditions. The excavation was backfilled with a layer of plastic sample bags across the base, followed by a thick layer of archaeologically sterile white beach sands from the adjacent mangrove fringe and topped with the sediments that had passed through the sieve (see Chapter 3 for a detailed discussion of the standard excavation methods employed at all sites).

Cultural deposit and stratigraphy

Excavation revealed a 85cm thick deposit of dense shell resting on well-rounded beach sands containing occasional pieces of shell and degraded pumice. A veneer of fragmented shell encountered at the top of the deposit may be attributed to a combination of cultural and environmental factors, such as cessation of cultural deposition, trampling, weathering and/or the effects of fire and the recent invasion of exotic grasses, shrubs and cattle (see Robins and Stock 1990). Although occasional cultural material was recovered between 90–100cm below the top of the mound, systematic and continuous use of the site is not evidenced until the base of the shell mound deposit at around 85cm, roughly contiguous with the surrounding natural ground surface. The lowermost excavated deposits between 100–117cm were culturally-sterile, indicating that initiation of midden formation can be dated with confidence. The rate of site accumulation calculated using the methods outlined by Stein et al. (2003) indicate an extremely rapid rate of site formation of 23.3cm/100 years.

The deposit can be divided into four major stratigraphic units (SUs) based on sediment colour and texture (Figs 6.5–6.6, Table 6.1). Few sediments occur at all in SU1–2 which are overwhelmingly dominated by the shell fraction. The dark greyish brown to yellowish brown sediments of the upper shell deposit are assumed to derive from a combination of aeolian transport of sediments trapped by an exposed irregular shell surface and subsequent percolation of sediments down the shell profile. Occasional tree roots and sometimes large non-artefactual rocks were observed throughout the deposit. Acidity (pH) values are highly alkaline and decrease with depth from 10.0 at the surface and throughout the dense shell zone to 8.0 below the shell zone.

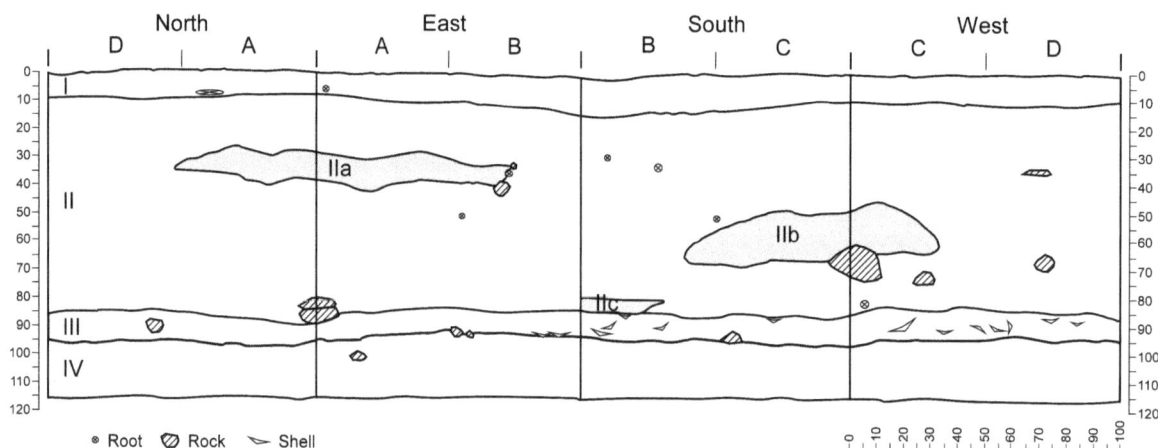

Figure 6.6 Stratigraphic section, Seven Mile Creek Mound, Squares A–D.

Table 6.1 Stratigraphic Unit descriptions, Seven Mile Creek Mound, Squares A–D.

SU	DESCRIPTION
I	Extends across the entire square, with an average depth of 8–9cm and a maximum depth of 13cm below the surface. The unit comprises compact shell with a minor component of fine angular poorly-sorted dark to very dark greyish brown (10YR-3/2) humic sediments. The SU is penetrated by fibrous grass roots, particularly across its upper margin. Cultural materials are dominated by fragmented oyster shell, but also include mud ark, hairy mussel, hercules club whelk, nerite, stone artefacts, charcoal and fish bone. pH values are highly alkaline (9.5–10.0).
II	Extends across the entire square with a maximum thickness of 82cm and a maximum depth of 90cm below the surface. It comprises a compact shell matrix. Sediments are poorly consolidated with colour ranging from dark greyish brown (10YR-4/2) to yellowish brown (10YR-5/4). This unit includes several areas distinctive in either fragmentation (SUIIa), limited taxa representation (SUIIb), or burning (SUIIc). A few tree roots up to 20mm in diameter occur in Squares B and C. Cultural materials consist of whole shells with patches of fragmented shell, dominated by oyster with small quantities of hairy mussel, mud ark, scallop, hercules club whelk, nerites as well as stone artefacts, charcoal and fish bone. Occasional rocks and shells with oyster bases attached are also present. pH values are highly alkaline (9.0–10.0).
IIa	An extremely compact lens of fragmented shell exposed for almost the entire length of the east section (Squares A and B) and the northeast half of Square A. The unit has a maximum thickness of 10cm and maximum depth of 41cm below the surface. Otherwise the unit is the same as SUII. The deposit is dominated by oyster and hairy mussel.
IIb	A compact lens of mud ark valves exposed across the southwest corner of Square C. The unit has a maximum thickness of 18cm and maximum depth of 68cm below the surface. Otherwise the unit is similar to SUII with the presence of many articulated mud ark valves and a noticeable reduction in the representation of other taxa.
IIc	Layer of burnt shell and charcoal fragments exposed only across the southeast corner of Square B towards the base of SUII. The unit has a maximum thickness of 5cm and maximum depth of 83cm below the surface. Otherwise the unit is the same as SUII.
III	Extends across the entire square with a maximum thickness of 11cm and a maximum depth of 98cm below the surface. The unit comprises poorly consolidated fine subrounded poorly-sorted brown (10YR-4/3) sands underlying the dense shell deposit. SUIII is differentiated from SUII by the dominance of sandy sediments rather than shell in the matrix. Cultural material includes occasional shell, rocks, charcoal and small pieces of pumice. The pH values are alkaline (8.0–8.5), decreasing with depth.
IV	Extends across the entire square, with a maximum thickness of at least 24cm and maximum depth of at least 117cm below the surface. The base of this unit was not reached. It comprises medium rounded well-sorted poorly consolidated light olive brown (2.5Y-5/4) sands with extensive mottling of yellow sands. This SU is culturally-sterile with occasional degrading rocks and pumice. The pH values are alkaline (8.0–8.5), decreasing with depth.

Radiocarbon dating and chronology

Eight radiocarbon determinations were obtained for the deposit from Square A (Table 6.2). Four conventional radiocarbon dates were obtained on articulated *A. trapezia* valves which were plotted *in situ* during excavation. These shell samples were paired with associated single fragments of blocky charcoal which were dated by accelerator mass spectrometry (AMS) to investigate local marine reservoir conditions (see Chapter 4). Single pieces of charcoal were selected to avoid combining fragments representing possibly separate events (e.g. Ashmore 1999). Calibration calculations employed a ΔR correction value of –155±55 (see Chapter 4). On the basis of the x-ray diffraction analysis, three shell samples (Wk-8325, Wk-8347 and Wk-8455) derived from SUII 13–20cm below ground surface were rejected as they contained recrystallised material. The charcoal date NZA-12272 is clearly at odds with the other seven age determinations available and indicates a lack of association with the paired shell sample (Wk-8324) from the same depth. This small charcoal sample (<0.1g) is derived from SUI close to the top of the deposit (7–10cm) and may have been introduced through percolation into the midden profile. It is therefore rejected from further consideration. The lower two shell dates (Wk-8327 and Wk-8328) appear inverted, but in fact overlap within one standard deviation. Although occasional cultural material was recovered up to 10cm below the lowest dated sample much of this material was probably pushed into the loosely consolidated sandy sediments from the base of the mound, suggesting that the determination of NZA-12118 dates the initiation of deposition at this part of the site. Given the homogeneity of SUI, Wk-8324 from the base of SUI provides a determination that may be relied upon for dating the termination of occupation. The ^{14}C and AMS determinations display a firm stratigraphically sound sequence overall, suggesting extremely rapid accumulation over a period of about 350 years with initial occupation around 3,947 cal BP and abandonment shortly after 3,608 cal BP. A standard chi-square test to determine statistically significant differences in the determinations (Ward and Wilson 1978) indicates that all four conventional shell determinations are indistinguishable at the 95% confidence level.

Table 6.2 Radiocarbon dates from the Seven Mile Creek Mound, Square A (see Appendix 1 for full radiometric data for each determination).

SQUARE	XU	DEPTH (cm)	LAB. NO.	SAMPLE	δ^{13}C (‰)	^{14}C AGE	CALIBRATED AGE/S
A	4	6.8–10.4	NZA-12272	charcoal	-26.0±0.2	1260±80	1293(1171)955
A	4	7.1	Wk-8324	*A. trapezia*	-0.9±0.2	3540±80	3850(3608)3372
A	13	39–43.6	NZA-12117	charcoal	-25.7±0.2	3500±60	3886(3693)3569
A	13	40.4	Wk-8326	*A. trapezia*	-0.8±0.2	3610±70	3919(3684)3462
A	20	67.8–71.5	NZA-12273	charcoal	-23.4±0.2	3570±60	3979(3829,3786,3780)3639
A	20	67.8	Wk-8327	*A. trapezia*	-1.2±0.2	3780±60	4140(3904)3688
A	26	88.7–92.2	NZA-12118	charcoal	-27.8±0.2	3660±60	4137(3957,3952,3924,3921,3910)3725
A	26	88.2	Wk-8328	*A. trapezia*	-0.5±0.2	3750±60	4089(3867)3652

Stratigraphic integrity and disturbance

Several lines of evidence suggest that the deposit exhibits a high degree of stratigraphic integrity. The structural properties of the matrix itself, dominated by closely interlocking whole shells, would prevent significant post-depositional movement of all but very small objects. The three subunits identified as SUIIa, IIb and IIc are linear and roughly parallel to the major stratigraphic units, indicating that they were probably formed on a relatively flat surface (Fig. 6.6). The clear stratigraphic disjunction at the base of the deposit unambiguously demarcates the base of the cultural deposit. No other obvious signs of disturbance, such as burrows, voids, major roots, root casts or similar stratigraphic features, were encountered during excavation. The radiocarbon sequence shows a regular age-depth relationship (with the exception of NZA-12272, see above).

An unusual feature of the deposit is the high incidence of articulated bivalves. Attenbrow (1992:16) noted that articulated shells rarely occur in shell middens, yet 22 articulated *A. trapezia* were encountered in Square A and 158 in the 1m² excavation as a whole. Articulated specimens of the fragile mussel *T. hirsutus* were also recorded during excavation. As noted earlier, articulated specimens are frequently viewed as evidence for the natural origin of sediments in both archaeological (e.g. Claassen 1998:74; O'Connor and Sullvan 1994a; Sullivan and O'Connor 1993) and geomorphological (e.g. Dodd and Stanton 1981:304) studies. The rationale for this assumption is unclear, but probably derives from palaeoenvironmental research in which articulated bivalves preserved in growth position have been used extensively for sea-level and other palaeo-environmental reconstructions. For cultural deposits, however, there is no reason to expect that articulated bivalves will not be present in archaeological shell deposits (Attenbrow 1992).

Ethnographic data from the Anbarra of the Northern Territory indicate that *A. granosa*, a taxon closely related to *A. trapezia*, was prepared by either steaming, being placed directly in hot ashes or having a fire placed on top (Meehan 1982; Robins et al. 1998). It was noted in the previous chapter that such low intensity heating is unlikely to break the ligaments that hold the valves together at the hinge. In the absence of significant mechanical action on the valves after deposition, at least some intact articulated valves are likely to be preserved. The potential for articulated valve preservation is significantly enhanced when high rates of cultural and/or natural sedimentation rapidly emplace a matrix around the articulated valves to minimise further movement. Based on the evidence from the Seven Mile Creek Mound, the presence or absence of articulated bivalves should not be viewed as definitive in distinguishing cultural shell deposits from natural shell deposits. In fact the presence of some articulated specimens resulting from food refuse suggests that shells were deposited soon after consumption (i.e. fresh) before ligament deterioration rather than by secondary disposal. The large number of articulated *A. trapezia* encountered and their distribution throughout the deposit suggests little post-depositional movement in this part of the site.

Conjoin analysis of the *A. trapezia* valve assemblage reinforces the argument for high stratigraphic integrity of the deposits. A minimum number of conjoins of 61 was identified. Most pairs (55.74%) were separated by less than 5cm and over 80% were separated by less than 10cm. Only 3.28% were separated by more than 15cm (Fig. 6.7). The deposit exhibits a high degree of stratigraphic integrity, given that the maximum separation measurements of conjoins overestimate the actual separation distance between valves in many instances (see Chapter 5). The relatively small distance separating valve-pairs supports the radiocarbon chronology indicating that shell material accumulated during episode(s) of extremely rapid deposition (see Chapter 5 for detailed results of the conjoin analysis conducted on the assemblage).

A. trapezia valves are generally in good condition with low rates of fragmentation, fracturing or cracking suggesting that they were not exposed to extensive post-depositional mechanical damage. *A. trapezia* fragmentation rates are relatively low throughout the sequence (compare, for example, with the Mort Creek Site Complex, see Chapter 7) with an average of 24.5 NISP/100g with peaks of 60.6 and 51.4 at the base and top of the deposit respectively. Zones of low fragmentation are often associated with areas with high numbers of conjoining *A. trapezia* valves.

Laboratory methods

Owing to the large quantity of excavated materials, detailed analysis to date has focussed on a single square (Square A). Field observations and preliminary processing of the other three squares indicate broad consistency in the contents of the excavated material. Results of the analysis of Square A are presented below (see Chapter 3 for a detailed discussion of the standard laboratory methods employed at all sites). Further summary results are available in Appendix 4.

Cultural materials

Invertebrate remains

Fifty taxa of shellfish weighing 135,794.8g were identified, consisting of 24 marine gastropods, 20 marine bivalves, five terrestrial gastropods and one freshwater bivalve (Table 6.3). Forty-four taxa are relatively rare, each contributing less than 0.2% of the shell assemblage by weight. The shell deposit is dominated by rock oyster (*S. glomerata*), comprising 78.7% by weight (Fig. 6.8), followed by hairy mussel (*T. hirsutus*) (8.8%) (Fig. 6.9), mud ark (*A. trapezia*) (8.5%) (Fig. 6.10), scallop (*Pinctada albina sugillata*) (1.4%) (Fig. 6.11), hercules club whelk (*Pyrazus ebeninus*) (1.1%) and nerite (*Nerita balteata*, syn. *Nerita lineata*) (1.1%). The assemblage exhibits relatively high diversity

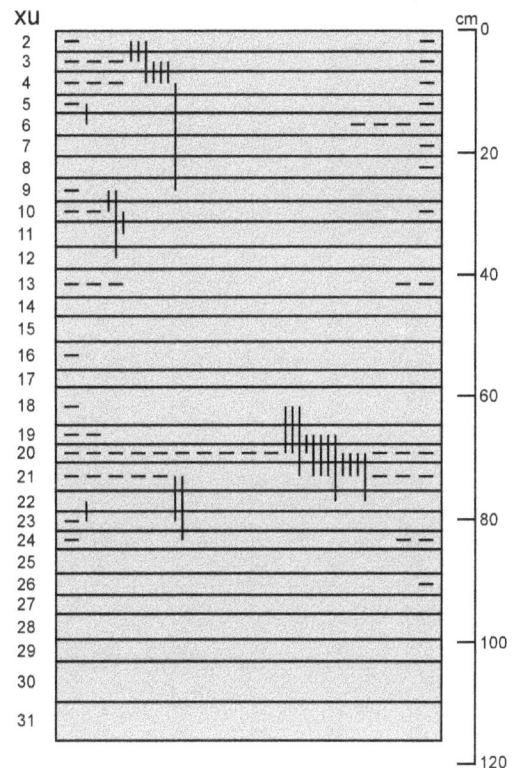

Figure 6.7 Distribution of identified *A. trapezia* valve-pairs (n=61), Seven Mile Creek Mound, Square A. An additional 22 valve-pairs encountered as articulated specimens during excavation are shown as short horizontal lines down the right hand side of the figure. Line termination points indicate the vertical mid-points of the excavation units from which conjoining valves were recovered. Short horizontal lines indicate valve-pairs identified within excavation units. Not to scale on the horizontal axis. See Chapter 5 for details of methods.

with a calculated Shannon-Weaver Function (H′) of 1.596 and Simpson's Index of Diversity (1–D) of 0.587. The relative proportions of the four main taxa remain relatively stable throughout the deposit except for an increase in the relative abundance of hairy mussel in XU4–5 and mud ark in XU2–3 and 19–21 (Fig. 6.16). Crustaceans are represented throughout the cultural assemblage, consisting of mud crab (*Scylla serrata*) weighing a total of 404.5g and barnacle (*Balanus variegatus*) weighing 130.7g.

The bulk of the non-artefactual stone material (20,916.4g; Fig. 6.15) is thought to have entered the site attached to rock platform species such as oysters, as is evidenced in the large number of oysters found still attached to rocks and other shells such as *P. ebininus* and *Velacumantus australis* in the deposit (5,158.5g). Most of the marine gastropods in the assemblage are less than 15mm in height with many less than 5mm, such as *Zafra avicennia*, *Epitonium* sp., *Bembicium nanum*, *Littoraria* sp., *Mitra* sp., *Pseudoliotia* sp., *Metaxia* sp. and *Herpetopoma atrata*. Similarly, many of the marine bivalves are less than 10mm in length, such as *Venericardia* sp., *Corbula crassa*, *Acropsis deliciosa*, *Arcopsis symmetrica*, *Tellina* sp. and *Trapezium sublaevigatum*. Many of the smaller taxa present, particularly many of the gastropods, probably entered the deposit attached to larger shellfish taxa. Even if these small taxa may have been intentionally collected (see Rowland 1994), they clearly did not make a major contribution to the overall diet. These small shellfish, however, reveal potentially important data about environmental zones exploited and foraging strategies.

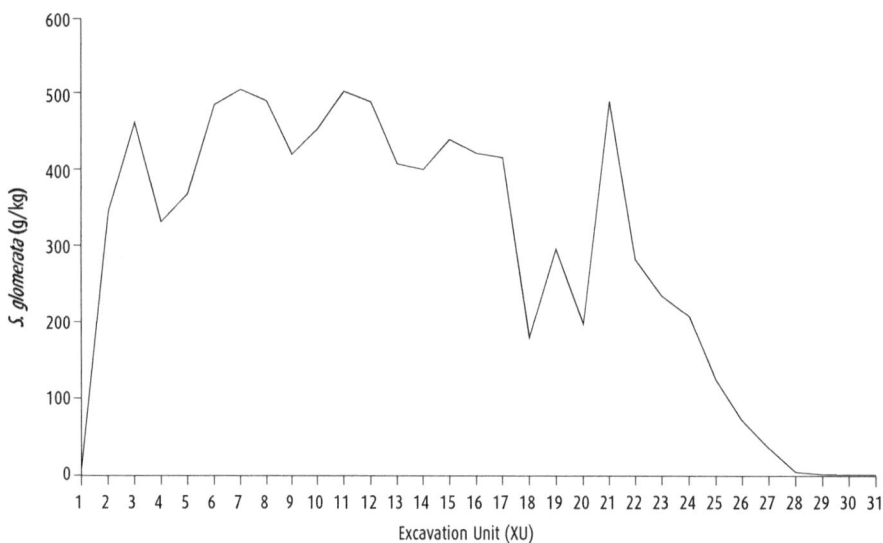

Figure 6.8 Abundance of oyster (*S. glomerata*).

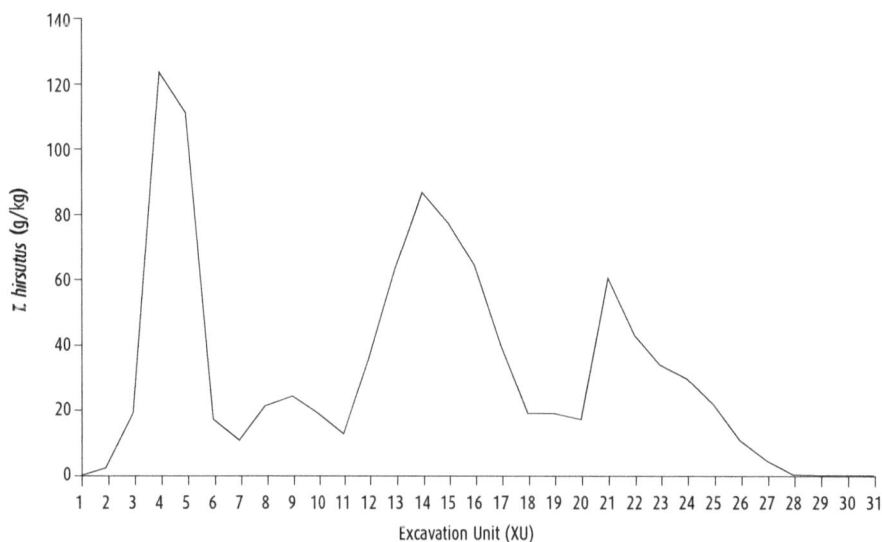

Figure 6.9 Abundance of hairy mussel (*T. hirsutus*).

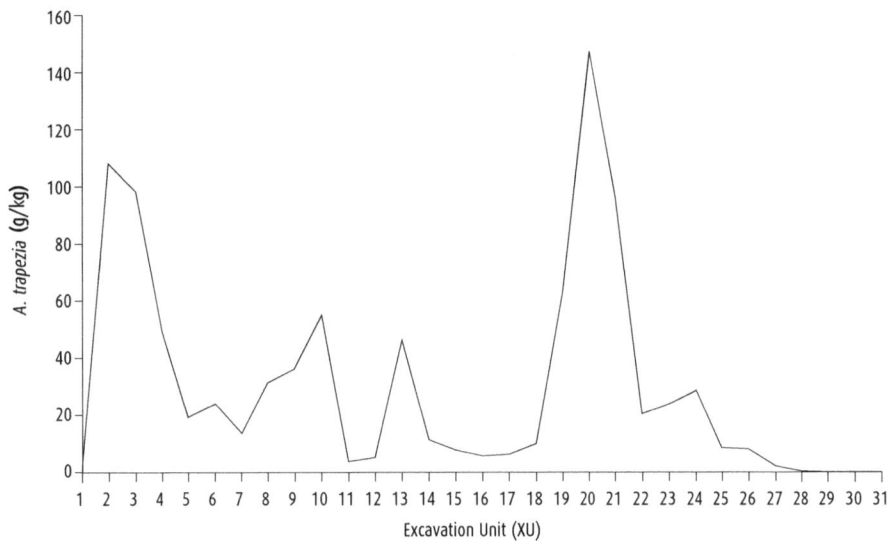

Figure 6.10 Abundance of mud ark (*A. trapezia*).

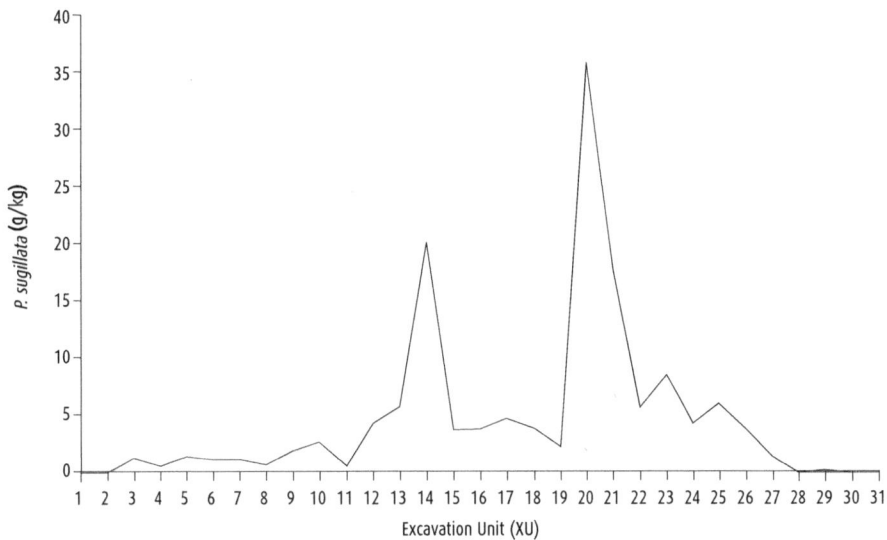

Figure 6.11 Abundance of scallop (*P. sugillata*).

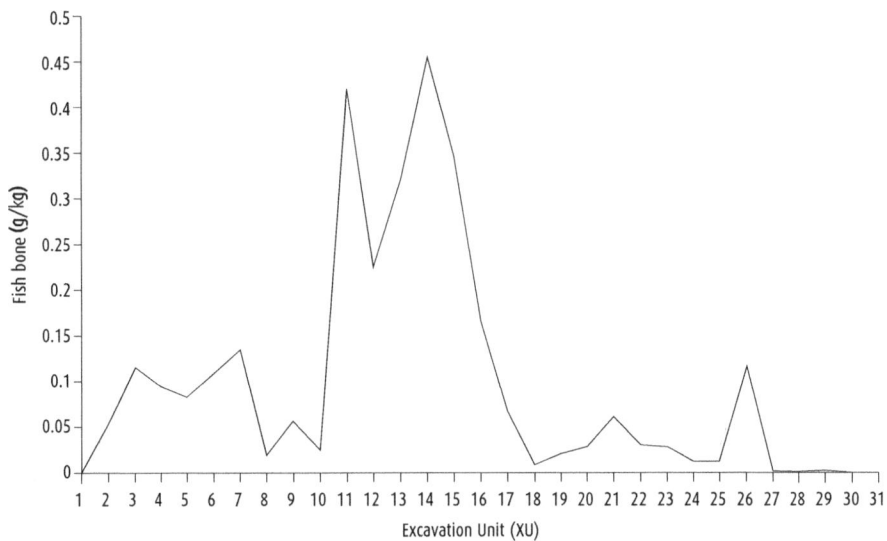

Figure 6.12 Abundance of fish bone.

Table 6.3 Presence/absence of shellfish identified in the Seven Mile Creek Mound, Square A.

FAMILY	TAXON	1	2	3	4	5	6	7	8	9	10	11	12	13	14	15	16	17	18	19	20	21	22	23	24	25	26	27	28	29	30	31	TOTAL (g)
														MARINE BIVALVIA																			
Anomiidae	*Anomia trigonopsis*				X																												5.9900
Arcidae	*Anadara trapezia*	X	X	X	X	X	X	X	X	X	X	X	X	X	X	X	X	X	X	X	X	X	X	X	X	X	X	X	X				11489.1296
Cardiidae	*Acrosterigma vertebratum*			X								X																					20.8000
Carditidae	*Venericardia* sp.			X	X																												0.1759
Chamidae	*Chama fibula*				X	X	X								X	X	X	X						X	X		X						141.4335
Corbulidae	*Corbula crassa*									X				X			X																4.8640
Mytilidae	*Trichomya hirsutus*	X		X	X	X	X	X	X	X	X	X	X	X	X	X	X	X	X	X	X	X	X	X	X	X	X	X	X	X	X	X	11899.4321
Noetiidae	*Arcopsis deliciosa*													X	X																		0.2351
Noetiidae	*Arcopsis symmetrica*																			X		X								X			0.6131
Ostreidae	*Saccostrea glomerata*	X		X	X	X	X	X	X	X	X	X	X	X	X	X	X	X	X	X	X	X	X	X	X	X	X	X	X	X	X	X	106833.4143
Pteriidae	*Pinctada albina sugillata*			X	X	X	X	X	X	X	X	X	X	X	X	X	X	X	X	X	X	X	X	X	X	X	X	X	X	X			1882.9924
Tellinidae	*Tellina* sp.																						X										0.0999
Trapeziidae	*Trapezium sublaevigatum*										X	X	X	X	X		X	X	X		X	X		X	X	X							2.7089
Ungulinidae	*Felaniella subglobosa*														X																		0.0321
Veneridae	*Antigona chemnitzii*																					X											0.3196
Veneridae	*Dosinia tumida*														X																		1.8271
Veneridae	*Gafrarium australe*					X	X	X	X	X	X	X	X	X	X	X		X		X	X	X	X	X	X	X	X	X	X				20.4363
Veneridae	*Irus* sp.					X	X	X	X	X	X	X	X	X	X	X	X	X	X		X	X	X	X	X	X	X	X					26.6503
Veneridae	*Placamen* sp.								X	X		X											X										1.7165
Veneridae	*Venerid* sp.					X																	X	X	X								2.7148
												MARINE GASTROPODA																					
Batillariidae	*Pyrazus ebininus*	X	X	X	X	X	X	X	X	X	X	X	X	X	X	X	X	X	X	X	X	X	X	X	X	X	X	X					1487.8767
Batillariidae	*Velacumantus australis*	X	X	X	X	X	X	X	X	X	X	X	X	X	X	X	X	X	X	X	X	X	X	X	X	X	X	X	X				167.5934
Cerithiidae	*Cerithiid* sp.			X	X		X									X			X				X										2.9232
Cerithiidae	*Cerithium* sp.		X		X	X		X		X								X	X						X	X	X	X					4.0566
Cerithiidae	*Clypeomorus bifasciata*	X	X	X	X	X	X		X	X	X	X	X	X	X	X	X	X	X	X	X		X	X	X	X	X	X					35.0240
Columbellidae	*Zafra avicennia*			X	X	X	X			X	X	X	X	X									X	X	X		X	X					0.5156
Costellariidae	*Vexillum* sp.	X		X	X																					X	X	X					3.6403
Epitoniidae	*Epitonium* sp.																							X			X	X					0.0281
Fasciolariidae	*Latirus* sp.	X							X			X					X		X	X		X	X	X	X								13.5168
Fissurellidae	*Diodora ticaonica*								X				X			X	X							X			X						0.5322

continued over

Table 6.3 continued

FAMILY	TAXON	1	2	3	4	5	6	7	8	9	10	11	12	13	14	15	16	17	18	19	20	21	22	23	24	25	26	27	28	29	30	31	TOTAL (g)
													MARINE GASTROPODA continued																				
Lottiidae	*Acmaeid* sp.				X	X	X	X		X	X				X	X		X	X		X	X	X										0.5253
Littorinidae	*Bembicium nanum*	X		X	X			X					X	X	X	X	X		X	X	X	X		X		X	X	X			X		23.5524
Littorinidae	*Littoraria* sp.			X	X	X	X	X				X				X	X	X	X	X	X	X	X		X	X	X						3.3904
Mitridae	*Mitra* sp.					X												X							X								0.0037
Muricidae	*Bedeva paivae*	X		X	X	X	X	X	X	X	X	X	X	X	X	X	X	X	X	X	X	X	X	X	X	X	X	X					77.2569
Muricidae	*Morula marginalba*					X															X	X											15.1567
Nassariidae	*Nassarius burchardi*															X							X	X			X						0.5650
Neritidae	*Nerita balteata*	X		X	X	X	X	X	X	X	X	X	X	X	X	X	X	X	X	X	X	X	X	X	X	X	X	X	X		X		1467.6258
Neritidae	*Nerita squamulata*											X	X			X				X	X	X	X	X	X		X						2.2910
Planaxidae	*Planaxis sulcatus*			X	X								X	X	X		X		X	X	X	X											4.8336
Skeneidae	*Pseudoliotia* sp.																X																0.0233
Triphoridae	*Metaxia* sp.					X	X							X	X		X						X		X		X	X					0.1790
Triphoridae	*Subulphora* sp.			X	X			X		X			X	X	X	X	X	X			X	X	X	X	X	X	X						1.0017
Trochidae	*Herpetopoma atrata*			X	X			X	X	X	X	X	X	X	X	X	X		X		X	X	X	X	X	X	X						7.3648
Trochidae	*Thalotia* sp.	X							X	X	X	X	X	X	X	X	X	X					X	X	X		X	X					7.0767
													TERRESTRIAL GASTROPODA																				
Camaenidae	*Figuladra* sp.	X		X	X	X	X	X	X	X	X	X	X	X	X	X	X	X	X	X	X	X	X	X	X	X	X	X	X	X			116.0877
Camaenidae	*Trachiopsis mucosa*			X	X				X																								9.4184
Pupillidae	*Pupoides pacificus*	X		X	X	X	X	X	X	X	X	X	X	X	X	X	X	X	X	X	X	X	X	X	X	X	X					X	0.6877
Subulinidae	*Eremopeas tuckeri*	X		X	X	X	X	X	X	X	X	X	X	X	X	X	X	X	X	X	X	X	X	X	X	X	X	X			X		4.2172
													FRESHWATER BIVALVIA																				
Corbiculidae	*Corbicula australis*																						X	X	X	X							2.4568

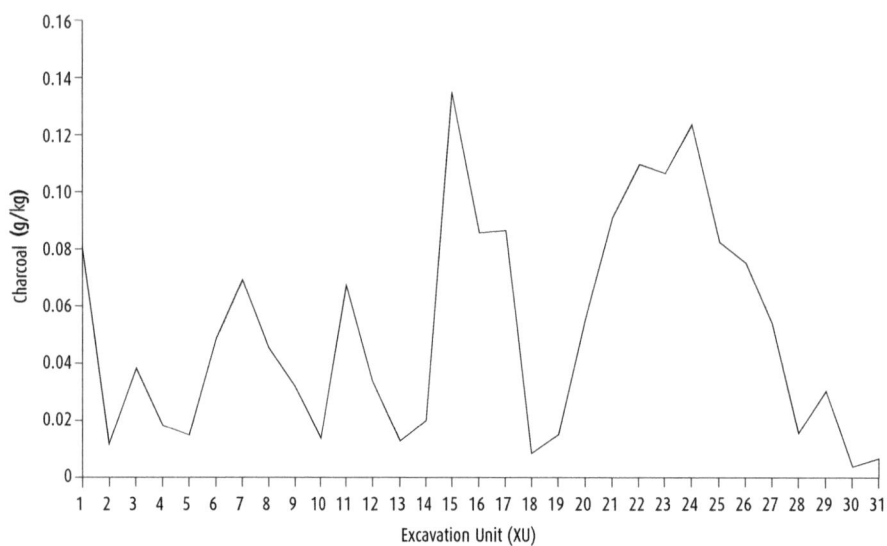

Figure 6.13 Abundance of charcoal.

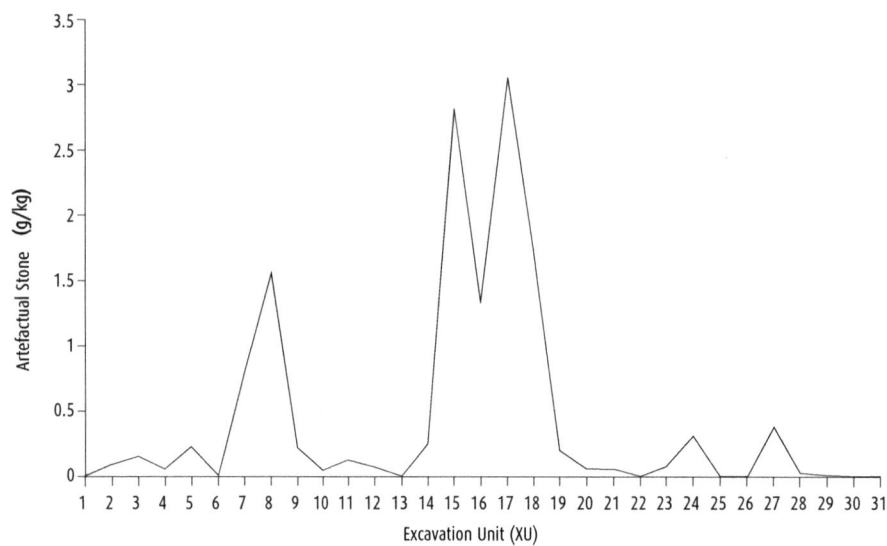

Figure 6.14 Abundance of artefactual stone.

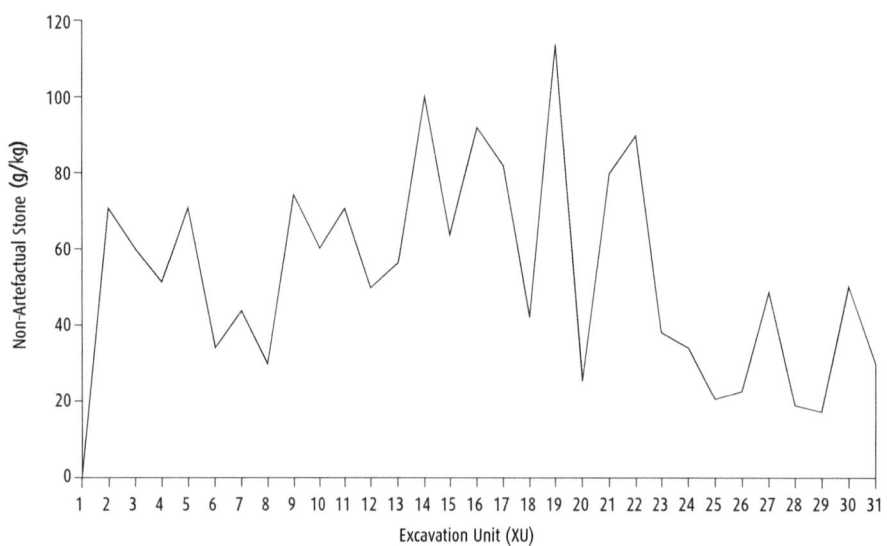

Figure 6.15 Abundance of non-artefactual stone.

Table 6.4 Metrical data for intact and broken (with umbo) *A. trapezia* valves from the Seven Mile Creek Mound, Square A.

XU	MEAN LENGTH			MEAN HEIGHT			MEAN WIDTH			MEAN WEIGHT			MEAN HINGE		
	n	mm	±	n	mm	±	n	mm	±	n	g	±	n	mm	±
2	38	40.6	4.5	47	35.6	4.1	61	14.8	1.9	37	13.0	4.5	52	26.0	2.9
3	66	39.9	5.5	86	27.3	10.5	90	14.6	2.1	61	11.6	4.28	79	25.1	3.4
4	41	38.9	5.0	44	33.4	4.6	48	14.1	2.1	39	10.5	4.4	44	24.3	3.3
5	12	37.7	3.3	13	33.2	3.9	14	15.8	4.0	12	10.4	3.4	14	23.9	2.8
6	18	41.4	3.6	19	35.0	3.6	19	15.4	1.8	18	12.8	4.0	17	25.5	2.7
7	8	40.2	5.6	9	35.4	4.5	10	15.9	1.7	8	13.2	4.7	8	25.1	2.7
8	24	42.2	4.3	24	35.5	2.8	25	15.7	1.7	24	13.9	3.1	22	26.0	2.6
9	31	40.2	4.6	31	34.0	3.7	32	15.0	1.2	31	12.1	3.3	31	24.5	3.0
10	39	39.9	3.3	40	34.5	2.9	40	14.8	1.2	39	11.8	2.6	39	24.3	2.2
11	4	34.8	10.7	4	29.5	9.6	4	12.7	4.4	4	9.0	5.5	4	22.0	5.4
12	7	31.7	12.8	7	25.6	11.1	7	11.3	5.2	6	7.4	6.2	6	20.2	8.3
13	39	42.8	3.4	39	36.9	3.3	39	15.3	1.7	38	13	3.6	36	25.8	2.6
14	8	44.5	5.2	8	37.7	4.4	8	16.7	2.1	8	14.7	4.9	8	25.9	4.0
15	7	37.9	10.0	7	32.0	8.6	7	14.9	4.2	7	11.6	6.6	7	23.9	5.4
16	6	38.1	6.8	6	31.6	6.2	6	13.2	2.4	6	9.7	3.9	6	23.9	4.0
17	5	42.2	2.3	5	37.7	1.5	5	15.1	0.9	5	13.4	2.4	5	26.6	2.1
18	11	43.3	7.8	11	38.0	6.8	11	15.6	3.0	11	14.5	6.9	11	26.8	5.0
19	46	45.5	3.9	50	39.9	3.8	50	16.5	1.9	44	16.7	5.3	45	27.5	2.7
20	130	44.9	4.3	130	39.2	4.1	131	16.3	1.9	123	15.7	4.5	122	27.2	2.8
21	75	44.9	5.6	76	40.0	4.8	78	16.4	2.2	66	16.5	4.8	69	28.1	3.3
22	21	36.4	4.9	24	31.4	4.0	24	13.3	1.8	21	8.2	2.9	22	22.8	2.6
23	21	39.0	8.6	21	33.6	7.4	22	14.3	3.0	19	12.4	6.1	21	24.9	4.1
24	21	43.7	6.2	22	37.7	6.3	22	15.7	2.7	19	14.5	6.7	21	26.3	3.6
25	5	44.6	8.7	5	38.3	8.4	5	15.9	3.0	5	17.1	9.3	5	27.5	4.8
26	6	44.4	5.3	7	37.6	4.0	8	15.4	2.5	6	14.5	4.5	4	27.9	4.3
Total	689	41.4	0.9	735	36.1	0.8	766	15.2	0.4	657	12.4	0.8	698	25.5	0.6

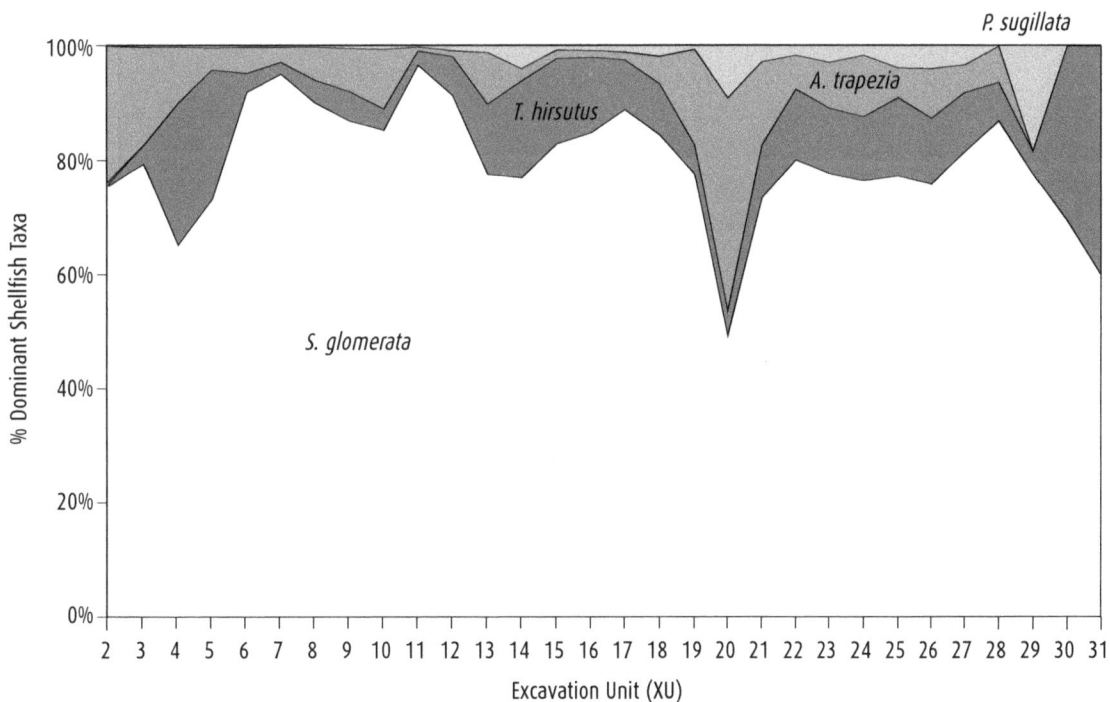

Figure 6.16 Relative contribution of dominant shellfish taxa. Note that XU1 is excluded as it contained negligible quantities of shell (see Appendix 4).

The relative homogeneity in the diversity and abundance of the dominant taxa throughout the deposit suggests no major changes in foraging targetting strategies or resource availability in the area during the period of site use. The entire suite of shellfish taxa recovered is consistent with shallow estuarine conditions and the presence of intertidal and subtidal flats containing a range of rock/shell debris beds, seagrass and mudflats. The only apparently anomalous taxon identified is the freshwater bivalve *Corbicula (Corbiculina) australis*, which populates coastal rivers and streams, although incidental gathering of this taxon from shell debris beds seems a reasonable assumption. The four species of terrestrial gastropods also indicate the broad similarity of the current vegetation structure to that obtaining during the period of site occupation.

Very intensive localised shellfishing has the potential to skew the demographic structure of mollusc assemblages in intertidal communities and reduce the availability of large specimens (Swadling 1976). This was investigated using *A. trapezia* as an indicator taxa. There is no significant change in the mean size of *A. trapezia* throughout the deposit as measured by five attributes (length, height, width, weight and hinge length) (Table 6.4). The mean length of *A. trapezia* does not fall below 31.7mm in any excavation unit, with a combined weighted mean length calculated on 689 valves of 41.4±0.9 (χ^2=7.2197, df=24, p≤0.05) and a terminal mean of 40.6±4.5mm for XU2 (XU1 was the shallow surface covering of turf). Inglis (1992) found that *A. trapezia* attains a length of 20–30mm within 12 months with large individuals (>40mm) growing less than c.1mm/year up to 70–80mm. The majority of individuals in this assemblage are, therefore, over 12 months of age and probably several years of age, discounting the hypothesis that site abandonment is related to overpredation of *A. trapezia*.

Vertebrate remains

Fish bone is present throughout the cultural deposit, totalling 34.4g consisting of 1,346 pieces of bone and a NISP of 54 (Fig. 6.12). A total MNI of 37 was calculated by totalling the MNI for each excavation unit. The weight of bone identified to taxon was 1.6g, giving an identification rate of 4.75% (Table 6.5). Identified taxa in descending order of abundance include flathead (Platycephalidae), whiting (Sillaginidae), Sparidae (including bream, *Acanthopagrus australis*) and mullet (Mugilidae) (Table 6.6). Size-classing of vertebrae showed that 69% have a centrum diameter of 3mm or less. These represent very small fin fish. Some larger fish are represented by vertebrae from XU9–15 (see Vale 2002 for further details). Twenty-five pieces of bone weighing 4g could not be assigned to a fish skeletal element. The small size of these specimens and the lack of diagnostic attributes prevented identification to taxon.

Of the six vertebral samples subject to DNA analysis, only one returned a positive fish-like polymerase chain reaction (PCR) product, although this extract did not produce a product when sequenced (Hlinka et al. 2002). Taphonomic factors are thought to be responsible for the low amplification success rate (see Hlinka et al. 2002 for further details).

Stone artefacts

Stone artefacts are distributed throughout the cultural deposit between XU2–29 (Fig. 6.14). A total of 207 stone artefacts weighing 169.1g was identified in Square A (Table 6.7). Seven stone artefacts were plotted *in situ* during excavation, with the remainder recovered from the sieve residue. Virtually the entire assemblage is manufactured on quartz (n=199) with occasional granodiorite (n=4), greywacke (n=2), chert (n=1) and silcrete (n=1). Four flakes were identified, with the remainder consisting of flaked pieces. All raw materials are available in the immediate vicinity of the site, with a range of rock types available on estuarine gravel beds adjacent to the site. Most artefacts are extremely small, with an average maximum dimension of 13.1mm and average weight of 0.8g.

Other remains

A range of other materials was recovered from the site. Small nodules of red ochre totalling 1.5g were recovered from XU12–15. Small pieces of pumice totalling 32.2g were recovered from the bottom half of the deposit (XU13, 20–31). Charcoal is represented in very small quantities throughout the deposit, totalling 18.8g (Fig. 6.13). The presence of charcoal in culturally-sterile sediments below the shell deposit suggests the possibility that some of the other charcoal represented in the assemblage may be natural.

Table 6.5 Fish bone abundance, Seven Mile Creek Mound, Square A.

XU	NUMBER SPECIMENS	TOTAL WEIGHT (g)	NISP	WEIGHT NISP (g)	MNI	% IDENTIFIED BY WEIGHT
1	0	0	0	0	0	0
2	14	0.5106	0	0	0	0
3	47	1.2786	3	0.0495	2	9.70
4	74	1.1613	1	0.0115	1	0.90
5	53	0.9165	2	0.0217	1	0.02
6	98	1.3380	5	0.0760	3	5.68
7	60	1.4591	5	0.5054	4	34.64
8	16	0.2120	0	0	0	0
9	27	0.6474	2	0.0314	2	4.85
10	18	0.2565	0	0	0	0
11	169	4.8736	6	0.1938	3	3.98
12	115	2.2661	3	0.0755	2	3.33
13	61	3.5652	8	0.1072	6	3.01
14	176	4.7367	10	0.1443	5	3.05
15	132	3.9176	6	0.2088	5	5.33
16	52	2.0026	0	0	0	0
17	47	0.7844	0	0	0	0
18	8	0.1657	0	0	0	0
19	22	0.2808	0	0	0	0
20	12	0.4353	0	0	0	0
21	69	0.8567	0	0	0	0
22	16	0.3883	2	0.0591	2	15.22
23	17	0.3532	0	0	0	0
24	14	0.1496	1	0.1496	1	100.00
25	7	0.1879	0	0	0	0
26	18	1.5749	0	0	0	0
27	2	0.0260	0	0	0	0
28	1	0.0113	0	0	0	0
29	1	0.0299	0	0	0	0
30	0	0	0	0	0	0
31	0	0	0	0	0	0
Total	1346	34.3858	54	1.6338	37	4.75

Table 6.6 Fish bone taxonomic representation, Seven Mile Creek Mound, Square A.

TAXON	NISP	MNI	WEIGHT (g)	XUs
Platycephalidae	21	15	0.94	6–7, 9, 11–15, 22, 24
Sillaginidae	16	11	0.25	3, 5–7, 9, 11, 13–15, 22
Sparidae	13	6	0.22	3–4, 6–7, 12–14
Mugilidae	4	3	0.06	11, 13, 15

Table 6.7 Stone artefacts from the Seven Mile Creek Mound, Square A.

RAW MATERIAL	ARTEFACT TYPE	NUMBER	WEIGHT (g)	XUs
Quartz	Flake	2	10.2785	7, 8
Quartz	Flaked Piece	197	125.6980	2-12, 14-21, 23, 27-28
Granodiorite	Flake	2	8.3893	24, 27
Granodiorite	Flaked Piece	2	23.8782	17, 23
Greywacke	Flaked Piece	2	0.1824	4, 20
Chert	Flaked Piece	1	0.6016	4
Silcrete	Flaked Piece	1	0.0789	29
Total	–	207	169.1069	–

Discussion

The Seven Mile Creek Mound was deposited over a period of some 350 years between c.3,950–3600 cal BP. Deposits represent broad utilisation of intertidal and near-shore resources, including extensive shellfishing, crabbing and fin fishing. Shell remains are dominated by oyster, with varying quantities of hairy mussel, mud ark and scallop. Although fish remains occur throughout the deposit, they are infrequent and represent very small fish.

There are no clear stratigraphic features within the main shell unit (SUII) of Square A, with the exception of SUIIa. However, the internally consistent radiocarbon chronology and analysis of the shell and sediment fractions suggest that rapid deposition at the site occurred in a sequence of events rather than by continuous occupation. The distribution of overlapping *A. trapezia* conjoin sets indicates that the site can be divided into at least two major sequential accumulation events (XU2–12, XU18–24), separated by zones that are also possibly associated with these events (Fig. 6.7).

The Seven Mile Creek Mound does not appear to have been used as a camp site, but rather functioned as a refuse pile similar to that documented for the Bayley Point 3 shell mound in the Gulf of Carpentaria (Robins et al. 1998). The absence of significant oxidation and fragmentation of *A. trapezia* suggests no significant exposure to heating either before or after discard (Robins and Stock 1990). Further, no hearths or ashy sediments were observed in the deposit and charcoal is virtually absent. These observations provide strong support for a model of offsite preparation and consumption of shellfish (and other food) with secondary discard on the mound (Meehan 1982:86; Robins et al. 1998:120). The creation of the mound is therefore likely to result from intentional discard rules operating at the site.

Hiscock (2001:141) has suggested that mound-building in northern Australia is related to patterns of reduced residential mobility. This does not appear to be the case at the Seven Mile Creek Mound. The site differs in a number of respects from shell mounds reported across northern Australia (e.g. Bailey 1994; Beaton 1985). The site is an isolate (as are the other two shell mounds known in the region) and not part of a group of mounds or even located adjacent to other sites. Oyster dominates a shellfish assemblage comprising a broad range of intertidal taxa, whereas northern Australian mounds are mono-specific accumulations of *A. granosa* (e.g. Robins et al. 1998). The mound is not coeval with any other dated deposits in the area, pre-dating the next oldest site (the Mort Creek Site Complex) by some 300 years. A single date from a disturbed section of the Round Hill Creek Mound of 1,910±42 BP (Wk-10090) suggests that mound formation in the region may have occurred independently through time and space. The episodic nature of site occupation and the absence of coeval coastal assemblages raises the possibility that occupation of the Seven Mile Creek Mound was part of an inland-based settlement-subsistence system rather than one based on permanent occupation of the coast (this idea is explored further in Chapter 14).

The cessation of mound-building and abandonment of the Seven Mile Creek Mound can be ascribed to a combination of possible factors: (1) changes in resource availability, caused by either environmental change or overexploitation; (2) changes in subsistence strategies and/or settlement behaviour, related to either environmental and/or cultural factors; (3) changes in discard rules away from mounding behaviour to less visible forms of deposit with lower survival potential; and/or (4) changes in land-use through alterations to regional settlement strategies. These arguments have been advanced previously in north and northwest Australia to explain the abandonment of shell mound sites dominated by *A. granosa* around 1,000 BP. Hiscock (1997) and O'Connor (1999), for example, have both linked mound abandonment to habitat change caused by mangrove invasion. At the Seven Mile Creek Mound, with the possible exception of catastrophic storm-based environmental change, the diversity of estuarine taxa represented in the mound and the lack of change in shellfish size throughout the sequence suggest the presence of a stable estuary that was not suffering from overexploitation. Although discard rules may have changed, an extensive survey and dating program has failed to find coeval sites indicating a change in site form. These data therefore lend some support to the idea that site abandonment was linked to changes in regional settlement strategies that were not principally reliant on coastal resources.

The age and contents of the site also have implications for understanding regional palaeoenvironments. The basal age of the site is congruent with recent models of sea-level change on the Queensland coast suggesting a Holocene stillstand of +1.65m at c.5,500 BP until c.3,700 BP, when sea-levels dropped to approximately modern values (Larcombe et al. 1995). The low elevation landform that the Seven Mile Creek Mound is situated on was probably overtopped by the mid-Holocene sea-level highstand, dating highstand on the southern Curtis Coast to before 4,000 BP. Site contents indicate that productive estuaries were in place by this time, with a broad range of commensal species representing shallow intertidal and subtidal rock/shell debris, mudflats and seagrass beds. Results from the Seven Mile Creek Mound therefore support the findings of other recent research suggesting that Aboriginal groups were resident in some regions of the Queensland coast at the time of local sea-level stabilisation (e.g. Barker 1996; McNiven 1991a), contrary to models of mid-Holocene coastal abandonment proposed by Beaton (1985).

Broader implications

Coastal archaeological sites older than 3,000 BP are not common on the Queensland coast (see Ulm et al. 1995; Ulm and Reid 2000, 2004 for a review). Although several rockshelters containing evidence for marine resource exploitation in the Whitsunday Islands (Barker 1996) and Princess Charlotte Bay areas (Beaton 1985) date to around or before the mid-Holocene, only two open sites on the Queensland coast have evidence of focussed marine resource exploitation pre-dating 3,000 BP. The Hope Island site on the Coomera River, dated on charcoal to 4,350±220 BP (Beta-20799), contains abundant shell remains, although fish bone is apparently absent (Walters et al. 1987). Mazie Bay on North Keppel Island is dated on charcoal to 4,274±94 BP (NZA-456) but quantities of shell and fish bone are only present in deposits dated to shortly before 3,000 BP (Rowland 1999).

In all other open coastal sites in Queensland pre-dating 3,000 BP (n=7) faunal remains are either entirely absent, represented in minute quantities or restricted to deposits dated to the last 3,000 years. This is the case at Wallen Wallen Creek (Neal and Stock 1986), New Brisbane Airport (Hall and Lilley 1987), Teewah Beach Site 26 (McNiven 1991), King's Bore Sandblow Site 97 (McNiven 1992a), Eurimbula Site 1 (Ulm et al. 1999a), Mort Creek Site Complex (see Chapter 7) and Townsville Common (Kelly 1982), although the cultural status of this last site is equivocal. The absence of faunal remains from these early deposits is commonly attributed to ephemeral occupation, taphonomic considerations and/or recovery strategies (Ulm 2002a). The presence of fish bone, crab shell and a variety of shellfish remains throughout the sequence enables data from the Seven Mile Creek Mound to contribute uniquely to understandings of Aboriginal use of late Holocene coasts during the period 3.5–4ka.

Walters (1986, 1989, 1992a, 1992b) has consistently argued that marine fishing was only incorporated as a regular feature of subsistence-settlement systems in southeast Queensland after

2,000 BP. Data from the Seven Mile Creek Mound provide the first unequivocal evidence for significant fish bone discard dating earlier than 3,500 BP. The Seven Mile Creek Mound is the oldest dated shell mound in Queensland, pre-dating the earliest mounds at Weipa (Bailey et al. 1994; Stone 1992) and Princess Charlotte Bay (Beaton 1985; David and Lourandos 1997) by over 1,000 years and pre-dating the Booral Shell Mound (Frankland 1990) by over 700 years.

Summary

The Seven Mile Creek Mound was briefly and intensively occupied by Aboriginal hunter-gatherers for a period of approximately 350 years between 3,950–3,600 years ago. Mound construction appears to have occurred episodically and involved secondary disposal of food refuse collected from adjacent intertidal flats and prepared and consumed offsite. The mound is unique in the region and exhibits more similarity in form, content and chronology to the occasional mounds recorded further to the south than the very large mono-specific mounds which feature in the archaeological record above the Tropic of Capricorn. The short duration and focussed nature of cultural discard at the site provides critical chronological resolution of Indigenous occupation compared with the long-term, spatially-diffuse Aboriginal occupation sites which dominate the archaeology of the region. The data enable a significant and detailed understanding of the late Holocene social, cultural and natural environment of the southern Curtis Coast which may now be explored.

7

Mort Creek Site Complex

Introduction

This chapter reports the results of archaeological excavations undertaken at the Mort Creek Site Complex. Previous surveys and test excavations undertaken by Lilley et al. (reported in Carter et al. 1999) in 1995 revealed the potential of the site to contribute to a regional understanding of early coastal occupation with cultural deposits dated to before 2,000 BP. These excavations and subsequent analyses demonstrated a complex history of landscape formation, with interfingering natural and cultural shell deposits in some areas of the site. The excavations reported below were part of a detailed reappraisal of the site complex in 1998 which sought to determine its antiquity of occupation and better characterise its content and structure through controlled excavations and recording of deposits.

Site description and setting

The Mort Creek Site Complex is located on the west bank of Mort Creek, on the west coast of Rodds Peninsula (Latitude: 24°00′45″S; Longitude: 151°37′45″E) (Fig. 7.1). Natural and cultural shell deposits extend discontinuously over an area of about 6ha characterised by a complex of beach ridges, cheniers, shell middens and tidal inlets. The deposits are associated with a probable stone-walled tidal fishtrap and a possible stone monolith (Figs 7.2–7.3). This entire area is referred to collectively as the Mort Creek Site Complex. The deposits front the shallow, open waters of Rodds Harbour to the south and west and a large area of mangroves and intertidal flats to the east. Mort Creek is a minor estuary of Rodds Harbour, which comprises an extensive estuarine system sheltered from full oceanic conditions by Rodds Peninsula. According to Olsen (1980a:18), Rodds Harbour has no brackish zone, despite periodic freshwater inflow from Worthington Creek at its southeast extremity. Seagrasses and soft corals occur at the mouth of Mort Creek. The lower two-thirds of Mort Creek are dominated by dense stands of spotted mangroves (*Rhizophora stylosa*), with a low shrubland of yellow mangroves (*Ceriops tagal*), fringing saltflats to the north of the site and patches of tall grey mangroves (*Avicennia marina*) interspersed in the spotted mangrove forest in the northern half of the estuary. The main area of shell deposits is vegetated by microphyll vine forest with emergent Moreton Bay ash (*Corymbia tessellaris*). The exposed southern margin fronting Rodds Harbour is fringed by black she-oak (*Allocasuarina littoralis*). The understorey is frequently closed with introduced weeds including prickly pear (*Opuntia stricta*) and lantana (*Lantana camara*). Intertidal flats adjacent to the site include areas of muddy and shell/rocky debris substrates which support dense populations of telescope mud whelk (*Telescopium telescopium*) and occasional mud ark (*Anadara trapezia*) and razor clam (*Pinna bicolor*).

A low chenier ridge extends southwest beyond the mainland into Rodds Harbour, terminating in a rounded area vegetated with spotted mangroves (Fig. 7.3). Shell ridges observed in the area generally have a northeast-southwest orientation. A narrow projection of land marked on Bedwell et al.'s 1870 chart running over a kilometre south-southwest from the current Spit End may have been a western extension of such shell deposits (Fig. 7.1). This feature may have exerted considerable structural control over patterns of sedimentation to the east during its existence.

Evidence for non-Indigenous use of the site area is limited. There has been a permanent European presence on the peninsula since the late 1890s and extensive cattle grazing from the 1920s (Buchanan 1999:76). There is virtually no material evidence for activities associated with European occupation in the site area, with the exception of introduced weeds and a four wheel drive track which transects the site. Dams and fencelines are common features in other parts of Rodds Peninsula. The Queensland Parks and Wildlife Service (QPWS) has breeched some dams and demolished residential structures associated with the former grazing lease at Richards Point, although the remains of terraced gardens, exotic plantings and scatters of glass and metal remain. The Rodds Peninsula Section of Eurimbula National Park has been closed to public vehicle access since gazettal in October 1990 (QDEH 1994), limiting recent use of the area. Mort Creek is frequented today by recreational crabbers, with access by small boat from the nearby townships of Turkey Beach and Tannum Sands.

A possible stone-walled tidal fishtrap was identified during surveys on the western margin of Mort Creek, approximately 50m east of the excavation grid for Squares A–D (Fig. 7.2). The oyster-encrusted rocks appear to be anthropogenic extensions of the larger boulder outcrops which extend towards the creek from the area of The Granites excavation (see below). The rocks on the intertidal flats are much smaller than those under the canopy of the adjacent mangrove fringe. The portability of these smaller rocks and the absence of larger boulders in this area suggest that the rocks were transported to extend oyster habitats and/or to form a stone-walled trap. The feature consists of several tiers of rocks, with lower tiers visible through the top of the mangrove muds. The rocks form two low, linear banks which are raised above the level of the surrounding flats by accumulating muds. Although the two arcs curve towards each other, they do not meet. It is possible that the eastern end was at times enclosed by saplings or nets as documented in other parts of southeast Queensland (Petrie 1904:72–3). The southern arc appears to be broader than the northern one. The feature is not visible on low level aerial photography, discounting this avenue for investigating the antiquity of the structure. Oyster leases were established at nearby Bustard Head by 1909 (Buchanan 1999:76), but no records were found for such activities in Mort Creek. The abundance of fish remains recovered from the adjacent cultural deposits provides circumstantial evidence for an Indigenous origin of the feature. The structure is thus tentatively identified as an Indigenous stone-walled tidal fishtrap.

A large standing block of microgranite is located adjacent to the four wheel drive track c.25m west of the excavation grid (Fig. 7.2). In this area, the basal microgranite bedrock underlying the ridge is exposed and several large boulders of microgranite sit on the exposed rock. One of these large roughly triangular-shaped stones has been placed in an upright position. As no other information for the origin of the stone monolith is currently available, its attribution to either an Indigenous or European origin remains uncertain. Horton (1994:1029) classifies such features as stone arrangements.

Previous investigations

Shell deposits in the area were initially reported by Burke (1993) as sites CC-067 and CC-068 during a heritage management study of the Curtis Coast and were described as an Aboriginal site of 'high significance' (Burke 1993:Table 17). On the Queensland Environmental Protection Agency (EPA) Site Index Form completed for site CC-067, Burke notes that 'this shell midden appears to be interspersed with a natural beach ridge. It was quite difficult to determine if the midden was real or natural, it seems to me that it is probably a mixture of both'.

Test excavations and augering were subsequently undertaken under the auspices of the Gooreng Gooreng Cultural Heritage Project in January 1995. In previous publications the site has been called 'Rodds Peninsula' (Lilley et al. 1996) and the 'Rodds Peninsula Site Complex' (Carter 1997) but was renamed the Mort Creek Site Complex (MCSC) to distinguish it from other major site complexes recorded on Rodds Peninsula (e.g. the Pancake Creek Site Complex; see Chapter 8). The site is registered on the EPA's Indigenous Sites Database as KE:A41 and Queensland Museum Scientific Collection Number S866.

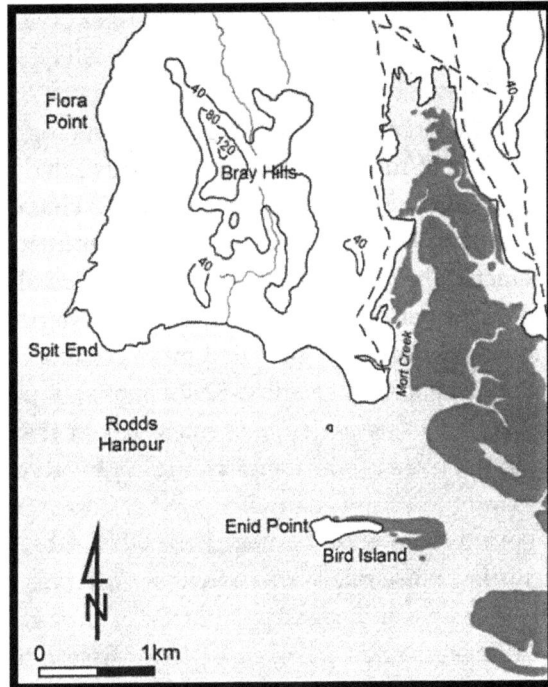

Figure 7.1 The Mort Creek catchment area. Dark grey shading indicates the general extent of mangrove, saltflats and claypans. Dashed lines denote 4WD tracks.

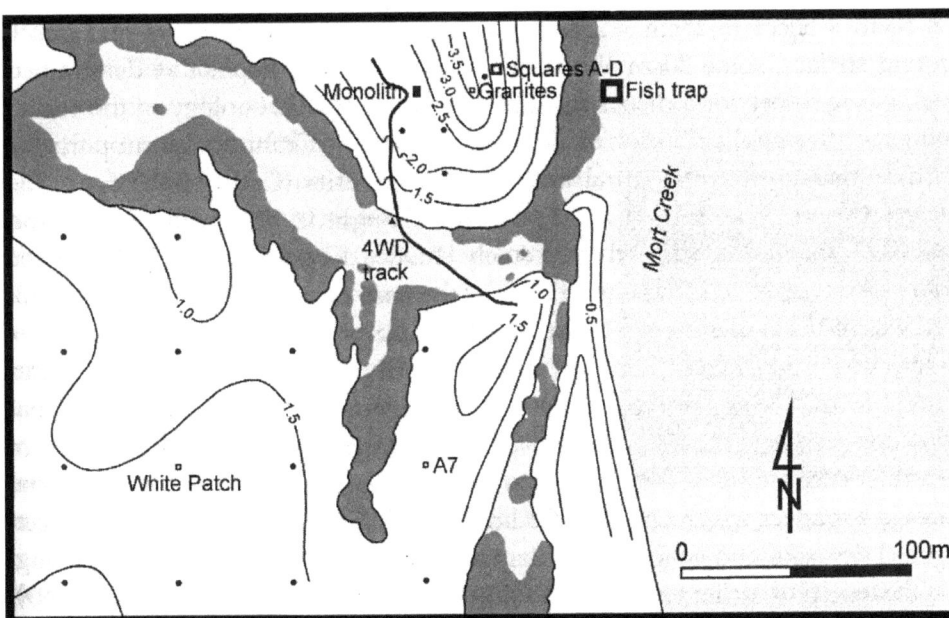

Figure 7.2 Topographic map of the Mort Creek Site Complex. Contours are in 0.5m intervals. Dark grey shading indicates the general extent of mangroves. Dots indicate auger test holes (not all shown).

In the initial field season, three 50cm × 50cm test pits were excavated to recover samples of cultural and natural marine shell deposits as part of an investigation of site formation processes in the area. In addition, a grid of 38 × 75mm auger holes were drilled across the site at 50m intervals to delineate the subsurface distribution of shell deposits. The first excavation was conducted in an area named 'White Patch' owing to the abundance of highly fragmented surface shell. The second excavation was undertaken beside auger hole number seven ('A7'), where a dense shell layer some 10–15cm thick was located approximately 20cm below ground surface in a sandy area with virtually no surface shell. The third excavation was placed in an area with abundant surface shell near 'The Granites' named because of microgranite boulders at this location (Fig. 7.2).

The White Patch excavations revealed a densely packed, highly fragmented shell deposit attributed to chenier development. It is characterised by a large range of species (>72 shell taxa), comprising juvenile shells and micro-molluscs and a general absence of charcoal, bone and stone artefacts. The Granites excavation revealed shell midden (>9 shell taxa) including stone artefacts and burnt fish bone overlying a chenier deposit (>62 shell taxa) resting on microgranite bedrock. The excavation at A7 revealed more complex sediments, suggesting the interfingering of cultural and natural shell deposits (>52 shell taxa). A possible shell artefact manufactured on the right valve of a *Antigona chemnitzii* was recovered at the base of the dense shell layer. Although the heavily scalloped edge is not typical of natural bivalve fracture patterns, detailed microscopy and chemical screening for use-wear and organic residues failed to confirm its cultural status (Culbert 1996). The augering demonstrated that there were substantial subsurface shell deposits over the entire area, including those parts where surface shell was largely absent.

Analysis (including studies of foraminifers) and radiocarbon dating of the excavated materials revealed a complex site formation history, with interfingering cultural and natural shell deposits in some areas of the site (Carter 1997; Carter et al. 1999; Lilley et al. 1999; Lilley et al. 1996). Probable cultural deposits at The Granites were dated to 2,335 cal BP (Wk-3941) – a relatively early date in the context of other sites on the Queensland coast (see Ulm et al. 1995). The resolution available for The Granites test excavation does not enable a high level of confidence to be placed in these findings. There are inconsistencies in the recorded elevations for the critical excavation units at the presumed contact point between cultural and natural deposits. XU11, the unit spanning this transition, was also disturbed by augering from the base of XU10, when it was assumed that the base of the deposit had been reached. Somewhere around XU11 or soon after, the excavation area was decreased from 50cm × 50cm to 25cm × 25cm and excavation was terminated at a depth of c.70cm below ground surface, some 20cm above the base of the shell deposit as determined by subsequent augering (see below for a discussion of problems in the chronology of the earlier excavations). The intra-specific size distribution of shells in XU11 of The Granites also supports the view that this unit comprises a mix of natural and cultural deposits (Carter 1997:85–6). The presence of stone artefacts and bone in both XU11 (the unit thought to be mixed midden and chenier deposit) and XU12 (the unit concluded to be solely chenier deposit) further indicates the presence of a mixed deposit (Carter n.d.:8), suggesting that the base of cultural deposits were not reached before the pit was reduced to a 25cm × 25cm sampling area.

Radiocarbon dates from the chenier deposits at White Patch and the base of The Granites indicate an overlap in the formation of natural and cultural shell deposits at the site, suggesting that Aboriginal occupation occurred in this area while the local landscape was in a significant state of flux. The aim of the further excavations at the site was to increase the sample of excavated material from The Granites area in a manner which ensured maximum resolution in discriminating between cultural and natural shell deposits, and enabled confident sampling of cultural material for dating. For further details on the results of earlier test excavations see Carter (n.d., 1997), Carter et al. (1999), Lilley et al. (1999) and Lilley et al. (1996).

Nearby sites include a probable stone-walled tidal fishtrap (KF:A12) at Richard's Point, c.7.5km to the north-northwest (Fig. 2.12). No significant cultural deposits are associated with this feature, although any deposits which may have occurred there are likely to have been disturbed by heavy mineral sandmining activities in this area in the 1970s. Although occasional stone artefacts and low density shell scatters have been encountered across the peninsula (see Chapter 2), the only other major site recorded is the Pancake Creek Site Complex, some 19km to the southeast (see Chapter 8; see also Burke 1993).

Excavation methods

A single 1m² pit divided into four 50cm × 50cm separately excavated squares (Squares A-D) was excavated to bedrock between 28 October and 5 November 1998 (Figs 7.2, 7.4–7.5). The excavation grid was situated in an open, level area approximately 8–10m northeast of the original Granites excavation. Although the precise location of the earlier excavations was not established, Squares A-D were located on the same low, sandy ridge to maximise the recovery of cultural deposits based on the high density cultural materials encountered earlier in this area. Excavation proceeded in shallow, arbitrary excavation units averaging 3.1cm in depth and 10.6kg in weight. Excavation ceased at a maximum depth of 68.4cm below ground surface after microgranite bedrock had been exposed over the entire base of the excavation area (Fig. 7.4). A total of 81 XUs was removed, distributed as follows: Square A (20 XUs), Square B (22 XUs), Square C (19 XUs), Square D (20 XUs). A total of 881.7kg of sediment was excavated. Excavated sediments were gently dry-sieved through 3mm screens onto a plastic tarpaulin located 10m northeast of the excavation to prevent contamination of underlying sediments. Stone (n=35), bone (n=14), carapace (n=6), charcoal (n=6), shell (n=4) and pumice (n=3) specimens encountered *in situ* during excavation were plotted three-dimensionally. The excavation was backfilled with a layer of green plastic sample bags across the base, followed by c.60l of

Figure 7.3 Chenier ridge at the Mort Creek Site Complex extending into Rodds Harbour. Facing southwest (Photograph: Ian Lilley).

Figure 7.4 General view of completed excavation, Squares A-D. Note continuous microgranite bedrock across the base of the excavation. Facing north.

Figure 7.5 General view of completed excavation, showing the section of Squares C-D. Note shell layer across the upper 20cm of the deposit. Facing east.

culturally-sterile white beach sands from the beach bordering Mort Creek and finally the sediments that had passed through the sieve (see Chapter 3 for a detailed discussion of the standard excavation methods employed at all sites).

Cultural deposit and stratigraphy

Excavation revealed approximately 65cm of sediments overlying microgranite bedrock (Fig. 7.5). Large quantities of shellfish remains, dominated by mud ark (*A. trapezia*), were recovered from a shell layer across the upper 20cm of the deposit. Remains of dugong (*Dugong dugon*) and turtle, probably loggerhead (*Caretta caretta*), were recovered towards the middle of the deposit immediately below the shell layer. Occasional shell and stone artefacts were recovered to bedrock, including several stone artefacts lying on the bedrock. Fish bone was recovered from every excavation unit. Two small areas of natural shell deposit were encountered at the base of the excavation in bedrock crevices clearly separated from the deposits containing cultural material. Bedrock was exposed over the entire base of the 1m^2 excavation with natural shell deposits clearly distinguishable at the base of the profiles in the extreme southwest and northwest corners of the pit. The rate of site accumulation calculated using the methods outlined by Stein et al. (2003) indicate a relatively slow overall rate of site formation of 2.66cm/100 years.

The deposit can be divided into five major stratigraphic units (SUs) on the basis of sediment colour and texture (Table 7.1, Fig. 7.6). The reddish brown sediments which dominate the deposit appear to derive from weathering of the surrounding and underlying microgranite. Fragments of non-artefactual microgranite are abundant towards the base of the deposit and are assumed to derive from *in situ* weathering of the bedrock. Acidity (pH) values are neutral to slightly alkaline throughout (7.0–8.0).

The occurrence of chenier material at the base of the western extent of the excavation suggests that the current excavations may have encountered the margin of a chenier which is more concentrated in the ridge area of the original Granites excavation to the southwest. Although both occurrences of chenier material appear to be infilling natural depressions in the bedrock, they occur at different elevations in the present pit: in the southwest corner it is 55–60cm below ground surface while in the northwest corner it is 62–68cm. Chenier material was encountered in The Granites test excavation at a depth of c.43cm. Assuming that all of this material belongs to the same sedimentary unit, the difference in relative depth below ground surface in the occurrence of the chenier material between the two excavations areas can probably be accounted for by topographic variations in the underlying bedrock. The differences in elevation between chenier material in Squares A and B, however, suggest that there has been erosion of the chenier deposit in the area, with remnants retained in crevices in the bedrock (see below for a proposed chronology of chenier formation).

Radiocarbon dating and chronology

Twelve radiocarbon determinations were obtained for the deposits, including the seven dates obtained for the initial excavations. Three dates were obtained from A7, two from The Granites, two from White Patch and five from Squares A–D. Ten dates were obtained on *A. trapezia* samples, one on charcoal and one on mixed shell (Table 7.2). The charcoal date (Wk-7458) was paired with two associated *A. trapezia* samples (Wk-6987 and Wk-7836) to investigate local marine reservoir conditions (see Chapter 4). Calibration calculations presented in Table 7.2 employ a ΔR correction value of +12±58. On the basis of the x-ray diffraction analysis (XRD), two shell samples (Wk-3939a

Table 7.1 Stratigraphic Unit descriptions, Mort Creek Site Complex, Squares A–D.

SU	DESCRIPTION
I	Extends across the entire square with an average depth of 5cm and a maximum depth of 10cm below ground surface. The unit comprises angular to subangular loosely consolidated surface sands which are brown (7.5YR-4/2) to very dark brown in colour (7.5YR-2.5/3). Occasional tufts of grass penetrate this surface layer with numerous small, fibrous roots. Cultural materials include occasional whole and fragmented mud ark shell. pH values are slightly alkaline (7.5).
II	Extends across the entire square with a maximum thickness of 40cm and a maximum depth of 42cm below the surface. It comprises a dense shell layer across its upper half surrounded by a more consolidated fine subrounded and poorly-sorted sandy matrix which is a dark reddish brown (5.0YR-3/2) in colour. The soils are very organic containing abundant shell materials, stone artefacts, complex root systems and fibrous rootlets as well as occasional large pieces of marine mammal and marine reptile bone. Shell and bone cultural material appear to be well-preserved. pH values are slightly alkaline to alkaline (7.5–8.0).
III	Extends across the entire square with a maximum thickness of 23cm and a maximum depth of 55cm below the surface, tapering in thickness towards the northwest corner of the excavation. The unit continues to bedrock in the eastern third of Square D. The sandy matrix has decreased to a loosely consolidated consistency in this layer, while the colour has changed slightly to a very dark grey (5.0YR-3/1) grading to dark reddish brown (5.0YR-3/2). The matrix is defined by generally less abundant quantities of degraded shell material and some small roots. This layer appears to slope towards the northwest section. pH values are slightly alkaline to alkaline (7.5–8.0).
IV	Extends across the entire square and contacts bedrock over most of this area with the exception of the eastern third of Square D where SUIII continues to bedrock and the small areas of SUV (see below). The unit has a maximum thickness of 27cm and a maximum depth of 75cm below the surface where it infills crevices in the bedrock. Distinctly different from the previous units, this layer is moist and well-consolidated. The matrix appears a dark reddish brown to black colour (5.0YR-3/2 to 5.0YR-2.5/1), and contains many coarse sediments. The layer also contains numerous fragments of degraded microgranite, probably derived from *in situ* weathering of the underlying bedrock. Occasional heavily weathered shell fragments and stone artefacts were noted, while the majority of the excavation has been exposed to bedrock at this depth. pH values for this SU remain slightly alkaline to alkaline (7.5–8.0).
V	Discrete thin layer of shell fragments and small whole shells of various taxa in a very dark grey (7.5YR-3/1) sandy matrix resting on bedrock encountered in the two extreme western corners in the excavation pit (southwest corner of Square A and northwest corner of Square B). The layer represents numerous species of shell and was identified in the field as chenier (i.e. natural marine shell deposited by water action). pH values remain alkaline (8.0).

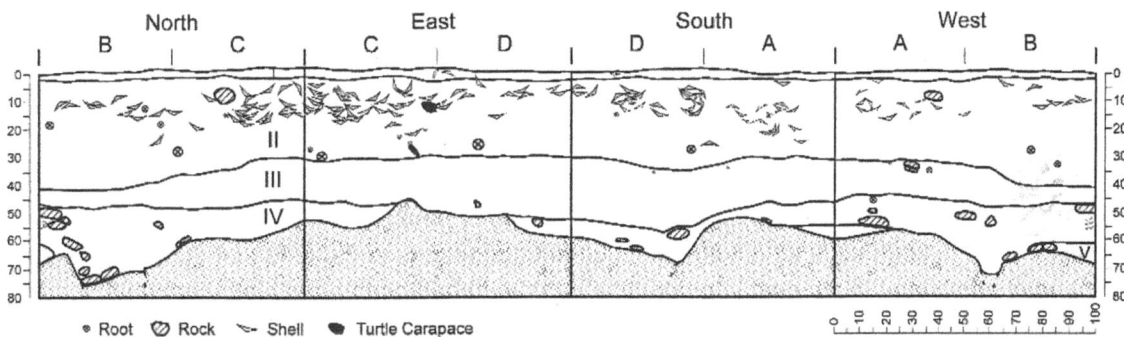

Figure 7.6 Stratigraphic section, Mort Creek Site Complex, Squares A–D.

and Wk-3939b) derived from 18.5–28.3cm below ground surface at The Granites were rejected as they contained recrystallised material.

It has been argued elsewhere that The Granites XU11C date of 3,065 cal BP (Wk-3940) related to the surface of a buried chenier ridge while The Granites XU11M determination of 2,335 cal BP (Wk-3941) dates a veneer of midden material lying directly on top of the chenier (Carter et al. 1999; Lilley et al. 1999). Although the dated shell specimens were excavated as part of a single unit, they were separated into cultural and natural specimens on the basis of colour staining and the colour and texture of the matrix adhering to the shell specimens (Lilley et al. 1996:39). The mixed status of XU11 is evident in the range of shellfish taxa recovered (n=17), including very small individuals and taxa such as *Bembicium auratum* and *Bendeva hanleyi* which are unlikely to be deliberately targetted for collection (Carter et al. 1999:93, 99–102). Like the shell from the White Patch chenier, it was argued that The Granites chenier was characterised by pink-tinged shell and clean yellow sand, whereas shell from the midden deposit was defined by a lack of pink colouration of the shell and by the fine, dark, organic sediment adhering to it (Lilley et al. 1999). The criteria employed for separating cultural from natural shell specimens is called into question by results from Squares A–D, where unambiguously cultural shell and bone materials are stained a dark reddish brown (5.0YR-3/2) colour, probably associated with local sediment supply relating to weathering of the microgranite bedrock than natural versus cultural formation processes. A second set of problems in accepting the accuracy of the Wk-3940 determination is raised by the sample composition. Wk-3940 consisted of a 66.7g marine shell sample comprising 84 fragments of shell from 17 taxa. Oyster comprised 22.2g of the sample. Owing to the fact that not all 84 fragments in the sample were screened for recrystallisation the apparently large age difference between the samples in XU11 might be attributed to some components of the dated sample being recrystallised, rather than a significant time lapse between cessation of chenier formation and Aboriginal occupation of this area of the site. This caveat aside, the date is broadly consistent with indirect dates for chenier formation from Squares A–D (see below).

The two White Patch determinations (Wk-3942 and Wk-3943) date the chenier deposit southwest of The Granites, suggesting that it was forming while the lower Granites midden was being deposited on the surface of the chenier possibly formed centuries earlier. The dates obtained from A7 indicate formation of this deposit between c.2,400–2,800 cal BP. The apparent inversion of the date from XU9 (Wk-3938) may simply indicate rapid formation of the deposit as all three determinations overlap at the two sigma calibrated age-range. The calibrated age-ranges of the three dates from the shell layer in SUII of Squares A–D (Wk-7458, Wk-7836 and Wk-6987) overlap at one standard deviation indicating an extremely rapid accumulation of this unit. Thus the radiocarbon dates suggest an overlap in the formation of cultural and natural shell deposits in the study area. The termination date provided by Wk-7458, while not from the actual surface of the deposit, is from the middle of SUII indicating that this part of the site was probably abandoned not long after c.1,850 cal BP. The dates are thus in sequence overall, suggesting first occupation shortly before c.3,300 cal BP and abandonment after 1,900 years ago.

Two periods of chenier-building are dated at the site. The earliest episode ceases around 3,310–3,065 cal BP as defined by the contact point between cultural and natural deposits at The Granites (see caution above) and the basal dates for sediments overlying the chenier units in Squares A–D. A later episode of chenier formation to the south and west occurred between 2,353–2,059 cal BP, as evidenced by the White Patch deposits and a dated *in situ* chenier layer mid-way down the A7 profile. A lower chenier deposit identified in XU13–14 of A7 (Carter et al. 1999:94) may coincide with the earlier episode of chenier deposition mentioned above. These dates broadly coincide with major periods of chenier formation identified at 3,550 BP and 2,500 BP at Broad Sound, 250km to the north, where they have been linked to phases of decreased sediment supply (Cook and Mayo 1977; Cook and Polach 1973).

Stratigraphic integrity and disturbance

Several lines of evidence suggest that the deposit exhibits a high degree of stratigraphic integrity. The virtually continuous dense layer of *A. trapezia* valves across the four excavation squares c.10–20cm below ground surface caps the underlying sediments. The presence of bedrock at the base of the deposit demarcates the maximum extent of cultural materials. No burrows or voids were encountered during excavation. The radiocarbon date sequence shows a regular age-depth relationship. In addition, several conjoining *A. trapezia* valves were noted during excavation of SUII in both Squares A and D. There is also a predictable shell decay profile with highly weathered specimens recovered from the base of the deposit and relatively well-preserved specimens from the upper deposit. The major source of post-depositional disturbance appears to be the presence of numerous roots throughout the deposit. Abundant fibrous roots occur to a depth of c.40cm with a zone of larger roots between 20–30cm. Occasional larger roots occur to bedrock (Fig. 7.12).

Table 7.2 Radiocarbon dates from the Mort Creek Site Complex (see Appendix 1 for full radiometric data for each determination).

SQUARE	XU	DEPTH (cm)	LAB. NO.	SAMPLE	δ^{13}C (‰)	^{14}C AGE	CALIBRATED AGE/S
A7	4	18–20.2	Wk-5602	*A. trapezia*	-0.3±0.2	2880±50	2769(2688)2356
A7	6	22.6–26.7	Wk-3937	*A. trapezia*	0.1±0.2	2930±60	2845(2712)2446
A7	9	32.4–37	Wk-3938	*A. trapezia*	0.1±0.2	2720±60	2694(2353)2195
Granites	11M	45.5–52.1	Wk-3941	*A. trapezia*	-0.2±0.2	2680±60	2657(2335)2143
Granites	11C	45.5–52.1	Wk-3940	mixed shell[a]	0.7±0.2	3260±70	3320(3065)2823
WP	4	12.8–18.4	Wk-3942	*A. trapezia*	0.6±0.2	2440±80	2314(2059)1826
WP	10	37.6–44.8	Wk-3943	*A. trapezia*	-0.5±0.2	2570±60	2387(2269)2002
C	6	11.3–15.8	Wk-7458	charcoal	-26.5±0.2	1970±80	2057(1875)1633
C	6	11.3–15.8	Wk-7836	*A. trapezia*	-1.4±0.2	2320±50	2108(1915)1730
C	7	15.8–18.1	Wk-6987	*A. trapezia*	-1.5±0.2	2260±50	2026(1855)1682
C	18	53.6–56.4	Wk-6988	*A. trapezia*	-1.1±0.2	3380±90	3462(3235)2928
B	19–20	52.8–59.8	Wk-6986	*A. trapezia*	-1.6±0.2	3430±140	3633(3310)2870

a Mixed shell consisting of *Saccostrea* sp., *Polynices* sp., *Nerita chamaeleon*, *Placamen calophyllum*, *Fragum hemicardium*, *Gafrarium australe*, *Cymatium* sp., *Corbula* sp., *Antigona chemnitzii*, *Trisidos tortuosa*, *Tapes dorsatus*, *Meropesta* sp., *Pinctada* sp., *Trichomya hirsutus*, *Bembicium auratum*, *Calthalotia arruensis* and *A. trapezia*.

Table 7.3 Identified *A. trapezia* conjoin sets, Mort Creek Site Complex, Square C.

CONJOIN SET	XU		MIN. SEPARATION (cm)	MAX. SEPARATION (cm)	MID-POINT (cm)	± (cm)
	L	R				
Set 1	5	5	0	3.02	1.51	1.51
Set 2	2	3	0	5.58	2.79	2.79
Set 3	6	6	0	4.48	2.24	2.24
Set 4	6	6	0	4.48	2.24	2.24
Set 5	3	5	2.74	6.62	4.68	1.94
Set 6	4	4	0	2.74	1.37	1.37
Set 7	3	6	3.02	11.10	7.06	4.04
Set 8	6	6	0	4.48	2.24	2.24
Set 9	6	6	0	4.48	2.24	2.24
Set 10	4	3	0	3.60	1.80	1.80

Conjoin analysis of the *A. trapezia* assemblage shows that at least the shell layer toward the top of the deposit has excellent integrity. Out of a total dataset of 595 measured intact and broken valves, 470 were discarded from consideration in the conjoin analysis owing to an absence of hinge length or valve length and width, indicating the presence of valve damage (especially marginal damage). This left 125 relatively intact valves for consideration in the conjoin analysis, using the methods described in Chapter 5. A total minimum number of 10 *A. trapezia* conjoins was identified from Square C. Most pairs (n=7) were separated by 5cm or less and nine were separated by 10cm or less. Only one conjoin was separated by over 10cm (Fig. 7.7, Table 7.3). The shell lens therefore appears to exhibit a high degree of stratigraphic integrity, especially when it is considered that the maximum separation measurements of conjoins tends to overestimate the actual separation distance between valves in many instances (see Chapter 5). Although the distribution of identified conjoins largely reflects the vertical distribution of *A. trapezia* in the deposit, the 10 bivalve conjoin sets identified between 0–16cm below ground surface support the impression gained from the radiocarbon chronology that the shell layer accumulated during episode(s) of rapid deposition.

The *A. trapezia* assemblage is generally in poor condition, with a high ratio of broken to intact valves (11:1) and high rates of fragmentation, suggesting that the shells have been exposed to sustained heating after initial discard (see Chapter 5) and/or mechanical damage from treadage. *A. trapezia* fragmentation rates are relatively high throughout the sequence, with an average of 129.3 NISP/100g (compare, for example, with the average for the Seven Mile Creek Mound of 24.5

XU

| 2 |
| 3 |
| 4 |
| 5 |
| 6 |
| 7 |
| 8 |
| 9 |
| 10 |
| 11 |
| 12 |
| 13 |
| 14 |
| 15 |
| 16 |
| 17 |
| 18 |
| 19 |

cm 0

20

40

60

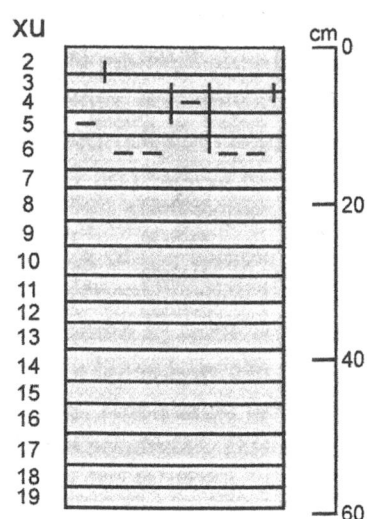

Figure 7.7 Distribution of identified *A. trapezia* valve-pairs (n=10), Mort Creek Site Complex, Square C. Line termination points indicate the vertical mid-points of the excavation units from which conjoining valves were located. Short horizontal lines indicate valve-pairs identified within excavation units. Not to scale on the horizontal axis. See Chapter 5 for details of methods.

NISP/100g) with peaks of 428.6 and 193.9 at the base and top of the deposit respectively. The high rates of fragmentation at the base of the deposit are attributed to an expected decay profile, while those at the top may be related to trampling. Zones of relatively low fragmentation are associated with the centre of the shell layer of SUII, where the matrix seems to have protected valves from extreme mechanical and chemical destruction.

Laboratory methods

Owing to the large quantity of excavated materials, detailed analysis of only a single square has been completed to date (Square C). Field observations and preliminary processing of the other three squares indicate a broad homogeneity in the contents of the excavated material. Results of the analysis of Square C are presented below, with reference to finds in other squares where relevant. Preliminary sorting of sieve residues from Squares C and D was conducted prior to wet sieving to remove marine fish vertebral components for DNA analysis and to prepare samples for radiocarbon dating (see Chapter 3 for a detailed discussion of the standard laboratory methods employed at all sites). Further summary results are available in Appendix 4.

Cultural materials

Invertebrate remains

Ten taxa of shellfish weighing 8,748.8g were identified in the Square C assemblage, consisting of four marine bivalves, two marine gastropods and four terrestrial gastropods (Table 7.4). The shell deposit is dominated by mud ark (*A. trapezia*) comprising 96.1% of the shell assemblage by weight (Fig. 7.8), followed by rock oyster (*Saccostrea glomerata*) (3.1%) (Fig. 7.9). The remaining eight taxa are relatively rare in the deposit, each contributing less than 1% of the shell assemblage by weight (Fig. 7.10). The assemblage exhibits relatively low diversity with a calculated Shannon-Weaver Function (H') of 0.272 and Simpson's Index of Diversity (1–D) of 0.119. Virtually all of the shell was recovered from the shell layer located across the top 20cm of the deposit, with occasional *A. trapezia* encountered to bedrock.

A similar range of taxa was recovered in the upper, unequivocally cultural, units of the nearby The Granites excavation. Carter (1997:89–90) found that *A. trapezia* comprised 85% of total MNI for The Granites excavation.

There is no significant change in the mean size of *A. trapezia* throughout the deposit as measured by five attributes (length, height, width, weight and hinge length) (e.g. length: $\chi^2=0.8055$, df=6, $p \le 0.05$) (Table 7.5). The mean length of *A. trapezia* does not fall below 38.4mm in any excavation unit which contains intact or broken valves, with a combined weighted mean calculated on 67 valves of 42.7±2 and a terminal mean of 43.4±7mm for XU2 (XU1 was the thin surface covering of turf).

Table 7.4 Presence/absence of shellfish identified in the Mort Creek Site Complex, Square C.

FAMILY	SPECIES	1	2	3	4	5	6	7	8	9	10	11	12	13	14	15	16	17	18	19	TOTAL (g)
								MARINE BIVALVIA													
Arcidae	*Anadara trapezia*	X	X	X	X	X	X	X	X	X	X	X	X	X	X	X	X	X	X		8408.9132
Mytilidae	*Trichomya hirsutus*						X	X	X											X	0.7151
Ostreidae	*Saccostrea glomerata*	X	X	X	X	X	X	X	X	X		X	X			X				X	275.5522
Pteriidae	*Pinctada albina sugillata*						X														0.3129
								MARINE GASTROPODA													
Batillariidae	*Pyrazus ebininus*					X	X	X	X												56.2353
Neritidae	*Nerita balteata*							X	X	X		X	X								4.6073
								TERRESTRIAL GASTROPODA													
Camaenidae	*Figuladra* sp.		X	X	X	X	X	X	X												2.1178
Camaenidae	*Trachiopsis mucosa*		X											X	X						0.1502
Pupillidae	*Pupoides pacificus*	X	X	X				X													0.1476
Subulinidae	*Eremopeas tuckeri*		X	X																	0.0429

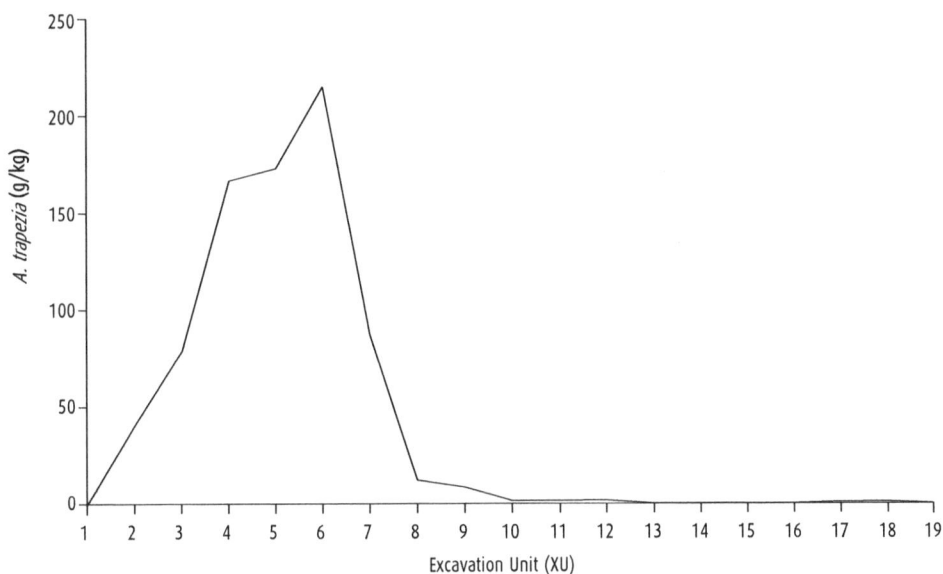

Figure 7.8 Abundance of mud ark (*A. trapezia*).

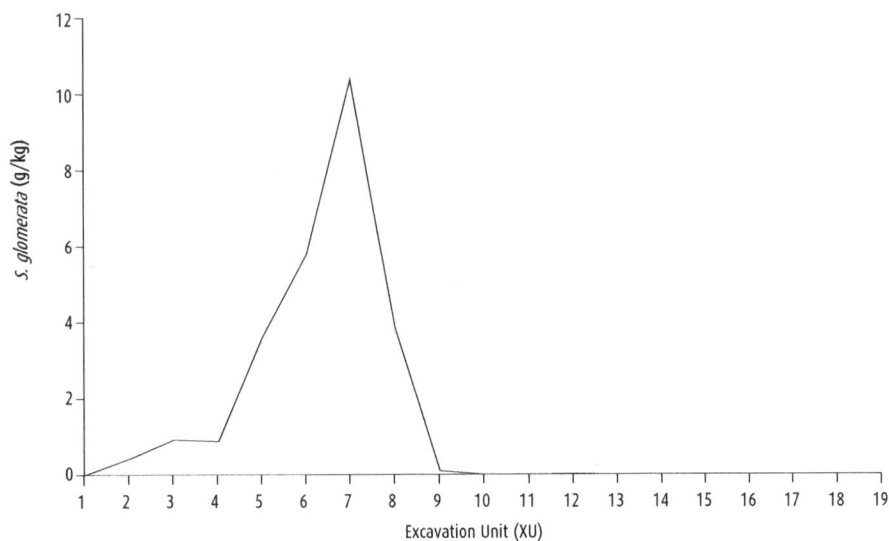

Figure 7.9 Abundance of oyster (*S. glomerata*).

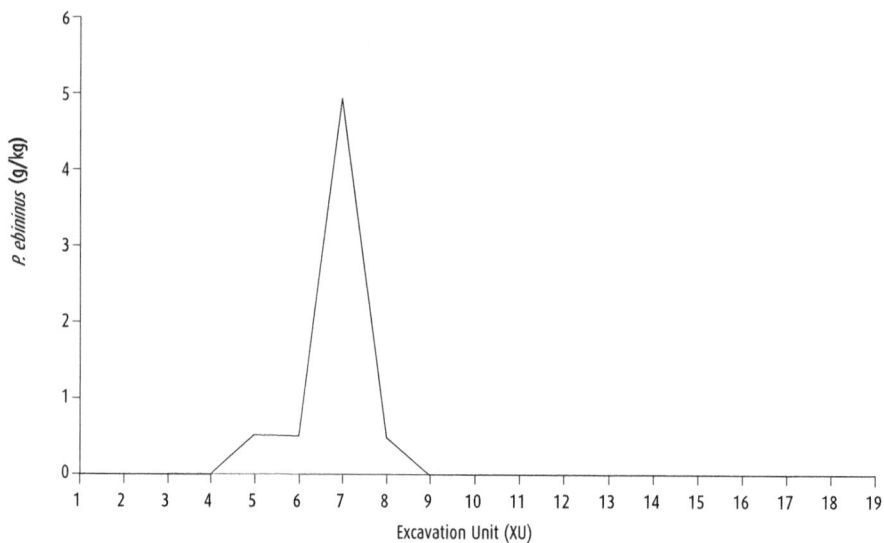

Figure 7.10 Abundance of whelk (*P. ebininus*).

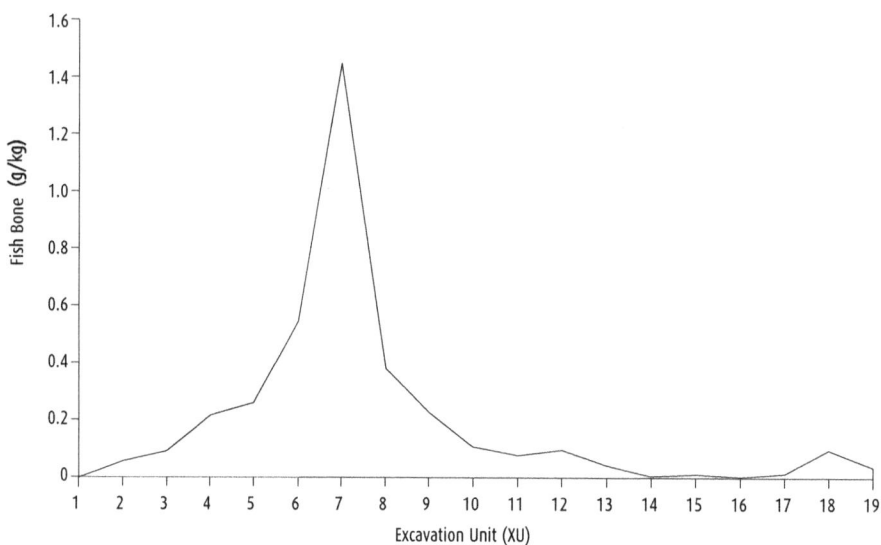

Figure 7.11 Abundance of fish bone.

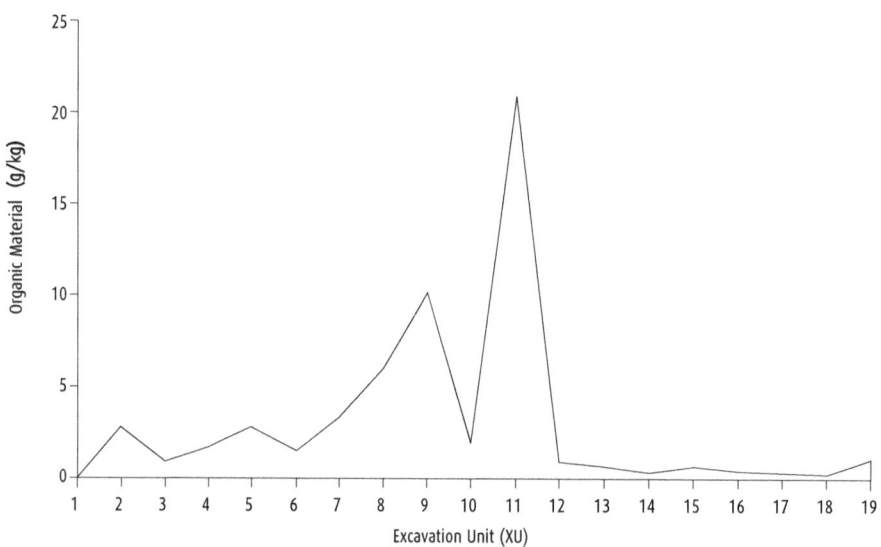

Figure 7.12 Abundance of organic material.

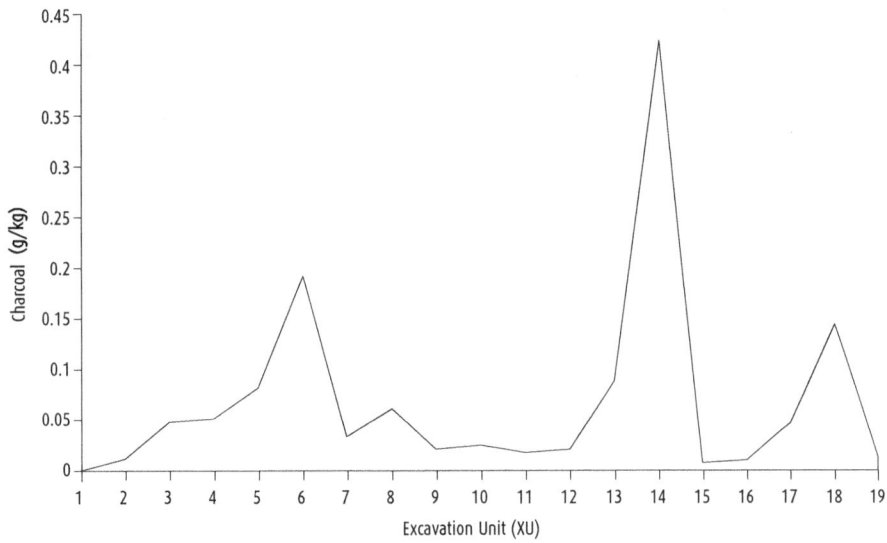

Figure 7.13 Abundance of charcoal.

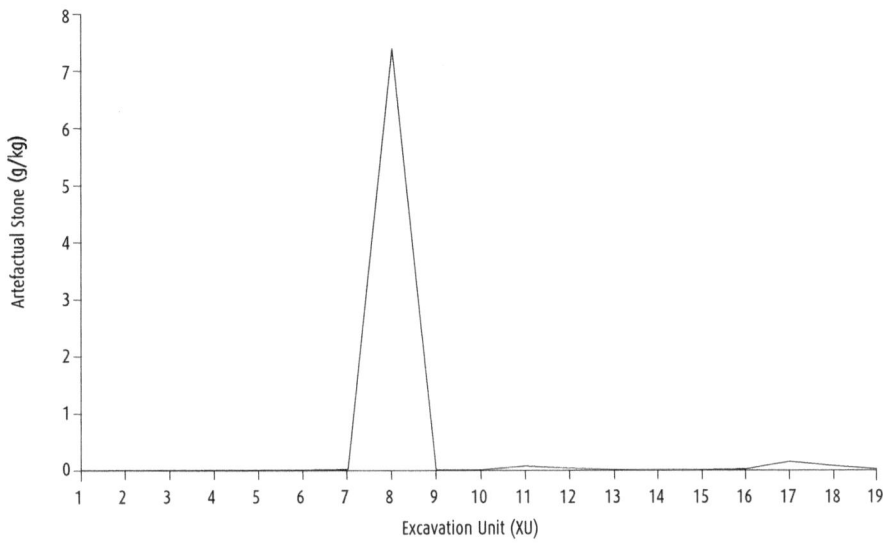

Figure 7.14 Abundance of artefactual stone.

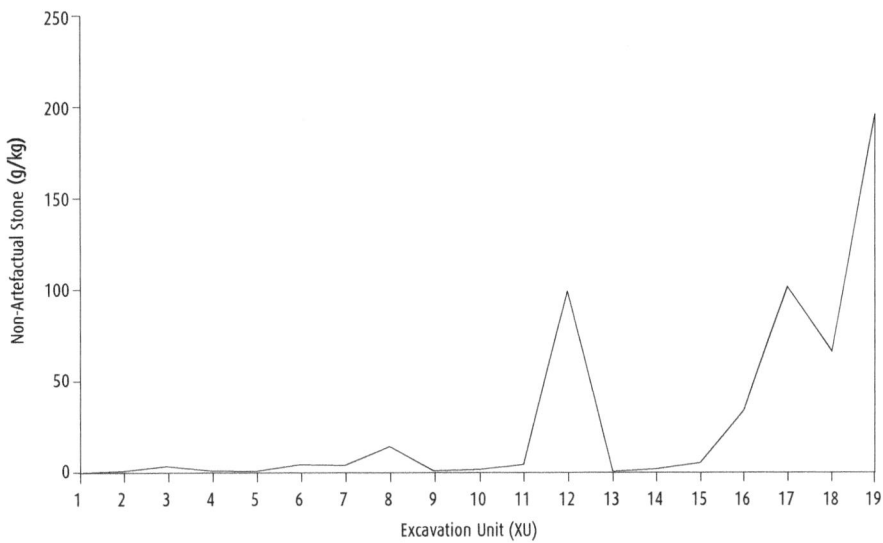

Figure 7.15 Abundance of non-artefactual stone.

Table 7.5 Metrical data for intact and broken (with umbo) *A. trapezia* valves from the Mort Creek Site Complex, Square C.

XU	MEAN LENGTH			MEAN HEIGHT			MEAN WIDTH			MEAN WEIGHT			MEAN HINGE		
	n	mm	±	n	mm	±	n	mm	±	n	g	±	n	mm	±
2	9	43.4	7.0	17	36.0	4.3	33	14.2	1.7	8	13.8	5.3	21	27.3	3.5
3	13	42.1	4.4	21	37.2	4.3	52	14.9	2.1	12	13.0	3.6	39	28.7	2.8
4	11	43.6	3.6	21	37.7	3.6	108	15.3	2.3	8	13.2	4.0	68	28.6	2.8
5	3	45.6	6.4	1	41.0	6.0	114	14.8	2.2	3	18.4	6.3	60	29.0	3.4
6	19	45.2	3.8	44	38.2	3.4	195	15.3	1.9	13	15.7	3.6	141	28.4	2.4
7	8	43.6	6.5	15	38.6	4.7	46	14.8	1.5	3	15.3	2.8	32	29.0	2.5
8	4	38.4	5.8	5	33.1	2.0	19	14.1	2.2	3	7.1	1.0	13	26.8	2.8
9	0	0	0	0	0	0	4	12.5	1.9	0	0	0	2	21.5	0.7
10	0	0	0	0	0	0	1	10.0	0	0	0	0	1	19.0	0
11	0	0	0	0	0	0	1	12.2	0	0	0	0	1	23.0	0
12	0	0	0	0	0	0	1	15.1	0	0	0	0	0	0	0
Total	67	42.7	2.0	134	35.9	1.3	574	14.5	0.7	50	9.0	0.8	378	23.6	0.6

Vertebrate remains

Fish bone is present throughout the cultural deposit, consisting of 1,635 pieces of bone totalling 38.4g and a NISP of 34 (Table 7.6, Fig. 7.11). A total MNI of 21 was calculated by summing the MNI for each excavation unit. The weight of bone identified to taxon was 4g, giving an identification rate of 10.5% (Table 7.6). Identified taxa in descending order of abundance include Sparidae, flathead (Platycephalidae), whiting (Sillaginidae) and catfish (Ariidae) (Table 7.7). Size-classing showed that 71% of all vertebrae have a centrum diameter of 3mm or less. These represent very small fin fish. Some larger fish are represented by single vertebrae recovered from XU5 and XU9. Three Sparidae otoliths from XU7 range in size from 12.4–21.5mm which represent very large fish. Comparative Sparidae otoliths from bream (*Acathopagrus australis*) with total lengths of 305mm and 365mm had otoliths measuring 9.8mm and 11.1mm in length respectively. In addition, a Platycephalidae dentary fragment from XU7 measured 6.9mm at the symphysis, much larger than a comparative Platycephalidae dentary with a symphysis length of 4.3mm for a 418mm fish (see Vale 2002, 2004 for further details). There is a discordance between the presence of at least several large individual fish represented by cranial elements (dentary and otoliths) and the very few vertebrae recovered above 3mm in centrum diameter. This pattern may relate to differential representation of skeletal elements caused by butchering practices on large fish, with post-cranial elements discarded elsewhere. Although fish bone occurs in every unit, it is most abundant in units where shell is also abundant (compare Figs 7.8 and 7.11).

Of the 16 vertebral samples subject to DNA analysis, only one returned a positive fish-like polymerase chain reaction (PCR) product, although this extract did not produce a product when sequenced (Hlinka et al. 2002). Taphonomic factors are thought to be responsible for the low amplification success rate. In particular, microscopic examination revealed that all samples had been penetrated by plant roots and showed signs of discolouration consistent with burning (see Hlinka et al. 2002 for further details).

Sixty-eight pieces of bone weighing 26.2g could not be assigned to a fish skeletal element. The small size of these specimens and the lack of diagnostic attributes generally prevented identification to taxon. However, positive identification of bone elements at a similar depth towards the base of SUII in adjacent squares as dugong (*D. dugon*) and turtle, probably loggerhead (*C. caretta*), suggest that many of the unidentified small bone fragments in Square C derive from these taxa. Dugong remains were recovered from Squares A–C in association with the lower half of the shell layer in SUII between 13.6–21.3cm below ground surface. Although the turtle carapace fragments recovered from between 24.5–35.2cm in Squares B and D are associated with occasional *A. trapezia* valves, they are clearly located below the major shell layer.

Table 7.6 Fish bone abundance, Mort Creek Site Complex, Square C.

XU	NUMBER SPECIMENS	TOTAL WEIGHT (g)	NISP	WEIGHT NISP (g)	MNI	% IDENTIFIED BY WEIGHT
1	0	0	0	0	0	0
2	36	1.0086	0	0	0	0
3	64	0.9123	0	0	0	0
4	97	2.1781	2	0.1433	1	6.93
5	90	2.3407	5	0.2710	4	11.44
6	272	7.0749	7	0.1520	2	2.35
7	313	10.7277	7	2.6287	5	24.45
8	332	6.7944	8	0.1633	5	2.76
9	82	1.9631	1	0.3738	1	20.65
10	74	1.1594	0	0	0	0
11	73	0.9109	2	0.0508	1	8.79
12	69	0.8443	0	0	0	0
13	28	0.4571	1	0.0245	1	7.32
14	8	0.0957	0	0	0	0
15	13	0.1390	0	0	0	0
16	5	0.0587	0	0	0	0
17	18	0.2461	0	0	0	0
18	49	1.1300	1	0.2353	1	36.13
19	12	0.3115	0	0	0	0
Total	1635	38.3525	34	4.0427	21	10.50

Stone artefacts

Stone artefacts are distributed throughout the cultural deposit between XU7–19 (Fig. 7.14). A total of 13 stone artefacts weighing 135.7g was identified in Square C (Table 7.8). Two of these items were plotted *in situ* during excavation with the rest recovered from the sieve residue. Virtually the entire assemblage is manufactured on quartz (n=10) with the remainder microgranite (n=3). The microgranite artefacts include a hammerstone exhibiting impact-pitting and flaked pieces. All raw materials are available in the immediate vicinity of the site or in the adjacent Bray Hills. Most artefacts are extremely small, with an average maximum dimension of 15.3mm and average weight of 10.4g. Two of the largest artefacts recovered from the excavation, a quartz core and flaked piece from Square B, were located immediately above bedrock

Francis (1999) analysed a single small (6.3g) quartzite flaked piece (FS60) from Square B, XU21, for residues using incident-light microscopy (see Loy 1994 for a discussion of techniques). In addition to rootlets, sand grains and spores attributed to post-depositional processes, Francis (1999) described a concentration of white crystalline raphides along the edge of the tool suggesting that it was used to scrap or cut plant tissue containing raphides.

Table 7.7 Fish bone taxonomic representation, Mort Creek Site Complex, Square C.

TAXON	NISP	MNI	WEIGHT (g)
Sparidae	19	10	2.95
Platycephalidae	8	6	0.98
Sillaginidae	4	4	0.09
Ariidae	2	1	0.02
Total	33	21	4.04

Table 7.8 Stone artefacts from the Mort Creek Site Complex, Square C.

RAW MATERIAL	ARTEFACT TYPE	NUMBER	WEIGHT (g)	XUs
Quartz	Flaked Piece	10	2.2	7–8, 11–12, 16–19
Microgranite	Flaked Piece	2	1.9	17–18
Microgranite	Hammerstone	1	131.7	8
Total	–	13	135.7	–

Other remains

A range of other materials was recovered from the site. Small nodules of red ochre totalling 1.5g were recovered from XU12–15. Small pieces of pumice totalling 32.2g were recovered from the bottom half of the deposit (XU13, 20–31). Charcoal, totalling 15.8g, is represented in very small quantities throughout the deposit (Fig. 7.13). The presence of charcoal in culturally-sterile sediments below the shell deposit suggests that some of the other small quantities of charcoal represented in the assemblage may derive from natural rather than cultural deposition.

Discussion

The Mort Creek Site Complex accumulated over a period of about 1,400 years between c.3,300–1,900 cal BP. The first 1,000 years of site use is characterised by intermittent low intensity occupation with subsistence focussing on fin fish with occasional hunting of dugong and turtle and incidental shellfishing. Results support Carter's (1997:107) conclusion that ephemeral occupation of the area, including chenier ridges, occurred while the local landscape was in a significant state of flux. Around 2,000 BP the intensity of use of the site dramatically increased, with a layer of shell material dominated by *A. trapezia* and fish remains deposited over a large area between 1,800 and 2,000 BP. Current evidence points to the site being little used since that time.

Like the nearby Seven Mile Creek Mound, reduction of use or abandonment of the Mort Creek Site Complex does not appear to be linked to changes in resource availability, caused by either environmental change or overexploitation. The range of estuarine taxa represented in the assemblage and the lack of change in *A. trapezia* size throughout the sequence suggest the presence of a stable estuary that was not suffering from overexploitation. These data therefore lend support to the idea that site abandonment was linked to alterations in regional settlement strategies that were not exclusively based on coastal resources.

The Mort Creek Site Complex is one of the most scientifically significant sites investigated on the southern Curtis Coast during the course of this study. In addition to its antiquity, it is the only site which has evidence for the procurement of large marine animals such as dugong and turtle. Rodds Harbour currently supports the largest dugong population along the Curtis Coast (QDEH 1994:66), with an estimated minimum population of 300±95 (Marsh and Saalfeld 1989). Minnegal (1982:21) notes that dugong are the largest animals hunted by coastal Aboriginal populations in northern Australia, with adult weights up to 420kg (Heinsohn 1991). Despite several detailed eighteenth and nineteenth century ethnohistoric accounts of dugong and turtle hunting in southeast Queensland (e.g. Backhouse 1843; Colliver and Woolston 1978; Fairholme 1856; Flinders 1814; MacGillivray 1852; Petrie 1904:24–5, 65–9, 82–3; Watkins 1891), archaeological remains are relatively rare and very recent. At the site of St Helena Island in Moreton Bay, dugong bone was recovered from the top 15cm of the deposit, which Alfredson (1984:73) dates to the mid-to-late nineteenth century. On Moreton Island, several bone fragments have been tentatively identified as dugong at the Little Sandhills site, although the radiocarbon date indicates a probable post-contact chronology (Robins 1983, 1984). Similarly at Wallen Wallen Creek on Stradbroke Island, dugong remains were identified in the late Holocene deposits, but the chronology of the remains is unclear from the reported data (Neal and Stock 1986). Walters (1979:47) identified nine specimens of dugong bone from Trench B at Toulkerrie on Moreton Island, although no provenance details are available and the entire analysed assemblage appears to date from the last 400 years (see Ulm 2002a:86–7). The secure dating of dugong remains to before 2,000 BP provides some of the earliest, if not the earliest, evidence for dugong procurement activities in southeast Queensland.

Data from the Mort Creek Site Complex provide evidence for a well-developed marine economy including fin fish, shellfish, dugong and turtle before 2,000 BP and possibly up to 3,300

BP. A similar antiquity for these activities is indicated by other recent research (see Ulm 2002a for an overview) and further undermines the model proposed by Walters (1986, 1989, 1992a, 1992b) that marine fishing was only incorporated as a regular feature of subsistence-settlement systems in southeast Queensland in the last 2,000 years. For the Waddy Point Rockshelter 1 on Fraser Island, McNiven et al. (2002:15) has recently demonstrated that fish bone is only consistently represented in the faunal assemblage after shell densities exceed 9–10g/kg of deposit. He (1991a:21; McNiven et al. 2002:15) has suggested that the survival of fish bone in southeast Queensland deposits may be correlated with the occurrence of shell, with the presence of shellfish remains providing a protective matrix and altering the chemical properties of the sedimentary matrix towards conditions conducive to fish bone preservation. At the Mort Creek Site Complex fish bone is represented throughout the deposit, including excavation units with shell densities well below 1g/kg of deposit. However, there is generally a positive relationship between the abundance of fish bone and shellfish remains, particularly in the upper deposit. As McNiven et al. (2002:15) warned, whether this indicates patterns of subsistence or patterns of bone survivorship remains to be determined, although the presence of fish bone in the lower deposits of the Mort Creek Site Complex indicates that fish bone may survive under certain conditions without an accompanying shell matrix.

Results from the Mort Creek Site Complex indicate that care must be taken in excavation, analysis and interpretation of the archaeology of the area owing to the co-occurrence of cultural and natural shell deposits. Interpretations would greatly benefit from detailed geomorphological investigation of the area. Further dating of terminal deposits would be useful to confirm the time of abandonment. Successful dating of chenier material encountered along the basal western margin of Squares A–D would confirm the antiquity of this unit and help to better define periods of chenier formation in the area. As Carter (1997) noted, further application of foraminifers analysis to all deposits would also benefit the discrimination of cultural from natural deposits. Ultimately, excavation of a larger area across the ridge close to The Granites and Squares A–D would help resolve stratigraphic relationships between natural and cultural deposits and identify further variability in the structure of the site.

Summary

The Mort Creek Site Complex was occupied for at least 1,400 years between c.3,300–1,900 cal BP. An initial period of low intensity occupation was followed by a brief more intensive use of the site around 2,000 years ago. The site appears to have been abandoned soon after this period of rapid deposition. Like the Seven Mile Creek Mound, the evidence from the Mort Creek Site Complex suggests activities and site functions that have no obvious parallels with other sites investigated on the southern Curtis Coast dating to the last 1,500 years. The occupation and subsequent abandonment of these two sites appears to be associated with systems of land-use that changed in the last 2,000 years. These patterns will be elucidated through the description of investigations undertaken at other sites.

Pancake Creek Site Complex

Introduction

This chapter reports the results of archaeological excavations undertaken at the Pancake Creek Site Complex. The large size, linear structure and recent chronology of this site are features common to other major archaeological deposits identified on the lower margins of the main estuaries in the region. Similar sites investigated in this study include the Ironbark Site Complex on Middle Creek (Chapter 9) and Eurimbula Site 1 on Round Hill Creek (Chapter 12). The excavations and analyses reported in this chapter demonstrate that although the cultural deposits at the Pancake Creek Site Complex are of a relatively low density, the large extent and shallow chronology of the deposits indicate that a large quantity of material was discarded over a relatively short period of time.

Site description and setting

The Pancake Creek Site Complex is a large multiple-component linear shell midden located on the north bank of Pancake Creek, on the south coast of Rodds Peninsula (Latitude: 24°02′25″S; Longitude: 151°43′00″E) (Fig. 8.1). Extensive shell midden is exposed as a thin, discontinuous layer between 20–30cm below the ground surface in two low (c.1–3m high) erosion banks which border Pancake Creek and are separated by a small (c.100m long) unnamed mangrove-lined tidal inlet, oriented roughly north-south (Fig. 8.2). Shell is visible on or near the frontal beach ridge, often in crab burrow spoil, east to a large area of tidal flats southwest of Pancake Point. The present configuration of Pancake Point and its immediate hinterland appears to be very recent, with southeasterly progradation of the area evident in aerial photographs dating to 1948. Abundant shell is also evident at the base of the erosion bank and in the adjacent intertidal zone. Large trees in growth position on the beach up to 15m south of the present erosion bank reflect significant bank recession in this area. Densities of shell appear to decrease dramatically in the eastern section of the site complex, although this may simply be a function of differential visibility as there is no erosion bank in this part of the site and the height of the frontal dune is much lower. The inland extent of the shell was not determined but surface shell was observed up to 100m inland from the erosion bank and varies widely along its c.1.5km creek frontage. The entire area of shell exposure is referred to as the Pancake Creek Site Complex.

Vegetation over the area of the shell deposits is dominated by red bloodwood (*Corymbia intermedia*) tall woodland with shrub/heath mid-stratum and cloudy teatree (*Melaleuca dealbata*) open forest (Fig. 8.2). The understorey is generally open, with occasional introduced weeds including prickly pear (*Opuntia stricta*). Mangrove vegetation in Pancake Creek is dominated by dense stands of spotted mangroves (*Rhizopora stylosa*), with patches of grey mangroves (*Avicennia marina*) and yellow mangroves (*Ceriops tagal*) in the minor estuaries/inlets in the middle and at either end of the site area (Olsen 1980a:18). A bank of live hard coral, unique in the region, occurs in Pancake Creek towards Bustard Head (Olsen 1980a:18). Extensive sandy to muddy intertidal flats adjoin the site, supporting populations of hercules club whelk (*Pyrazus ebininus*) and oyster (*Saccostrea glomerata*) attached to mangrove substrates. Small rock/shell debris beds occur at the top margin of the subtidal zone, exhibiting a wide range of shellfish taxa including hairy mussel (*Trichomya hirsutus*). At least some of the shell present in these debris beds is likely to have originated from the adjacent eroding middens.

Material evidence for non-Indigenous use of the site area is limited. Although shipping on the southern Curtis Coast increased throughout the 1850s, the establishment of a permanent settlement at Gladstone saw Pancake Creek used as a regular anchorage only in the late 1860s with the construction of the Bustard Head Lightstation. Permanent navigation beacons were erected in the creek in 1883, reflecting the increased use of the area as a safe anchorage. By the early 1900s, oyster leases were being worked at Bustard Head and *bêche-de-mer* harvested in Pancake Creek (Buchannan 1999:76). In 1899, the Kettlewell family built a slab hut at Pancake Point immediately east of the site complex (Buchannan 1999:76), marking the first recorded permanent non-Indigenous presence on Rodds Peninsula. In 1993 Christine Burke noted a squat in the same location as scattered surface shell material at the site. This structure has since been removed by the Queensland Parks and Wildlife Service (Denis Dray, Queensland Parks and Wildlife Service, pers. comm., 2000). It is unclear whether the structure recorded by Burke (1993) is the 'Kettlewell' hut referred to by Buchannan (1999), as squats constructed by professional and recreational fishing people are common in the area. A designated Queensland Parks and Wildlife Service (QPWS) camping area (accessible to the public only by boat) is currently situated towards the western end of the visible shell exposure.

This site was originally recorded by Burke (1993) as six separate sites (CC094–CC099). Burke noted shell deposits and occasional stone artefacts covering an area of at least 22,500m² and up to 50m inland. These original descriptions were conflated into five sites when registered by the Queensland Environmental Protection Agency (KE:A44–KE:A48). The site was briefly inspected during 1994 as part of the Gooreng Gooreng Cultural Heritage Project (Lilley et al. 1997). This initial inspection suggested that all of the apparently discrete exposures of shell recorded by Burke would more usefully be considered as parts of a single site complex. The site was subsequently registered as Queensland Museum Scientific Collection Number S865. Burke (1993:Table 17) identified part of the Pancake Creek Site Complex (registered as KE:A46) as one of only 10 sites of 'extremely high significance' documented for the 160km stretch of coast between Round Hill Head and Raglan Creek at Port Alma. This assessment was made on the basis of the site's Aboriginal significance, research potential (reflected in the presence of charcoal and depth of deposit), the large size and good condition of the site and the fact that it was atypical of middens found in the area as it contained a quite dense cultural layer (Burke 1993:Table 17). Burke (1993:Table 18, Table 19) also noted that parts of the site (registered as KE:A45 and KE:A46) were threatened by tidal bank erosion and recommended monitoring to establish the rate of erosion (Fig. 8.3). To date no such monitoring program has been implemented and it remains uncertain to what extent bank recession has impacted on the integrity of the site.

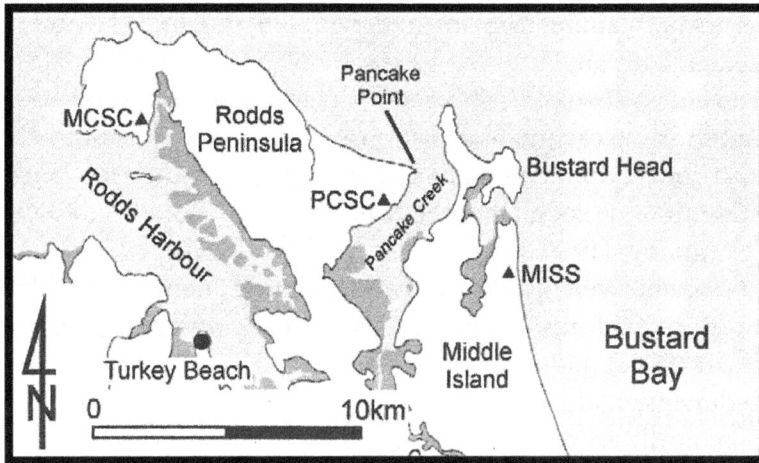

Figure 8.1 The Pancake Creek catchment area showing the location of the Pancake Creek Site Complex. Dark grey shading indicates the general extent of mangrove, saltflats and claypans.

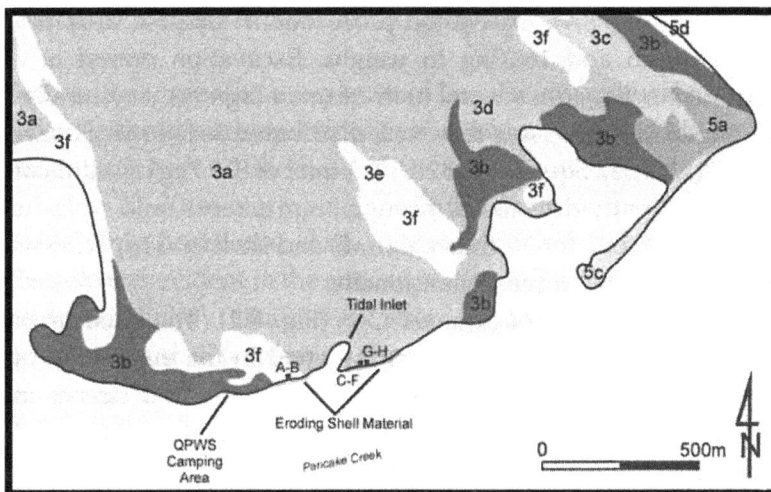

Figure 8.2 Detail of the site area showing the location of excavated squares and beach ridge vegetation units. 3a=*Corymbia intermedia* tall woodland with shrub/heath midstratum; 3b=*C. tessellaris, C. intermedia* tall to very tall woodland with *Livistona decipiens/Melaleuca dealbata* and shrub understorey; 3c=*L. decipiens/Melaleuca* tall forest/tall open forest with emergent *Eucalyptus tereticornis/C. intermedia*; 3d=*L. decipiens* open forest; 3e=*Acacia julifera/A. flavescens* tall shrubland; 3f=*M. dealbata* open forest; 5a=*C. tessellaris* woodland with understorey of low microphyll vine thicket; 5c=*Casuarina equisetifolia* low open woodland; 5d=Tall open shrubland (windsheared vegetation) (after QDEH 1997).

Excavation methods

Three areas were selected, coinciding with the densest exposures of shell observed in section in the erosion bank. A total of 8 × 50cm × 50cm pits (Squares A–H) were excavated to form 2 × 1m × 50cm trenches (Squares A–B and Squares G–H) and a single 1m² (Squares C–F). A total of 1,728.8kg of sediment was excavated at the site. All squares were excavated to 60–70cm below ground surface. Excavations were conducted between 5 November and 20 November 1998. The excavation squares were situated close to the visible erosion bank profile to facilitate reference to stratigraphic features during excavation.

Squares A–B were situated adjacent to the first major exposure of shell encountered to the east of the QPWS camping area. In this area, *in situ* shell was visible as a layer c.20cm below ground surface adjacent to a deflation area containing large quantities of shell, dominated by mud ark. The excavation grid was oriented at right angles to the erosion bank and situated c.190cm northwest of the edge of the bank. The erosion bank in this area is oriented northwest-southeast. Excavation proceeded in shallow, arbitrary excavation units averaging 3.67cm in depth and 12.73kg in weight. Excavation ceased at a maximum depth of 68.32cm below ground surface after several units of unambiguously culturally-sterile sediments had been removed. A total of 37 XUs was removed, distributed as follows: Square A (19 XUs), Square B (18 XUs). A total of 471.1kg of sediment was excavated. Excavated sediments were gently dry-sieved through 3mm screens onto a plastic tarpaulin located 10m northwest of the excavation to prevent contamination of underlying strata. Stone (n=4), charcoal (n=12) and shell (n=14) specimens encountered *in situ* during excavation were plotted three-dimensionally.

Squares C–F were located 250m northeast of Squares A–B (Fig. 8.2). There is evidence for extensive erosion in this area, with large tree roots protruding from the erosion bank and the stump of a large tree in growth position located on the beach 8.5m from the current erosion face. Oyster is the dominant taxon visible in both the *in situ* cultural layer and material eroded down the bank, with minor representation of mud ark and whelk. The excavation grid was situated 40–45cm northwest of the edge of the erosion bank (Fig. 8.4). Excavation proceeded in shallow, arbitrary excavation units averaging 4.72cm in depth and 16.03kg in weight. Excavation ceased at a maximum depth of 61.7cm below ground surface after several units of unambiguously culturally-sterile sediments had been removed. A total of 51 XUs was removed, distributed as follows: Square C (13 XUs), Square D (13 XUs), Square E (12 XUs), Square F (13 XUs). A total of 817.7kg of sediment was excavated. Excavated sediments were gently dry-sieved through 3mm screens onto a plastic tarpaulin located 5m west of the excavation. Stone (n=3), charcoal (n=1) and shell (n=1) specimens encountered *in situ* during excavation were plotted three-dimensionally.

Squares G–H were located c.20m northeast of Squares C–F (Fig. 8.2). The shell layer exposed in the erosion bank in this area is continuous with that targetted in the excavation of Squares C-F although the shell layer visible in section appears to be less dense and deeper in extent. A 1m × 0.5m excavation grid was oriented at right angles to the erosion bank and situated c.50–60cm northwest of the bank margin. The erosion face here is slightly slumped and undercut in this area. Excavation proceeded in shallow, arbitrary excavation units averaging 3.7cm in depth and 12.22kg in weight. Excavation ceased at a maximum depth of 67.56cm below ground surface after several units of unambiguously culturally-sterile sediments had been removed (Fig. 8.5). A total of 36 XUs was removed, distributed as follows: Square G (18 XUs), Square H (18 XUs). A total of 440kg of sediment was excavated. Excavated sediments were gently dry-sieved through 3mm screens onto a plastic tarpaulin located 5m east of the excavation. Charcoal (n=1) and shell (n=2) specimens encountered *in situ* during excavation were plotted three-dimensionally.

A layer of plastic sample bags was placed over the base of all excavated squares and c.50l of sterile yellow sand from the adjacent beach placed in the base of each excavation area. The remainder was backfilled with the sediments which passed through the 3mm sieve. A subsequent inspection of the site in early 2000 revealed that further bank recession has caused the erosion bank to migrate further north, actually exposing the area of the Squares C–F and G–H excavations in the erosion bank section.

Figure 8.3 Eroding bank in the vicinity of Squares G–H showing displaced trees and shell. Facing southwest.

Figure 8.4 General view of completed excavation at Squares C–F. Facing south.

Figure 8.5 Close-up view of concentration of *A. trapezia* valves encountered in Square H, XU9, at a depth of 26–30cm. These shells date to around 600 years ago and include a single valve conjoin. Facing southeast.

Cultural deposit and stratigraphy

Excavations revealed a sequence of low density cultural material, concentrated between 20–30cm below ground surface, deposited in beach ridge sands overlying culturally-sterile sediments continuing beyond the base of excavations at 60–70cm (Table 8.1). The concentration of shell between 20–30cm corresponds with the depth of the shell layer observed in the adjacent erosion bank. Only occasional shell was encountered below this zone, reinforcing the impression of discreteness evident in the erosion section. All sediments comprise grey-brown sands, with slightly acidic to slightly alkaline pH values throughout (6.5–7.5). Squares A–B exhibited only small quantities of shell (consisting almost entirely of oyster), despite the proximity of the excavation to a dense exposure of shell on the adjacent erosion bank. Large quantities of charcoal were recovered, particularly towards the base of the excavation. This material is thought to be largely of natural origin. The deposit can be divided into four major stratigraphic units (SUs) on the basis of sediment colour and texture (Table 8.2, Fig. 8.6). The upper unit is dominated by humic material with roots common in the upper two units.

Squares C–F revealed a fairly continuous layer of shell across the excavation c.20cm below the surface. The deposit can be divided into five main stratigraphic units (Table 8.3, Fig. 8.7).

Squares G–H revealed several apparently discrete concentrations of whelk and mud ark up to 42cm below ground surface, with occasional shell recovered below these features. The deposit can be divided into five stratigraphic units (Table 8.4, Fig. 8.8). SUIII contains virtually all the shell recovered, again coincident with the depth of the shell layer observed in the erosion section.

With the exception of Square A, all pits excavated immediately adjacent to the erosion bank (Squares C, F, G) exhibited much higher densities of cultural material than squares located on the inland margin of excavation grids. This raises the possibility that the densest deposits of midden material located towards the creek margin of the site complex have been removed by erosion.

Table 8.1 Pancake Creek Site Complex, Squares A–H: summary excavation data and dominant materials.

SQUARE	XUs (#)	DEPTH (cm)	WEIGHT (kg)	SHELL (g)	BONE (g)	CHARCOAL (g)	ARTEFACTS (g)	STONE (g)	ORGANIC (g)
A	19	68.32	247.75	120.58	0.01	252.84	0.03	468.71	387.34
B	18	67.40	223.30	293.81	0.49	220.27	0.02	149.85	557.38
C	13	61.70	190.30	1918.53	0.14	158.22	0	109.61	400.00
D	13	60.64	204.70	663.11	0	93.68	12.82	16.72	789.00
E	12	58.64	196.70	609.88	0	171.15	0	62.93	474.20
F	13	59.92	226.00	1200.95	2.30	281.75	0.18	45.47	608.80
G	18	65.64	212.30	1274.08	0	201.73	0	217.31	643.16
H	18	67.56	227.70	535.92	0.01	338.55	0.43	65.01	354.99
Total	124	–	1728.75	6616.86	2.95	1718.19	13.48	1135.61	4214.87

Table 8.2 Stratigraphic Unit descriptions, Pancake Creek Site Complex, Squares A–B.

SU	DESCRIPTION
I	Extends across the entire trench with an average depth of 5cm and a maximum depth of 10cm below the surface. The unit comprises dry, loosely consolidated fine poorly-sorted subrounded light grey (7.5YR-7/1) sediments. A thick fibrous mat of humic material occurs across the upper extremity of the unit. Small quantities of fragmented shell and charcoal are present. pH values are slightly acidic (6.5).
II	Extends across the entire trench with a maximum thickness of 30cm and a maximum depth of 36cm below the surface. It consists of fine, poorly-sorted subrounded grey (7.5YR-5/1) sandy sediments which are interspersed with numerous roots and fine rootlets. This unit is moister than the previous one and therefore more consolidated. Occasional shell (dominated by oyster and mud ark) occurs throughout, with some whole shell encountered towards the base of the unit. Small blocky fragments of charcoal are common. pH values are slightly acidic to acidic (5.5–6.5), with values increasing with depth.
III	Extends across the entire trench with a maximum thickness of 33cm and a maximum depth of 55cm below the surface. The sediments change distinctly at this interface of stratigraphic units. Sediments comprise light brownish grey (10YR-6/2) to brown (10YR-5/3) sands. Sediments are moist and well-consolidated. Concentrations of blocky charcoal and scattered fragments are abundant (especially in Square A). Shell is virtually absent from this unit. Small quantities of non-artefactual stone dominate the sieve residues. pH values are slightly acidic (6.5).
IV	Unit extends across the entire trench with a maximum thickness of 22cm and maximum depth of at least 68cm below the surface. The base of this unit was not reached. Sediments are well-consolidated, well-sorted, fine and subrounded. The unit exhibits distinctive mottling with the greyish brown (10YR-5/2) frequently interspersed with more yellow sediments. This mottling may relate to burrowing or root penetration. This unit appears to be culturally-sterile with occasional non-artefactual stone, pumice and large quantities of charcoal recovered (especially in Square A). Occasional roots occur throughout. pH values remain slightly acidic (6.5).

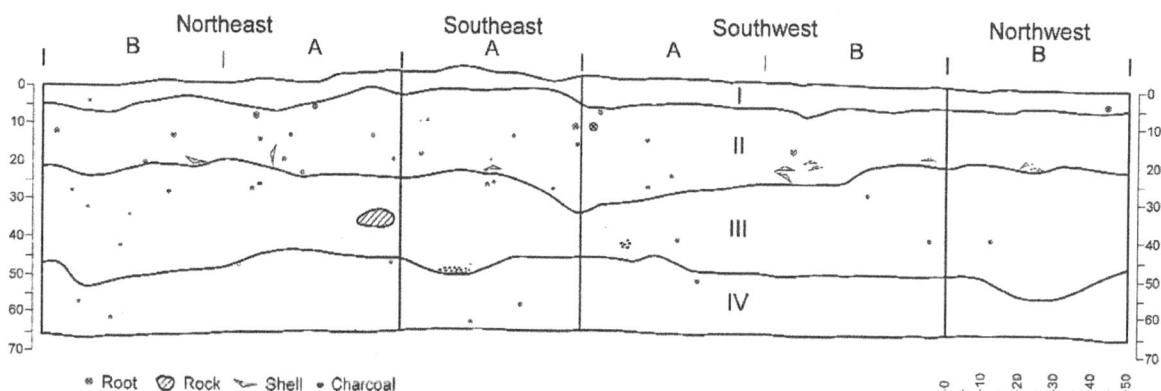

Figure 8.6 Stratigraphic section, Pancake Creek Site Complex, Squares A–B.

Table 8.3 Stratigraphic Unit descriptions, Pancake Creek Site Complex, Squares C–F.

SU	DESCRIPTION
I	Extends across the entire square with an average depth of 4cm and a maximum depth of 14cm below the surface. The unit consists of dry, loosely consolidated fine poorly-sorted subrounded light grey (10YR-7/1) sediments. Occasional tufts of grass penetrate the surface with numerous small fibrous roots. Large quantities of charcoal and occasional shell fragments are present. pH values are neutral to slightly acidic (6.5–7.0).
II	Extends across the entire square with a maximum thickness of 26cm and a maximum depth of 28cm below the surface. The unit comprises fine, poorly-sorted subrounded greyish brown (10YR-5/2) sediments with mottling to lighter shades of grey. Occasional oyster shell fragments occur throughout, with whole shell exposed at the base of the unit. Small blocky fragments of charcoal are common. pH values are neutral to slightly acidic (6.0–7.0).
IIa	Concentration of shell, charcoal and stone exposed across the southwest of Squares C and D. The unit has a maximum thickness of 14cm and maximum depth of 28cm below the surface. The concentration of shell in this unit is restricted specifically to the centre of the southwest section. The matrix is poorly-consolidated, but otherwise similar to SUII.
III	Extends across the entire square with a maximum thickness of 18cm and a maximum depth of 39cm below the surface. Sediments are poorly consolidated and remain greyish brown (10YR-5/2), exhibiting some mottling to lighter grey and yellow sand in places. Sediments are fine, poorly-sorted and subrounded. Shell is abundant across the upper margin of this unit with numerous roots and rootlets evident. Scattered blocky charcoal is common. Several stone artefacts were noted. pH values are alkaline to slightly acidic (6.5–8.0).
IV	Extends across the entire square with a maximum thickness of 17cm and a maximum depth of 53cm below the surface. The unit comprises mottled, light brownish grey (10YR-6/2) sediments which are generally fine, poorly-sorted and subrounded. Large roots occur throughout the matrix. Shell is virtually absent from this unit. Occasional large blocky charcoal fragments occur. pH values are slightly acidic to slightly alkaline (6.5–7.5).
V	Unit extends across the entire square with a minimum thickness of 18cm and maximum depth of at least 62cm below the surface. The base of this unit was not reached. The matrix remains light brownish grey (10YR-6/2) to light grey (10YR-7/1), but mottles to yellow, brown grey and white. This unit appears to be culturally-sterile with occasional non-artefactual stone, pumice and numerous pieces of blocky charcoal recovered. Occasional roots occur throughout. A void from an animal burrow was encountered immediately below the base of excavations. pH values are neutral to slightly alkaline (7.0–7.5).

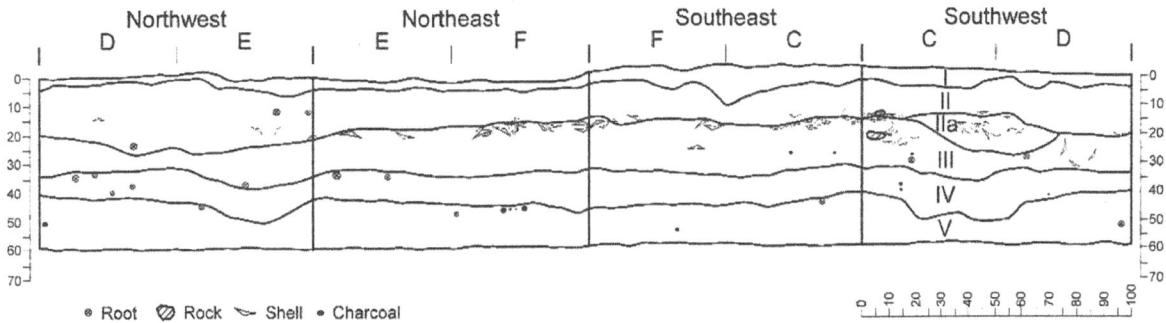

Figure 8.7 Stratigraphic section, Pancake Creek Site Complex, Squares C–F.

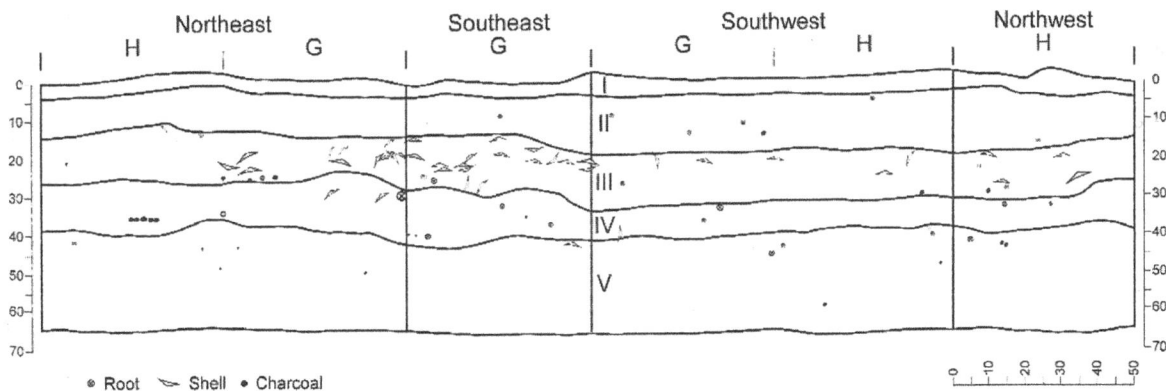

Figure 8.8 Stratigraphic section, Pancake Creek Site Complex, Squares G–H.

Table 8.4 Stratigraphic Unit descriptions, Pancake Creek Site Complex, Squares G–H.

SU	DESCRIPTION
I	Extends across the entire trench with an average depth of 5cm and a maximum depth of 7cm below the surface. The unit consists of dry, loosely consolidated, fine, poorly-sorted subrounded light grey (10YR-7/1) sediments. Occasional tuffs of grass penetrate the surface with numerous small fibrous roots. Occasional shell and charcoal fragments are present. pH values are slightly acidic to acidic (6.0–6.5).
II	Extends across the entire trench with a maximum thickness of 17cm and a maximum depth of 22cm below the surface. The unit comprises fine, poorly-sorted subrounded light brownish grey (10YR-6/2) sediments. Occasional oyster shell fragments and charcoal blocks occur throughout. Numerous roots and rootlets are evident. Occasional small pieces of ironstone were recovered. pH values are slightly acidic (6.0).
III	Extends across the entire trench with a maximum thickness of 17cm and a maximum depth of 36cm below the surface. Sediments are moist and well-consolidated and remain light brownish grey (10YR-6/2). Sediments are fine, well-sorted and subrounded. This unit is defined by the abundance of shell throughout, and contains virtually all the shell recovered from the excavation. Numerous roots and rootlets continue to be abundant. Scattered blocky charcoal is common. Several stone artefacts were noted. pH values are neutral to slightly acidic (6.0–7.0).
IV	Extends across the entire trench with a maximum thickness of 16cm and a maximum depth of 44cm below the surface. The unit comprises moist greyish brown (10YR-5/2) sediments, mottling to light brown and white. Sediments are fine, well-sorted and subrounded. Large roots occur throughout the matrix. Occasional shell was recovered but at much reduced densities to SUIII. Large pieces of blocky charcoal are abundant. pH values are neutral (7.0).
V	Unit extends across the entire trench with a minimum thickness of 28cm and maximum depth of at least 67cm below the surface. The base of this unit was not reached. The matrix remains greyish brown (10YR-5/2) to brown (10YR-5/3), with distinct mottling noted. Sediments are moist and well-consolidated. This unit appears to be culturally-sterile with occasional non-artefactual stone, pumice and numerous pieces of blocky charcoal recovered. Occasional roots occur throughout, but much less numerous than SUIV. pH values are neutral (7.0).

Radiocarbon dating and chronology

Six radiocarbon dates are available for the deposit, comprising one from Squares A–B, two from Squares C–F and three from Squares G–H. All are conventional determinations with five obtained on samples of *Anadara trapezia* and one on charcoal (Table 8.5). A shell/charcoal pair (Wk-6992/Wk-6993) was obtained from a discrete concentration of *A. trapezia* valves in Square H, XU8 (Fig. 8.5). It exhibited an apparent age difference of 100 ^{14}C years with ΔR= –259±137 years (see Chapter 4 for further discussion of this pair). The large error estimate associated with this value makes it indistinguishable from the generalised local open ocean value of ΔR= +10±7. The paired sample consisted of a valve of *A. trapezia* tightly packed with blocky charcoal fragments, suggesting a secure stratigraphic association. This site is also located close to the broad entrance of the creek, which currently exhibits good tidal flushing and no obvious large terrestrial water input (Olsen 1980a). It is possible that terrestrial CO_2 dissolved in water stored in the beach ridges lining the north bank of the creek (possibly with long residence times) may have been introduced into shellfish beds in the intertidal zone. Although the large error estimate associated with the ΔR value introduces considerable ambiguity, and only a single paired result is available for this estuary, a ΔR= –259±137 years is adopted as a first approximation in calibrating radiocarbon dates obtained on marine shell from the site.

All of the dated samples come from the approximate level of the shell layer observed in the erosion bank. The calibrated ages are broadly consistent, all overlapping at 2σ, and suggest that the major period of occupation at the site occurred between 500–700 cal BP. The apparent range of the three dates from a depth of 25cm in Square G–H, XU8, may be associated with rapid formation of the deposit, with both the radiocarbon and calibrated ages overlapping at one standard deviation. There is little evidence for major use of the site over the last 500 years, with observed surface shell associated with deflation surfaces, disturbed areas and crab burrows, suggesting a subsurface origin. It is possible that more recent deposits located seaward of those investigated were removed by erosion.

Table 8.5 Radiocarbon dates from the Pancake Creek Site Complex (see Appendix 1 for full radiometric data for each determination).

SQUARE	XU	DEPTH (cm)	LAB. NO.	SAMPLE	$\delta^{13}C$ (‰)	^{14}C AGE	CALIBRATED AGE/S
A	9	22	Wk-7837	*A. trapezia*	−1.1±0.2	670±50	750(523)280
E	7	25	Wk-6989	*A. trapezia*	−0.1±0.2	870±130	1051(667)414
F	6	20	Wk-6990	*A. trapezia*	−0.4±0.2	630±70	726(504)254
G	8	25	Wk-6991	*A. trapezia*	0.5±0.2	740±60	865(564)316
H	8	25	Wk-6992	*A. trapezia*	−0.3±0.2	800±80	921(635)406
H	8	25	Wk-6993	charcoal	−26.8±0.2	700±140	910(651,576,574)467

Stratigraphic integrity and disturbance

Several lines of evidence indicate that the deposits exhibit a reasonable degree of stratigraphic integrity. Recovered cultural materials are consistent with the distribution observed in section in the erosion bank. The consistent radiocarbon dates for the excavated material and its narrow vertical range lend further support to the idea that most, if not all, cultural material derives from a shell layer located 20–30cm below the modern ground surface. The fact that little shell was recovered outside this zone suggests that there has been little movement of materials within the deposit.

Conjoin analysis of the *A. trapezia* assemblage was limited by the small number of excavated valves and their generally poor preservation. Out of a total dataset of 53 measured broken valves, 24 were discarded from consideration owing to an absence of hinge length. This left 29 valves to be considered in the conjoin analysis, distributed as follows: Squares A–B (12 valves); Squares C–F (9 valves); Squares G–H (32 valves). Methods proceeded as described in Chapter 5. A single conjoin was identified. The conjoining valves came from the shell concentration encountered in Square H, XU9, located 26–30cm below ground surface (see Fig. 8.5). The identification of this conjoin suggests that this concentration of *A. trapezia* exhibits considerable integrity and was probably discarded as part of a single event.

Root penetration and crab burrowing (probably the smooth-handed ghost crab, *Ocypode cordimanus*, which is common at the site today) appear to be the major sources of post-depositional disturbance in all squares. Fragments of oyster shell were recovered from the basal excavation units of both Squares C–F and G–H in a mottled unit which appears to be otherwise culturally-sterile. The mottled appearance of this stratigraphic unit indicates that these are likely to be infilled crab burrows or even root voids. Only a single active burrow was encountered at the base of Square D, where a large cavity (c.15cm in diameter) exposed after excavation had ceased indicated a burrow originating from the erosion bank.

Laboratory methods

Owing to the relatively low density of cultural material recovered from the Pancake Creek Site Complex all squares were analysed to maximise the available sample (see Chapter 3 for a detailed discussion of the standard laboratory methods employed at all sites). In the sections below, the results from all squares are summarised although only the data from Square G is illustrated in Figures 8.9–8.16. This approach has been adopted to minimise repetition. Square G has been selected as it yielded the densest cultural material, the general distribution of which is broadly consistent across the site. Further summary results for all excavated squares are available in Appendix 4.

Cultural materials

Invertebrate remains

Thirty-four taxa of shellfish weighing a total of 6,617g were recovered from Squares A–H, consisting of 12 marine bivalves, 17 marine gastropods, four terrestrial gastropods and a freshwater bivalve (Table 8.6). The shell deposit is dominated by rock oyster (*S. glomerata*), making up 78.5% of the shell assemblage by weight (Fig. 8.9), followed by mud ark (*A. trapezia*) (10.5%) (Fig. 8.10) and hercules club whelk (*P. ebininus*) (10%) (Fig. 8.11). The remaining 31 taxa are relatively rare, each contributing less than 0.5% of the shell assemblage by weight. The assemblage exhibits relatively high diversity with a calculated Shannon-Weaver Function (H') of 1.21 and Simpson's Index of Diversity (1–D) of 0.445. Virtually all of the shell was recovered from between 20–30cm below ground surface. Occasional small shell fragments were recovered to the base of excavations and are thought to have been displaced by crab burrowing (see above). Many of the smaller taxa represented, particularly many of the gastropods, are likely to have entered the deposit attached to larger shellfish taxa.

The vertical distribution of specific taxa shows distinct patterning suggestive of focussed species selection. In Square G (Figs 8.9–8.13), there are distinct almost mono-specific concentrations of mud ark (XU12–13) and scallop (XU8–10), while oyster appears to co-occur with hairy mussel and whelk (XU5–7). The most diverse shellfish assemblages are consistently recovered from units dominated by rock oyster. Oyster was recovered attached to both rock (Square A) and shell (Square E) platforms, indicating that at least some oyster gathering occurred on debris beds such as that currently located at the top of the subtidal zone. Excavation units exhibiting high densities of oyster are coincident with high densities of non-artefactual stone, raising the possibility that much of the latter entered the site attached to rock platform species. Oysters occupy habitats favoured by other taxa such as debris beds and the aerial prop roots of mangroves where associated taxa can be incidentally harvested and thus enter cultural deposits.

The range of shellfish recovered indicate gathering activities focussed on the broad intertidal zone adjacent to the site. The small quantity of the freshwater bivalve *Corbicula (Corbiculina) australis* recovered from Square G, which inhabits coastal rivers and streams, probably reflects incidental gathering from shell debris beds. Although shell sizing was undertaken on the shell assemblage the small sample makes results unreliable and so they are not discussed further.

Vertebrate remains

Occasional fish bone was recovered totalling 1.28g and consisting of 15 pieces of bone. A single specimen was recovered from Square A, four from Square C and 10 from Square F. This material was highly fragmented and none could be identified to taxon. A small vertebrae was recovered from Square H, XU11, which may originate from a small reptile (see Vale 2004 for further details).

Stone artefacts

Stone artefacts were recovered from each excavation area. Twelve stone artefacts weighing 13.4g were identified (Table 8.7). All of these were recovered from the sieve residue. The assemblage is manufactured on a variety of materials, including microgranite (n=5), quartz (n=4), rhyolitic tuff (n=2) and one unknown rock type. All artefacts are flaked pieces. Although no rock occurs naturally in the beach ridge system where the Pancake Creek Site Complex is situated, microgranite and quartz are available nearby in the hills in the centre of Rodds Peninsula or on Bustard Head across Pancake Creek. The nearest recorded source of rhyolitic tuff is at the Ironbark Site Complex on Middle Creek, some 11.5km southeast. Most artefacts are extremely small, with an average maximum dimension of 9.5mm and average weight of 1g.

Table 8.6 Presence/absence of shellfish identified in the Pancake Creek Site Complex, Squares A–H.

FAMILY	TAXON	A	B	C	D	E	F	G	H	TOTAL (g)
				MARINE BIVALVIA						
Arcidae	*Anadara trapezia*	X	X	X	X	X	X	X	X	681.8826
Chamidae	*Chama fibula*				X					1.9244
Donacidae	*Donax deltoides*		X							0.0046
Mytilidae	*Trichomya hirsutus*		X	X	X	X	X	X		10.1011
Noetiidae	*Arcopsis symmetrica*							X		0.1648
Ostreidae	*Saccostrea glomerata*	X	X	X	X	X	X	X	X	5190.3799
Pteriidae	*Pinctada albina sugillata*	X	X	X	X	X		X	X	30.6987
Tellinidae	*Tellina* sp.			X	X	X	X	X	X	1.1518
Veneridae	*Gafrarium australe*						X			0.8874
Veneridae	*Irus* sp.	X		X						0.0795
Veneridae	*Placamen* sp.							X		0.0551
Veneridae	*Venerid* sp.						X			0.5171
				MARINE GASTROPODA						
Batillariidae	*Pyrazus ebininus*		X	X	X	X	X	X	X	663.1323
Batillariidae	*Velacumantus australis*		X			X	X		X	1.4083
Cerithiidae	*Cerithiid* sp.		X	X				X	X	0.5652
Cerithiidae	*Cerithium* sp.				X					0.0490
Cerithiidae	*Clypeomorus bifasciata*			X	X		X	X		0.5137
Costellariidae	*Vexillum* sp.			X						0.4337
Cypraeidae	*Cypraea* sp.							X		1.7554
Lottiidae	*Acmaeid* sp.			X			X	X	X	0.4323
Littorinidae	*Bembicium nanum*			X		X	X			0.2794
Littorinidae	*Littoraria* sp.						X			0.0845
Muricidae	*Bedeva paivae*						X	X		0.2465
Muricidae	*Morula marginalba*		X	X	X	X	X	X	X	14.8585
Nassariidae	*Nassarious burchardi*			X						0.0278
Neritidae	*Nerita squamulata*			X						0.9361
Planaxidae	*Planaxis sulcatus*					X				0.0716
Trochidae	*Herpetopoma atrata*			X			X			0.8750
Trochidae	*Thalotia* sp.			X	X		X	X		1.9965
				TERRESTRIAL GASTROPODA						
Camaenidae	*Figuladra* sp.		X	X	X	X	X	X	X	4.9309
Camaenidae	*Trachiopsis mucosa*	X	X	X	X	X	X	X	X	5.7908
Pupillidae	*Pupoides pacificus*			X	X	X		X		0.2765
Subulinidae	*Eremopeas tuckeri*			X	X	X	X			0.2863
				FRESHWATER BIVALVIA						
Corbiculidae	*Corbicula australis*							X		0.0684

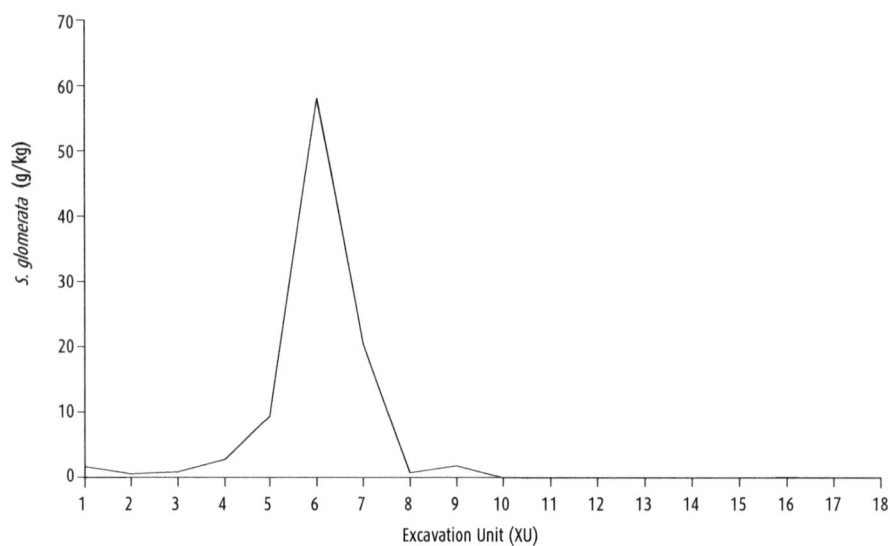

Figure 8.9 Abundance of oyster (*S. glomerata*).

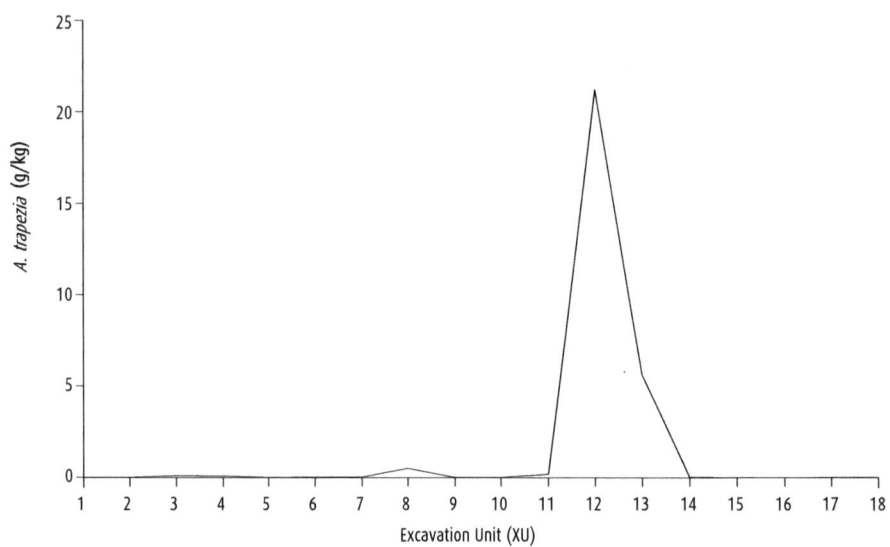

Figure 8.10 Abundance of mud ark (*A. trapezia*).

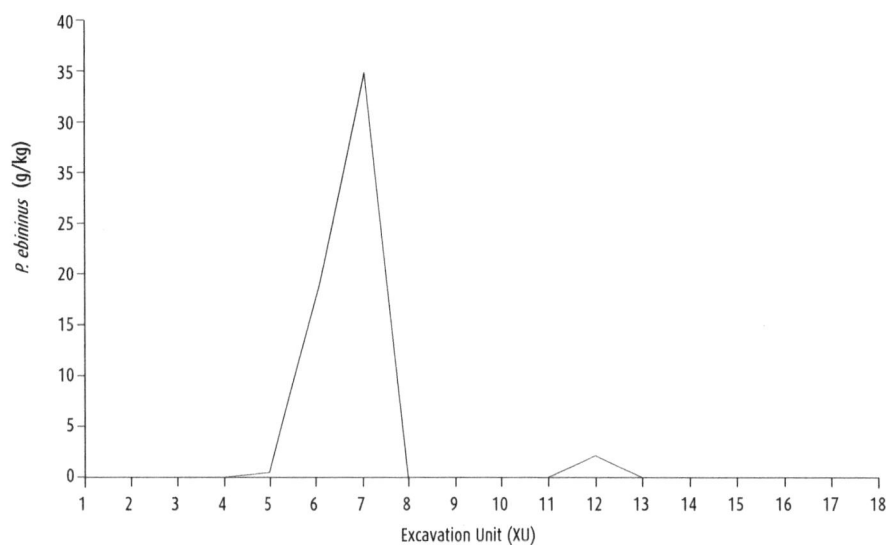

Figure 8.11 Abundance of whelk (*P. ebininus*).

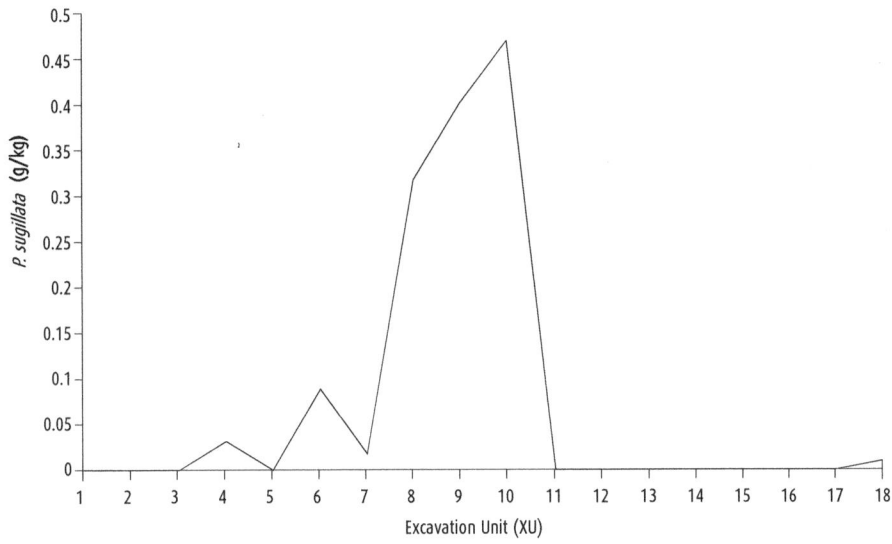

Figure 8.12 Abundance of scallop (*P. sugillata*).

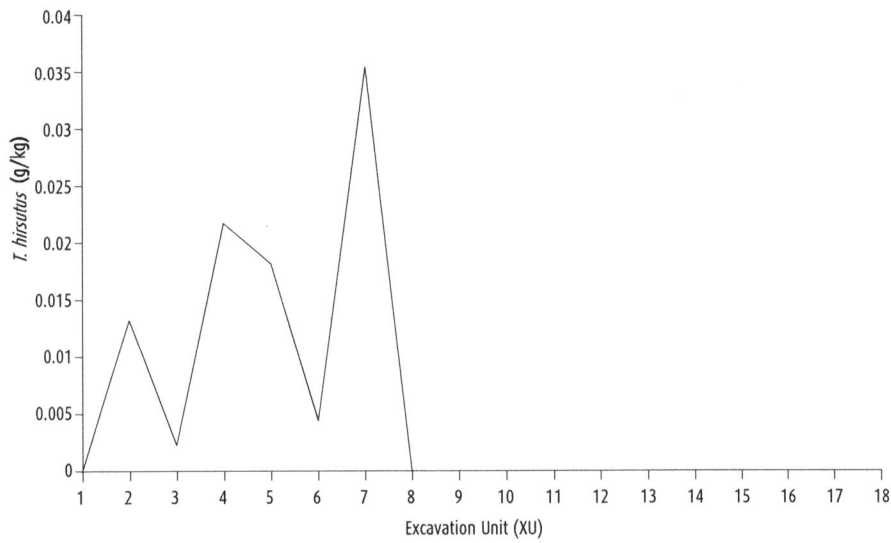

Figure 8.13 Abundance of hairy mussel (*T. hirsutus*).

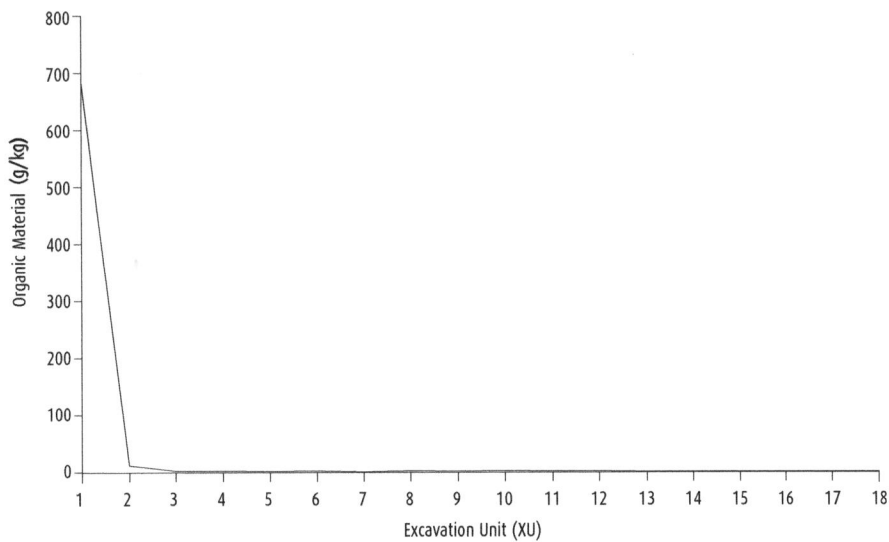

Figure 8.14 Abundance of organic material.

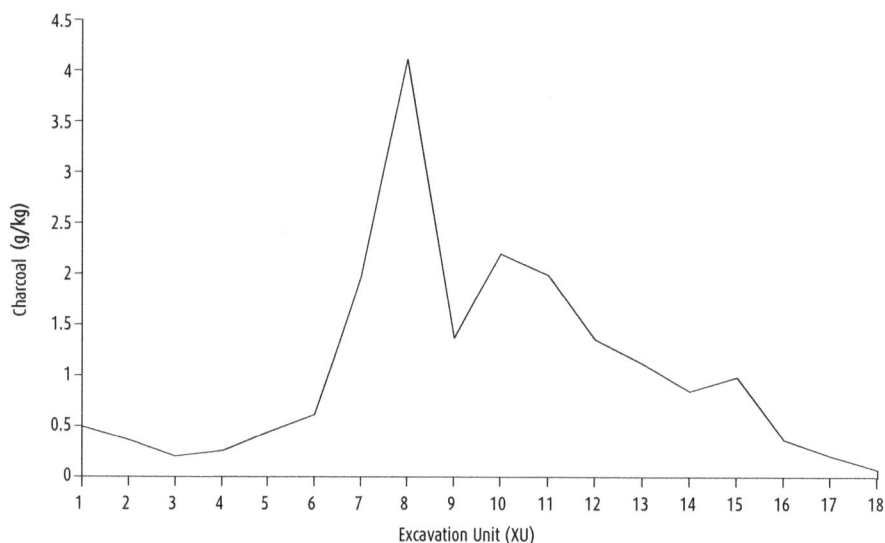

Figure 8.15 Abundance of charcoal.

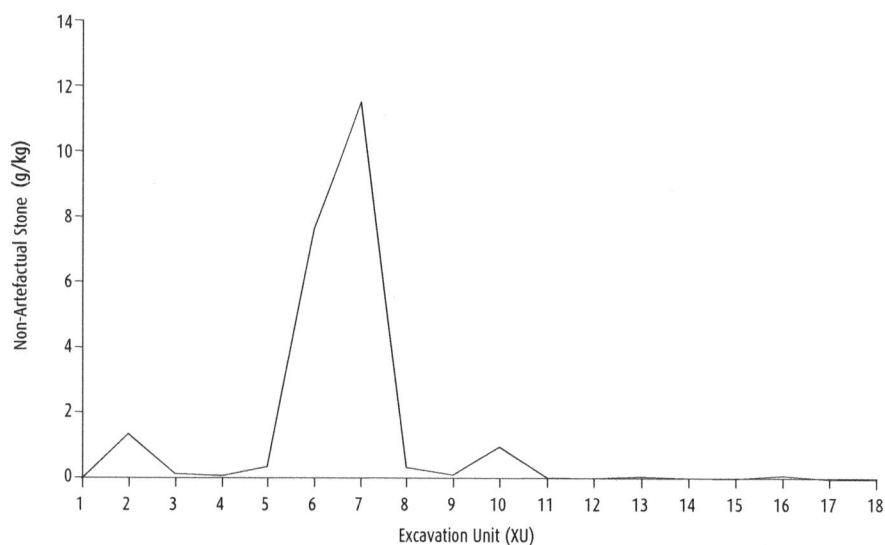

Figure 8.16 Abundance of non-artefactual stone.

Table 8.7 Stone artefacts from the Pancake Creek Site Complex, Squares A–H.

RAW MATERIAL	ARTEFACT TYPE	NUMBER	WEIGHT (g)	SQUARES
Quartz	Flaked Piece	4	0.1	A, B, D, H
Microgranite	Flaked Piece	5	12.9	D, F, H
Rhyolitic Tuff	Flaked Piece	2	0.2	D, H
Unknown	Flaked Piece	1	0.2	F
Total	–	12	13.4	–

Other remains

Abundant blocky charcoal was recovered from throughout all squares (1,718.2g), including the basal deposits which are otherwise culturally-sterile (Fig. 8.15). This material almost certainly derives from bushfires as it is not directly associated with any shell or other definitively cultural material. Pumice, totalling 56.4g, occurs throughout the deposits but is most abundant in the basal units. Small quantities of coral were recovered from Squares C and G (8.1g). It could have entered the deposit incidentally as part of debris attached to platform shellfish, been collected deliberately or entered the site naturally.

Discussion

Excavation revealed highly variable distribution of cultural material across the site area. Only minor quantities of shellfish were recovered from Squares A and B despite dense exposures of mud ark and oyster on the adjacent erosion bank. Squares C–F and G–H, on the other hand, revealed quantities of shellfish remains dominated by mud ark (*A. trapezia*) and oyster (*S. glomerata*), with lesser quantities of hercules club whelk (*P. ebininus*) and minor representation of other shellfish species. Variations in the vertical distribution of various species was also noted, which may relate to patterns of predation, mobility or resource availability.

The excavations and analyses reported above demonstrate that although the cultural deposits at the Pancake Creek Site Complex are generally of a relatively low density (compared, for example, to those at the Seven Mile Creek Mound), the extent and recent chronology of the deposits indicate that a large quantity of cultural material was discarded over a relatively short period of time. To put this into perspective, we can extrapolate from the figures available from the analysed excavations to the estimated site area. Burke (1993) estimated the deposits to be in excess of 22,500m^2. Excavation of Squares A–H indicated that the density of shell recovered varied from 482.32g/m^2 (Square A) to 7,674.12g/m^2 (Square C). Even if the lowest density figure is taken as a basis for calculations, the entire site area is likely to exhibit a minimum of 10,850kg of shell. These figures become more plausible when it is considered that an unknown quantity of possibly denser deposits once located seaward of the current site has been removed by erosion. If the known chronology can be extrapolated to the deposits as a whole, the entire deposit may have accumulated over a period of less than 500 years between 300–800 years ago.

The low density and highly variable structure of such extensive cultural deposits raise critical sampling issues. The linear form of many sites and their exposure along eroding creek sections mitigated this to some degree, although it was shown in some squares (A–B) at the Pancake Creek Site Complex that apparent concentrations of material in the eroding section were not encountered in excavated deposits situated <1m inland.

This general site form appears to be the principal manifestation of settlement and subsistence systems in place over the last 1,000 years in the region, and indeed for southeast Queensland generally (e.g. Sandstone Point, Toulkerrie, NRS, Maroochy River Mouth, Corroboree Beach). The results from this site should act as a cautionary tale for assessments of the scientific significance of sites which rely too heavily on the density of visible cultural material to establish significance values.

Summary

The Pancake Creek Site Complex was occupied for a period of at least 200 years between c.300–800 cal BP. There is little evidence for major use of the area over the last 500 years although more recent deposits which may have been located seaward of those investigated may have been lost to erosion. Low density deposits dominated by shellfish occur over a very large area, emphasising the need to examine sites in context. The patterning and chronology of site use at the Pancake Creek Site Complex appear to be associated with a more recent phase of land-use than that evident at the Seven Mile Creek Mound and Mort Creek Site Complex.

9

Ironbark Site Complex

Introduction

This chapter describes archaeological investigations of a major stone quarry/shell midden complex in the approximate centre of the southern Curtis Coast study region. In addition to an extensive, low density linear midden common to the archaeological record of the region, the Ironbark Site Complex features evidence for major stone quarrying and reduction activities. The site has yielded evidence for Aboriginal use from around 1,500 years ago into the post-contact period. Survey and excavation methods are outlined and the results of excavations are presented. Analyses demonstrate that activities at the site were focussed on the reduction of rhyolitic tuff, with edge-ground hatchets one of the key manufacturing products. Geochemical and archaeological studies indicate that the quarry was a major raw material source in the region, with artefacts from nearby sites and beyond found to be geochemically indistinguishable from rock available at the Ironbark Site Complex.

Site description and setting

The Ironbark Site Complex is a large multiple-component stone quarry/shell midden site complex located on the lower southern bank of Middle Creek (Latitude: 24°07′45″S; Longitude: 151°46′30″E) (Fig. 9.1). The site has three broad components: an exposed concentration of flaked rhyolitic tuff covering an area of c.1,000m^2 along a section of narrow beach separating the mangrove fringe from an eroding frontal ridge; a discontinuous exposure of shell and stone artefacts which covers at least 150,000m^2 across low transgressive dune ridges adjacent to the creek; and small isolated exposures of shell, flaked bottle glass and a large baler shell artefact associated with a stand of cycads (*Cycas megacarpa*) on an elevated ridge immediately inland of the quarry (Fig. 9.2). Artefact densities vary across the quarry from 9,500 to less than 10 artefacts/m^2. The entire site is estimated to contain more than one million stone artefacts. Stone artefacts and shell occur discontinuously both in section and along the base of a low erosion bank bordering the creek, stretching some 1,373m between the two elevated headlands from the Queensland Parks and Wildlife Service (QPWS) camping area in the east, to the western tip of the embayment c.100m west of the quarry and in the adjacent intertidal zone in the mangrove fringe (Fig. 9.2). Large dead trees in growth position on the fringing beach and the presence of many dead trees indicate significant bank recession across the deposits adjacent to the creek margin. Tidal action and storm-surge are probably responsible for much of the bank erosion, although recent cattle grazing has exacerbated the problem as evidenced by cattle faeces close to the bank. Driftwood has also been washed into piles along the fringing beach by tidal action with several piles pushed up against the base of the erosion bank, particularly in the area of the quarry. Driftwood deposition suggests that most, if not all, of the exposed quarried material is occasionally impacted by large tides and/or storm-surge activity. Occasional scatters of shell and flaked stone were also observed on four wheel drive tracks up to 1km south of the site complex, although it is unclear whether these exposures are associated with those visible in the erosion bank and close to the creek margin or part of the low density background scatter across the region.

Rhyolitic tuff outcrops on the headlands at either end of the embayment. Exposures in the area of the QPWS camping area in the east do not appear to have been quarried. The western headland, however, exhibits evidence for stone reduction, with rhyolite boulders (both modified and unmodified) on the beach fringe extending for 100m before the beach continues. This exposed rock is divided roughly in two by a group of large boulders which originates in the main elevated hill to the south and continues down to and below the water's edge, creating a break in the mangrove fringe (Fig. 9.3). The eastern half of the quarry and adjacent bank are protected from the direct impact of wave and wind action by a low intertidal sand bank covered with mangroves which runs parallel to the mainland for about 500m (Figs 9.2, 9.7). The western half of the quarry is close to the northwestern extent of this sand bank and is exposed to a much higher energy tidal environment, as evidenced by the presence of coarser angular sands in the intertidal zone and heavier weathering of exposed rock (Fig. 9.7). The less weathered material on the eastern half of the quarry appears to have been targetted for reduction. Densities of modified stone decrease markedly to the west. The southern margin of the quarry exposure adjacent to the bank exhibits many large boulders. Flaked artefacts are evident on the surface of and in crevices between these boulders, and it is conjectured that the artefacts may have deflated from erosion bank deposits receding to the south with sediment, shell, lighter artefacts and other cultural materials removed by tidal action. As part of this process some smaller artefacts may have become lodged in the interstices between rocks and other artefacts on the gentle slope towards the creek, partly explaining the abundance of smaller material on the parts of the quarry closest to the creek.

Large modified and apparently unmodified blocks of rhyolitic tuff (some individual blocks weighing over 50kg) occur between the rocky headlands both along the narrow sandy beach separating the low erosion bank from the mangrove fringe, and on the muddy substrate in the intertidal zone of the mangrove fringe itself. The location of these artefacts away from the natural rock occurrence is thought to be the result of a combination of deliberate transport and progressive recession of bank deposits containing cultural material, leaving lag deposits of heavier material in the intertidal zone. There is no higher ground from which cobbles could have rolled down, and the low energy estuary does not appear to have sufficient energy to move large (>5kg) stones under normal conditions.

Vegetation over the Ironbark Site Complex is distinctively zoned by substrate. Elevated land on the western headland is covered by open woodland dominated by eucalypts (*Eucalyptus intermedia, E. acmenioides, E. umbra*), with black she-oak (*Allocasuarina littoralis*) on the fringes of the zone closest to the creek margin. The understorey includes native grasses, grass trees (*Xanthorrhoea* sp.) and cycads (*C. megacarpa*). The low sandy ridges and swales between the western and eastern headlands support open forests of melaleucas (*Melaleuca quinquenervia*), eucalypts (*E. tereticornis*) and swamp box (*Lophostemon suaveolens*), with cloudy teatree (*M. dealbata*) and weeping cabbage palm

(*Livistonia decipiens*) dominating the base of swales. A small area of closed, dry rainforest occurs on the southeastern half of the dune system, and includes many traditional Aboriginal food sources such as burdekin plum (*Pleiogynium timorense*), bumpy ash (*Flindersia schottiana*), brown pine (*Podocarpus elatus*) and native cherry (*Exocarpus latifolius*). The more exposed eastern headland supports eucalypts (*E. tessalaris*), burdekin plum and wattle (*Acacia aulacocarpa*). Mangrove vegetation adjacent to the site comprises a dense fringe of spotted mangroves (*Rhizophora stylosa*), with grey mangroves (*Avicennia marina*) dominating the low sand bank in the creek parallel to the mainland (Olsen 1980a:18). Extensive sandy to muddy intertidal flats adjoin the site, supporting populations of

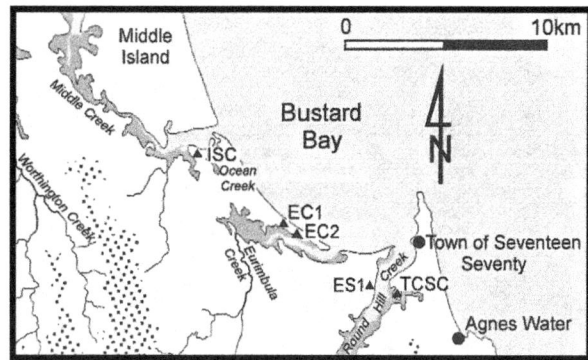

Figure 9.1 The Middle Creek catchment area showing the location of the Ironbark Site Complex (ISC) and nearby excavated sites (EC1=Eurimbula Creek 1; EC2=Eurimbula Creek 2; ES1=Eurimbula Site 1; TCSC=Tom's Creek Site Complex). Dark grey shading indicates the general extent of mangrove, saltflats and claypans. Dotted shading indicates land above 200m. Solid dots indicate local population centres.

hercules club whelk (*Pyrazus ebininus*), telescope mud whelk (*Telescopium telescopium*), and oyster (*Saccostrea glomerata*) attached to mangroves. No oysters were observed on the rhyolitic tuff of the quarry exposure, which has both intertidal and subtidal components, suggesting that the rock is not a preferred substrate for these shellfish.

Evidence for European use of the immediate area is limited. Although the Bustard Head Lightstation 11km to the north had a permanent presence from 1867, it was not until 1907 that cattle were farmed on Middle Island, across the creek from the Ironbark Site Complex, with Katherine Bowton officially taking up the 3,432ha Red Hill grazing lease in 1917 (Buchanan 1999:64). Anecdotal evidence reported by Buchanan (1994) suggests that the Bowton family, resident on Middle Island from 1907 to 1977, regularly interacted with local Aboriginal people. The Bowton homestead was located close to the northern bank of Middle Creek, just 3km upstream of the Ironbark Site Complex. During a visit to the homestead in the mid-1970s, Buchanan (1994:102) was shown a collection of 'Aboriginal grinding stones, stone cutting implements and woven baskets'. The site area is likely to have been logged in the late nineteenth century, with a sawmill and loading jetty in operation near the mouth of nearby Eurimbula Creek, 8km to the southeast, by 1867 (Buchanan 1999:33; Growcott and Taylor 1996:65–6). Eurimbula Creek Station was established in the area of the sawmill in 1868 (QDEH 1994:79), although the Eurimbula run was not officially surveyed and leased until 1878 (Growcott and Taylor 1996:65).

The Ironbark Site Complex area is situated on a Special Purposes Reserve jointly controlled by the Queensland Environmental Protection Agency (EPA) and Department of Natural Resources, Mines and Energy. Middle Creek is a popular recreational fishing area, with access from a designated QPWS camping area with an unofficial boat ramp which cuts through midden deposits at the eastern end of the site complex (Fig. 9.2). Surveys confirmed that material culture associated with these contemporary fishing and camping activities is largely confined to the eastern end of the site complex and along a narrow 4WD access track which runs along the inland margin of visible surface midden materials before turning south. Occasional recent human defecation sites are located to the west of the camping area and along the margins of the 4WD track. Except for occasional flotsam washed onto the fringing beach through the mangroves, the only other object found during the present project was a 20 gallon 'BP' brand oil drum placed upright next to a large teatree at the western edge of the teatree swamp, just inland of Squares L–M.

The site was first recorded in September 1996 by the author during systematic pedestrian surveys focussing on open beaches and the lower reaches of major creeks in the area, conducted as

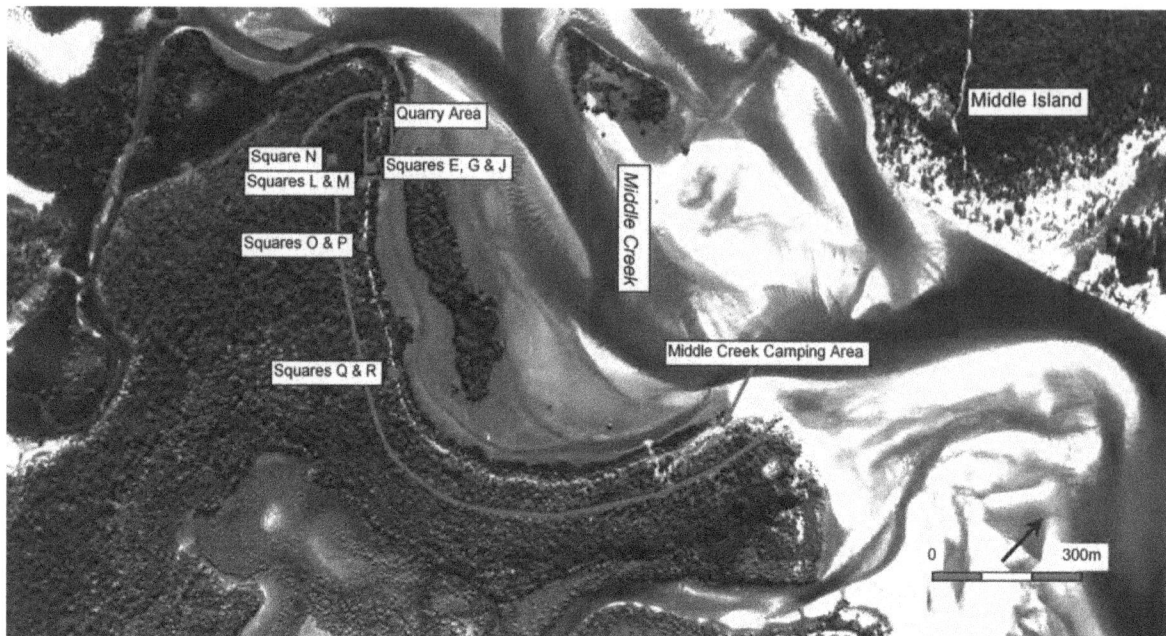

Figure 9.2 Aerial view of the Ironbark Site Complex, showing the maximum extent from the creek of shell and stone exposed at the surface (heavy line) and the general location of excavation squares (after BPA Run 15D/79, 30 July 1996). Based on data provided by the Department of Natural Resources and Mines, Queensland 2006, which gives no warranty in relation to the data (including accuracy, reliability, completeness or suitability) and accepts no liability (including without limitation, liability in negligence) for any loss, damage or costs (including consequential damage) relating to any use of the data.

part of the Gooreng Gooreng Cultural Heritage Project (GGCHP) (see Lilley et al. 1997). Adopting an encounter-and-record strategy, the site was originally described and recorded as five separate sites corresponding to major concentrations of midden and flaked stone encountered along the creek margin from east to west (originally designated the GGCHP Site Numbers CC26, CC27, CC28, CC29 and CC41; see Lilley et al. 1997). Subsequent and more intensive survey of the area confirmed earlier speculation that these separate 'sites' were in fact different exposures of a single, large, multiple-component site complex. The site is registered on the EPA's Indigenous Sites Database as KE:B07–KE:B10 and KE:B22 (corresponding to the original five separate recordings above) and as Queensland Museum Scientific Collection Number S863.

The geologically distinctive rhyolitic tuff that has been quarried at this site is restricted in distribution to coastal headlands and near-coastal outcrops in a 40km long coastal zone between Wreck Rock in the south and Middle Creek in the north, on which the Ironbark Site Complex is situated. The Ironbark quarry is located within an outcrop of rhyolitic tuff with ignimbrite flows that infilled a palaeotopographic low with undifferentiated and heavily weathered granites of Permian to Triassic age (Ellis and Whitaker 1976). Stephen Cotter (Cooperative Research Centre for Landscape Evolution and Mineral Exploration, University of Canberra, pers. comm., 2000) suggested that this rhyolitic tuff is Triassic in age. The rocks are aphyric, containing minor sanidine and/or quartz micro-phenocrysts, and dark green, partly devitrified glass. Lithic fragments of trachyte and rhyolite are noticeably absent from this suite of flow units compared to other outcrops of Agnes Water Volcanics along the coastline, with a greater proportion of glassy material (Stephen Cotter, Cooperative Research Centre for Landscape Evolution and Mineral Exploration, University of Canberra, pers. comm., 2000; for further details see Ulm et al. 2005). Despite systematic examination of virtually all coastal occurrences of rhyolitic tuff, the Ironbark Site Complex is the only location where evidence for significant Aboriginal procurement has been documented, even though virtually every site recorded in the region exhibits artefacts manufactured on this material.

Reid (1998) analysed a sample of the lithic assemblage excavated from the Ironbark Site Complex quarry (Square E) and adjacent middens (Square M), demonstrating a level of standardisation of the reduction sequence on several technological and descriptive indices. On this basis, Reid (1998) suggested that the quarry was primarily a place of manufacture of edge-ground hatchets. The Ironbark quarry was not used exclusively for the production of hatchets, however, as numerous flaked pieces, cores and flakes sourced to the quarry have been recovered from sites throughout the region, making it difficult to separate quarried hatchet blanks from stone reduced for other end-products (Dickson 1981:34). Significantly, nine bifacially flaked edge-ground hatchets from various locations in central Queensland and manufactured from rhyolitic tuff like that found at the Ironbark Site Complex have been located in museum collections and during field surveys. One hatchet was found at the Ironbark site during surveys c.100m east of the quarry exposure (Fig. 9.12). Five hatchets held by the local Miriam Vale Shire Historical Society Museum in Agnes Water were collected from Lowmead (45km from the Ironbark Site Complex), Kalpowar (79km), Moondoondah (25km), Miriam Vale (30km) and Bororen (31km). Further south, hatchets held by the Queensland Museum were collected from Bundaberg (100km), Gin Gin (97km) and Sharon (96km). Laser ablation inductively coupled plasma mass spectrometry (LA-ICP-MS) was used in an attempt to provenance the hatchets to particular outcrops of rhyolitic tuff on the basis of trace element geochemistry. Preliminary results confirm that all hatchets identified as rhyolitic tuff exhibit a similar geochemical signature. Moreover, this geochemistry can be correlated with the background samples from the Ironbark Site Complex, the only major rhyolite quarry known in the region (for further details see Ulm et al. 2005).

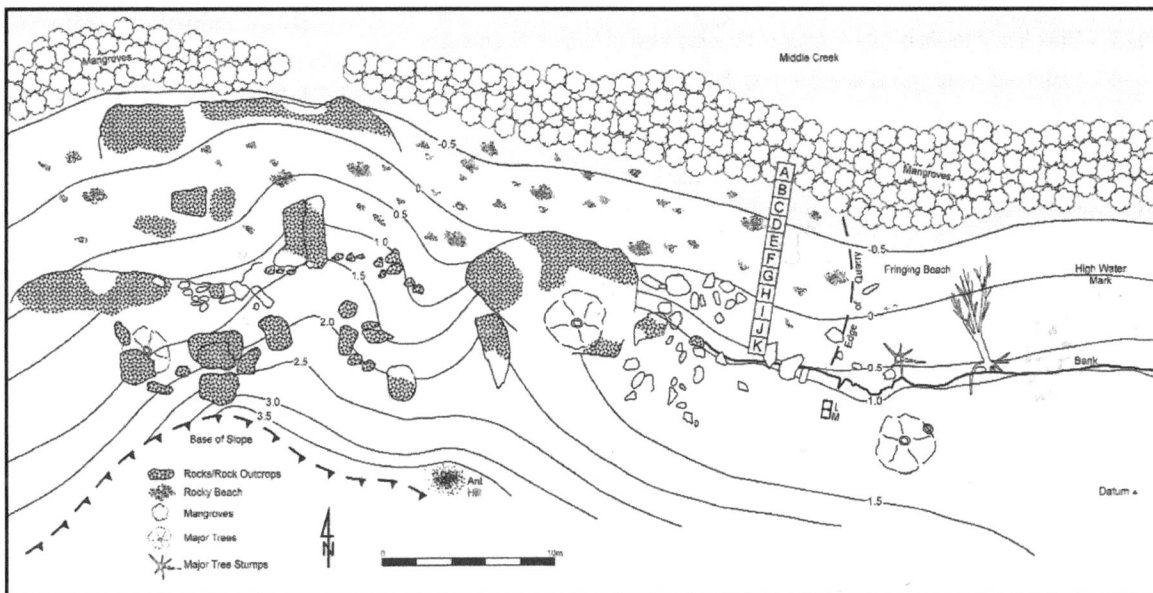

Figure 9.3 Site plan of Ironbark Site Complex, showing area of Squares A–M. Contours are in 0.5m intervals. Rock outcrops and low density artefactual material continue for a further c.50m southwest of the limits of this plan.

Excavation methods

Owing to the extent of the site and resource constraints prohibiting large open area excavation, a strategy was adopted whereby a broad area could be characterised through a program of limited excavations. In total, 4.75m^2 was excavated. Three 1m × 1m squares (E, G and J) were excavated along a north-south transect on the quarry to obtain a sample of material to characterise onsite reduction. A further 1m × 50cm trench excavated as contiguous 50cm × 50cm squares (L–M) was located on the bank adjacent to the quarry in an attempt to recover organic materials to date the initiation of quarrying and increase the sample of flaked stone artefacts (Fig. 9.3). Two 1m × 50cm trenches excavated as contiguous 50cm × 50cm squares (O–P and Q–R) were excavated into the midden east of the quarry (Figs 9.5–9.6) to sample areas not immediately associated with it. A single 50cm × 50cm square (N) was excavated on the ridge inland of the quarry (Fig. 9.4) to sample

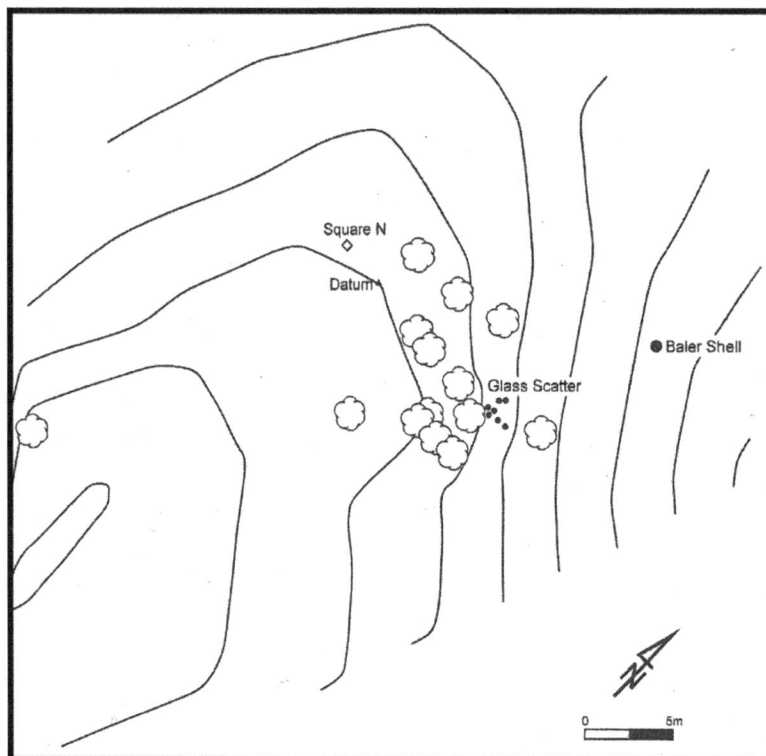

Figure 9.4 Site plan of Ironbark Site Complex, showing area of Square N. Contours are in 0.5m intervals. Only cycads over 50cm in trunk height are shown.

shell associated with the cycad grove and glass artefacts. Over 1,225.4kg of sediment was excavated at the site (this figure excludes material removed from Squares E, G and J on the exposed quarry, see below). Excavations were conducted between 12 January and 13 February 1998. Excavation areas were mapped in detail. The large area over which cultural material is distributed prohibited detailed mapping of the entire site. All sections were photographed, drawn and described except Squares E, G and J on the quarry, which were only photographed and described.

To sample quarry materials, an excavation grid was established as an 11m long and 1m wide transect extending from the mangrove fringe (Square A) to the erosion bank (Square K) (Fig. 9.3). Three squares (E, G and J) were excavated. This approach was adopted to sample an apparent separation of larger material located further upslope on the quarry exposure and smaller material downslope towards the creek. The excavation aimed to test if these observations represented discard of smaller materials downslope and/or post-depositional sorting of the assemblage by periodic tidal inundation. A portable wooden frame comprising eleven 1m × 1m squares was constructed to facilitate excavation of the quarry (Fig. 9.8). This was necessary as the lower part of the excavation grid adjacent to Middle Creek was periodically inundated by the tide. Traditional string lines were not practicable because chain arrows could not penetrate the rocky substrate and string lines moved with tidal action. A sealant was applied to the wooden frame to minimise warping caused by exposure.

Conventional excavation methods using trowel and brush could not be used to remove the first stratigraphic unit of Squares E, G and J as it comprised of layered stone with no intervening sediment (Fig. 9.8). Rather than imprecisely drawing the thousands of pieces of stone onto excavation forms, they were recorded photographically to expedite recovery. The method involved using a large colour photograph of each excavation unit of each square taken in plan, or close to plan, view. All material larger than 30mm was plotted three-dimensionally, assigned unique field specimen (FS) and object numbers and bagged individually. Individual object numbers were written directly onto the photographic print using a 0.3mm waterproof marker (Fig. 9.8). Given the uneven nature of sections formed solely of rock, a general protocol was adopted to remove artefacts and rocks that were more than half-way into the square. Artefactual and non-artefactual material smaller than 30mm was removed with tweezers. This material was bagged according to excavation unit, assigned a field specimen number and weighed.

Conventional excavation methods were generally followed once the sediments underlying the rock were encountered (see Chapter 3 for a detailed discussion of the standard excavation methods employed at all sites). Excavated sediments were wet-sieved through 3mm mesh in the

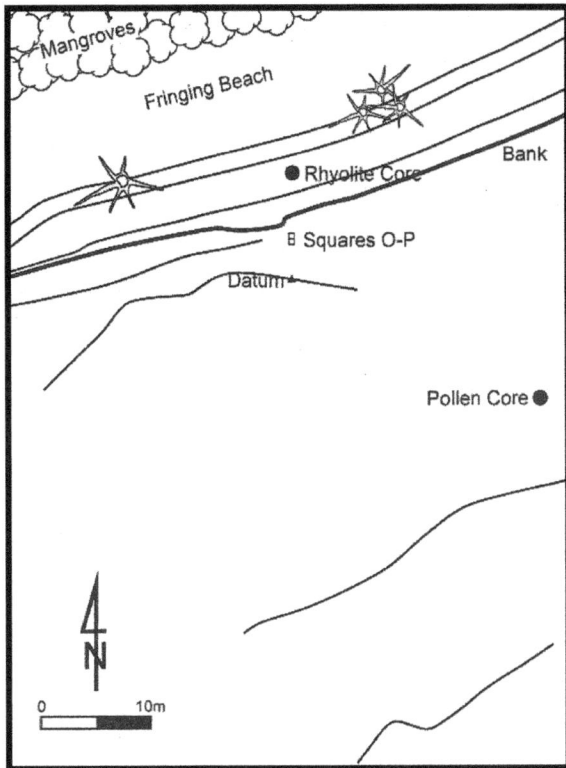

Figure 9.5 Site plan of Ironbark Site Complex, showing area of Squares O-P. Contours are in 0.5m intervals. Tree stumps are shown on the fringing beach.

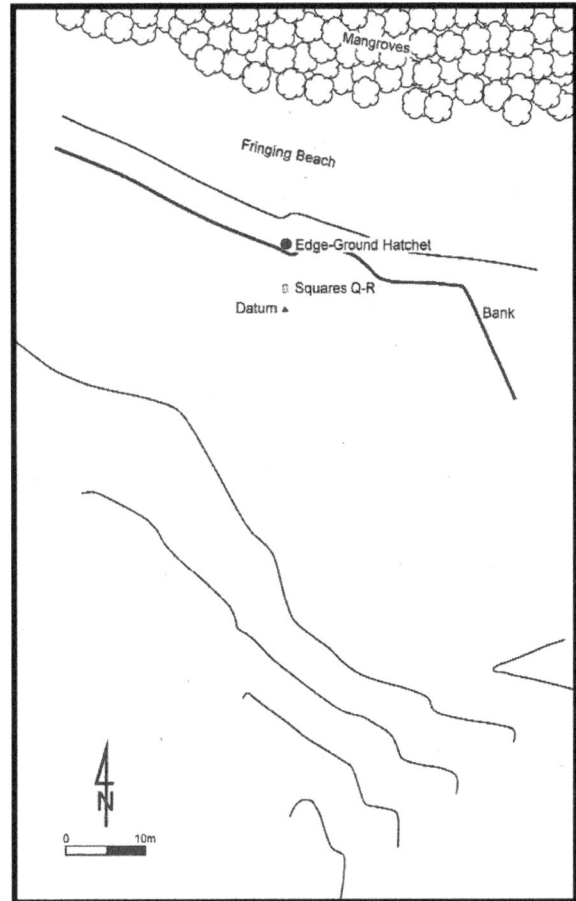

Figure 9.6 Site plan of Ironbark Site Complex, showing area of Squares Q-R. Contours are in 0.5m intervals.

adjacent estuary, because the moisture content of the matrix prevented effective dry-sieving. Sediment samples (c.200g) were taken for each excavation unit from the material that passed through the 3mm sieve prior to wet sieving. Large boulders encountered towards the base of Squares E, G and J were checked for modification before being plotted, removed, weighed and eventually returned as backfill (see below). Excavation of Square E ceased at a maximum depth of c.50cm below surface, when artefactual material was no longer recovered. Stone (n=1,754) and shell (n=4) objects encountered *in situ* during excavation were plotted three-dimensionally. Squares L–R were excavated conventionally.

Squares L–M were situated on the low bank immediately adjacent to the quarry (Fig. 9.3). In this area, *in situ* stone artefacts were visible as a concentrated layer c.30cm below ground surface in the erosion bank, with flaked rhyolitic tuff and a single oyster valve on the surface. The excavation trench was oriented at right angles to the erosion bank and situated 50cm south of the edge of the bank (Fig. 9.3). The erosion bank in this area is oriented east-west and undercut by wave action to a depth of c.30cm, preventing excavation closer to the eroding section. Recent bank recession in this area is evidenced by a large dead tree in growth position 2m from the current erosion section (Fig. 9.3). Excavation proceeded in shallow, arbitrary excavation units averaging 4.55cm in depth and 10.18kg in weight. Excavation ceased at a maximum depth of 68.24cm below ground surface after unmodified boulders of rhyolitic tuff covered the entire base of the excavation area (Fig. 9.11). A total of 30 XUs was removed, distributed as follows: Square L (13 XUs), Square M (17 XUs). A total of 305.3kg of sediment was excavated. Excavated sediments were gently dry-sieved through 3mm screens onto a plastic tarpaulin located 10m southeast of the excavation to prevent contamination to underlying strata. Stone (n=180), pumice (n=17), charcoal (n=5) and shell (n=1) specimens encountered *in situ* during excavation were plotted three-dimensionally.

Square N was located 100m southwest of Squares L–M (Fig. 9.2). Archaeological materials in this area are focussed on a relatively flat ridge located c.100m inland of the quarry (Fig. 9.4). The ridge trends northeast-southwest, is approximately 20m in elevation, and contains two isolated exposures of shell (dominated by oyster), a scatter of bottle glass and a large baler shell artefact associated with a cluster of over 80 cycads (*C. megacarpa*). This cycad species is restricted in distribution to a small area of central Queensland centred on Miriam Vale and including the Bustard Bay area (Hill 1992). Some cycad plants in the group suggest considerable antiquity, with trunk heights of up to 3m. A 50cm × 50cm square was located over one of the shell exposures. Excavation proceeded in shallow, arbitrary excavation units averaging 3.46cm in depth and 8.8kg in weight. Excavation ceased at a maximum depth of 27.7cm below ground surface after culturally-sterile sediments had been reached. A total of 8 XUs was removed comprising 70.6kg of sediment. Excavated sediments were gently dry-sieved through 3mm screens onto a plastic tarpaulin located 10m west of the excavation. Stone (n=112) and shell (n=51) specimens encountered *in situ* during excavation were plotted three-dimensionally. Glass was collected using tweezers wrapped in plastic cling wrap, which was changed between samples to avoid cross-contamination of any residues. Samples of near-surface sediments were taken from the location of the glass scatter to allow analysis of background starch and cellulose.

Squares O–P were located 150m southeast of Squares L–M (Fig. 9.2). A layer of oyster shell was exposed in the low erosion section in this area adjacent to a large block of rhyolite (with flaking evident around its top margin) on the fringing beach. Immediately inland of the frontal dune is an oblong-shaped teatree swamp c.30m wide (north-south). The land rises gently to the south of the swamp, where shell material is visible on the surface in disturbed areas around the base of trees and in crab burrow spoil, indicating a probable subsurface origin. There is evidence for extensive erosion of the deposits fronting the creek, with the stumps of several large trees in growth position located on the fringing beach several metres away from the current erosion face (Fig. 9.5). The excavation trench was situated 50cm south of the edge of the erosion bank. Excavation proceeded in shallow, arbitrary excavation units averaging 3.7cm in depth and 12.94kg in weight. Excavation ceased at a maximum depth of 65.7cm below ground surface after several units of unambiguously culturally-sterile sediments had been removed. A total of 32 XUs was removed, distributed as follows: Square O (13 XUs), Square P (19 XUs). A total of 414kg of sediment was excavated. Excavated sediments were gently dry-sieved through 3mm screens onto a plastic tarpaulin located 10m south of the excavation. Shell (n=85), charcoal (n=12), crustacean carapace (n=9), stone (n=9) and pumice (n=2) specimens encountered *in situ* during excavation were plotted three-dimensionally.

Squares Q–R were located 250m southeast of Squares O–P (Fig. 9.2) and adjacent to the location of an edge-ground hatchet which was recorded in the erosion section in October 1996 during the original recording of the site complex (Fig. 9.12). By January 1998 the hatchet had eroded out of the bank onto the fringing beach. The erosion bank in this area has slumped and is slightly rounded so the 1m x 0.5m excavation grid was located on the top of a low ridge c.3m south of and at right angles to the erosion bank (Fig. 9.6). Excavation proceeded in shallow, arbitrary excavation units averaging 3.6cm in depth and 11.77kg in weight. Excavation ceased at a maximum depth of 67.06cm below ground surface after several units of culturally-sterile sediments had been removed. A total of 37 XUs was removed, distributed as follows: Square Q (17 XUs), Square R (20 XUs). A total of 435.5kg of sediment was excavated. Excavated sediments were gently dry-sieved through 3mm screens onto a plastic tarpaulin located 10m west of the excavation. Shell (n=41), charcoal (n=16) and stone (n=5) specimens encountered *in situ* during excavation were plotted three-dimensionally.

At the completion of the excavations a layer of plastic sample bags was placed over the bases of Squares L–M, N, O–P and Q–R, which were then backfilled with sediments which had passed through the 3mm sieve, and culturally-sterile yellow sands from the beach fringing Middle Creek. Large dead tree branches were placed over the top to discourage goanna burrowing in the backfill. Squares E, G and J were backfilled with the large non-artefactual boulders which had been removed during the course of excavation.

Cultural deposit and stratigraphy

Excavation revealed a highly variable distribution of cultural material across the site (Table 9.1). While large quantities of stone artefacts and minimal shell were recovered from the quarry and adjacent bank, excavations to the east of the quarry revealed a low density sequence of shell with only occasional small stone artefacts.

Squares E, G and J consisted almost entirely of flaked stone, with occasional shell and crab remains assumed to be of recent origin given the presence of colouration and/or flesh still attached to many specimens. The deposit can be divided into three major stratigraphic units (SUs) (Table 9.2). The surface (SUI) is dominated by large pieces of stone with abundant smaller stone material directly underneath (Fig. 9.8). Many of the smaller pieces of stone were oriented vertically along their longitudinal margin, suggesting an origin higher in the deposit. This pattern may be the result of size-sorting of the deposit by wave action, resulting in smaller material working its way down in the rock matrix. Alternatively, it could represent functional differences in the use of space for different reduction activities (see below). Lower excavation units contained abundant flaked material under 30mm in maximum dimension which is not visible on the present quarry surface. The base of the exposed quarry deposit comprised large boulders embedded in coarse sands (SUII). The boulders appeared to have been flaked *in situ*: although flaked material was recovered around their margins, no material was found directly beneath them on removal. Basal deposits also contained small, rounded gravels presumably derived

Figure 9.7 General view of the western side of the quarry sloping into Middle Creek, showing a massive core in centre foreground. Middle Island in background. Facing north.

Figure 9.8 Surface of Square E, XU1, showing photographic recording method. Facing north.

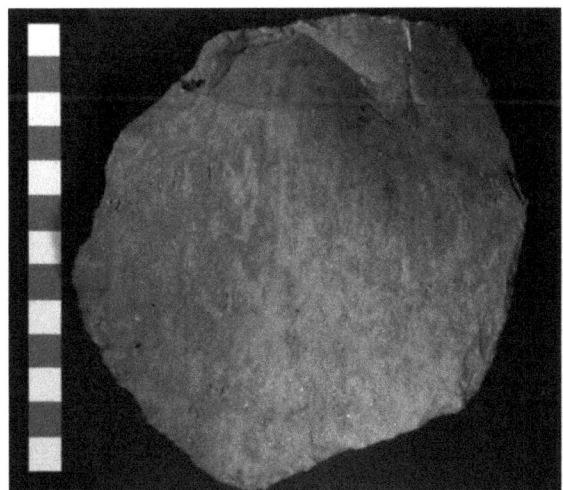

Figure 9.9 Large flake (FS54) manufactured on non-local banded rhyolite recovered from the surface of Square G on the quarry. Note the heavy edge-damage along the distal margin. Scale=1cm (Photograph: Paul Aurisch).

Figure 9.10 Water-rounded microgranite hammerstone (FS88) exhibiting impact-pitting recovered from the surface of Square J on the quarry. Several similar artefacts were noted eroding from adjacent midden deposits. Bustard Head, some 11km to the north is the nearest source of this raw material. Scale=1cm (Photograph: Paul Aurisch).

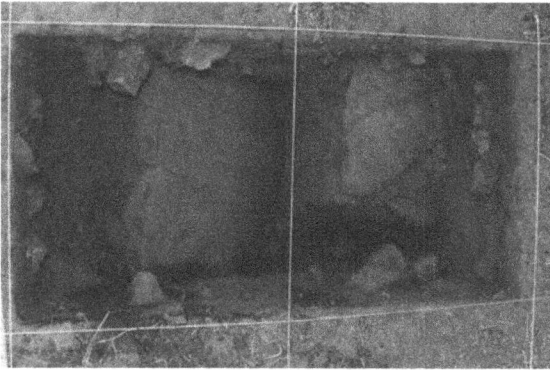

Figure 9.11 General view of completed excavation, Squares L–M. Facing southeast.

Figure 9.12 Close-up view of edge-ground hatchet (FS2747) (Photograph: Paul Aurisch).

from long-term tidal deposition of sediments. In Square E, the closest excavation square to the creek, the coarse, infilling sands gave way to dark grey mangrove muds and clays with depth (SUIII). This pattern suggests that the original rock outcrop consisted of rounded boulders stacked on top of one another with an infilling of sands and gravels in the interstices between rocks in all but the upper 30cm of deposit.

Sediments at Squares L–M were similar to those encountered at Square N, but quite different from those excavated at Squares O–P and Q–R. The Squares L–M deposit can be divided into four major SUs on the basis of sediment colour and texture (Table 9.3, Fig. 9.13). The matrix is basically brown throughout, with SUs differentiated largely on the basis of texture, and markedly different levels of consolidation throughout the deposit. Large quantities of stone artefacts were recovered from SUII, with occasional stone artefacts recovered to the base of excavations, when large boulders were exposed across the basal units. pH values are slightly acidic (6.0) to acidic (5.5) throughout, which may partially account for the absence of shell and bone material in this part of the deposit. The location of Squares L–M at the base of a steep hill suggests that local sediments derive primarily from eroding soils on the surface of the rhyolitic debris flow which forms the substrate of the elevated ground (Fig. 9.3). This pattern contrasts with the excavations to the east, which encountered the quartz-dominated sandy sediments of the low transgressive dunes.

Angular rhyolitic gravels were encountered in Square N c.10cm below the surface and excavations ceased at 27.7cm (Table 9.4, Fig. 9.14). Most of the cultural material was recovered from the thin layer of soil overlying the rock, with occasional shell pieces located in the interstices of the lower rocky unit. Cultural materials were dominated by oyster (*S. glomerata*), with some crustacean carapace, nerite (*Nerita balteata*) and stone artefacts. The shallow and limited distribution of cultural materials suggests that all material is likely to be contemporaneous. The pH values range from 7.5 at the top of the sequence to 6.0 in the basal sterile sediments.

Limited quantities of shell, dominated by oyster, were recovered from the shallow bank deposits east of the quarry exposure in Squares O–P and Q–R consistent with the low density shell observed in the erosion bank. Most shell was recovered from 20–30cm below ground surface. Occasional small shell fragments were recovered to the base of excavations and are thought to have been displaced by crab burrowing (see above). All sediments comprise brown to yellowish-brown sands. The pH values are generally slightly acidic (6.0), but range from neutral (7.0) to highly acidic (4.0). Despite being separated by 250m, Squares O–P and Q–R share a similar stratigraphic sequence. These deposits can be divided into six major SUs (Tables 9.5–9.6, Figs 9.15–9.16). The upper units are dominated by humic material and numerous roots. Squares O–P yielded a greater range of cultural materials than Squares Q–R, including fish otoliths and crustacean carapace fragments.

Table 9.1 Ironbark Site Complex, Squares E–R: summary excavation data and dominant materials.

SQUARE	XUs (#)	DEPTH (cm)	WEIGHT (kg)	SHELL (g)	BONE (g)	CHARCOAL (g)	ARTEFACTS (g)	STONE (g)	ORGANIC (g)
E	11	16.02	NA	NA	0	0	143359.2[a]	NA	0
L	13	68.24	142.95	3.92	0	33.12	3058.45	2538.88	965.84
M	17	68.10	162.30	0	0	21.87	8941.10	1672.87	606.88
N	8	27.70	70.60	940.79	0.02	8.81	5.99	52177.60	1193.50
O	13	52.72	179.50	345.64	0.63	54.42	3.31	195.91	988.60
P	19	65.66	234.50	334.37	0.24	78.91	51.29	275.86	1119.44
Q	17	67.06	213.60	173.00	0	21.14	216.17	80.39	2738.80
R	20	66.24	221.90	46.53	0	24.34	1.21	29.25	1310.20
Total	107	–	1225.35	1844.25	0.89	242.61	155636.72	56970.76	8923.26

a   Only includes artefacts over 30mm in maximum dimension (n=869).

Table 9.2 Stratigraphic Unit descriptions, Ironbark Site Complex, Square E.

SU	DESCRIPTION
I	Exposed rhyolitic tuff rock matrix consisting of both artefactual and non-artefactual stone. Approximately 96% of this material is artefactual. SU depth varies across the square owing to the uneven nature of the large rock matrix. No sediments are present in this SU.
II	Extends across the entire square to a depth of c.19cm. A rocky matrix with moist, dark yellowish-brown (10YR-4/4) coarse sands. Approximately 50% of the rock is artefactual. pH values are alkaline (8.5).
III	Extends across the entire square to a depth of at least 40cm below the surface of the quarry. Compact dark grey (1N-6/6) clay matrix containing largely non-artefactual gravels and boulders. The few artefacts recovered from the unit are positioned vertically along the longitudinal margin in the cracks between larger rocks. Mangrove roots are visible in the underlying unexcavated matrix. pH values remain alkaline (8.5).

Table 9.3 Stratigraphic Unit descriptions, Ironbark Site Complex, Squares L–M.

SU	DESCRIPTION
I	Extends across the entire trench with an average depth of c.10cm and a maximum depth of 17cm below the surface. The unit comprises humic, loosely consolidated, dark greyish brown (10YR-4/2) sediments which are poorly-sorted, fine and subrounded. It contains many fine fibrous and some larger roots. Small pieces of pumice are present. pH values are slightly acidic (6.0).
II	Extends across the entire trench with a maximum thickness of 20cm and a maximum depth of 35cm below ground surface. More consolidated than SUI, with visible clumping of sediments. Matrix remains fine and sandy with dark greyish brown (10YR-4/3) sediments. Several large roots encountered, with fewer fine rootlets. This SU contains the majority of stone artefacts recovered from the deposit. Small pieces of pumice are present. pH values are slightly acidic (6.0).
III	Extends across the entire trench with the exception of an area of Square M where a large rhyolitic boulder encountered at the base of SUII interrupts the unit. The SU has a maximum thickness of 17cm and a maximum depth of 48cm below ground surface. It consists of fine sandy brown (10YR-5/3) sediments mottling in places to yellowish brown (10YR-5/4). Some large roots present. Less pumice present than SUII. pH values are slightly acidic to acidic (5.5–6.0).
IV	Extends across the entire trench, except where interrupted by boulders, with a base defined by rhyolitic boulders. The unit has a minimum thickness of 33cm and a minimum depth of 72cm where it infills crevices between boulders. The SU may extend further in depth in the continuing thin interstices between boulders where it was not possible to excavate. The unit exhibits moist, well-sorted, fine subrounded sandy brown (10YR-4/3) sediments mottling in places to dark yellowish brown (10YR-4/4) sediments. The unit is less consolidated than SUIII. No pumice is present. Few fine rootlets are present. pH values remain slightly acidic (6.0).

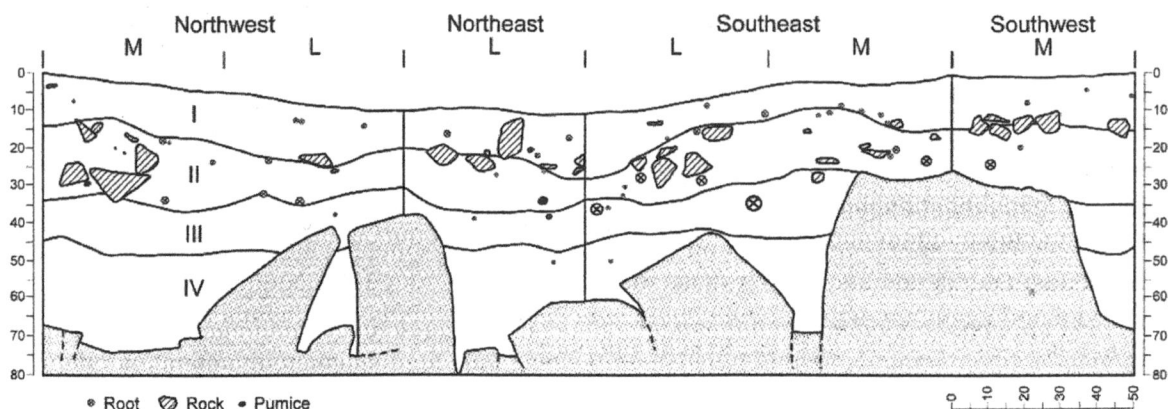

Figure 9.13 Stratigraphic section, Ironbark Site Complex, Squares L–M.

Table 9.4 Stratigraphic Unit descriptions, Ironbark Site Complex, Square N.

SU	DESCRIPTION
I	Extends across the entire pit with an average depth of 3cm and a maximum depth of 6cm below the ground surface. The unit comprises a thin, loosely consolidated matrix capped by a dark organic humic layer containing many fibrous roots. Occasional tufts of grass penetrate the surface with numerous small fibrous roots. Sediments are dark greyish brown in colour (10YR-4/2) and are poorly-sorted, coarse and subangular. Includes some shell (dominated by oyster) which appears to be concentrated on the surface, roots up to 1cm in diameter and rhyolitic tuff gravels. pH values are slightly alkaline (7.5).
II	Extends across the entire pit with a maximum thickness of 8cm and a maximum depth of 13cm below the surface. The loosely consolidated sediments continue to be dark greyish brown (10YR-4/2) although the matrix is slightly finer with poorly-sorted medium and subangular sediments predominating. Some larger roots are present in this unit (up to 2cm in diameter), although there is less organic material in general than in SUI. Shell material is common at the top of this unit, but decreases in abundance with depth. Occasional small stone artefacts and pieces of blocky charcoal also occur. Most of the cultural material (including shell, crustacean carapace and stone artefacts) recovered from the square was located at the interface of SUI/II. pH values are neutral (7.0).
III	Extends across the entire pit with a maximum thickness of 11cm and a maximum depth of 18cm below the surface. Non-artefactual rocks make up the majority of this unit. Sediment colour and texture remains consistent with SUII: dark greyish brown and loosely consolidated. Only very small shell and charcoal fragments were recovered with large quantities of small subangular gravels. There are few roots and the sediment is extremely dry and dusty. pH values are slightly acidic (6.0).
IV	Unit extends across the entire trench with a minimum thickness of 16cm and a minimum depth of at least 30cm below the surface. The base of this unit was not reached. As in SUIII, non-artefactual coarse gravels make up the majority of the unit. Sediments are poorly consolidated and lighter than previous SUs grading from brown (10YR-5/3) to pale brown (10YR-6/3). The matrix continues to be poorly-sorted, medium and subangular. There is evidence for minor bioturbation in this unit in the form of ant burrowing. There are few roots present and only minute fragments of oyster and charcoal were recovered. pH values remain slightly acidic (6.0).

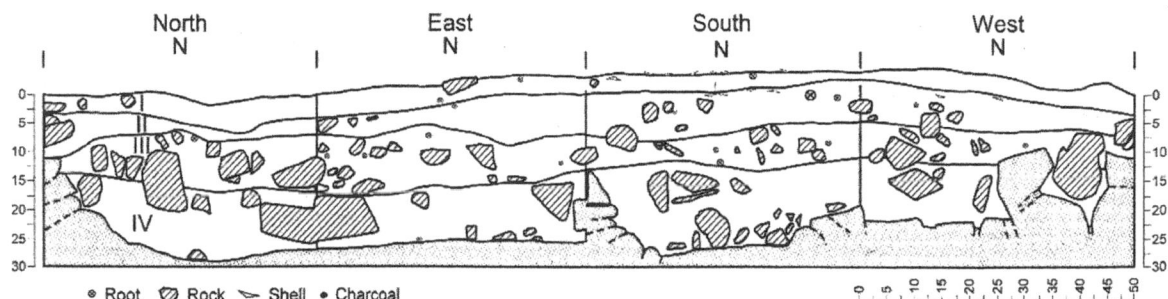

Figure 9.14 Stratigraphic section, Ironbark Site Complex, Square N.

Table 9.5 Stratigraphic Unit descriptions, Ironbark Site Complex, Squares O–P.

SU	DESCRIPTION
I	Only extends across the southern half of Square P to a depth of approximately 3cm below ground surface. The unit comprises a dense mat of humic material including rootlets, bark, leaves and small fragments of charcoal. Sediments are dark greyish brown (10YR-4/2) and are well-sorted, fine and subrounded. This sediment matrix was found to be consistent with that of the lower SUs. pH values are neutral (7.0).
II	Extends across the entire trench with a maximum thickness of 8cm and a maximum depth of 13cm. In Square O and the northern half of Square P this unit extends from the surface. It consists of loosely consolidated brown (10YR-5/3) sandy sediments. Numerous small rootlets and a small number of larger roots (up to 3cm in diameter) are present as well as some humic matter, especially towards the upper margin of the SU. Shell (dominated by oyster), charcoal and pumice are present, particularly towards the base of the unit. pH values are slightly acidic (6.0).
III	Extends across the entire trench with a maximum thickness of 11cm and a maximum depth of 22cm. Brown (10YR-5/3) to greyish brown (10YR-5/2) sediments with numerous roots ranging from <1–3cm in diameter. Occasional shell encountered throughout. pH values are slightly acidic (6.0–6.5).
IVa	Extends across the entire trench, with the exception of a small area in the northeast section (see SUIVb), with a maximum thickness of 17cm and a maximum depth of 32cm. Sediments comprise a loosely consolidated yellowish brown (10YR-5/4) sand. Shell is most abundant in this SU, especially in Square P. Small roots and fine hair rootlets are numerous. pH values are neutral to slightly acidic (6.5–7.0).
IVb	Well-defined dark greyish brown (10YR-4/2) sandy matrix with numerous minute charcoal fragment inclusions and occasional shell restricted to the northeast corner of Square O. The unit has a maximum thickness of 17cm and a maximum depth of 33cm below the surface. The matrix is poorly consolidated, but otherwise similar to SUIVa. pH values are slightly acidic (6.5).
Va	Extends across the entire trench with a maximum thickness of 12cm and a maximum depth of 42cm. Very similar to SUIV but graded to a slightly lighter yellowish brown (10YR-5/4) with more roots. This SU contains small quantities of shell and occasional pumice and charcoal. pH is slightly acidic (6.5).
Vb	Extends across the entire trench. Although the base of this SU was not reached in Square O it is assumed to have a maximum thickness of c.20cm and a minimum depth of c.60cm. Comprises light yellowish brown (10YR-6/4) loosely consolidated sediments including a small number of roots. Contains occasional artefacts, non-artefactual stone and pumice. pH values are slightly acidic (6.0–6.5).
VI	Extending across the entire base of Square P with a minimum thickness of 12cm and a maximum depth of at least 67cm below the surface. The base of this unit was not reached and was only encountered in Square P owing to deeper excavation, but it is assumed to continue across the base of Square O. Sediments comprise loosely consolidated very pale brown (10YR-7/4) sands and appear to be culturally-sterile with occasional non-artefactual stone, pumice and charcoal fragments. Few roots occur in this unit. pH values are slightly acidic (6.5).

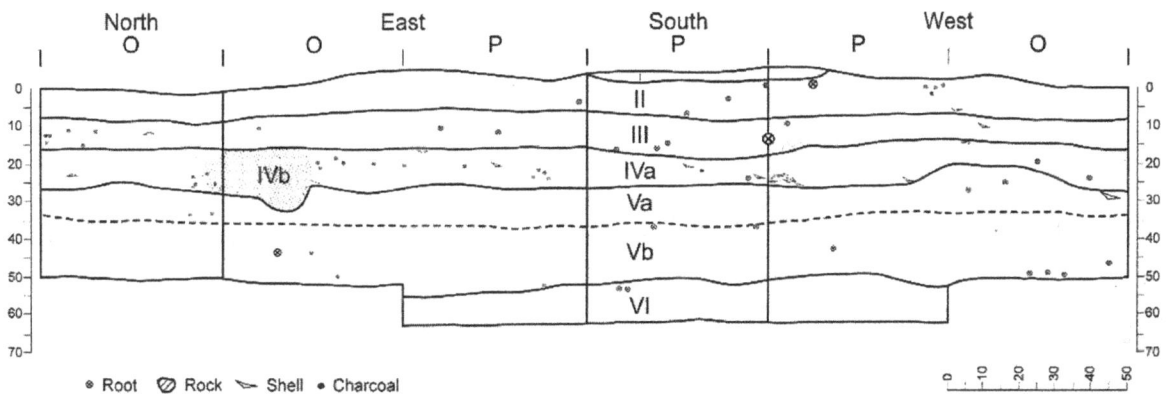

Figure 9.15 Stratigraphic section, Ironbark Site Complex, Squares O–P.

Table 9.6 Stratigraphic Unit descriptions, Ironbark Site Complex, Squares Q–R.

SU	DESCRIPTION
I	Extends across the entire trench with an average depth of 3cm and a maximum depth of 5cm below the surface. Thin unit of fibrous, humic material and roots in a poorly-sorted, fine, subrounded matrix mottling from brown (10YR-5/3) to dark yellowish brown (10YR-4/3). pH values are slightly acidic (6.0).
II	Extends across the entire trench with a maximum thickness of 15cm and a maximum depth of 18cm. It comprises loosely consolidated yellowish brown (10YR-5/4) sediments with some fibrous roots and occasional large roots. Contains occasional shell (dominated by oyster), artefacts, charcoal and pumice. pH values are slightly acidic (6.0–6.5).
III	Extends across the entire trench with a maximum thickness of 14cm and a maximum depth of 28cm. Sediments are very loosely consolidated and brown (10YR-5/3) to pale brown (10YR-6/3). pH values are slightly acidic (6.5).
IV	Extends across the entire trench with a maximum thickness of 24cm and a maximum depth of 52cm. Comprises brown (10YR-5/3) sediments. More consolidated and fewer roots than SUIII. Includes occasional stone artefacts, pumice, charcoal and minute shell fragments. pH values range from slightly acidic to acidic (5.0–6.0).
V	Extends across the entire trench with a maximum thickness of 20cm and a maximum depth of 62cm. Unit comprises loosely consolidated yellowish brown (10YR-5/4) sediments with few roots. Includes some minute stone artefacts and shell fragments, with pumice nodules and small charcoal fragments common throughout the unit. pH values are slightly acidic (6.0–6.5).
VI	Extends across the entire base of the trench with a minimum thickness of 14cm and a minimum depth of at least 70cm below the surface. The base of this unit was not reached. Comprises loosely consolidated pale brown (10YR-6/3) to very pale brown (10YR-7/4) sediments with few roots. This unit appears to be culturally-sterile with pumice nodules and occasional fragments of charcoal common. pH values are slightly acidic (6.0–6.5).

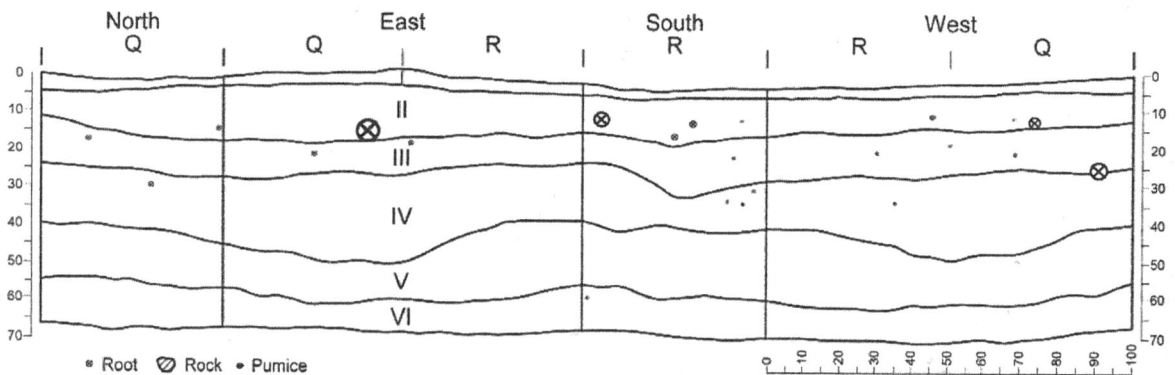

Figure 9.16 Stratigraphic section, Ironbark Site Complex, Squares Q–R.

Radiocarbon dating and chronology

Eight radiocarbon dates were obtained from the site (Table 9.7). Five conventional radiocarbon dates were obtained on charcoal and two on *Anadara trapezia* valves. In addition, a single accelerator mass spectrometry (AMS) date was obtained on organic material from a pollen core from the teatree swamp adjacent to Squares O–P. A single shell/charcoal paired sample (Wk-8558/Wk-8557) was obtained from Square P to investigate local marine reservoir conditions (see Chapter 4 for details). The samples consisted of whole *A. trapezia* valves and charcoal fragments collected from the 3mm sieve residue. Although the shell/charcoal pair returned an apparent age difference of 390 ^{14}C years, a ΔR value could not be calculated because the charcoal age is modern at one sigma. This places the determination in a recent segment of the radiocarbon calibration time-scale where significant uncertainties in ^{14}C activity are introduced by large-scale fossil fuel combustion beginning in the late nineteenth century. In the absence of a determined estuarine ΔR value for Middle Creek, the local open water value of ΔR= +10±7 is adopted as the default value for the calibration calculations presented in Table 9.7.

In the absence of datable organic material on the quarry, three radiocarbon dates were obtained for Squares L–M on the adjacent bank, where stratified deposits containing both flaked stone artefacts and charcoal occurred. The date of 1,519 cal BP (Wk-6361) is associated with the lowest artefactual material in Square M and provides a date for initiation of use of the bank adjacent to the quarry as a discard area, and by implication, provides a minimum age for quarrying at the site. The date for these activities, however, is not necessarily synchronous with initiation of reduction at the quarry itself. The two other dates of c.600 cal BP (Wk-6359) and 1,288 cal BP (Wk-6360) correspond to synchronous peaks in the abundance of charcoal and stone artefacts in Square M. The simple age-depth relationship indicates at least two major phases of deposition. The apparently rapid sedimentation evident in the bottom two-thirds of the deposit is exaggerated by the small volume of sediment excavated between the large basal boulders and corresponds with very low rates of stone artefact discard. The determination of 1,288 cal BP (Wk-6360) dates the first major peak in artefact density and is at a depth where sediments have infilled the underlying crevices between major boulders to form a relatively flat surface. From this date onwards this bank area appears to have been regularly used for stone artefact discard. Dates from Squares O–P and R suggest that the bank deposits adjacent to the quarry largely date to the last 500 years. Evidence for older deposits in this area may have been removed by erosion.

Indirect age estimates for quarrying can also be derived by establishing the presence of Ironbark Site Complex quarrying products in other dated sequences in the region. Although rhyolitic tuff does not appear regularly in regional stone artefact assemblages until the last 1,500 years (see Chapter 14), stone artefacts manufactured from rhyolitic tuff geochemically consistent with the Ironbark Site Complex raw material have been recovered from the basal levels of Square E2 at Eurimbula Site 1, located some 11km to the southeast, dating to 3,020±70 BP (Wk-3945), raising the possibility that the quarry was in use at a much earlier date (Ulm et al. 1999a).

Table 9.7 Radiocarbon dates from the Ironbark Site Complex (see Appendix 1 for full radiometric data for each determination). * = assumed value only.

SQUARE	XU	DEPTH (cm)	LAB. NO.	SAMPLE	δ^{13}C (‰)	^{14}C AGE	CALIBRATED AGE/S
M	4	5.4–10.5	Wk-6359	charcoal	-26.9±0.2	650±60	669(626,600,560)517
M	9	22.9–28.1	Wk-6360	charcoal	-25.7±0.2	1400±60	1351(1288)1173
M	17	60–69.3	Wk-6361	charcoal	-26.2±0.2	1640±150	1865(1519)1260
O	9a	27.4	Wk-8556	*A. trapezia*	-0.5±0.2	910±55	608(509)442
P	7	16.3	Wk-8557	charcoal	-26±0.2	200±140	508(271,210, 197,194,146,15,3)0*
P	7	17.6	Wk-8558	*A. trapezia*	-0.3±0.2	590±60	317(254)0*
R	9	17.5–20.4	Wk-10964	charcoal	-26.8±0.2	290±89	506(297)0*
Core	–	25–30	OZD-756	organics	-25*	215±55	310(275,174,149,10,4)0*

In addition to the radiocarbon assays, an assemblage of bottle glass associated with Aboriginal activities also provides evidence for the chronology of site use. Only one (FS186) of the seven pieces of glass could be confidently dated. The bottle fragment has the base mark of the Australian Glass Manufacturers. Bottles with this base mark were manufactured between AD 1900 and AD 1915 (Boow 1991:180; Errol Beutel, Queensland Museum, pers. comm., 1999). Another glass specimen (FS182) is dated to between AD 1820 and AD 1870 on the basis of the absence of a mould seam and the presence of a deep push-up with ridges made by the pontil (Boow 1991:26). A more precise date for manufacture cannot be assigned to this bottle, although its close association with the base confidently dated to AD 1900–AD 1915 suggests that the entire assemblage dates to the early twentieth century.

A low-lying ephemeral swamp dominated by weeping cabbage palm (*L. decipiens*) and melaleuca (*M. quinquenervia*) occurs in the base of a swale immediately inland of the frontal dune in the area of Squares O–P, running parallel to the creek bank. The basin sediments of this feature were sampled using a simple method of pile-driving a 3m length of 50mm diameter aluminium pipe into the approximate centre of the swamp, c.30m southeast of Squares O–P, to obtain a sample of the organic deposits. The core penetrated to a depth of 1.26m below ground surface and revealed a band of rich organic sands to c.30cm, underlain by sterile beach sands. The determination of 215±55 BP (OZD-756) dates the base of the organic layer and provides a minimum age for swamp formation, suggesting that this feature is of very recent origin.

Future dating efforts could make use of cosmogenic radionuclide dating (10B and 26Al) to provide a chronological framework for when the rhyolitic tuff present at the quarry was exposed and thus became available for Aboriginal utilisation. A large, worked boulder (see Fig. 9.7) provides a good candidate for dating the exposure time for the upper surfaces. The assumption is that the boulder, like the rest of the outcrop at the quarry, was transported to its present position by a rock debris flow, and that during that transport the rock was split, exposing fresh surfaces. It is this fresh face which has been knapped. The age determination obtained for the timing of the exposure of this fresh face would provide indications, firstly, of the probable timing of the debris flow and indirectly the timing at which the surface became available for Aboriginal resource procurement. Large nodules of sea-rafted pumice recovered from stratified bank deposits (especially Squares L–M) may also be suitable for indirect dating using geochemical source characterisation (see Ward and Little 2000).

In summary, radiocarbon determinations from the frontal dune excavations indicate that cultural deposits began accumulating by 1,640±150 BP (Wk-6361) with indirect evidence for site use by 3,020±70 BP (Wk-3945). Bottle glass dated to AD 1900–AD 1915, assigned an Aboriginal behavioural origin (see below), establishes a *terminus post quem* for use of the site.

Stratigraphic integrity and disturbance

Several lines of evidence suggest that the stratified deposits generally exhibit good stratigraphic integrity. Although stone on the quarry has been displaced to an unknown degree by tidal action, cultural materials from the adjacent bank deposits are consistent with the pattern of distribution observed in section in the erosion bank. This pattern suggests that the extant bank deposits exhibit integrity, although of major concern is the differential representation of cultural deposits caused by bank recession. The presence of large tree stumps in growth position on the mangrove fringe indicates significant recent erosion, while the presence of large artefacts in the intertidal mangrove fringe up to 15m from the present erosion section suggests an erosional regime of some antiquity. Interpretation of aerial photographs from the late 1940s is currently underway in an attempt to quantify the rate of recent erosion. Anecdotal evidence from local crabbers suggests that there has been a major restructuring of the creek mouth over the last 30 years.

Conjoin analysis of the *A. trapezia* assemblage was limited by the small number of valves present in excavations and their generally poor preservation. Twelve measured intact and broken valves were considered in the conjoin analysis, distributed as follows: Square O (10 valves); Square P (2 valves). Methods proceeded as described in Chapter 5. A single conjoin was identified. The conjoining valves both came from the shell concentration encountered in Square O, XU9, located c.30cm below ground surface. The close vertical proximity of these valves suggests that material in this zone of the deposit has experienced limited post-depositional movement.

After tidal erosion and root penetration, ghost crab (*Ocypode cordimanus*) and goanna (*Varanus punctatus*) burrowing appear to be the major source of post-depositional disturbance across the site. Although no voids or other unambiguous evidence for burrowing were encountered during excavation, ghost crabs and goannas are common in the area. A goanna was observed burrowing into the backfill of Squares L–M on several occasions. Goannas are known to have caused significant disturbance to shell deposits elsewhere (see Roberts 1991).

Laboratory methods

Owing to the large quantity of excavated materials from the quarry, detailed analysis of Square E only has been completed to date. Field observations and preliminary processing of Squares G and J indicate a broad homogeneity in the composition of all three pits, although Square E yielded the most artefactual material. All other excavated squares (Squares L–R) were analysed to maximise the available sample (see Chapter 3 for a detailed discussion of the standard laboratory methods employed at all sites). Results from all squares are summarised below, although only selected data are illustrated in Figures 9.17–9.24. Further summary results for all excavated squares are available in Appendix 4. Stone artefacts from Squares E and M were analysed in detail by Reid (1998) who used 22 attributes to investigate the reduction sequence at the quarry. These attributes were designed to identify the primary stages of reduction, focussing, for example, on the presence of cortex. These attributes were applied to all of the artefacts larger than 30mm from Squares E and M (Reid 1998). For further details on methods, see Reid (1998). Use-wear and residue analyses were conducted on the bottle glass assemblage in the Archaeological Sciences Laboratory, University of Queensland, using standard procedures outlined by Loy (1994).

Cultural materials

Invertebrate remains

Nineteen taxa of shellfish weighing 1,844.3g were recovered from Squares L–R, consisting of nine marine bivalves, eight marine gastropods and two terrestrial gastropods (Table 9.8). The shell is dominated by rock oyster (*S. glomerata*), comprising 74.4% by weight (Figs 9.19–9.20), followed by scallop (*Pinctada albina sugillata*) (8.9%) and mud ark (*A. trapezia*) (8.5%). The remaining 16 taxa are relatively rare, each contributing less than 3% by weight. The assemblage exhibits relatively high diversity with a calculated Shannon-Weaver Function (H') of 1.633 and Simpson's Index of Diversity (1–D) of 0.65. With the exception of a single piece of oyster found on the surface of Square L, no shell was recovered from the 305.3kg of deposit excavated from Squares L–M adjacent to the quarry. Square N yielded the most shell despite the small volume of the excavation, with 940.8g, consisting of 886.9g of oyster. The range of shellfish taxa recovered indicates gathering focussed on the intertidal zone and creek margins adjacent to the site. Although shell sizing was undertaken, the small sample makes results unreliable and they are not discussed further.

Some 23.8g of mud crab (*Scylla serrata*) carapace are present in Squares N and O–P. Virtually all of this material was recovered from Square O, XU4, approximately 10cm below ground surface. The assemblage was dominated by the hard, decay-resistant claw tips, providing another indication that acidic soil preservation conditions may have adversely affected the representation of other crustacean carapace components.

Table 9.8 Presence/absence of shellfish identified in the Ironbark Site Complex, Squares L–R.

FAMILY	TAXON	L	M	N	O	P	Q	R	TOTAL (g)
				MARINE BIVALVIA					
Arcidae	*Anadara trapezia*			X	X	X		X	156.1525
Chamidae	*Chama fibula*					X			0.6770
Donacidae	*Donax deltoides*					X			2.1302
Mytilidae	*Trichomya hirsutus*			X	X	X			23.0983
Ostreidae	*Saccostrea glomerata*	X		X	X	X	X	X	1372.7084
Pteriidae	*Pinctada albina sugillata*			X	X	X	X		164.7145
Veneridae	*Gafrarium australe*				X				2.7673
Veneridae	*Placamen* sp.					X			32.7395
Veneridae	*Venerid* sp.				X				0.1644
				MARINE GASTROPODA					
Batillariidae	*Pyrazus ebininus*				X	X			6.8144
Lottiidae	*Acmaeid* sp.			X					0.2285
Littorinidae	*Bembicium nanum*			X					1.9586
Littorinidae	*Littoraria* sp.				X				0.0187
Muricidae	*Bedeva paivae*				X				0.1294
Muricidae	*Morula marginalba*				X				0.8771
Neritidae	*Nerita balteata*			X					48.6466
Planaxidae	*Planaxis sulcatus*			X					0.0553
				TERRESTRIAL GASTROPODA					
Camaenidae	*Trachiopsis mucosa*			X	X	X	X	X	29.8380
Subulinidae	*Eremopeas tuckeri*			X		X	X	X	0.5263

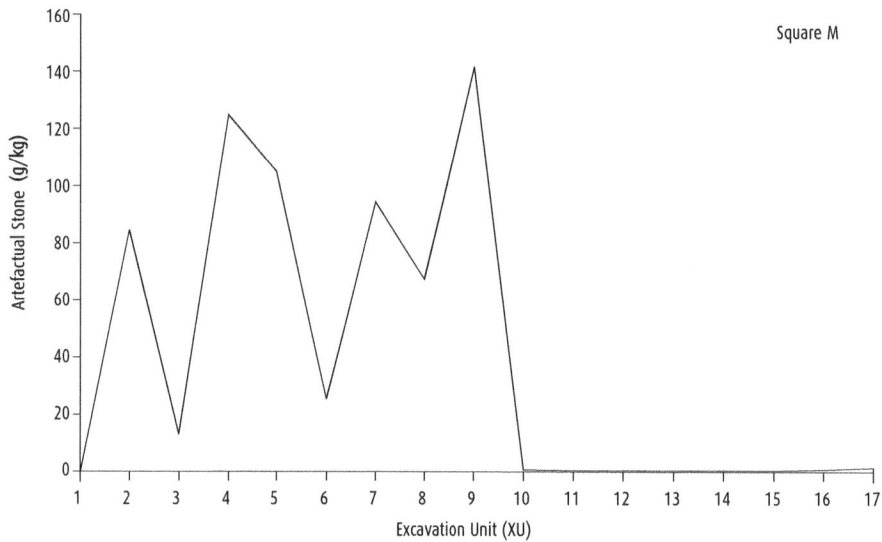

Figure 9.17 Abundance of artefactual stone.

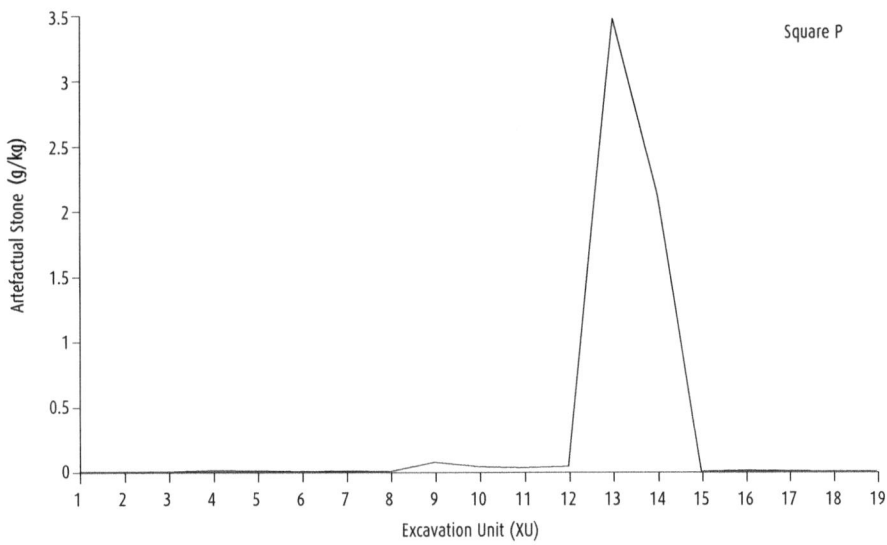

Figure 9.18 Abundance of artefactual stone.

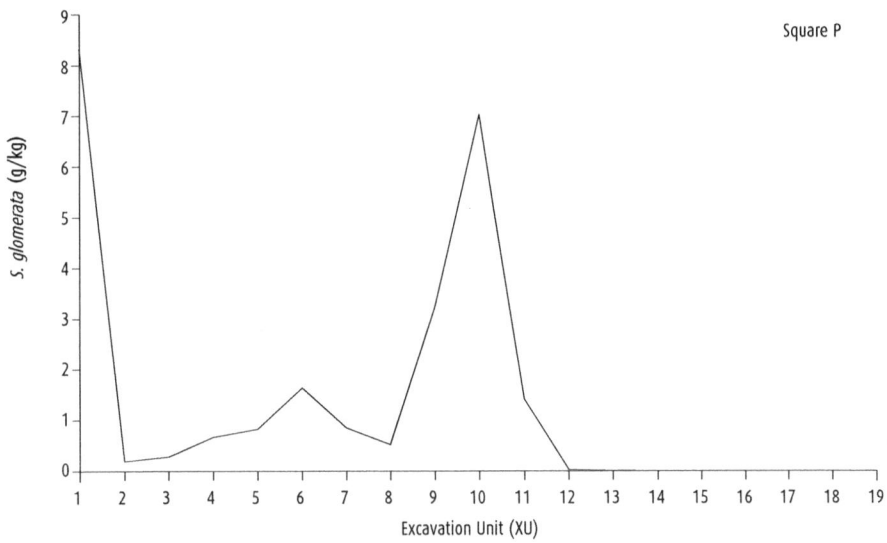

Figure 9.19 Abundance of oyster (*S. glomerata*).

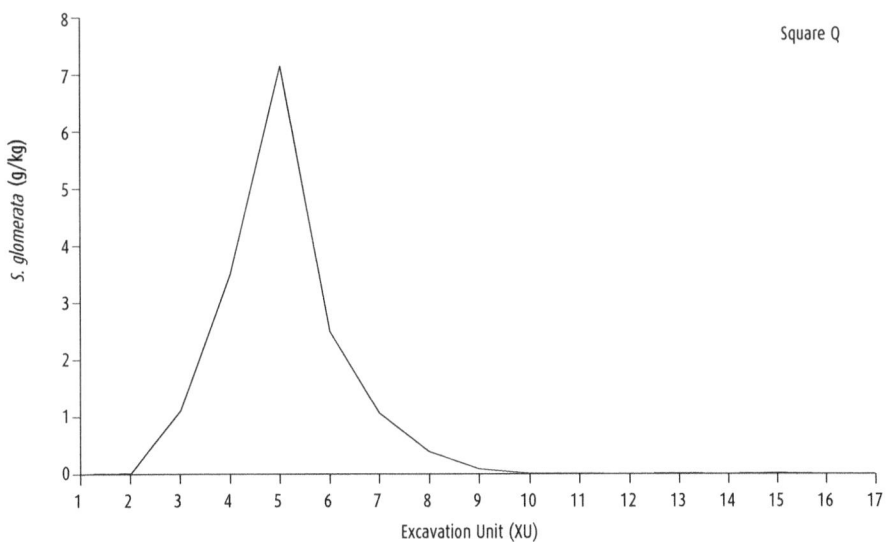

Figure 9.20 Abundance of oyster (*S. glomerata*).

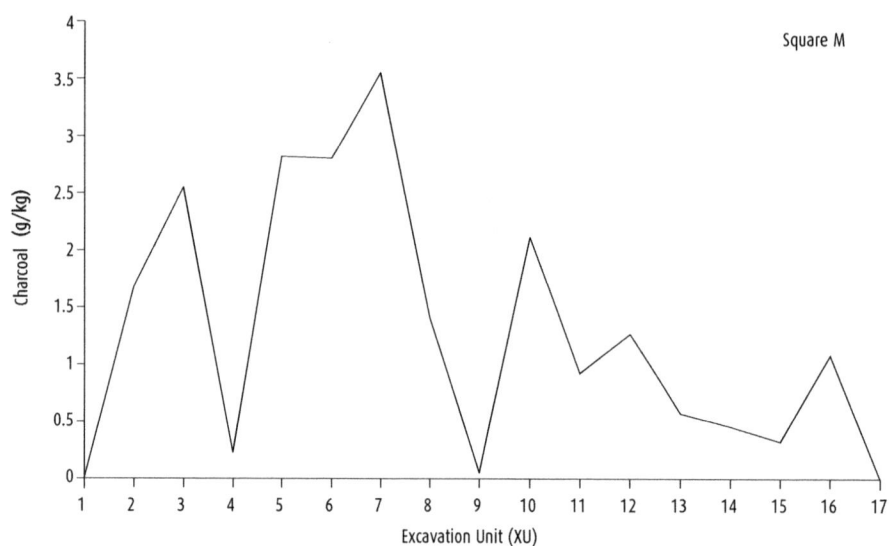

Figure 9.21 Abundance of charcoal.

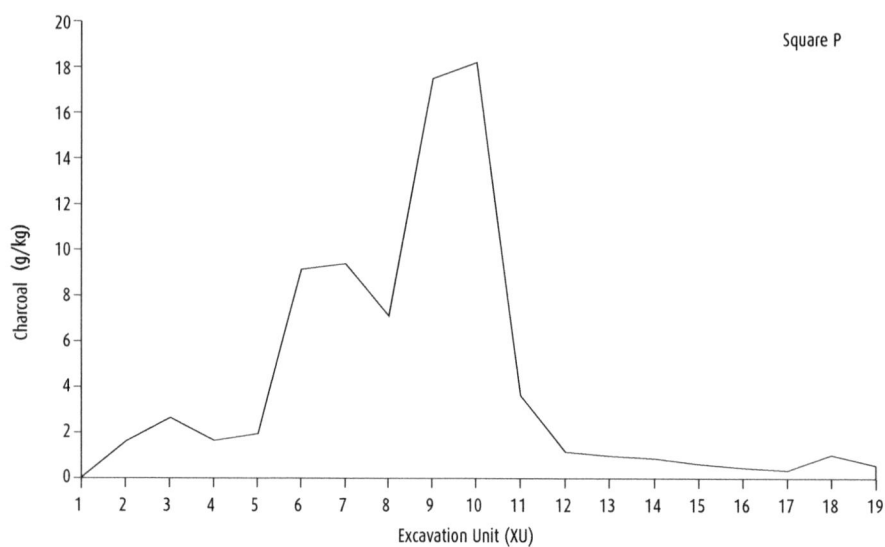

Figure 9.22 Abundance of charcoal.

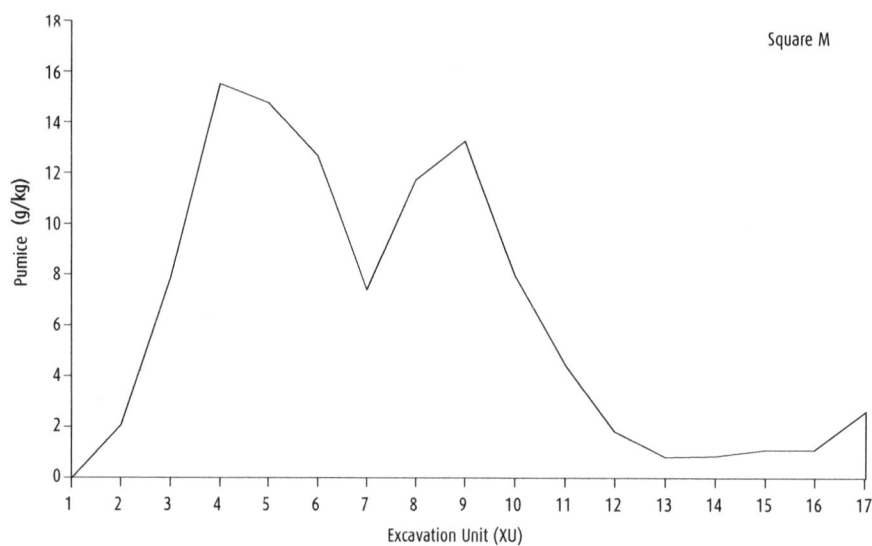

Figure 9.23 Abundance of pumice.

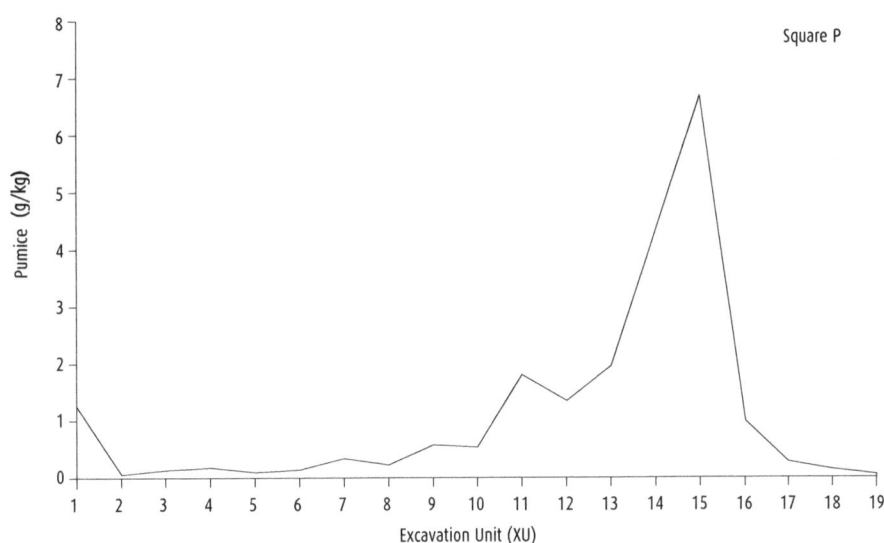

Figure 9.24 Abundance of pumice.

Baler shell artefact

The baler shell (*Melo amphora*) artefact recovered near the cycad grove in the vicinity of Square N is the first of its type found in the region. Although badly deteriorated, the shell exhibits evidence for intentional modification. An ovulate area approximately 10cm long and 5cm wide appears to have been removed from the ventral surface of the final whorl. Fragments of shell have been removed from the ventral margin, either intentionally or during use, to form a jagged edge. Despite intensive survey of the area where the artefact was recovered, no other fragments of baler shell were located, suggesting that shell working took place elsewhere. Roth (1904:29) describes three major forms of modification of *Melo* sp. shells as 'water-carriers' in Queensland: on the east coast north of Bowen *Melo* sp. shells have the ventral surface of the last whorl, the spire and columella completely removed to form an open basin; in the Whitsunday Island region an area of the ventral surface of the final whorl is removed to allow insertion of the hand to hold the columella as a handle; and along the Gulf of Carpentaria the last whorl of the shell is pierced for insertion of the thumb during transport. The present example is similar to the form described by Roth for the Whitsunday Island region. Similar examples are held by the Queensland Museum indicating a general distribution of this form throughout central Queensland.

Vertebrate remains

Bone is extremely rare in all sampled deposits at the Ironbark Site Complex owing to the acidic sediments. Five pieces of bone were identified as fish, weighing 0.8895g, with a NISP of three. A total MNI of three was calculated. The three pieces of bone identified in Squares O–P were all lefthand otoliths identified as Sparidae. These otoliths range in size from 9.9–11.8mm in length which equate to modern comparative collection fish lengths of c.305–365mm. The weight of bone identified to taxon was 0.8738g, giving an identification rate of 98.23% (Table 9.9) (see Vale 2004 for further details).

Table 9.9 Fish bone abundance, Ironbark Site Complex, Squares N, O and P.

SQUARE	XU	NUMBER SPECIMENS	TOTAL WEIGHT (g)	NISP	WEIGHT NISP (g)	MNI	% IDENTIFIED BY WEIGHT
N	5	2	0.0157	0	0	0	0
O	6	1	0.2948	1	0.2948	1	100
O	8	1	0.3395	1	0.3395	1	100
P	6	1	0.2395	1	0.2395	1	100
Total	–	5	0.8895	3	0.8738	3	98.23

Stone artefacts

Stone artefacts were recovered from all excavated squares (Table 9.10). Large quantities of stone artefacts were recovered from Square E, with 869 artefacts above 30mm in maximum dimension weighing 143.4kg. Only three of these artefacts were not rhyolitic tuff, comprising one basalt and two silcrete artefacts. All of the rhyolitic tuff is assumed to derive from the Ironbark exposure itself. The predominance of rhyolitic tuff is not surprising given the proximity of the raw material source. Few formal artefacts were identified in the assemblage. The low silica content and presence of large inclusions in the raw material greatly diminish the predictability of flaking the material. Since virtually all of the excavated stone is rhyolitic tuff, with a similar specific gravity, weight is a useful proxy for artefact size. A plot of weight of artefacts more than 30mm versus depth shows a clear pattern of size-sorting in Square E deposit (Fig. 9.25). Heavier artefacts are located at the top of the deposit while the abundance of lighter material increases with depth. This pattern is likely the result of tidal action transporting smaller material into interstices in the deposit.

Elsewhere at the quarry, a large flake (FS54) manufactured on banded rhyolite was recovered from the surface of Square G (Fig. 9.9). This artefact was found between two large boulders with its platform facing upwards, suggesting that the object was intentionally cached. Heavy edge-damage and polish are evident along the distal margin. Banded rhyolite is not available locally and no likely sources of this material have been identified during geological reconnaissance of the wider region (Stephen Cotter, Cooperative Research Centre for Landscape Evolution and Mineral Exploration, University of Canberra, pers. comm., 2000). In Square J a water-rounded microgranite hammerstone (FS88) exhibiting impact-pitting on the margins was recovered from the surface (Fig. 9.10). Similar artefacts were noted eroding from adjacent midden deposits. Bustard Head, 11km to the north, is the nearest source of this raw material.

A total of 4,178 stone artefacts weighing 12,277.5g was recovered from the bank deposits of Squares L–R (Table 9.10). The assemblage is manufactured only on rhyolitic tuff (n=4,165) and quartz (n=13). The assemblage is overwhelmingly dominated by small flaked pieces, with occasional flakes, broken flakes and cores only identified in Squares L–M. Most artefacts are extremely small, with an average weight of 2.9g. Size-classing of artefacts less than 30mm in maximum dimension from Squares E and M was undertaken by Reid (1998:70–2). Results show that <30mm material is dominated by very small (<10mm) artefacts identified as flaking by-products in experimental studies. Furthermore, Reid (1998:74–5) found that c.60% of artefacts recovered from Square E and c.50% of artefacts recovered from Square M exhibited either primary or secondary cortex, indicating initial stages of reduction consistent with quarrying activities. These data are consistent with a pattern of initial reduction on the exposed quarry itself, with later stages of artefact production undertaken on the adjacent bank. The large volume of very small flakes as the by-product of these activities support this conclusion (for further details see Reid 1998).

Lamb (2003) examined one large artefact (FS3166) for organic residues which was collected from the erosion bank c.20m west of Squares O–P. This artefact is manufactured on rhyolitic tuff and exhibits a roughly triangular cross-section and distinct bevelling along one margin. Lamb (2003) found starch grains and plant fibres concentrated on the bevelled edge of the tool including cooked, partially cooked and otherwise damaged starch grains indicating probable processing of both cooked and raw plants.

Glass artefacts

Seven bottle glass fragments were recovered from the surface near the cycads on the ridge south of the quarry near Square N. The glass assemblage consists of three bottle bases and four small body sherds, one of which conjoins to a base. The recovery of glass at a long-term Aboriginal site some distance from known early European population centres supports the inference that it was discarded by Aboriginal people. This inference is strengthened by the fact that only incomplete

Table 9.10 Stone artefacts from the Ironbark Site Complex, Squares E, L–R.

| | RHYOLITIC TUFF | | | | | | | | QUARTZ | | TOTAL | |
| | FLAKE | | FLAKED PIECE | | CORE | | OTHER[a] | | FLAKED PIECE | | | |
SQUARE	#	(g)	#	(g)	#	(g)	#	(g)	#	(g)	#	(g)
E[b]	7	1078.1	867	112720.9	22	29438.7	1	121.3	0	0	869	143359.0
L	7	407.8	1347	1950.7	7	691.3	2	8.3	2	0.50	1365	3058.5
M	1	257.8	2640	7138.3	2	1313.4	1	233.7	0	0	2644	8943.2
N	0	0	8	3.9	0	0	0	0	2	2.1	10	6.0
O	0	0	59	3.3	0	0	0	0	1	<0.1	60	3.3
P	0	0	49	50.5	0	0	0	0	4	0.7	53	51.2
Q	0	0	26	216.4	0	0	0	0	2	0.1	28	216.5
R	0	0	16	1.1	0	0	0	0	2	0.1	18	1.2
Total	15	1743.7	5012	122085.1	31	31443.4	4	363.3	13	3.6	5047	155638.9

a   Includes broken flakes and hammerstones.

b   Only includes artefacts over 30mm in maximum dimension.

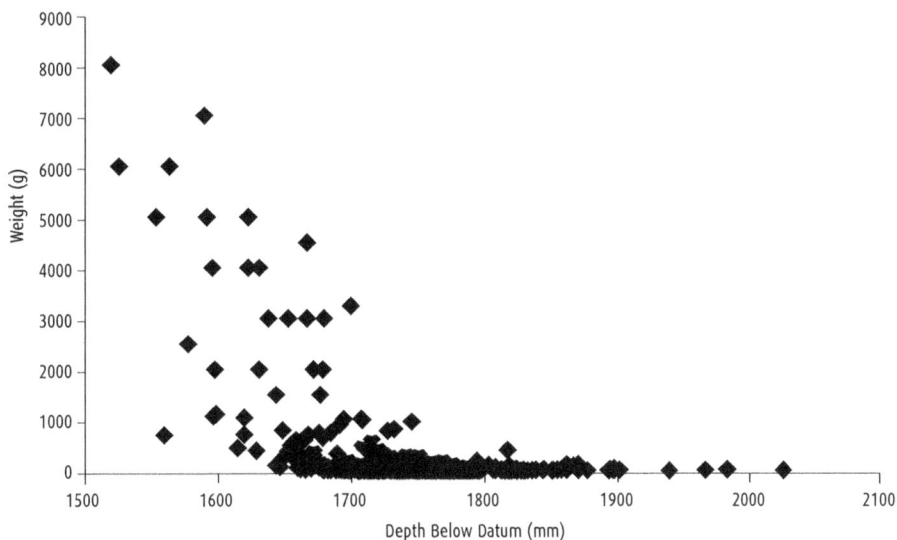

Figure 9.25 Weight-depth distribution of individually provenanced artefacts greater than 30mm in maximum dimension, Square E (n=784).

bottles were recovered. The presence of three bases in the assemblage suggests intentional selection and transport of the thicker bottle bases to the site. This pattern has been documented elsewhere in Australia, where thick bottle bases were targetted for acquisition while thinner neck, shoulder and body sherds were frequently discarded (Allen and Jones 1980; Freeman 1993). In the absence of unambiguous signs of intentional modification of the glass, the assemblage was inspected for use-wear and residues.

All seven glass objects were initially examined for use-wear and residues using an Olympus® metallurgical incident-light microscope under low level (<500×) magnification (see Loy 1994 for a discussion of techniques). Although starch grains were observed on the surface of all seven glass objects, only two sherds (FS181 and FS187) exhibited quantities of starch and use-wear features suggestive of systematic use in plant processing. Masses of large starch grains (>20µm) are present on the surface of FS187 directly behind the working edge. The other sherd (FS181) has starchy, soft plant tissue and fibres as well as deep scoring on the ventral surface. Both artefacts exhibit starch grains of greater average size and in much greater densities than occur in the surface and near-surface sediments at the site (for further details see Ulm et al. 1999b).

In a further study, Vernon (1999) intensively examined three of the glass objects (FS183, 184, 187) for use-wear and residues. All three artefacts were found to be covered in starch and an opalised, latex, frosted film, with a gritty, textured appearance. Residues were found set back from the cutting edge, and comprise large quantities of starch grains and cellulose. Structural elements include bark and insect casings, seeds and micro-hyphae colonised by cyanobacteria. Sections of xylem (secondary parenchymal wall thickening) were also found, which are common to all angiosperms (flowering plants). These elements are consistent with use of the artefacts on cycad seeds, tubers or mangroves (David Doley, Department of Botany, University of Queensland, pers. comm., 1999). The growth of fungi and microscopic plants on the surface of the glass artefacts had been promoted by the large quantities of starch residues. Using reference specimens, cycad (*C. megacarpa*) starch and sheets of tissue with large storage cells were identified in cross-section. Other residue elements found on the glass artefacts, such as bark, sap and starch with storage tissue, are consistent with tubers, indicating that the glass artefacts were used for a variety of plant processing activities.

Other remains

Scattered fragments of charcoal were recovered from all excavations (except Squares E, G and J on the quarry), with occasional large pieces of blocky charcoal totalling 242.6g (Figs 9.21–9.22). Pumice totalling 3,248.1g occurs throughout the bank deposits but is most abundant in the basal units of Squares L–M, which appear to have been more exposed to storm-surges than other investigated deposits. The quantity of pumice recovered from the bank excavations decreases from west to east (Figs 9.23–9.24). This pattern may be related to the greater protection afforded to these eastern deposits by the low sand bank located just offshore, although the antiquity of this feature is currently unknown (Fig. 9.7). Although pumice is known to have been used as an abrasive (Rowland 1994:121) the assemblage from the Ironbark Site Complex exhibits no obvious signs of use and is thought to have entered the site through a combination of wind and sea-rafting without causing major stratigraphic disturbance, evidenced in the consistent age-depth relationship of the radiocarbon dates for Squares L–M.

Discussion

Excavation revealed a highly variable distribution of cultural material across the site area. Analyses indicate that although the cultural deposits at the Ironbark Site Complex are generally low density, the extent and recent chronology of the majority of the deposits demonstrate that a large quantity of cultural material was discarded over a relatively short period of time. The volume of excavated stone artefacts manufactured on rhyolitic tuff indicates that stone extraction and reduction was the main activity undertaken throughout the period of site occupation, with the exception of the post-contact phase of occupation associated with the use of glass. Fishing, shellfishing and toxic (cycad) plant processing are also documented at the site. Although the Ironbark Site Complex was first used for raw material extraction before 1,500 years ago, intensive site use predominantly occured in the last 1,000 years. The presence of rhyolitic tuff (geochemically identical to the Ironbark quarry raw material) in the Eurimbula Site 1 assemblage by 3,000 cal BP suggests that the quarry was at least occasionally used prior to the last 1,500 years, although available data indicate that earlier phases of use were ephemeral. Systematic use of the quarry as an extraction and reduction site is clearly evidenced within the last 1,500 years although it is also possible that older deposits may have been removed by coastal erosion.

No shell or other food refuse was recovered adjacent to the quarry (Squares L–M). Although stone artefacts are visible in the erosion section in this area no shell was observed. The pH of soils in this area is relatively low (6.0), but shell is preserved in other parts of the site of a similar antiquity with similarly slightly acidic soils. There may be a functional explanation for the absence of food refuse in the immediate area of the raw material outcrop. This difference between midden and non-midden remains may represent a separation of activity areas. The initial stages of stone reduction centred on the raw material exposure and the adjacent bank (the area of Squares L–M) containing no shell remains. Domestic activities involving fish and shellfish use spread along the low, crescentic bank to the east (Squares O–P and Q–R) and west. Ross et al. (2003) observed a similar pattern on Moreton Island. They noted that the abundance of stone declines with distance from the Cape Moreton quarries with a concomitant increase in the density of shell, 'suggesting that close to the source of lithic material, site use was principally associated with tool manufacture, while further away from the quarries the focus of site use was subsistence activities' (Ross et al. 2003:78).

A key issue in situating the use of the Ironbark Site Complex within the broader context of the cultural chronology of the southern Curtis Coast is determining when the resource was first regularly incorporated into local land-use strategies. The great antiquity of the geological events which created the rhyolitic tuff debris flow pre-dates Aboriginal colonisation of the continent by at least several hundred million years, indicating that resource availability was not the determining factor in the initiation of exploitation or subsequent patterns of use. Use of the quarry is therefore more productively viewed as embedded within broader regional mobility patterns, as evidenced by the presence of contemporary sites exhibiting quarrying products. Geochemical sourcing to the Ironbark quarry of edge-ground hatchets distributed over a wide area of central Queensland, and the presence at the site of artefacts manufactured on non-local raw materials including banded rhyolite and microgranite, indicate that raw materials flowed both into and out of the site. The fact that artefacts manufactured on rhyolitic tuff sourced to the quarry are pervasive in excavated assemblages from local sites indicates that the quarry was a significant resource in local land-use strategies and an important node for regional exchange networks of Aboriginal groups along the southern Curtis Coast and adjacent hinterland. The commencement of systematic use of the quarry over the last 1,000 years is contemporaneous with periods of increased site use documented at other sites in the immediate area. The sourcing of numerous artefacts throughout the region to the quarry indicates a refocussing of land-use strategies on local raw materials, or localisation, which may be related to reduced patterns of mobility. These contentions are further developed in Chapter 14.

The presence of flaked bottle glass associated with evidence for continuity of pre-contact subsistence behaviours in the Middle Creek area is significant as it demonstrates continuities in Aboriginal occupation which is otherwise absent from the documentary record. This finding accords with Aboriginal oral historical data (Clarkson et al. n.d.). These data are consistent with McNiven's (1993, 1998; Courtney and McNiven 1998) findings on Fraser Island to the southeast, where clay pipes and flaked bottle glass artefacts have been recovered from the surface of 10 shell middens on the east coast of the island. The evidence from the Ironbark Site Complex also conforms to the evidence from other sites in southeast Queensland suggesting continuing Aboriginal use of places and landscapes into the post-contact period (see Lilley and Hall 1988; Neal 1984). Identification of starch grains adhering to the surface of glass artefacts also illustrates the potential of residue techniques in the study of contact period materials.

Cycads appear to have been widely used throughout central Queensland as a source of carbohydrate in the late Holocene (Beaton 1982; Beck 1985), with a variety of processing techniques adopted to remove toxins before consumption. Processing techniques typically take up to three to five days (Beck 1985). The identification of cycad processing at the Ironbark Site Complex in the early twentieth century indicates that complex food preparation was still undertaken well after a permanent European presence was established in the region. Logically, this evidence suggests that occupation of the site during this period was not ephemeral, but comprised a range of activities including shellfish gathering, artefact manufacture and use, and plant food processing.

The continuing use of occupation sites into the post-contact period may also contribute to our understanding of regional land-use trajectories. Whereas some sites such as the Seven Mile Creek Mound and the Mort Creek Site Complex were abandoned in the remote past, occupation at other sites was first initiated in the last two millennia and then maintained into the post-contact period, indicating historical continuities in the use of, and transmission of knowledge about, culturally important places. This suggestion is supported by findings at the Tom's Creek Site Complex (see Chapter 13) and is further discussed in Chapter 14.

Summary

The Ironbark Site Complex is unique in the regional archaeological record, being the only major lithic reduction site recorded for the coastal area between Elliot Heads near Bundaberg 100km to the south and South Molle Island in the Northumberland Island Group 530km to the north. The site was used for raw material extraction for the last 3,000 years, with evidence for regular use dating from 1,500 years ago, and shows continuity of traditional hunter-gatherer occupational strategies into the early post-contact period. Radiocarbon dates from the extensive midden deposits adjacent to the quarry indicate that the most intensive period of use of the site is probably archaeologically very recent, dating to within the last 1,000 years or so. However, it is likely that older deposits have been removed by coastal erosion. The ubiquity of stone artefacts manufactured on rhyolitic tuff at other sites in the region dating to the last 1,500 years suggests that the Ironbark Site Complex played a key role in provisioning local stone. As a major resource extraction site, changes in the use of the Ironbark Site Complex are important for understanding changes in Aboriginal settlement and subsistence patterns in the broader coastal region.

10

Eurimbula Creek 1

Introduction

This chapter reports archaeological excavations at the site of Eurimbula Creek 1, a small shell midden on the north bank of Eurimbula Creek. Small, stratified middens are rare on the southern Curtis Coast, where the archaeology is dominated by large linear deposits. Other recorded small middens are limited to surface contexts and/or exhibit signs of significant disturbance. Excavations and analyses presented in this chapter demonstrate that Eurimbula Creek 1 represents a single or limited number of occupation events over a relatively short interval in the recent past. The restricted range of activities represented at the site may indicate a specialised extraction function related to logistical patterns of local residential mobility.

Site description and setting

Eurimbula Creek 1 consists of a thin shell layer visible along a 3m section of a low erosion bank (c.40cm high) on the northern margin of Eurimbula Creek (Latitude: 24°09′54″S; Longitude: 151°49′02″E) (Figs 10.1–10.2). The site is situated on the western edge of a narrow projection of land separating Eurimbula Creek from the open ocean beach bordering Bustard Bay. Eurimbula Creek 1 is discrete, being visible only in the erosion bank as a single thin layer of shell dominated by rock oyster (*Saccostrea glomerata*) and pearl oyster (*Pinctada albina sugillata*) located c.18cm below the ground surface. Shell is concentrated over a 5m² area where it has deflated onto an erosion surface which gently slopes to a narrow band of saltgrasses bordering a dense mangrove fringe (Fig. 10.3). Despite intensive ground survey and good visibility, very little shell was located on the surface or the erosion face in the vicinity of the exposure, reinforcing the apparent discreteness of the site. A single valve of pearl oyster was observed on the surface of an inactive brush-turkey (*Alectura lathami*) mound located c.50m northwest of the site and surface shell densities away from the concentration of shell are low, in the order of one fragment/20m². The site is immediately

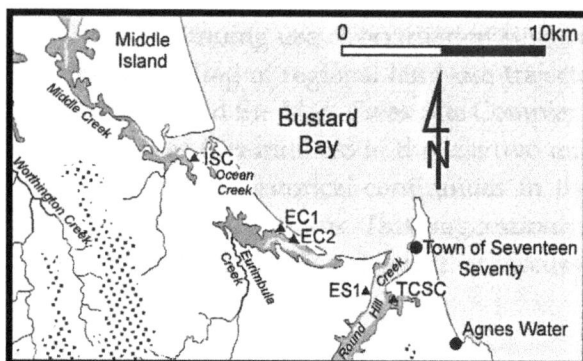

Figure 10.1 The Eurimbula Creek catchment area showing the location of Eurimbula Creek 1 (EC1) and nearby excavated sites (ISC=Ironbark Site Complex; EC2=Eurimbula Creek 2; ES1=Eurimbula Site 1; TCSC=Tom's Creek Site Complex). Dark grey shading indicates the general extent of mangrove, saltflats and claypans. Dotted shading indicates land above 200m. Solid dots indicate local population centres.

adjacent to extensive tidal flats to the west and southwest, with access to the main channel of Eurimbula Creek c.200m to the south. A deeply incised tidal inlet transects the frontal dune bordering Eurimbula Creek 100m east of the site. This ephemeral inlet circles around to the north of the site in a roughly northwest arc (Fig. 10.2). Many trees on the fringing beach have eroded out of bank deposits, indicating some bank recession along the creek. No tree stumps in growth position are evident on the creekward margin of the erosion bank, however, suggesting that the current phase of bank recession is recent. Cattle grazing has exacerbated bank erosion, as evidenced by cattle faeces and tracks close to the bank edge.

Dunes in the vicinity of the site support open forest dominated by eucalypts (*Eucalyptus tessalaris*), burdekin plum (*Pleiogynium timorense*), wattle (*Acacia aulacocarpa*) and red ash (*Alphitonia excelsa*). More diverse plant communities are also located nearby. Extensive swamps border the western and southern margins of Eurimbula Creek and include melaleucas (*Melaleuca quinquenervia*), eucalypts (*E. tereticornis*), swamp box (*Lophostemon suaveolens*), cloudy teatree (*M. dealbata*) and weeping cabbage palm (*Livistonia decipiens*). An area of closed dry rainforest also occurs on the lower southern margin of Eurimbula Creek comprising diverse vegetation including many Aboriginal bush food sources such as burdekin plum, bumpy ash (*Flindersia schottiana*), brown pine (*Podocarpus elatus*) and native cherry (*Exocarpus latifolius*). Mangrove vegetation adjacent to the site comprises a dense fringe of spotted mangroves (*Rhizophora stylosa*) along the northern margin of the creek with yellow mangroves (*Ceriops tagal*) and grey mangroves (*Avicennia marina*) tending to dominate the southern margin (Olsen 1980a:18). Extensive sandy to muddy intertidal flats adjoin the site, supporting a variety of molluscs dominated by gastropods, in particular the telescope mud whelk (*Telescopium telescopium*) which are common amongst the rearward spotted and grey mangroves (Shanco and Timmins 1975) (Fig. 2.8). Freshwater bivalves (*Alathyria pertexta* and *Velesunio ambiguus*) have also been collected from the nearby freshwater sections of Eurimbula Creek (Woodall et al. 1991).

Evidence for non-Indigenous use of the site area is limited. The land to the east of Eurimbula Creek is effectively an island: it is bordered to the east by Bustard Bay and to the west by Eurimbula and Ocean Creeks (Fig. 10.1). The small area of land between the upper reaches of the two creeks is of very low elevation and is frequently waterlogged, preventing easy vehicular access. Introduced weeds constitute the major evidence for non-Aboriginal occupation of the area, with

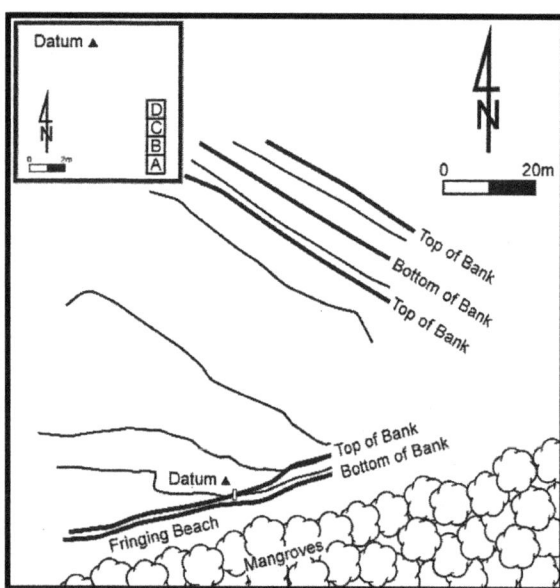

Figure 10.2 Site plan of Eurimbula Creek 1 area. Contours are in 0.5m intervals.

groundsel (*Baccharis halimifolia*) and lantana (*Lantana camara*) concentrated along the upper margins of the two creeks. The site is likely to have been logged in the late nineteenth century (see Chapter 9) and ballast associated with the early timber industry remains in the creek mouth. Eurimbula Creek 1 is located on a Special Purposes Reserve jointly administered by the Queensland Environmental Protection Agency (EPA) and Department of Natural Resources, Mines and Energy. Although Eurimbula Creek is closed to crabbing, line fishing is popular with recreational fishers who access the creek by small boat from a Queensland Parks and Wildlife Service camping area in Eurimbula National Park on the southern head of Eurimbula Creek. Thick mangroves bordering the north bank of Eurimbula Creek effectively prevent access to the site area from the creek itself.

The site was recorded on 2 October 1996 by the author during systematic pedestrian transect surveys focussing on the lower reaches of major estuaries in the study area conducted as part of the Gooreng Gooreng Cultural Heritage Project (GGCHP) (see Lilley et al. 1997). The site was originally designated as GGCHP Site Number CC37 and subsequently registered on the EPA's Indigenous Sites Database as KE:B19. It is registered as Queensland Museum Scientific Collection Number S231.

Several sites have been recorded in the Eurimbula Creek area. Large numbers of modified and apparently unmodified pieces of rhyolitic tuff occur at several sites on the mudflats and claypans of the upper reaches of both Eurimbula and Ocean Creeks at sites KE:B11, KE:B12 and KE:B16 (see Fig. 2.11, Appendix 2). Blocks of rhyolitic tuff have also been transported almost to the mouth of Eurimbula Creek at site KE:B21. The closest major outcrops of rhyolitic tuff are in the vicinity of the Ironbark Site Complex c.5km to

Figure 10.3 Excavations in progress at Squares A–D, showing cattle track on western (left) margin of the excavation. Facing north.

Figure 10.4 General view of completed excavation, Squares A–D, showing transect through erosion bank. Facing north.

Figure 10.5 General view of completed excavation, Squares A–D, showing transect through erosion bank. Facing east.

the northwest and the Tom's Creek Site Complex c.5km to the southeast. The location of rhyolitic tuff away from natural occurrences is thought to result from a combination of deliberate transport and recession of dune deposits containing cultural material leaving lag deposits of heavier material in the intertidal zone. The southern bank of Eurimbula Creek is yet to be surveyed in detail.

The small size and apparent discreteness of Eurimbula Creek 1 is unusual for archaeological sites in the region, which typically cover very large areas. Excavations aimed to help develop an understanding of activities undertaken at smaller sites and how these relate to activities undertaken at larger sites which dominate the archaeology of the region (see Chapters 8, 9, 12, 13). Additionally, the presence of pearl oyster (*P. sugillata*) in both the deflated and *in situ* deposits is unusual as it is uncommon on the surface of other recorded sites in the region and is not well-represented in excavated samples analysed to date.

Excavation methods

A single 2m × 40cm trench was marked out over the exposure at right angles to the erosion bank as 4 × 40cm × 50cm pits (Squares A–D) (Figs 10.3–10.5). Squares C and D were situated on top of the low erosion bank, Square B consisted of the bank margin and adjacent deflation surface, and Square A was wholly located on the deflation surface and contained no material considered to be in primary depositional context. The inclusion of Squares A and B, while lacking stratigraphic integrity, aimed to characterise the broad composition of the site by increasing the volume of recovered material. Excavations were conducted between 10–22 March 1999.

Excavation proceeded in shallow, arbitrary excavation units averaging 3.3cm in depth and 9.5kg in weight. Excavation ceased at an average maximum depth of 51.7cm (Square D) below surface after several units of culturally-sterile sediments had been removed. A total of 47 XUs was removed, distributed as follows: Square A (5 XUs), Square B (11 XUs), Square C (15 XUs), Square D (16 XUs). A total of 448.2kg of sediment was excavated. Excavated sediments were gently dry-sieved through 3mm screens onto a plastic tarpaulin located 5m north-northeast of the excavation grid (Fig. 10.3). XU1–3 of Square A were bulk sampled because their high moisture content prevented effective dry sieving. Shell (n=85), charcoal (n=5) and stone (n=1) specimens encountered *in situ* during excavation were plotted three-dimensionally. A layer of plastic sample bags was placed over the base of Squares A–D which were then backfilled with sediments which had passed through the 3mm mesh and with sands from the beach fringing Eurimbula Creek (see Chapter 3 for a detailed discussion of the standard excavation methods employed at all sites).

Cultural deposit and stratigraphy

Excavation reflected the structure of the site as observed in the eroding section, with a low density sequence of cultural material dominated by rock oyster (*S. glomerata*) and the fragile pearl oyster (*P. sugillata*) concentrated 10–25cm below surface. Small quantities of charcoal, pumice and non-artefactual stone were also recovered. Excavation demonstrated that the majority of cultural material is restricted to the level of the layer visible in the erosion bank and decreases with distance from the creek margin (Table 10.1).

As expected, virtually all of the shell from the deflation surface of Square A was recovered from the first two excavation units, within 6cm of the surface. Square B comprised a surface unit of leaf litter (XU1), a 10cm thick unit of sediment overlying the *in situ* shell layer (XU2), and the *in situ* shell layer itself (XU3–6). Below this level it was not possible to keep material from the deflation surface separated from the *in situ* cultural material in the bank with any confidence as there was no clear separation in the sediments. Parts of the erosion bank are also either slightly undercut or slope to the south to join the deflation surface.

Sediments comprise quartz-dominated sands of the low transgressive dunes which can be divided into two stratigraphic units (SUs) on the basis of colour and texture (Table 10.2, Fig. 10.6). Apart from the moisture-laden lower portion of the excavated profile, there is no clear variation in the sediments in the trench. Squares C and D and the northern third of Square B on the top of the erosion bank are capped by a shallow layer of matted roots and humic material (SUI). Virtually all cultural material was recovered from a narrow band across the top of SUII, with occasional shell fragments located lower in the unit. The shallow and limited distribution of cultural materials suggests that all material is roughly contemporaneous. The pH values range from acidic at the top of the sequence (5.5) to slightly alkaline (8.5) in the basal sterile sediments.

Table 10.1 Eurimbula Creek 1, Squares A–D: summary excavation data and dominant materials.

SQUARE	XUs (#)	DEPTH (cm)	WEIGHT (kg)	SHELL (g)	BONE (g)	CHARCOAL (g)	STONE (g)	ORGANIC (g)
A	5	13.66	52.10	245.31	0.02	4.12	40.88	42.28
B	11	41.94	81.50	499.85	0.07	12.69	155.66	253.19
C	15	46.82	152.10	288.18	0.02	23.89	389.25	1101.98
D	16	51.66	162.45	296.90	0.01	35.78	429.99	1353.80
Total	47	–	448.15	1330.24	0.12	76.48	1015.78	2751.25

Table 10.2 Stratigraphic Unit descriptions, Eurimbula Creek 1, Squares A–D.

SU	DESCRIPTION
I	Extends across the surface of Squares C and D and the northern third of Square B on the top of the erosion bank with an average depth of 7cm and a maximum depth of 10cm below the surface. It comprises a thin layer of matted fibrous roots and other organic matter including leaves, roots and bark. Sediments are extremely unconsolidated, consisting of dark brown (10YR-3/3), poorly-sorted, coarse, subangular particles. Pumice is present as small nodules and some charcoal occurs. Minute shell fragments are occasionally present. pH values are slightly acidic to acidic (5.5–6.0).
II	Extends across the entire trench with a minimum thickness ranging from 10–48cm and a maximum depth of at least 55cm below the surface. The base of this unit was not reached. Sediments are generally poorly consolidated, poorly-sorted, fine and subrounded to rounded. Moisture content of the sediments increases with depth, making the lower sediments appear darker in colour. Sediments are generally brown (10YR-4/3). A variety of shell taxa was recovered across the upper margin of the unit, including rock oyster and pearl oyster. Small blocky charcoal fragments and occasional large pumice nodules are common. Roots are present throughout, though less numerous and generally larger than SUI. pH values are slightly acidic to slightly alkaline (6.0–8.5).

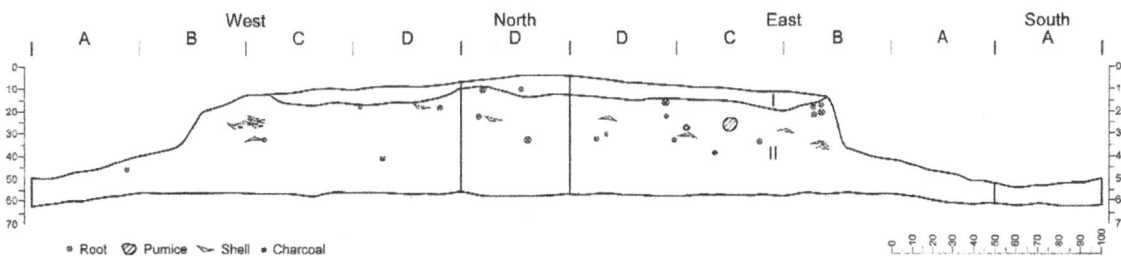

Figure 10.6 Stratigraphic section, Eurimbula Creek 1, Squares A–D.

Radiocarbon dating and chronology

A date for the site was obtained on a large blocky charcoal sample weighing 3.4g from Square C, XU6 (Table 10.3). The sample returned a value of 230±60 BP (Wk-7680) which equates to calibrated ages of 279, 171, 152 and 5 cal BP. These multiple intercepts are equally probable and are caused by short-term variation in atmospheric radioactive carbon activity in this segment of the radiocarbon calibration time-scale introduced by large-scale fossil fuel combustion beginning in the late nineteenth century. The absence of post-contact material culture (e.g. glass artefacts) in the assemblage, in evidence at other sites in the region, indicates a probable pre-European origin for the Eurimbula Creek 1 assemblage. In summary, the limited vertical distribution of recovered material and single radiocarbon date indicate that cultural deposits are very recent, probably being deposited in the 300 year period between AD 1652–AD 1950. The absence of post-contact materials narrows the probable chronology to before about AD 1900.

Table 10.3 Radiocarbon dates from Eurimbula Creek 1 (see Appendix 1 for full radiometric data).

SQUARE	XU	DEPTH (cm)	LAB. NO.	SAMPLE	δ^{13}C (‰)	^{14}C AGE	CALIBRATED AGE/S
C	6	14.9-18.3	Wk-7680	charcoal	-26.1±0.2	230±60	298(279,171,152,5)2

Stratigraphic integrity and disturbance

Several lines of evidence suggest that the stratified deposits exhibit good integrity. The location of excavated cultural materials is consistent with the pattern of distribution observed in section in the erosion bank. This pattern indicates that the extant bank deposits exhibit reasonable integrity, although of major concern is the differential representation of deposits caused by bank recession. As noted above, the absence of trees in growth position on either the fringing beach between the bank and mangrove forest or further south towards the creek suggests that the current phase of bank recession is of recent origin.

Evaluation of stratigraphic integrity using bivalve conjoin analysis was limited by the small number of *A. trapezia* valves in the shellfish assemblage. Only three intact and broken *A. trapezia* valves were recovered, one from Square A and two from Square D. Methods proceeded as described in Chapter 5. A single conjoin was identified. A right valve recovered from Square D, XU6, was found to conjoin with a left valve from Square D, XU10. This pair is separated by a minimum vertical distance of 8.6cm and a maximum of 15.2cm. This finding lends further support to the argument that the site represents a single or limited number of occupation events.

Although most shell was recovered 10–25cm below ground surface, occasional minute shell fragments were recovered to the base of excavations. These are considered unlikely to be in primary depositional context and are thought to have been displaced from the shell layer by crab burrowing and root penetration. Although no voids or other unambiguous evidence for burrowing was encountered during excavation, smooth-handed ghost crabs (*Ocypode cordimanus*) are common in the area.

Laboratory methods

Owing to the small size of the excavation and the relatively low density of cultural material recovered from Eurimbula Creek 1, all squares were analysed to maximise the available sample (see Chapter 3 for a detailed discussion of the standard laboratory methods employed at all sites). In the sections below, the results from all squares are summarised, although only selected data from Squares C and D are illustrated in Figures 10.7–10.12. This approach was adopted to emphasise results from the intact deposits rather than Squares A and B, which are impacted by tidal action. Further summary results for all excavated squares are available in Appendix 4.

Cultural materials

Invertebrate remains

Fourteen taxa of shellfish weighing 1,330.2g were recovered from Squares A–D, consisting of six marine bivalves, four marine gastropods and four terrestrial gastropods (Table 10.4). The shell assemblage is dominated by rock oyster (*S. glomerata*) comprising 86.7% of the shell assemblage by weight (Figs 10.7–10.8), followed by the fragile pearl oyster (*P. sugillata*) (7.5%), mud ark (*A. trapezia*) (3.1%) and hercules club shell (*Pyrazus ebininus*) (2.2%). The remaining 10 taxa are relatively rare, each contributing less than 1% of the shell assemblage by weight. The assemblage exhibits low diversity with a calculated Shannon-Weaver Function (H') of 0.507 and Simpson's Index of Diversity (1–D) of 0.21. A single pipi (*Donax deltoides*) valve was recovered from Square C, XU5. This finding of a non-estuarine open beach dweller is significant because it provides evidence from a stratified context for Aboriginal use of resources from the open beach located a minimum of 450m to the northeast (see Fig. 10.1). Square B yielded the most shell despite the small volume of the excavation, with 499.9g consisting of 451.8g of rock oyster. Shell in Squares C and D is concentrated between XU4–9, which equates to depths of 10–25cm. The range of shellfish taxa indicates gathering focussed on the intertidal zone and creek margins adjacent to the site. Shell sizing indicates that the shellfish assemblage is skewed towards larger sizes, consistent with a cultural origin for the deposit. The restricted range of taxa also suggests targetted collection. A small piece of crustacean carapace, probably mud crab (*Scylla serrata*), was recovered from Square A, XU1. This fragment may not derive from human behaviour, however, as it is situated in a deflation zone impacted by tidal inundation.

Vertebrate remains

Fourteen pieces of fish bone were recovered, weighing 0.13g. Three specimens were recovered from Square A, eight from Square B, two from Square C and one from Square D (Table 10.5). This material was highly fragmented and none could be identified to taxon. The specimen from Square A, XU1, is a vertebrae with a centrum diameter of 4.6mm. The cultural origin of this specimen is equivocal, however, given the deflated surface provenance of the sample in a zone of tidal influence. A single burnt cleithrum (the bone attached to the base of the pectoral fin) was recovered from Square B, XU6, suggesting a cultural origin (see Vale 2004 for further details).

Table 10.4 Presence/absence of shellfish identified in Eurimbula Creek 1, Squares A–D.

FAMILY	TAXON	A	B	C	D	TOTAL (g)
		MARINE BIVALVIA				
Arcidae	*Anadara trapezia*	X	X		X	41.3481
Donacidae	*Donax deltoides*			X		2.1999
Mytilidae	*Trichomya hirsutus*				X	0.0111
Ostreidae	*Saccostrea glomerata*	X	X	X	X	1153.3280
Pteriidae	*Pinctada albina sugillata*	X	X	X	X	100.6465
Veneridae	*Irus* sp.			X		0.1251
		MARINE GASTROPODA				
Batillariidae	*Pyrazus ebininus*		X	X	X	28.9469
Littorinidae	*Bembicium nanum*		X			0.0252
Neritidae	*Nerita balteata*	X				0.0999
Trochidae	*Thalotia* sp.	X				0.0714
		TERRESTRIAL GASTROPODA				
Camaenidae	*Figuladra* sp.			X	X	1.7902
Camaenidae	*Trachiopsis mucosa*	X				0.0252
Pupillidae	*Pupoides pacificus*	X				0.0415
Subulinidae	*Eremopeas tuckeri*		X		X	0.0265

Table 10.5 Fish bone abundance, Eurimbula Creek 1, Squares A–D.

SQUARE	XU	NUMBER SPECIMENS	TOTAL WEIGHT (g)	NISP	WEIGHT NISP (g)	MNI	% IDENTIFIED BY WEIGHT
A	1	1	0.0256	0	0	0	0
A	5	2	0.0149	0	0	0	0
B	4	1	0.0050	0	0	0	0
B	6	7	0.0695	0	0	0	0
C	6	2	0.0136	0	0	0	0
D	9	1	0.0053	0	0	0	0
Total	–	14	0.1339	0	0	0	0

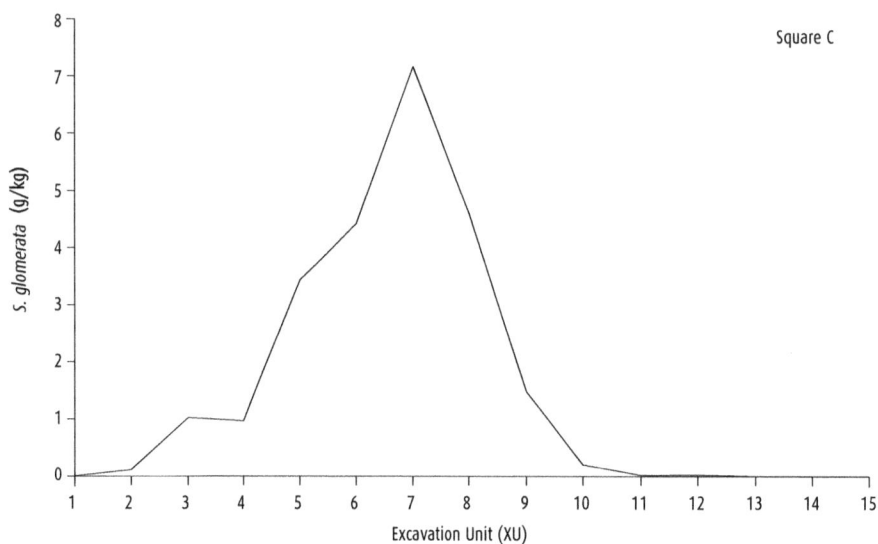

Figure 10.7 Abundance of oyster (*S. glomerata*).

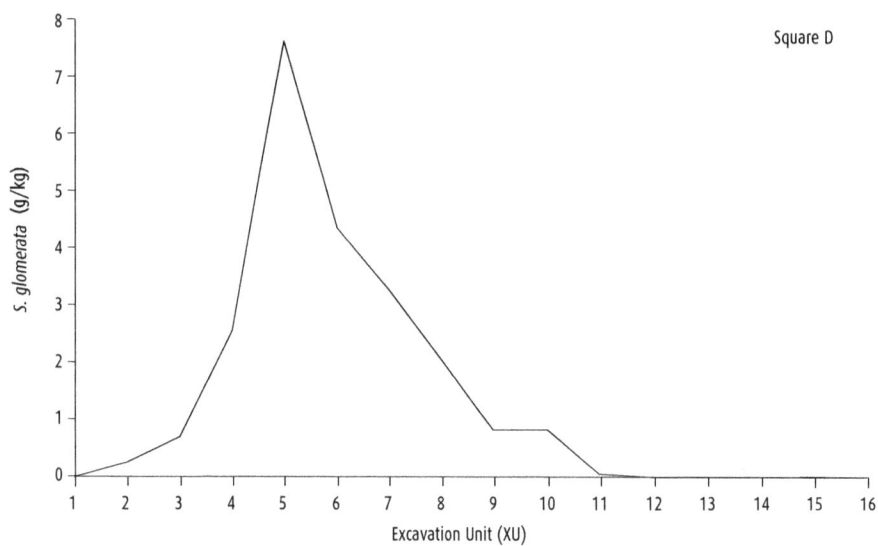

Figure 10.8 Abundance of oyster (*S. glomerata*).

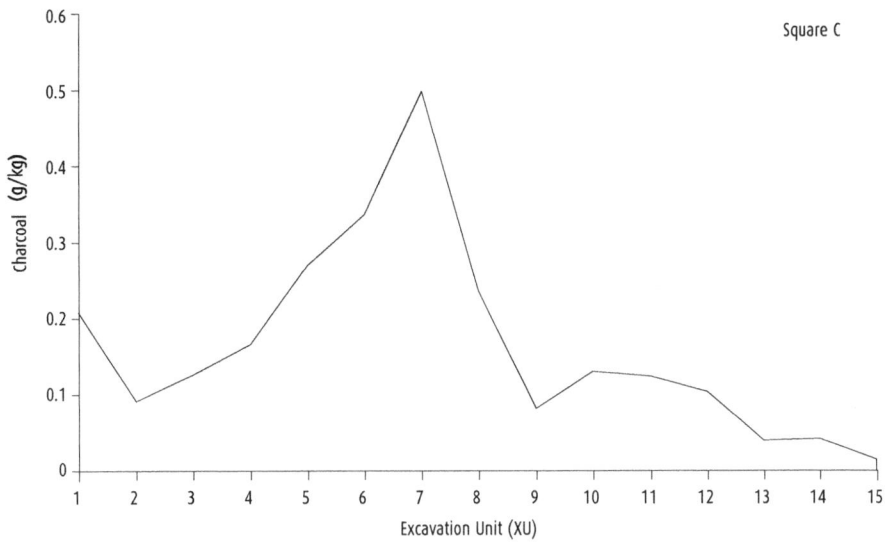

Figure 10.9 Abundance of charcoal.

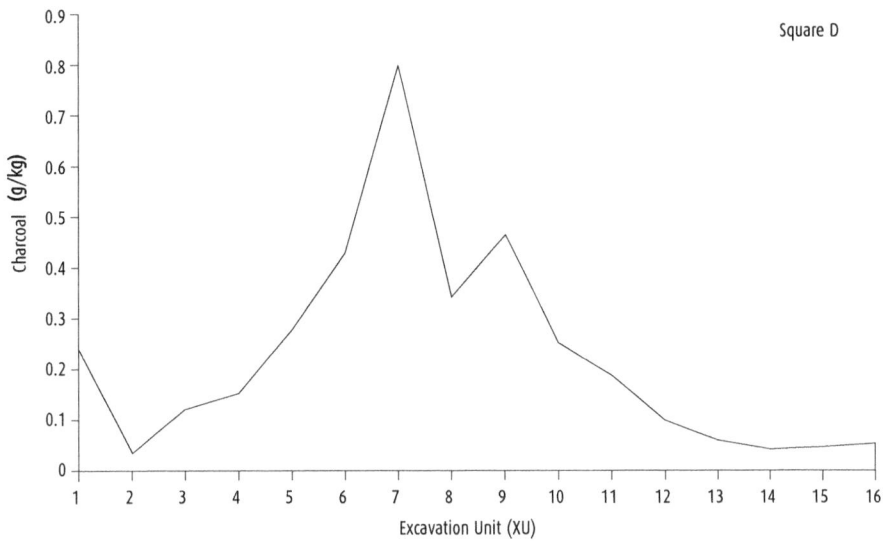

Figure 10.10 Abundance of charcoal.

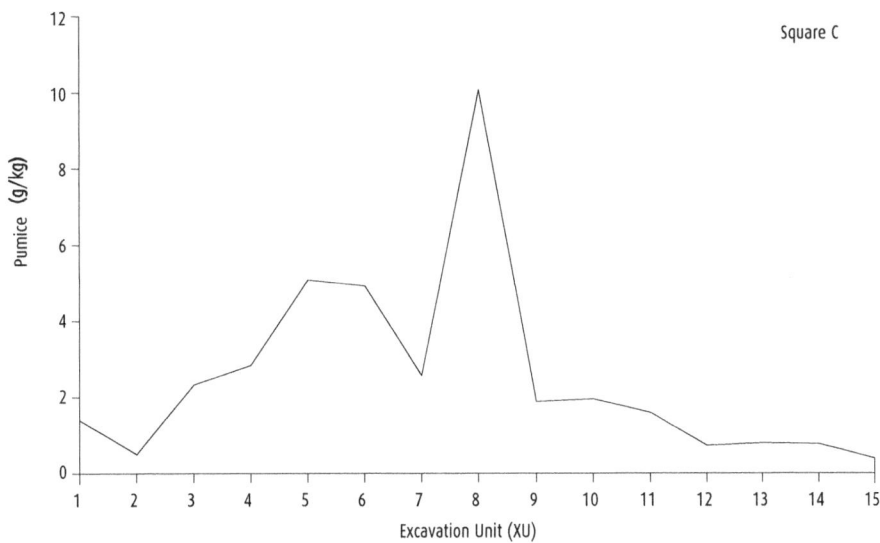

Figure 10.11 Abundance of pumice.

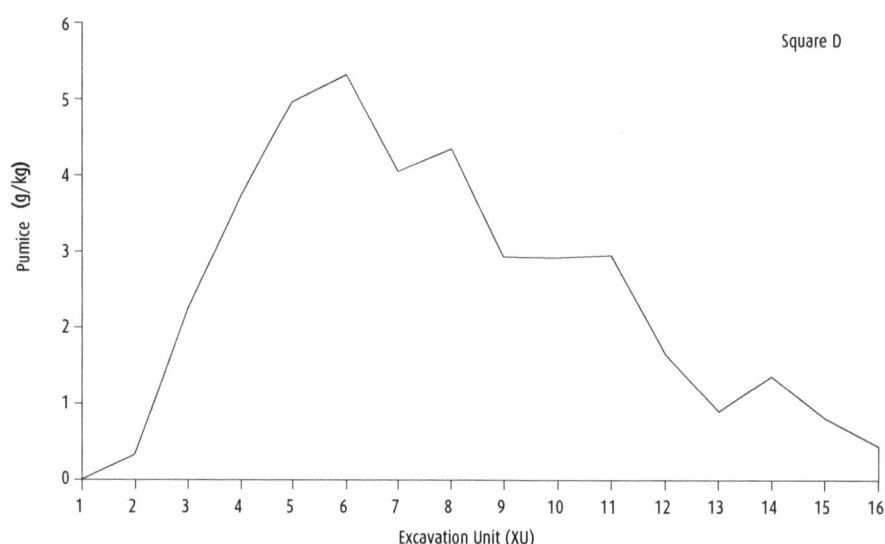

Figure 10.12 Abundance of pumice.

Other remains

Scattered fragments of charcoal totalling 76.5g were recovered from every excavation unit except the surface of Square B, with concentrations in the upper units coincident with the distribution of shell (Figs 10.9–10.10). Pumice totalling some 1,011.7g occurs throughout the bank deposits (Figs 10.11–10.12). Several large nodules of pumice in Square D account for most of the weight in the middle units of the deposit. Pumice is thought to have entered the site through a combination of wind and sea-rafting.

Discussion

Excavation at Eurimbula Creek 1 revealed a shallow, low density shell deposit consistent with observations of the material exposed in the erosion section. The apparent vertical and horizontal discreteness of the shell material, dominance of larger shellfish size-classes, absence of small shellfish and shell fragments, restricted range of shellfish taxa and presence of burnt fish bone support a cultural origin for the deposit. The concentration of shell material between 10–25cm and the single bivalve conjoin indicating a separation of contemporaneous cultural material of up to 15.2cm suggests that all recovered materials probably belong to a single or closely spaced series of small-scale deposition events.

Small-scale shellfishing and fishing are represented by the assemblage. The small extent of the site, low diversity of taxa and absence of evidence for stone artefact manufacture or maintenance is consistent with a site function as a 'dinner-time camp'. Meehan (1982:26) defined 'dinner-time camps' as 'small camp sites used during the middle of the day while people are engaged in hunting trips away from their home base. At these sites they cook and eat food that has been procured up to that time'. Bird and Bliege Bird's (1997) study of contemporary shellfish gathering among the Meriam of eastern Torres Strait concluded that small dinner-time camps might be expected during collective gathering activities taking place relatively far (average 1.9km) from the residential base. Meehan (1988) has noted that other factors also impact on the establishment of dinner-time camps, such as season and location of the home base. Dinner-time camps are commonly established on the foreshore fringe adjacent to the gathering area, where a large proportion (up to 75%) of food might be consumed. Meehan (1982:117) found that shellfish processing sites (as opposed to dinner-time camps) usually included only one taxon.

A key difference between long-term residential base camps and short-term dinner-time camps is the diversity of taxa targeted. Meehan's (1988) ethnographic observations documented that while 70 taxa were consumed at the base camp during April 1973, the 32 temporary dinner-time camps established during the same period only had an average of c.6.5 taxa/visit (data for all individual visits is not presented), with a range of 2–22 taxa. Meehan (1988) also noted the small size of dinner-time camps (maximum of 15m × 10m) versus base camps (200m × 100m) and the presence of manufacturing and maintenance activities at base camps. Together this evidence is consistent with ephemeral site creation during foraging activities focussed on the intertidal zone.

The recovery of a single pipi (*D. deltoides*) valve dated to the last 200 years is consistent with dated surface samples of pipi from the Middle Island Sandblow Site to the north (see Chapter 2), suggesting use of pipi over the last 550 years. Although pipi is not common in the area today, the volume of remains of this taxon on the Middle Island Sandblow Site and local oral history suggest it was more abundant at times in the recent past. It is thus curious that pipi is absent from excavated deposits on estuary margins dating to the last 500 years, despite close proximity to the ocean beach. At least three explanations are possible to account for this pattern. First, pipi may not have been actively targetted. Second, pipi were actively targetted and discarded in open beach environments prone to loss through erosion on the supratidal dune fringe. Third, pipi may have been occasionally consumed at temporary (dinner-time) camps on the open foreshore during foraging expeditions, with field processing of the remainder for transport of the meat only to nearby residential bases.

Summary

Eurimbula Creek 1 represents a different form of site from those which dominate the archaeological record of the southern Curtis Coast. It is one of the few sites investigated that does not appear to represent multiple occupation events spanning long periods of time. Radiocarbon dating indicates that the site was used within the last 200 years, contemporaneous with the period of greatest evidence for land-use in the region. These results provide an insight into patterns of resource procurement away from the large linear sites on lower creek margins that appear to have functioned as residential base camps.

11

Eurimbula Creek 2

Introduction

This brief chapter describes archaeological investigations at a small shell midden, Eurimbula Creek 2, on the north bank of Eurimbula Creek just southeast of the site of Eurimbula Creek 1 reported in Chapter 10. As noted in the previous chapter, small stratified middens are rare in the region. Investigations at Eurimbula Creek 2 were undertaken to explore the nature and chronology of these smaller assemblages. Excavations and analyses indicate that the site represents a single ephemeral occupation event in the recent past, probably pre-dating European invasion of the area. The limited range of remains suggest that the site was a temporary camp created during foraging activities, not a residential site.

Site description and setting

Eurimbula Creek 2 is located 600m southeast of Eurimbula Creek 1 and comprises a sparse scatter of shell spread over a 10m^2 area of mainly disturbed ground on the top of a low dune on the north bank of Eurimbula Creek (Latitude: 24°10'04"S; Longitude: 151°49'22"E) (Fig. 10.1). The site is located 3m above current tidal range and 39.5m north-northeast of the base of the dune abutting the fringing beach separating the frontal dune from the thick mangrove margin (Fig. 11.1). The surface expression of Eurimbula Creek 2 is discrete, with the majority of shell visible in a low ovular depression c.110cm long and c.75cm wide created by unidentified burrowing animals. Active smooth-handed ghost crab (*Ocypode cordimanus*) burrows are visible in the base of the depression although it is improbable that they are responsible for the entire disturbance, which covers a broader area than usually impacted by these crustaceans. The surface shell scatter is composed almost entirely of rock oyster lids and lid fragments (*Saccostrea glomerata*), with some common nerite (*Nerita balteata*) and the base of a telescope mud whelk (*Telescopium telescopium*). The latter is significant because although it abounds in the upper tidal reaches of mangrove

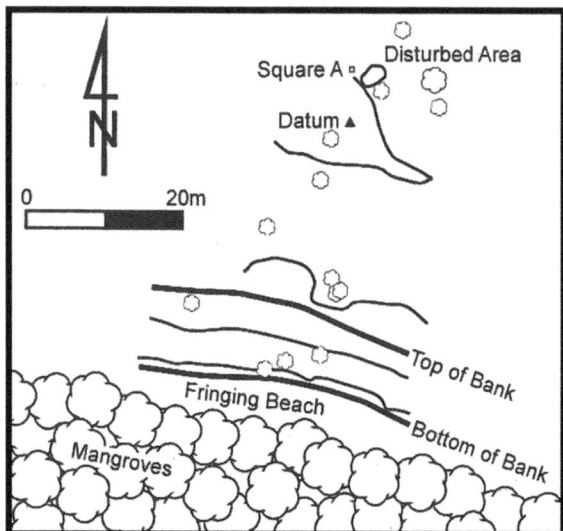

Figure 11.1 Site plan of Eurimbula Creek 2 area. Contours are in 0.5m intervals. Only major trees in the immediate area of the excavation are shown.

Figure 11.2 General view of location of Square A, showing disturbance zone to the northeast (rear left) of the excavation. Facing east.

Figure 11.3 General view of completed excavation, Square A, showing a large root protruding from the east section. Facing north.

Figure 11.4 General view of completed excavation, showing disturbance zone to the east (right) of the excavation. Facing north.

estuaries today (Shanco and Timmins 1975; Fig. 2.6), this taxon has not been observed on the surface of other shell middens recorded in the region nor recovered from excavations. Also, the apparent absence of mud ark (*Anadara trapezia*) is of note as it is the most common taxon in other sites recorded in the area. The surface exhibits a maximum shell density of 25 shell fragments/m^2. Despite intensive ground survey and good visibility, virtually no shell was located on the ground surface or in the erosion bank away from the identified exposure. Cattle activity in the site area is evidenced by a deeply worn track (c.30cm wide) along the top of the frontal dune.

The shell exposure is situated in a small clearing surrounded by dry vineforest thicket dominated by burdekin plums (*Pleiogynium timorense*) with occasional large eucalypts including an unusually large (264cm in circumference) tree located c.8m east of the site. The site is situated c.400m southwest of the open beach and c.100m west of a distinct thinning of the mangrove fringe, facilitating easier access to the extensive intertidal flats of Eurimbula Creek (Fig. 10.1). Refer to Chapter 10 for further details of vegetation, non-Indigenous impact and other recorded archaeological sites in the immediate area of Eurimbula Creek 2.

This site was recorded by the author on 2 October 1996 during Gooreng Gooreng Cultural Heritage Project (GGCHP) pedestrian transect surveys of the Eurimbula Creek northern bank (see Ulm and Lilley 1999). The site was originally designated as GGCHP Site Number CC38 and subsequently registered on the EPA's Indigenous Sites Database as KE:B20. It is registered as Queensland Museum Scientific Collection Number S232. The basic objective of excavation was to determine the presence and integrity of any subsurface deposits at the site. A higher-order objective was to sample the deposits to help develop an understanding of activities undertaken at smaller sites and how these relate to the larger sites which dominate the archaeology of the region.

Excavation methods

Owing to the small size of the site, a single 50cm x 50cm pit (Square A) was placed c.25cm from the western margin of the disturbed depression to assess the excavation potential of the site (Fig. 11.2). The test pit was located as close as practicable to the observable shell exposure, with the site datum established 670.5cm due south of the southwest corner of Square A (Fig. 11.1). As the excavation was situated on a dune ridge, bedrock was not reached, but excavation continued for 20cm below the last observed cultural material. Excavations were conducted between 22 March–1 April 1999.

Excavation proceeded in shallow, arbitrary excavation units averaging 3.1cm in depth and 11.2kg in weight. Excavation ceased at a maximum depth of 45.7cm below ground surface after several units of culturally-sterile sediments had been removed (Figs 11.3–11.4). A total of 167.2kg of sediment was excavated in 15 XUs. Excavated sediments were gently dry-sieved through 3mm screens onto a plastic tarpaulin located 5m southwest of the excavation. Three-dimensional plotting was undertaken for a single concentration of charcoal fragments (0.5g) encountered *in situ* during excavation of XU4. A layer of plastic sample bags was placed over the base of the completed excavation and then backfilled with sediments from the sieving station and with sands from the beach fringing Eurimbula Creek (see Chapter 3 for a detailed discussion of the standard excavation methods employed at all sites).

Cultural deposit and stratigraphy

Excavation reflected the low density and limited range of shellfish remains observed on the surface of the deposit. Only very sparse shell and scattered charcoal were recovered from the excavation (Table 11.1). The marine shell component comprises only two taxa, rock oyster (*S. glomerata*) and nerite (*N. balteata*), with two taxa of terrestrial gastropod the only other shell recovered (see below). Shell is concentrated between XU2–9 in the top 25cm of the deposit. Below this level only minute fragments of oyster shell were recovered and the abundance of charcoal falls off dramatically. The excavated sediments can be divided into three distinctive stratigraphic units (SUs) on the basis of colour and texture, with a third subunit across part of the surface (Table 11.2, Fig. 11.5). SUIa is thought to be spoil deriving from adjacent disturbance. Its sediments are consistent with those observed around the margins of the depression and the unit contains occasional shell and humic material. SUI appears to be the pre-disturbance surface and is partially overlain by SUIa along the eastern margin of the excavation. SUII contains the majority of shell and charcoal recovered in the excavation. This unit appears to be the source of all shell at the site. SUII grades into SUIII with depth, with a darkening of sediment and increased roundedness of particles. Shell fragments recovered from this unit are thought to derive from SUII. The shallow and limited distribution of cultural materials suggests that all material is likely to be roughly contemporaneous. The pH values are slightly acidic throughout (6.0–6.5).

Table 11.1 Eurimbula Creek 2, Square A: summary excavation data and dominant materials.

SQUARE	XUs (#)	DEPTH (cm)	WEIGHT (kg)	SHELL (g)	CHARCOAL (g)	STONE (g)	ORGANIC (g)
A	15	45.68	167.20	201.29	20.91	10.50	1252.69

Table 11.2 Stratigraphic Unit descriptions, Eurimbula Creek 2, Square A.

SU	DESCRIPTION
I	Extends across the entire square. This unit is exposed at the surface in the northwest corner and underlies SUIa along the eastern margin and southeast corner of the square. It has a maximum thickness of 7cm and a maximum depth of 14cm below the surface. Sediments are matted by humic material and comprise dark greyish brown (10YR-4/2), poorly-sorted, medium, subangular particles. Rock oyster and charcoal fragments are common. pH values are slightly acidic (6.0).
Ia	This unit overlies SUI along the eastern margin of the pit and is derived from the burrowing spoil from the surface of the adjacent disturbance zone. It has a maximum depth of 10cm below the surface. Sediments are matted by numerous fibrous roots. Sediments are brown (10YR-4/3). Rock oyster and charcoal are present. pH values are slightly acidic (6.0).
II	Extends across the entire trench with a maximum thickness of 20cm and a maximum depth of 27cm below the surface. Sediments are moist and loosely consolidated. Occasional large roots are present. Sediments are well-sorted, fine and subrounded to rounded and dark yellowish brown (10YR-4/4). Most shell and charcoal recovered derives from this unit. pH values are slightly acidic (6.5).
III	Unit extends across the entire square with a maximum thickness of 24cm and a maximum depth of at least 47cm below the surface. The base of this unit was not reached. Subsequent testing with a sand auger and observation of the nearby erosion bank profile indicates that this unit extends several metres. It comprises well-sorted, fine, rounded sediments which are dark yellowish brown (10YR-4/6). Sediments are moist and loosely consolidated. Roots are less numerous and generally smaller in diameter. Minute shell fragments recovered from this unit are thought to derive from higher in the profile. pH values are slightly acidic (6.5).

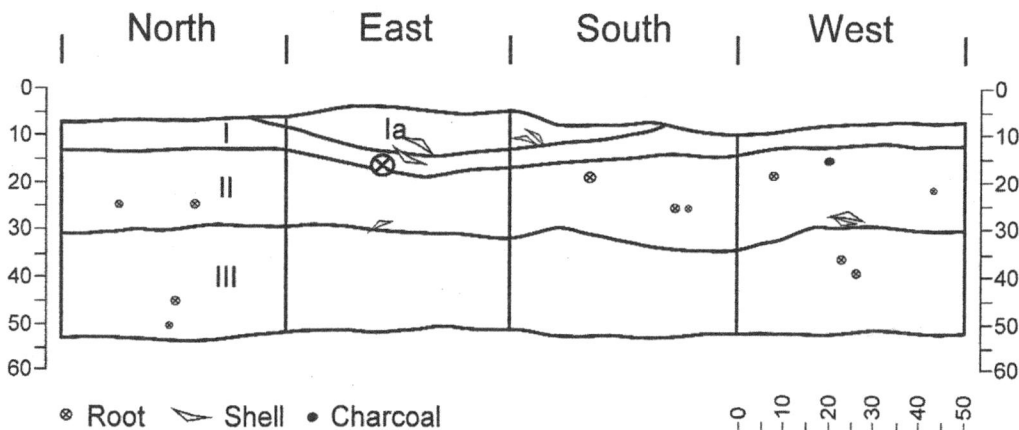

Figure 11.5 Stratigraphic section, Eurimbula Creek 2, Square A.

Radiocarbon dating and chronology

A conventional radiocarbon date was obtained for the site on a large blocky charcoal sample weighing 2.8g from XU6 (Table 11.3). The sample returned a value of 97.9±0.8% modern, which indicates an age of less than 200 years. Finite ages are problematic in this area of the calibration curve owing to high levels of variability in radiocarbon activity in the atmosphere caused by the onset of the industrial revolution and atmospheric testing of thermonuclear devices. Radiocarbon ages between 0 and 200 years could give calendar ages anywhere from AD 1750 to AD 1950 (Alan Hogg, Waikato Radiocarbon Dating Laboratory, pers. comm., 1999). The lack of post-contact material culture lends support to an argument for a pre-European origin for Eurimbula Creek 2. The near-surface context of the material and limited vertical distribution indicate that the cultural deposits are very recent, probably being deposited sometime between AD 1750–AD 1900.

Table 11.3 Radiocarbon dates from Eurimbula Creek 2 (see Appendix 1 for full radiometric data).

SQUARE	XU	DEPTH (cm)	LAB. NO.	SAMPLE	δ¹³C (‰)	% MODERN	CALIBRATED AGE/S
A	6	13.1–16.3	Wk-7689	charcoal	−25.7±0.2	97.9±0.8	modern (see text)

Stratigraphic integrity and disturbance

The Eurimbula Creek 2 deposits exhibit generally poor integrity. Burrowing near the excavation has dispersed subsurface shell material across a large area. The disturbance may have been caused by the brush-turkey (*Alectura lathami*) which are common in the area, although there are no mounds in the immediate vicinity of the site. Feral pigs (*Sus scrofa*) have also been recorded in adjacent areas. SUII is thought to be the source unit for all of the shell present at the site. This unit, at least in the area of excavation, does not show signs of significant disturbance (e.g. infilled burrows), although tree root penetration may well be responsible for some movement of shell material in the deposit. The similarity of shell diversity and density observed in the surface scatter with that encountered during excavation lends further support to the idea that the SUII deposits are *in situ*, with the scattered surface material originally deriving from a relatively small, discrete deposit. Unlike the deposits at the nearby Eurimbula Creek 1, the deposits at this site have not been impacted by water erosion, with the creek located some distance from the deposit and at a significantly lower elevation.

Although virtually all shell was recovered in the top 25cm of the deposit, occasional minute fragments were recovered to the base of excavations in SUII. These latter are considered unlikely to be in primary depositional context and have probably been displaced from the shell zone by crab burrowing and root penetration. Although no voids or unambiguous evidence for burrowing were encountered during excavation, ghost crab (*O. cordimanus*) burrows were observed in the base of the adjacent disturbed depression.

Laboratory methods

Laboratory methods followed the standard procedures employed at all sites (see Chapter 3). In the sections below, the results are summarised, although only selected data are illustrated in Figures 11.6–11.9. Further summary results are available in Appendix 4.

Cultural materials

Invertebrate remains

Four taxa of shellfish weighing 201.3g were recovered, consisting of one marine bivalve, one marine gastropod and two terrestrial gastropods (Table 11.4). The assemblage consists almost entirely of rock oyster (*S. glomerata*), comprising 96.8% of the shell assemblage by weight (Fig. 11.6), followed by common nerite (*N. balteata*) (3%) and the land snails *Trachiopsis mucosa* (<0.1%) and *Eremopeas tuckeri* (<0.1%). The assemblage exhibits very low diversity with a calculated Shannon-Weaver Function (H') of 0.173 and Simpson's Index of Diversity (1–D) of 0.08. This assemblage, together with the telescope mud whelk (*T. telescopium*) observed on the surface, are consistent with foraging strategies focussed on the mangroves fringing the estuary. The presence of the telescope mud whelk, which is absent from other midden assemblages but which dominates modern estuarine shellfish biomass, may point to a very recent origin for the proliferation of this species in the area, just pre-dating permanent European settlement in the area.

Other remains

Scattered fragments of charcoal, totalling 20.9g, were recovered from every excavation unit, with general abundance coincident with the distribution of shell (Figs 11.6–11.7). Small quantities of pumice, totalling 9.8g, occur throughout the bank deposits (Fig. 11.8). The small nodules of pumice recovered were most likely wind-transported from the tidal strand line of the nearby creek bank. Organic material decreases markedly below the leaf litter and humic-rich units at the surface of the deposit (Fig. 11.9).

Table 11.4 Presence/absence of shellfish identified in Eurimbula Creek 2, Square A.

XU	S. GLOMERATA (g)	(MNI)	N. BALTEATA (g)	(MNI)	T. MUCOSA (g)	(MNI)	E. TUCKERI (g)	(MNI)
1	4.66	1	0	0	0	0	0	0
2	38.60	6	6.12	1	0	0	0	0
3	12.00	1	0	0	0.09	2	0.08	2
4	28.10	2	0	0	0.07	2	0	0
5	33.40	5	0	0	0	0	0	0
6	31.50	2	0	0	0	0	0	0
7	3.91	1	0	0	0	0	0	0
8	17.40	1	0	0	0	0	0	0
9	18.00	3	0	0	0	0	0	0
10	1.91	1	0	0	0	0	0	0
11	1.16	0	0	0	0	0	0	0
12	0	0	0	0	0	0	0	0
13	0	0	0	0	0	0	0	0
14	0	0	0	0	0	0	0	0
15	4.29	0	0	0	0	0	0	0

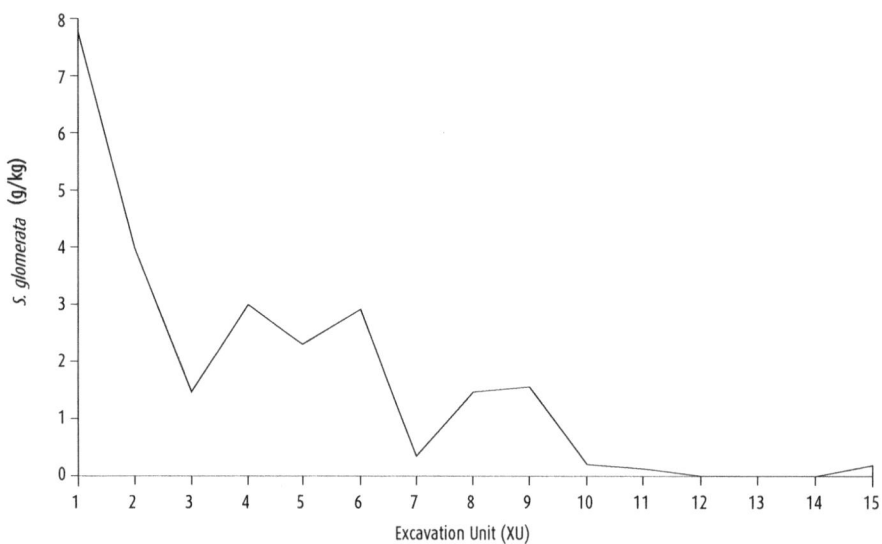

Figure 11.6 Abundance of oyster (*S. glomerata*).

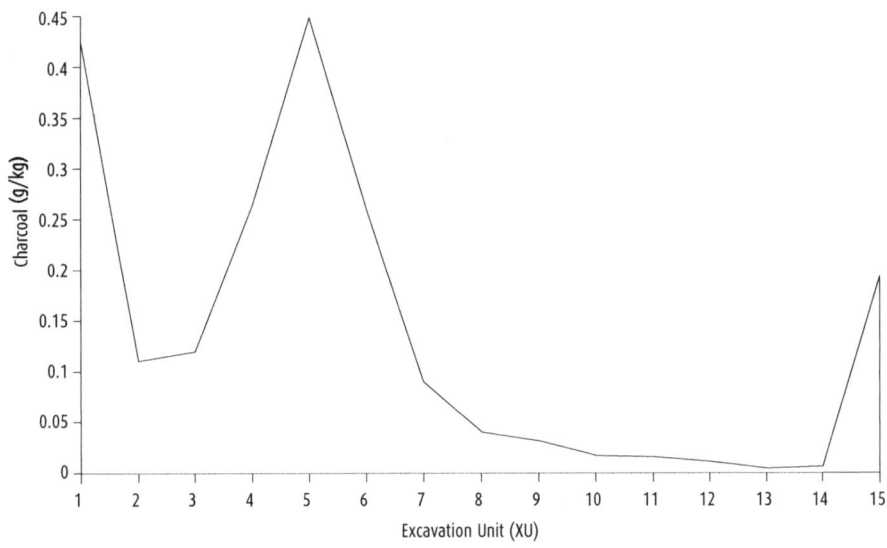

Figure 11.7 Abundance of charcoal.

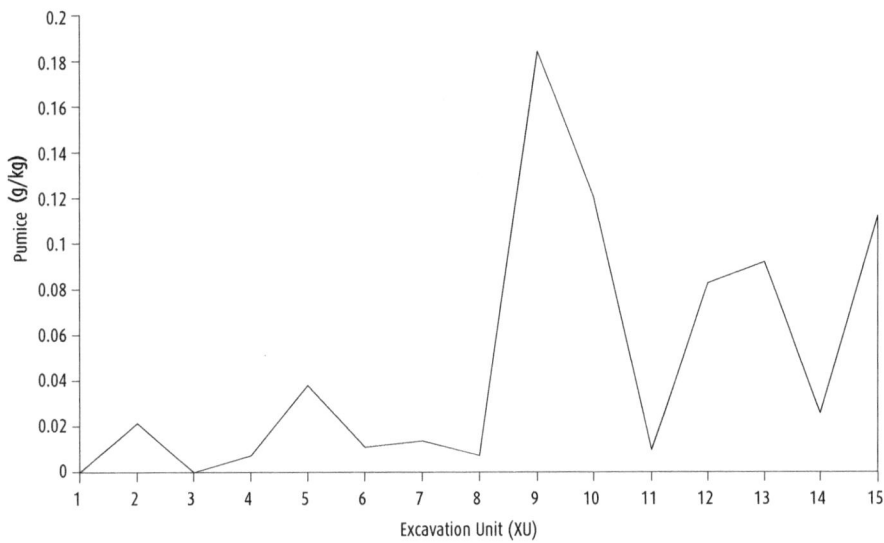

Figure 11.8 Abundance of pumice.

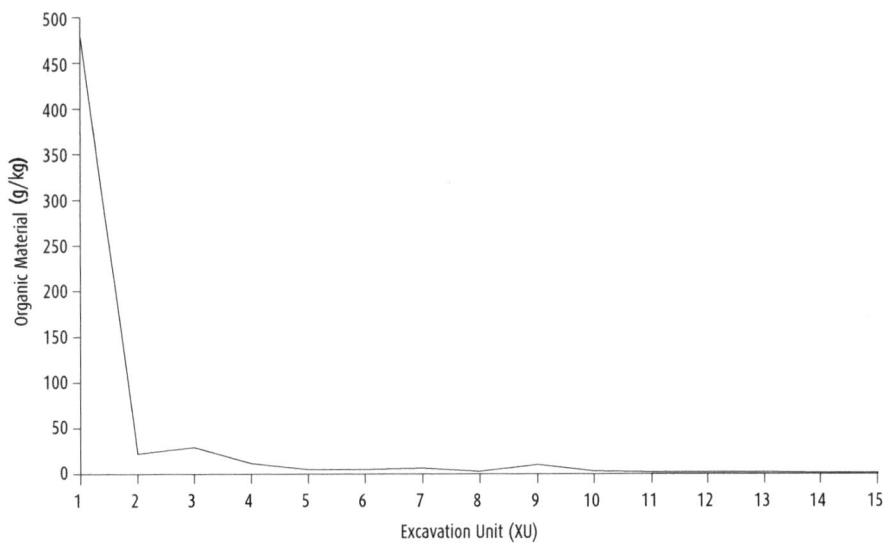

Figure 11.9 Abundance of organic material.

Discussion

Investigations at Eurimbula Creek 2 revealed a low density sequence of shell material located close to the surface. The density and diversity of the excavated shellfish assemblage is consistent with the material observed on the surface. Although the site exhibits poor integrity, all of the cultural material appears to derive from a shallow subsurface unit which dates to the last 200 years. The limited extent and composition of the assemblage suggest a general contemporaneity which is consistent with the creation of a temporary camp during foraging activities focussed on the adjacent upper intertidal zone. As at Eurimbula Creek 1, I suggest that the small-scale shellfishing and fishing represented by this assemblage are consistent with a site function as a dinner-time camp (*sensu* Meehan 1982:26) (see Chapter 10 for further discussion).

The recovery of a single telescope mud whelk base (*T. telescopium*) dating to the last 200 years may relate to local palaeoenvironmental changes and/or colonisation leading to the proliferation of this taxon observed in estuaries today. The presence of the telescope mud whelk late in the pre-European sequence and a general decline in the abundance of mud ark (*A. trapezia*) over the last 2,000 years is consistent with a model of small areas of mangroves being supplanted by dense fringing mangrove forests which characterise estuarine vegetation of the landscape today (see Chapter 2 for further discussion).

Summary

Eurimbula Creek 2 represents small-scale activities operating over a relatively short time-scale. Together with evidence from Eurimbula Creek 1, these investigations suggest a recent logistical pattern of land-use involving extraction and at least partial consumption of resources from estuaries located some distance from major archaeological sites identified as residential base camps. The age of this deposit is synchronous with the most intense period of occupation identified in the region and suggests that by this time complex patterns of localised settlement and subsistence strategies were in place.

12

Eurimbula Site 1

Introduction

This chapter reports the results of archaeological investigations at Eurimbula Site 1, a large site complex on the eastern margin of Eurimbula National Park. Test excavations conducted in 1995 (reported in Ulm et al. 1999a) revealed a low density cultural sequence with marked discontinuities in the distribution and antiquity of remains across the site complex. The densest and oldest deposits, dating to shortly before 3,000 BP, were located along the southern margin of the site with lower density deposits dating to the recent past across the northern two-thirds of the site. Densities of cultural material were also found to decrease markedly with distance from the creek. The excavations reported below sought to increase the sample of material available for analysis by targetting the concentrated near-creek deposits identified at the southern end of the site complex. Analyses of excavated material demonstrate extensive, low intensity site use from 3,000 BP into the historical period, with significant changes in both faunal and stone artefact assemblages over the last 1,500 years.

Site description and setting

Eurimbula Site 1 is a large, stratified midden complex intermittently exposed in a steep erosion section on the western bank of Round Hill Creek on the eastern boundary of Eurimbula National Park (Figs 12.1–12.2). The approximate centre-point of the site complex is located 4km southwest of Round Hill Head (Latitude: 24°11′32″S; Longitude: 151°51′45″E). The site is approximately 2km long (north-south) and up to 100m wide (east-west), although surface exposures of shell and stone artefacts are mainly confined to a 50m wide band parallel to the creek bank (Figs 12.3–12.4). The site thus covers a minimum area of 100,000m². It was formed on and in a series of low Holocene beach ridges and swales which run roughly parallel to the modern coastline forming Bustard Bay. These features were formed by the massive amounts of sand delivered to the coast by long-shore drift and long-shore transport of sediments on the continental shelf. Hopley (1985:76–7) defines this general area as a depositional coastline characterised by a series of beach ridges trailing northwards from the northern side of almost every estuary of note (see also Rowland 1987). A feature of these dunes is a band of dark cemented organic sands exposed in several locations in the erosion face approximately mid-way down the profile, commonly known as coffee rock in southeast Queensland (Fig. 12.5).

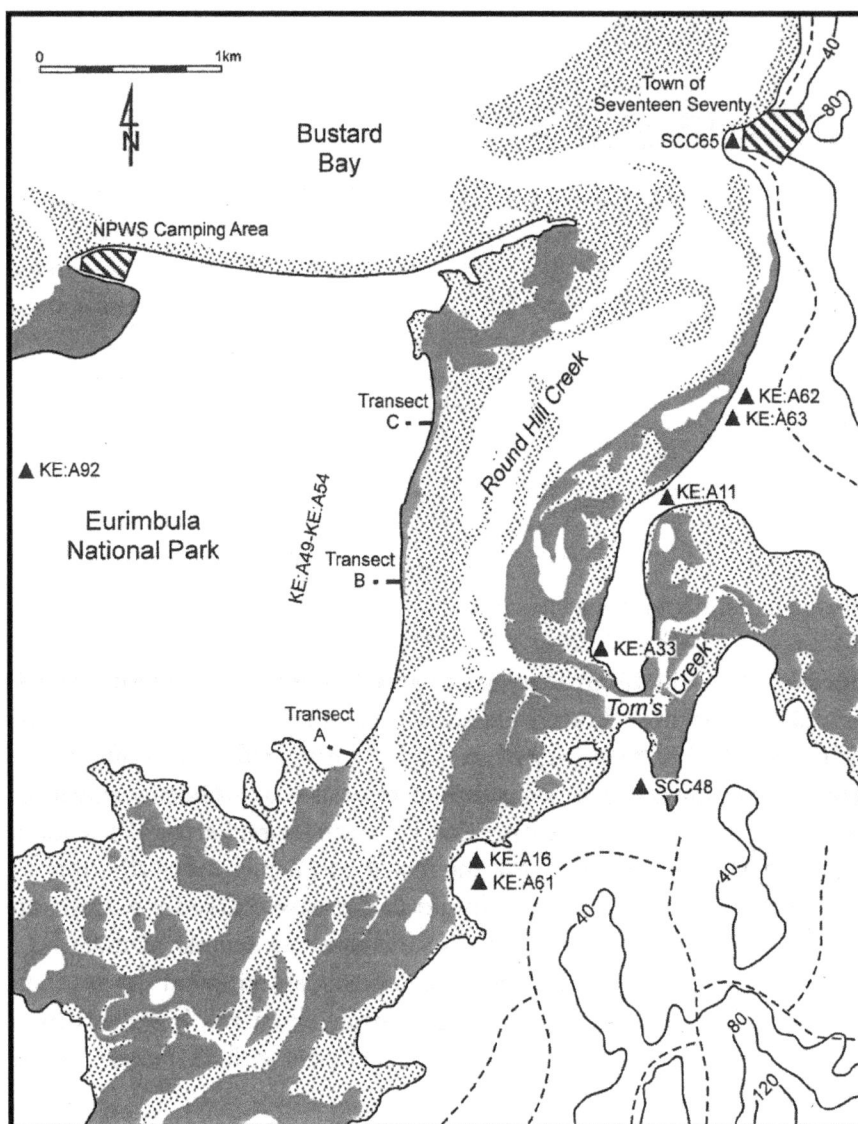

Figure 12.1 Lower reaches of Round Hill Creek, showing the location of Transects A, B and C at Eurimbula Site 1. Dark grey shading indicates the general extent of mangrove vegetation. Stippling indicates the general location of intertidal flats. Dashed lines are sealed roads.

Recent models proposed for southeast Queensland and northern New South Wales associate coffee rock formation with rising sea-levels in the mid-Holocene. It is thought that a subterranean salt wedge created by rising sea-levels interacts with organic material in coastal saltmarsh or back-barrier swamps to cement clastic material into humicretes or coffee rock (Maria Cotter, School of Human and Environmental Studies, University of New England, pers. comm., 2000). Therefore, the coffee rock exposed at Eurimbula Site 1 is likely to pre-date the Holocene highstand in the region dated to around 5,500 BP (see Chapter 2). Sediments located above this feature therefore post-date the mid-Holocene.

Large stone artefacts are common on or at the base of the erosion bank, having eroded out of *in situ* deposits located towards the top of the bank profile. The majority of these artefacts are manufactured on rhyolitic tuff. Occasional artefacts were also encountered on the fringing beach, suggesting significant bank retreat in these parts of the site, the deflation of larger stone artefacts and the complete removal of shell, charcoal and other lighter materials. The northern two-thirds of the site is bordered by a dense mangrove fringe which thins from north to south (Fig. 12.1). The mangrove fringe disappears altogether along the southern c.500m of the bank, coincident with an area of major recent bank retreat. The c.4m high creek bank in this area is very exposed: it abuts the main channel of Round Hill Creek and does not have a mangrove fringe to dissipate tidal surge (Figs 12.1, 12.3). The base of the bank in this area is littered with the remains of large trees which have eroded out of the bank. In places the bases of these trees occur up to 30m east of the current erosion bank in the subtidal zone, indicating significant bank recession. Few cultural materials have been observed along this section, possibly because they were removed by erosion. This observation accords well with previous findings revealing markedly decreased quantities of cultural material with distance from the creek margin (see below). A dense mangrove fringe commences at the extreme south of the erosion bank adjacent to large supratidal claypans and saltflats. Cultural material is abundant at the northern and southern extremities of the erosion bank where mangrove fringe development and the orientation of the creek channel may have reduced the impact of erosion, at least in recent times. At the northern end the frontal dune is set back slightly from the creek, offering further protection to deposits.

The cultural deposits border the extensive and shallow Round Hill Creek estuary, which comprises extensive intertidal and subtidal flats, samphire and claypan saltflats, seagrass beds and mangrove communities. Seagrass beds occur on the shallow banks in the middle reaches of the creek, concentrated near its confluence with Tom's Creek (Fig. 2.7). Although Olsen (1980a:17) referred to these beds as transitory seasonal features confined to the warmer months, they have been observed throughout the year during recent fieldtrips. Seagrass beds currently support a small population of mud arks (*Anadara trapezia*), with sometimes dense clumps of rock oysters (*Saccostrea glomerata*) and common nerites (*Nerita balteata*) attached to mangrove root substrates. Telescope mud whelks (*Telescopium telescopium*) are common amongst rearward mangroves.

Mangrove vegetation is dominated by spotted mangroves (*Rhizophora stylosa*) backed by yellow mangroves (*Ceriops tagal*) and grey mangroves (*Avicennia marina*) (Olsen 1980a:17). These vegetation communities appear to be persistent features in the local landscape with mangrove pollen dating to c.3,500 BP recovered from cored sediments underlying modern freshwater wetlands draining into Round Hill Creek (Maria Cotter, School of Human and Environmental Studies, University of New England, pers. comm., 2000). The position of the cored sediments suggests an expanded mid-Holocene Round Hill Creek estuary coincident with a sea-level highstand, with subsequent migration of vegetation zones associated with falling sea-levels. Cook (in Beaglehole 1968:256) and Banks (in Beaglehole 1963:65) noted many mangroves around the 'skirts' of the Round Hill Creek estuary during inspections on 23 May 1770. Shanco and Timmins (1975) noted saltmarsh backing the mangrove fringe comprising marine couch (*Sporobolus virginicus*), sea purslane (*Sesuvium portulacastrum*), and Australian seablite (*Sueada australis*) and other chenopods further inland. The seeds of marine couch are known to be an Aboriginal food source (Davie 1998:35). The site area is vegetated by tall open eucalypt forest (*Eucalyptus tereticornis, E. tessalaris, E. intermedia, Melaleuca dealbata*) with occasional cloudy teatree (*M. dealbata*) and weeping cabbage palm (*Livistonia decipiens*) more frequent towards the open beach. The understorey contains occasional concentrations of swamp fern (*Blechnum indicum*) and common bracken (*Pteridum esculentum*). The southern margin of the site adjacent to Transect A (Fig. 12.2) abuts a band of melaleuca (*M. quinquenervia, M. dealbata*) forest fringing the edge of the saltflats, with more diverse vegetation communities further to the south. An area of closed dry rainforest occurs <2km west of the site, comprising diverse vegetation including many Aboriginal bush food sources such as burdekin plum (*Pleiogynium timorense*), bumpy ash (*Flindersia schottiana*), brown pine (*Podocarpus elatus*) and native cherry (*Exocarpus latifolius*).

Evidence for non-Indigenous use of the site area is limited. There is no vehicular access within 2km of the creek bank containing the cultural deposits, with the only direct access by boat from the Town of Seventeen Seventy on the opposite side of the estuary. The site area is likely to have been selectively logged in the late nineteenth century, although a sawmill, established by 1867 around 3km northwest of the site, focussed on the hoop pine (*Araucaria cunninghamii*) located in the dry rainforest (Buchanan 1999:33; Growcott and Taylor 1996:65–6). A Queensland Parks and Wildlife Service camping area is located on Bustard Beach c.3km north-northwest of the site, although the focus of camper activities is on the open beach and recreational fishing of the adjacent Eurimbula Creek rather than the Round Hill Creek margin of the National Park. Surface survey of the area adjacent to the creek bank revealed occasional concentrations of recent (>AD 1970) beer bottles commonly associated with evidence for recent hearths and modern fishing equipment (e.g. fishing line, broken hand reels etc), suggesting that the area has been used in recent times for short-term camping by recreational fishers.

Numerous archaeological sites have been recorded around the margins of Round Hill Creek (see Fig. 2.11, Appendix 2), with a concentration of sites on the eastern bank of the creek opposite Eurimbula Site 1. Sites include a large shell mound dating to at least 1,600 cal BP (KE:A16) and a large site complex located at the junction of Tom's Creek and Round Hill Creek dating to at least c.1,500 cal BP (KE:A33; see Chapter 13). Other low density shell deposits recorded to the north of the concentration of sites around Tom's Creek (KE:A11; KE:A62–63) may be part of the same site complex, although disturbance associated with recent development activities has obscured the relationships between different exposures of this material.

Previous investigations

Eurimbula Site 1 is the only site on the southern Curtis Coast for which detailed ethnohistoric descriptions are available. The site is almost certainly that seen by Cook and Banks on 23 May 1770 when they explored the estuary by boat after a landing on the eastern bank earlier in the day. Both men made detailed dairy notes of the visit:

> As yet we had seen no people but saw a great deal of smook up and on the west side of the Lagoon which was all too far off for us to go by land excepting one; this we went to and found 10 small fires in a very small compass and some cockle shells laying by them but the people were gone. On the windward or South side of one fire was stuck up a little bark about a foot and a half high and some few pieces lay about in other places; these we concluded were all the covering they had in the night (Cook in Beaglehole 1968:256).

> Many large fires were made at a distance from us where probably the people were. One small one was in our neighbourhood, to this we went; it was burning when we came to it, but the people were gone; near it was left several vessels of bark which we conceivd were intended for water buckets, several shells and fish bones, the remainder I suppose of their last meal. Near the fires, for there were 6 or 7 small ones, were as many peices of soft bark of about the length and breadth of a man: these we supposd to be their beds: on the windward side of the fires was a small shade about a foot high made of bark likewise. The whole was in a thicket of close trees, defended by them from the wind; whether it was realy or not the place of their abode we can only guess. We saw no signs of a house or any thing like the ruins of an old one, and from the ground being much trod we concluded that they had for some time remaind in that place (Banks in Beaglehole 1963:67).

Other members of the landing party also reported the tail of a land animal at the camp to those that remained on the ship (Pickersgill in Bladen 1892:218). Parkinson (in Kippis 1814) noted this as 'the tail of a quadruped which we supposed might be a guanico [rat]'. Although Parkinson (in Kippis 1814) noted in passing that the fires might have been 'only an artifice of theirs to make us think they were numerous', the descriptions of Cook and Banks suggest small (still burning) fires, artefacts and food refuse consistent with a temporarily (and very recently) vacated camp site. Of particular relevance is the observation that the part of the site in active use was clearly located on the estuary rather than the open beach fronting Bustard Bay.

The site was briefly described by Godwin (1990), who noted the archaeological potential of the site as a large stratified deposit not common in the area. Burke (1993) subsequently recorded the deposits in more detail during a heritage management study of the Curtis Coast, identifying 20 separate sites comprising shell scatters and three linear middens. Burke (1993:Appendix 5) originally allocated these sites the field numbers CC-112A, CC-113A, CC-114–CC-131 and pre-allocated site numbers KE:A62–KE:A81, which were subsequently conflated into six sites when formally registered on the Queensland Environmental Protection Agency's (EPA) Indigenous Sites Database as KE:A49–KE:A54. In the site cards lodged with the EPA, Burke noted scattered mud ark, oyster and occasional club whelk in various densities and locales along the creek bank. Material was noted on the surface up to 40m from the creek bank and up to 30cm below the surface in the exposed erosion bank. A single stone artefact was recorded: a large, granitic core, which was thought to derive from the Round Hill Head headland. All of these deposits are referred to collectively here as Eurimbula Site 1. The Queensland Museum Scientific Collection Number is S864.

Test excavations were subsequently undertaken under the auspices of the Gooreng Gooreng Cultural Heritage Project (GGCHP) in April 1995 (reported in Ulm et al. 1999a). The major objective of these excavations was to establish the connection between the prograding beach ridge landforms and the deposition of cultural materials. In particular, data were collected to determine whether pre-European Aboriginal settlement patterns in the area were focussed on the estuary or the ocean beach; if the latter, the focus of settlement would be expected to be parallel to the orientation of the coastline and move northward as beach ridges developed in that direction.

During the initial field season, detailed survey of the erosion bank revealed quantities of shell and occasional stone artefacts which had fallen out of the bank owing to undercutting wave action (Fig. 12.3). Amongst the larger artefacts were several water-rounded microgranite hammerstones exhibiting impact-pitting. Although granites underlie the more recent rhyolitic tuffs which are common in the area, they rarely occur at or near the surface in the south of the study area. The closest documented surface occurrence of microgranite is at Bustard Head, some 20km to the northwest. Several large artefacts manufactured on sandstone and rhyolitic tuff were also noted. A number of these display possible bevelling along at least one margin and are roughly triangular in cross-section (Fig. 12.4). These artefacts appear morphologically similar to the 'bevelled-pounders' found further south, which are recognised as a formal tool type and functionally associated with processing of the root of the swamp fern (*B. indicum*) (Gillieson and Hall 1982; Higgins 1988; McNiven 1992b; Richter 1994). Although rhyolitic tuff is available on the east bank of Round Hill Creek at the junction of Tom's Creek 1km east, only two quarries have been identified: a minor extraction site on Round Hill Head 4km to the northeast (Rowland 1987), and the quarry described in Chapter 9. Sandstones are not common in the study area, indicating at least some transport of this material.

Three areas were selected for test excavation, located at the north, south and centre of the identified deposits (Fig. 12.1). Nine 50cm x 50cm test pits were excavated at 25m intervals along three transects placed approximately at right angles to the erosion face with three pits on each transect (Fig. 12.2). The test pits were placed across the site area in this way to characterise the broad patterns of variation in subsurface deposits. Excavations revealed an extensive, shallow, low

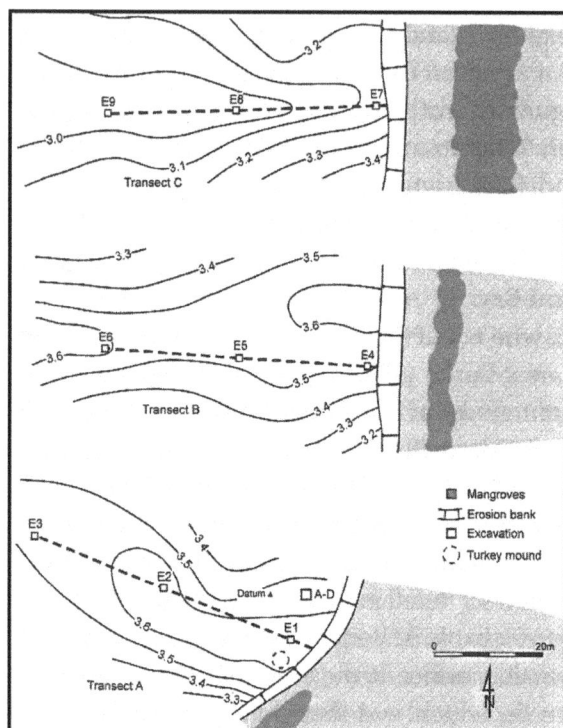

Figure 12.2 Location of test pits along Transects A, B and C at Eurimbula Site 1, showing topography in the immediate area of the transects.

Figure 12.3 General view of steep c.2m high erosion bank at the southern end of Eurimbula Site 1 fronting Round Hill Creek. Facing northwest.

Figure 12.4 Large stone artefact (FS1/2001) manufactured on rhyolitic tuff located mid-way down the erosion bank. Facing west.

density cultural sequence with shell, bone, stone artefacts and charcoal concentrated in the pits excavated along Transect A. Although small quantities of cultural material occur in the remaining pits, there appears to be a general decrease in quantity and diversity towards the sea. Squares E4 and E7, however, do contain substantial quantities of cultural remains in comparison to the other pits of Transects B and C. Cultural deposits were identified up to 50cm deep, with excavations terminating at a depth of c.70cm in culturally-sterile sediments. Shell was common in near-surface deposits with occasional concentrations up to 40cm below ground surface. Whole shells were not recovered from the basal excavated deposits, although occasional stone artefacts were noted in these units. Five radiocarbon dates were obtained, suggesting initial use of the site by 3,020±70 BP (Wk-3945) and probable abandonment in the contact period (Ulm et al. 1999a).

As surface observations had indicated, the two dominant mollusc species were rock oyster (*S. glomerata*) and mud ark (*A. trapezia*). The largest proportion of shell was recovered from Transect A, with Square E1 containing just over 2kg of oyster and mud ark combined. These two taxa exhibit a distinctly bimodal vertical distribution (Ulm et al. 1999a:Fig. 15). The earlier deposits show a dominance of mud ark, whilst the later units illustrate a shift towards exploitation of oyster. This trend was also apparent in Squares E2–E3 (Ulm et al. 1999a:Figs 16–17). Square E1 contained a clearly defined lens of *A. trapezia* located between 30–40cm below ground surface (Figs 12.6–12.7). The remaining squares contained only very small quantities of shell. Very small numbers of fish bone were also recovered. Bone from Squares E1–E9 was re-examined by Deborah Vale (School of Human and Environmental Studies, University of New England) as a component of the current project. Fish bone was identified only in Squares E1–E2 at the southern end of the site and was too fragmentary for identification (Table 12.1).

Stone artefacts (n=61) were concentrated at the southern end of the site in the vicinity of Transect A. A range of artefact types was represented, including flakes, flaked pieces and broken flakes as well as one backed artefact. Five raw materials are represented in the assemblage: quartz, quartzite, rhyolitic tuff, silcrete and a coarse sandstone. While quartz and rhyolitic tuff occur locally, the remaining raw materials are not common, suggesting the transport of stone into the area. Overall, rhyolitic tuff is the dominant raw material, although quartz is also well-represented. Significantly, there is a general pattern for artefacts manufactured on non-local raw material to be located towards the base of the cultural deposit. This pattern was noted in Squares E1–E3. Owing to the location of these raw material types in the excavations and based on the limited dating of the site, it seems likely that these artefacts are generally older than artefacts produced on local stone. This indicates a change in raw material focus in the local area and identifies a potentially important change in resource use that requires further investigation.

The primary objectives of further excavations in the vicinity of Square E1 were to: (1) expand the excavated sample to help understand the activities undertaken at the site; (2) recover data through controlled excavations which are directly comparable to other excavations conducted under the auspices of the

Table 12.1 Fish bone abundance, Eurimbula Site 1, Squares E1–E2.

SQUARE	XU	NUMBER SPECIMENS	TOTAL WEIGHT (g)
E1	2	12	0.13
E1	5	3	0.05
E1	6	16	0.19
E1	7	23	0.20
E1	8	3	0.10
E1	9	6	0.04
E1	10	11	0.05
E2	3	2	0.02
Total	–	76	0.78

Figure 12.5 Cleaned c.2.5m high section of erosion bank located along the southern third of the site, showing dark band of coffee rock mid-way down the profile. Facing west.

Figure 12.7 Close-up view of mud ark (*A. trapezia*) shell lens encountered during excavation, Square E1, XU10. Facing west.

Figure 12.6 General view of completed excavation, Square E1, showing shell lens mid-way down the western profile. Facing west.

Figure 12.8 General view of completed excavation at Squares A–D. Note large roots towards the top of the profile. Facing southwest.

Southern Curtis Coast Regional Archaeological Project (SCCRAP); and (3) collect further shell/charcoal paired samples, if possible, for radiocarbon dating to address issues of local marine reservoir effect in the Round Hill Creek estuary system. For further details on the results of earlier test excavations see Francis (1999), Lilley et al. (1996), Reid (1997) and Ulm et al. (1999a).

Excavation methods

A detailed surface examination of the entire lower western margin of Round Hill Creek was undertaken before final selection of the area to be excavated. Visibility away from creek margins was limited owing to dense vegetation cover, although erosion banks and clearings were examined in detail. The survey confirmed the results of previous studies, with scatters of surface shell and stone artefacts found to be concentrated at the southern end of the site and, more generally, in close proximity to the erosion bank. As in the earlier investigations, many large stone artefacts were located along the erosion bank. Ten of these artefacts were collected from the erosion bank for laboratory analysis. Square E1 from the 1996 excavations was relocated and backfill removed to allow plotting of the new excavation onto the same basemap as that used on the previous excavations. A 1m² excavation grid comprising four 50cm × 50cm pits (Squares A–D) was excavated between 21 February and 9 March 1999 (Fig. 12.2).

The excavation grid was situated in a level area 7.9m north-northeast of Square E1 and 160cm west of the top of the erosion bank (Latitude: 24°11′54″S; Longitude: 151°51′34″E). The new excavation was situated further away from the brush-turkey mound to minimise the potential for disturbance to the surface and near-surface cultural deposits. Although a near-surface date of 220±80 BP (Wk-5601) was obtained for Square E1, it is difficult to interpret owing to the proximity of the pit to the mound and the presence of large quantities of shell which have been scratched-up onto the surface of the mound. Fragments of shell and charcoal were present amongst the leaf litter on the surface both within the gridded area and in the immediate vicinity of the pit. A site datum was established 5.76m west of the centre western margin of the excavation grid. Excavation proceeded in shallow, arbitrary excavation units averaging 3.2cm in depth and 10.9kg in weight. Excavation ceased at a maximum depth of 80.5cm below ground surface after several units of culturally-sterile sediments had been removed (Fig. 12.8). A total of 98 XUs was removed, distributed as follows: Square A (25 XUs), Square B (24 XUs), Square C (26 XUs), Square D (23 XUs). A total of 1,069.5kg of sediment was excavated. It was gently dry-sieved through 3mm screens onto a plastic tarpaulin located 4m northeast of the excavation to prevent contamination of underlying sediments. Stone (n=23), charcoal (n=8) and shell (n=2) specimens encountered *in situ* during excavation were plotted three-dimensionally. The excavation was backfilled with a layer of plastic sample bags across the base, followed by a 15cm thick layer of sterile white sands from the beach at the base of the erosion bank and finally the sediments that had passed through the sieve (see Chapter 3 for a detailed discussion of the standard excavation methods employed at all sites).

Cultural deposit and stratigraphy

Excavation revealed approximately 55cm of sediments containing cultural material overlying culturally-sterile sands (Table 12.2). The pit yielded quantities of shellfish remains, dominated by rock oyster (*S. glomerata*) and mud ark (*A. trapezia*) with occasional hercules club whelk (*Pyrazus ebininus*) and common nerite (*N. balteata*), particularly from the upper 30cm of the deposit. Minute shell fragments and occasional small stone artefacts were recovered to the base of excavations. Fish bone was recovered from every excavation unit in the top 35cm, with occasional pieces present to the base of SUIII (Table 12.3). As the excavation was situated on a sandy dune, bedrock was not reached, although excavation continued for 20cm below the last *in situ* cultural material encountered. The rate of site accumulation calculated using the methods outlined by Stein et al. (2003) was relatively slow overall at some 1.42cm/100 years.

The deposit can be divided into four major stratigraphic units (SUs) on the basis of sediment colour and texture (Table 12.3, Fig. 12.9). The stratigraphy is not straightforward owing to mottling of parts of the deposit resulting from root penetration and burrowing which obscure the interface between units. Cultural materials were encountered *in situ* to the base of SUIII. Shell and stone recovered from the sieve residue below this level are thought to derive from higher up the sequence (see below). Mottling is a feature of SUIV, suggesting some penetration by roots and/or burrowing animals (see below). Acidity (pH) values are slightly acidic to neutral throughout (6.0–7.0). This stratigraphic sequence accords well with that described for the nearby Square E1 (Ulm et al. 1999a:108).

Table 12.2 Eurimbula Site 1, Squares A–D: summary excavation data and dominant materials. Data from Squares E1–E9 are included for comparison (after Ulm et al. 1999a:Appendix A–I).

SQUARE	XUs (#)	DEPTH (cm)	WEIGHT (kg)	SHELL (g)	BONE (g)	CHARCOAL (g)	ARTEFACTS (g)	STONE (g)	ORGANIC (g)
A	25	77.6	264.3	1459.6	6.6	118.6	2.9	18.0	3371.7
B	24	80.5	266.1	2096.4	7.6	133.8	25.9	65.4	2170.1
C	26	80.3	282.5	1919.3	8.7	96.1	4.6	66.3	2185.0
D	23	79.7	256.6	1192.0	5.1	282.9	132.8	58.0	1780.8
Subtotal	98	–	1069.5	6667.3	28.0	631.4	166.3	207.7	9507.6
E1	15	66.6	209.1	2451.0	3.0	194.0	16.1	NA	562.9
E2	9	51.1	161.0	177.6	0.5	159.5	43.9	NA	14.4
E3	9	47.7	143.5	211.6	0.1	94.2	4.0	NA	435.0
E4	6	29.5	100.0	1190.8	0.8	1.3	9.7	NA	13.7
E5	9	44.4	165.5	16.3	0	197.4	0	NA	155.0
E6	7	43.1	144.0	24.7	0	112.8	0	NA	425.9
E7	8	45.8	150.7	144.1	0	83.2	0	NA	320.4
E8	5	33.4	90.3	6.3	0	30.6	0	NA	99.5
E9	5	30.5	84.1	0	0	16.4	0	NA	1.6
Subtotal	73	–	1248.2	4222.4	4.4	889.4	73.7	NA	2028.4
Total	171	–	2317.7	10889.7	32.4	1520.8	240.0	207.7	11536.0

Table 12.3 Stratigraphic Unit descriptions, Eurimbula Site 1, Squares A–D.

SU	DESCRIPTION
I	Extends across the entire square with an average depth of 2.5cm and a maximum depth of 4cm below ground surface. The unit comprises greyish brown (2.5Y-5/2) medium subangular and poorly-sorted humic sands that are consolidated by a dense, matted fibrous root matrix. The surface of the unit is covered by leaf litter. Occasional tuffs of grass penetrate this surface layer with numerous small, fibrous roots. Cultural materials include occasional oyster and mud ark shell fragments, land snail and charcoal. pH values are slightly acidic (6.5).
II	Extends across the entire square with a maximum thickness of 28cm and a maximum depth of 32cm below the surface. It comprises fine subrounded and poorly-sorted dark greyish brown (10YR-4/2) to very dark greyish brown (10YR-3/2) sediments. The soils are humic and contain numerous roots up to 4cm in diameter. Large quantities of oyster and occasional mud ark, fish bone, stone artefacts and charcoal were recovered from the unit. pH values are slightly acidic (6.0–6.5).
III	Extends across the entire square with a maximum thickness of 36cm and a maximum depth of 58cm below the surface. The transition between SUII and SUIII is poorly defined owing to grading and mottling of the deposit from dark greyish brown (10YR-4/2)/brown (10YR-5/3) at the top to yellowish brown (10YR-5/4) at the base of the unit. Cultural material includes abundant mud ark (particularly in Squares B–C), with occasional oyster fragments, fish bone, charcoal and stone artefacts. Active tree roots are common in this unit, though fewer than SUII. pH values are slightly acidic to neutral (6.0–7.0), with an average of 6.5.
IV	Extends across the entire base of the excavation with a minimum thickness of 30cm and a maximum depth of at least 80cm below the surface. The base of this unit was not reached. Sediments comprise moist loosely consolidated light olive brown (2.5Y-5/4) to pale yellow (2.5Y-7/4) sediments with few roots. This unit appears to be culturally-sterile with abundant blocky charcoal (especially in Square D) and small pumice nodules. The several small stone artefacts and minute shell fragments recovered from this unit are thought to derive from the upper SUs. pH values are slightly acidic to neutral (6.0–7.0), with an average of 6.5.

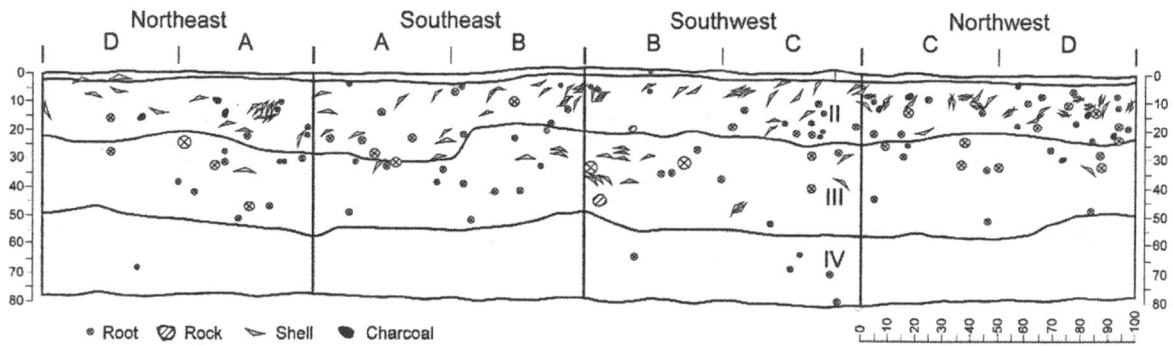

Figure 12.9 Stratigraphic section, Eurimbula Site 1, Squares A-D.

Radiocarbon dating and chronology

Twelve radiocarbon determinations have been obtained for the deposits, including the five obtained for the initial excavations (Lilley et al. 1996; Ulm et al. 1999a). As part of the present project, three further dates were obtained on samples from the initial excavations (one each from Squares E3, E4 and E7) to increase understanding of intra-site temporal variability. In total, three dates were obtained from Square E1, with one each from Squares E2–E4, two from Square E7 and four from Squares A–D. Seven dates were obtained on charcoal and five on *A. trapezia* samples (Table 12.4). A shell date from a discrete shell lens in Square E1, XU10 (Wk-3944), was paired with a charcoal sample from the same feature (Wk-5215) to investigate local marine reservoir conditions (see Chapter 4). Local reservoir effects may be a major factor in dating shell material from the Round Hill Creek estuary, as freshwater input from adjacent wetlands and incomplete isotopic exchange with the open ocean may have significantly altered radiocarbon activity within the estuary. The paired sample exhibits an apparent difference of 790 ^{14}C years (Table 12.4). The most probable explanation for this wide discrepancy is a lack of association between the shell and charcoal samples selected for dating. Although the discrete shell lens from which the samples derived appeared to be a secure stratigraphic context, it is possible that bulk sampling of the lens from the section resulted in contamination by more recent charcoal fragments. Alternatively, this apparent anomaly may be accounted for by mobilisation of small charcoal fragments in the matrix, as it is unlikely that densely packed shell valves with large surface areas such as that contained in the lens have moved far in the deposit (see Hughes and Lampert 1977). Calibration calculations for dates obtained on marine shell samples in Table 12.4 therefore employ a ΔR correction value of -305 ± 61, a provisional estimate based on a single shell/charcoal pair from the Tom's Creek Site Complex on the opposite side of Round Hill Creek (see Chapter 4 for details). It must be emphasised that this value was adopted as a first approximation, and less confidence can be placed in this figure than those available for other estuaries where more than one value has been determined.

Figure 12.10 plots the intercept closest to the mid-point of all calibrated ages against depth. All 12 samples show generally good concordance between age and sample depth. The linear regression only includes dates from Transect A, and excludes the paired samples (Wk-3944/Wk-5215) from Square E1 and the shell sample from Square E3 owing to uncertainties in sample provenance and marine reservoir offset. The regression shows a strong correlation between age and depth (r^2=0.9830) for these six determinations. Note that the origin intercept has not been forced through zero as it is not assumed that the surface represents 0 cal BP. Although straight line regressions can mask potentially significant variation in deposition rates (e.g. David and Chant 1995:377), they illustrate the general age-depth relationship of the deposit as a whole. All dates on *A. trapezia* samples shown in Figure 12.10 have been calibrated with the provisional ΔR= -305 ± 61.

The good concordance between the calibrated ages of Wk-8554, Wk-8553 and Wk-3944 and the expected general depth of these ages based on the linear regression lends further support to the broad validity of the ΔR employed.

Table 12.4 Radiocarbon dates from Eurimbula Site 1 (see Appendix 1 for full radiometric data for each determination). E* This date was undertaken on a sample of shell from a dense surface scatter adjacent to Square E7.

SQUARE	XU	DEPTH (cm)	LAB. NO.	SAMPLE	$\delta^{13}C$ (‰)	^{14}C AGE	CALIBRATED AGE/S
E1	5	9.5	Wk-5601	charcoal	-27±0.2	220±80	432(277,173,150,8,4)0
E1	10	35	Wk-3944	*A. trapezia*	-0.8±0.2	2390±60	2689(2347)2170
E1	10	35	Wk-5215	charcoal	-25.3±0.2	1600±160	1823(1416)1173
E2	9	50	Wk-3945	charcoal	-26.5±0.2	3020±70	3358(3205,3190,3162,3146, 3143,3086,3083)2948
E3	7	28.4-34.1	Wk-8553	*A. trapezia*	-0.6±0.2	1790±60	1869(1683)1479
E4	4	15-20	Wk-8554	*A. trapezia*	-0.9±0.2	560±55	619(493)317
E*	0	0	Wk-3946	*A. trapezia*	0±0.2	560±50	616(493)322
E7	5	18.8-24	Wk-8555	*A. trapezia*	-0.4±0.2	440±60	modern
A	5	9.7-12.4	Wk-10967	charcoal	-25±0.2	379±121	619(431,359,327)0
A	17	43.7-46.6	Wk-7688	charcoal	-25.5±0.2	2390±70	2710(2349)2158
B	12	34.4-38	Wk-10968	charcoal	-26±0.2	2218±126	2453(2282,2274,2151)1872
D	15	45.4-47.9	Wk-7687	charcoal	-24.7±0.2	2770±110	3158(2841,2829,2787)2547

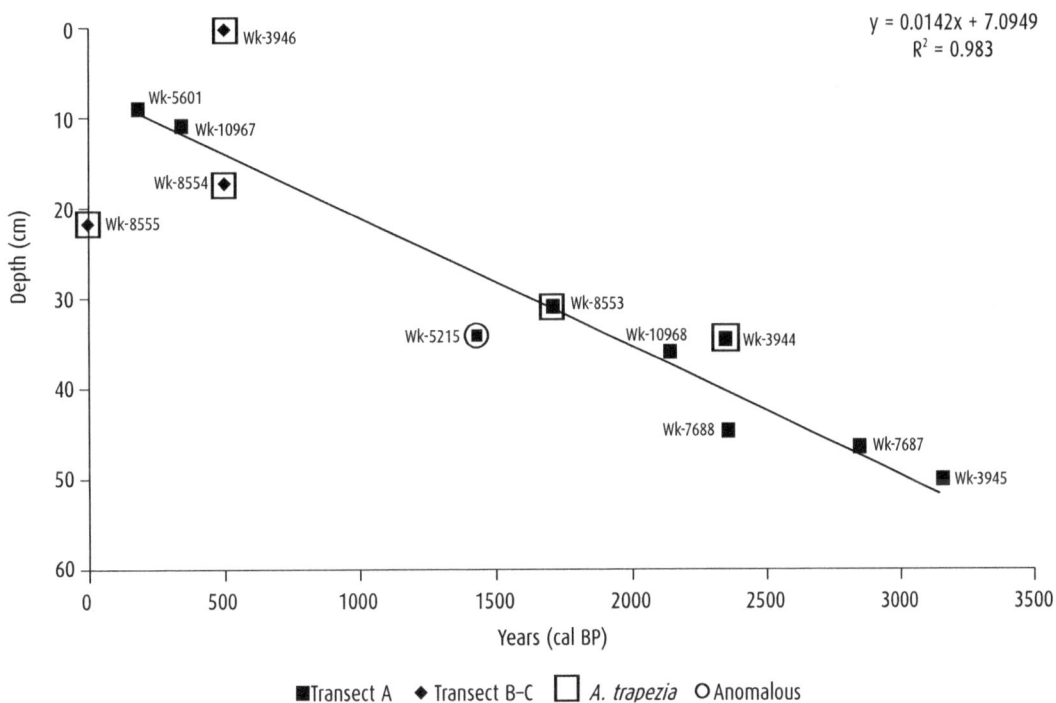

Figure 12.10 Age-depth relationship of all radiocarbon determinations obtained at Eurimbula Site 1 (n=12). The linear regression shown only includes the six dates obtained on charcoal samples from Transect A, Squares A–D, E1–E2.

Results from the two shell dates (Wk-3946 and Wk-8555) originating from and near Square E7 are problematic. The original date (Wk-3946) reported in Ulm et al. (1999a:Table 1) was undertaken on a sample collected from a dense exposure of *A. trapezia* on the erosion bank adjacent to Square E7. Wk-8555 was subsequently submitted to date a small concentration of *A. trapezia* in Square E7. It returned a modern calibrated result employing the ΔR= –305±61 value. The conventional radiocarbon dates (i.e. uncalibrated) of both determinations, however, overlap at two standard deviations, raising the possibility that they date the same or closely spaced events. The deflated context of the surface-collected sample adds further ambiguity. One possibility is that the surface material originated from the same stratum as the shell material dated from the square.

The dates from Squares A–D accord well with the earlier determinations obtained from Square E1, indicating that the top 10–15cm of the site dates to the last 500 years, with a concentration of cultural material between 30–40cm dating to 2,000–2,500 years ago, including the shell lens identified in Square E1. The basal undated cultural deposits probably belong to the 3,000–3,500 BP interval. The lowest stone artefact encountered *in situ* (see below) was recovered from a depth of 55.4cm, indicating that the cultural deposit pre-dates the lowest radiocarbon determination of c.2,800 cal BP (Wk-7687). These deposits are of broadly equivalent depth to the date of c.3,100 cal BP (Wk-3945) obtained from Square E2 (see Table 12.4) and may be of a similar antiquity. Dates from the basal units of Squares A–D therefore support the single date from Square E2, confirming occupation of the site by 3,000 BP. Samples dated from excavations conducted along Transects B and C indicate a much more recent chronology for the deposits north of Transect A, with widespread occupation apparently confined to the last 500 years. Dates from the top units of Squares E1 and E7 dating to the last 200–300 years combined with Cook's and Banks' ethnohistoric observations in AD 1770 suggest use of the site in the contact period. This scenario accords with evidence for post-contact occupation in the form of flaked bottle glass at the Tom's Creek Site Complex on the opposite side of Round Hill Creek. The dates are thus in sequence overall, suggesting first occupation shortly before c.3,000 cal BP and abandonment in the historical period (i.e. the last 150 years).

Stratigraphic integrity and disturbance

Several lines of evidence suggest that the deposit exhibits reasonable stratigraphic integrity. The sequence of radiocarbon dates is in order, with a regular age-depth relationship. There is also a predictable shell decay profile with highly weathered whole specimens recovered from the base of the deposit and relatively well-preserved specimens from the upper deposit. During excavation it was noted that small shell fragments recovered from the lower units were almost exclusively associated with small, rounded patches of dark grey sediment, which may indicate an origin from higher up the profile owing to burrowing and subsequent infilling. The sediments comprising these patches appear to derive from SUI (based on colour and grain characteristics). In contrast, stone artefacts encountered *in situ* in the lower units were found to be surrounded by yellowish brown sediments, indicating an *in situ* provenance in SUIII. The lowest *in situ* stone artefact recovered was from Square B, XU19, some 55.4cm below ground surface. This large flaked piece manufactured on rhyolitic tuff is 40.3mm in maximum dimension and weighs 14.3g. The broad surface area of this artefact and its association with the native yellowish brown sediments of SUIII suggest a valid stratigraphic relationship. In the initial series of excavations, several cavities were encountered in Square E3 along Transect A, presumably resulting from animal burrowing (Ulm et al. 1999a:108). A small burrow cavity was also noted in Square A, XU15–16, between 39.6–43.7cm in depth. The major source of post-depositional disturbance appears to be the presence of numerous roots throughout the deposit. Abundant fibrous roots occur to a depth of c.40cm with a zone of larger roots between 20–30cm (Fig. 12.16). The tuber of a convolvulus (*Ipomoea* sp.) was found in Square A, XU7, at a depth of 15.7–18.4cm. Tree roots are a feature of both the previous (Ulm et al. 1999a:108–10) and current excavations. Concentrations of shell encountered immediately above and below large, living roots suggest that they has been vertically displaced by root penetration and growth. However, conjoining *A. trapezia* valves found at a depth of 34.5–38cm in the adjacent Square B (see below) and conjoining fragments of non-artefactual stone found in Square D, XU13 suggest that the impact of burrowing can be very localised. Some evidence for insect burrowing was also noted in the form of a witchetty grub (*Cossidae* sp.) encountered some c.22–25cm below the surface in Square A, and active ant burrowing in the near-surface units of Square B, with burrows up to 2mm in diameter and to a depth of 5cm.

Conjoin analysis of the *A. trapezia* assemblage suggests that despite its relatively low density of cultural remains, the deposit has reasonable integrity. Out of a total of 95 measured intact and broken valves, 62 were excluded from consideration owing to an absence of hinge length or valve length and width, indicating the presence of valve damage (especially marginal damage). This left only 33 relatively intact valves for the conjoin analysis using the methods previously described in Chapter 5. A total minimum number of six *A. trapezia* conjoins was identified from Squares A–D. Most pairs (n=4) were separated by 6.5cm or less. A single conjoin had a maximum separation of just over 10cm (Table 12.5). Although this distribution largely reflects the main distribution of *A. trapezia* in the deposit, the six bivalve conjoin sets identified between 9.7–34.4cm below ground surface bracket some 25cm of the upper cultural deposit, supporting the impression gained from the radiocarbon chronology that shell and charcoal materials within the deposit are mostly *in situ*. This is also the zone of major tree root penetration, indicating that valves are generally closely associated despite visual impressions suggesting low site integrity.

The *A. trapezia* assemblage is generally in poor condition with a high ratio of broken to intact valves (15:1) and high rates of fragmentation. This indicates that the shells have probably been exposed to sustained heating after initial discard (see Chapter 5) and/or mechanical damage from treadage. Only the middle excavation units of Squares B and C had sufficient NISP of *A. trapezia* for calculation of fragmentation. Rates are relatively high in this part of the sequence, with an average of 234.7 NISP/100g with a range of 94.5–473.7.

Table 12.5 Identified *A. trapezia* conjoin sets, Eurimbula Site 1, Squares A–D.

CONJOIN SET	SQUARE/XU		MEAN DEPTH (cm)	MIN. SEPARATION (cm)	MAX. SEPARATION (cm)
	L	R			
Set 1	D/9	B/8	23.44	0.28	7.92
Set 2	A/6	A/5	12.69	0	6.02
Set 3	A/6	C/9	17.54	3.68	10.20
Set 4	B/6	D/6	15.40	0.24	6.48
Set 5	A/7	C/8	17.54	1.02	3.68
Set 6	B/11	B/11	32.43	0	3.94

Laboratory methods

Owing to the relatively low density of cultural material recovered from the site, all squares were analysed to maximise the available sample (see Chapter 3 for a detailed discussion of the standard laboratory methods employed at all sites). In the sections below, the results are summarised although the data from Square B is the main one illustrated in Figures 12.11–12.18. This approach has been adopted to minimise repetition. Further summary results for all excavated squares are presented in Appendix 4. Use-wear and residue analyses were conducted on selected stone artefacts in the Archaeological Sciences Laboratory, University of Queensland, using standard procedures outlined by Loy (1994).

Cultural materials

Invertebrate remains
Eighteen taxa of shellfish weighing 6,667.3g were recovered from Squares A–D, consisting of six marine bivalves, eight marine gastropods and four terrestrial gastropods (Table 12.6). The shell

deposit is dominated by rock oyster (*S. glomerata*), comprising 69.7% of the shell assemblage by weight (Fig. 12.13), followed by mud ark (*A. trapezia*) (29%) (Fig. 12.14). The remaining 16 taxa are relatively rare in the deposit, each contributing less than 1% of the shell assemblage by weight. The assemblage exhibits low diversity with a calculated Shannon-Weaver Function (H') of 0.715 and Simpson's Index of Diversity (1–D) of 0.308. The oyster and mud ark together with other taxa represented in small quantities in the shellfish assemblage, such as scallop (*Pinctada albina sugillata*), hercules club whelk (*P. ebininus*) and common nerite (*N. balteata*), suggest foraging strategies focussed on the mangrove fringe and adjacent intertidal and subtidal flats. The majority of shell was recovered from the upper deposit, with 87% of shell by weight occurring in the top c.30cm of the excavation.

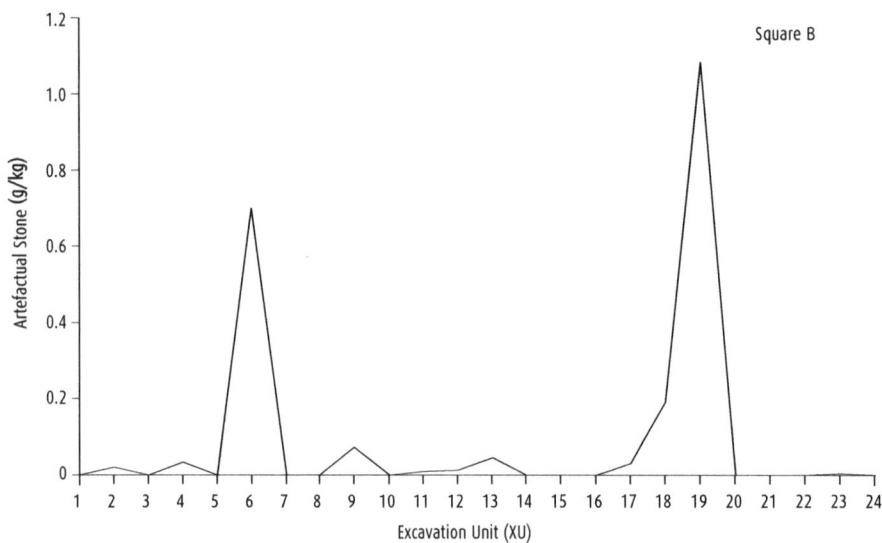

Figure 12.11 Abundance of artefactual stone.

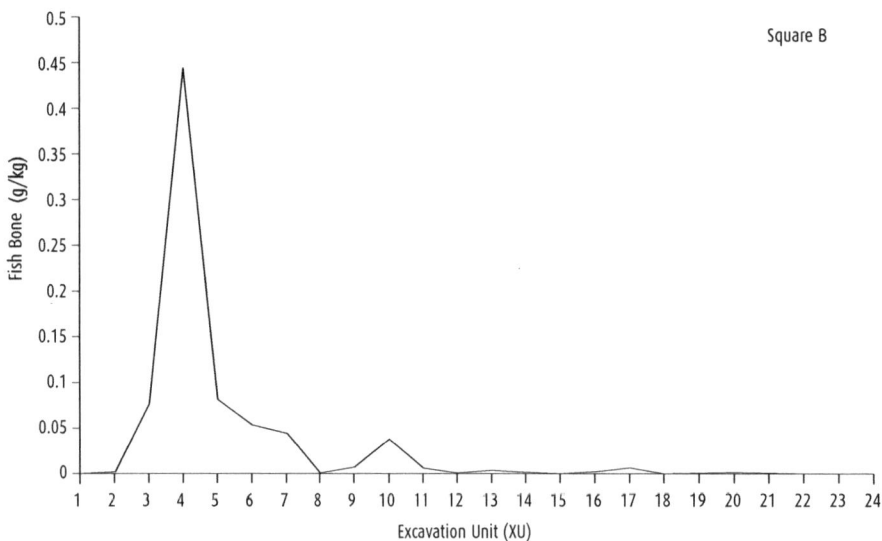

Figure 12.12 Abundance of fish bone.

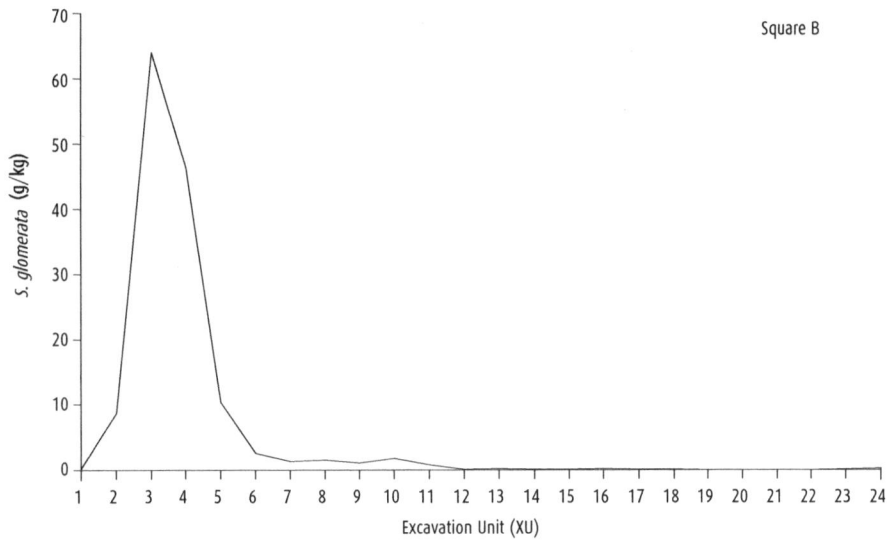

Figure 12.13 Abundance of oyster (*S. glomerata*).

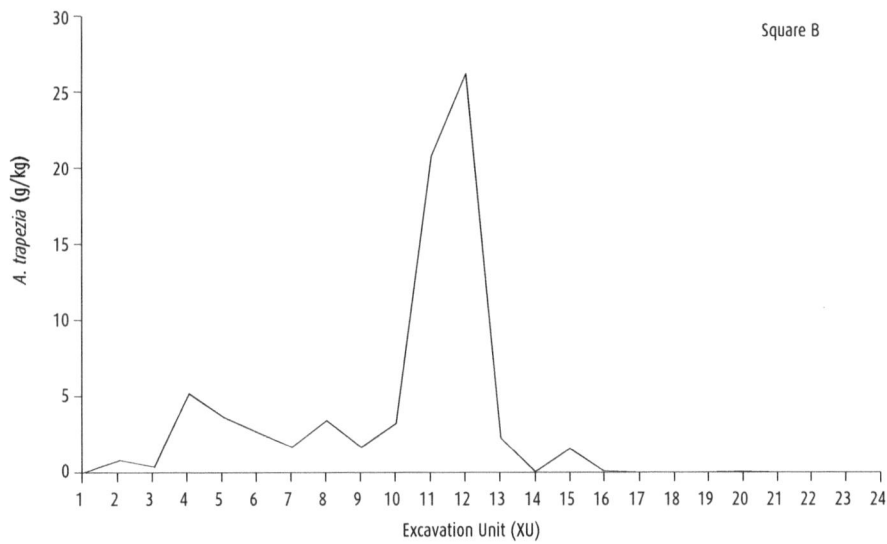

Figure 12.14 Abundance of mud ark (*A. trapezia*).

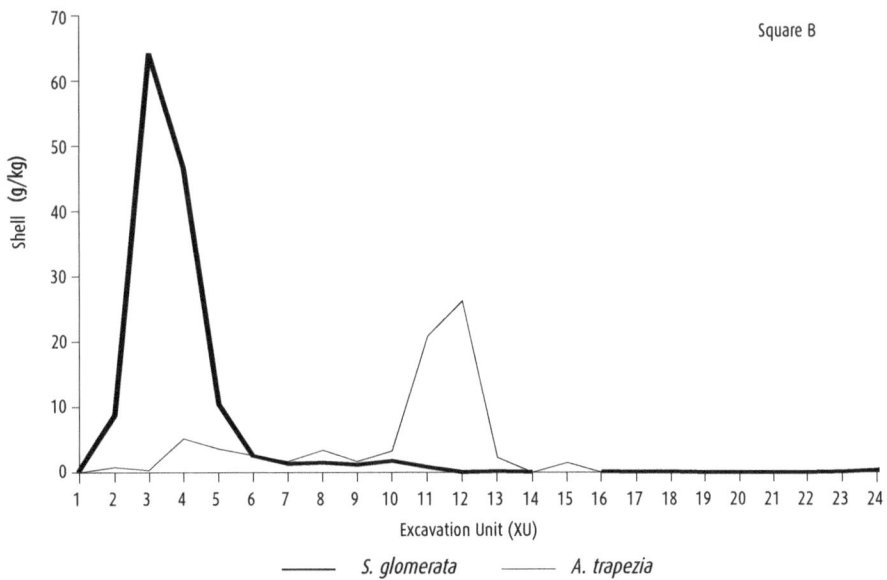

Figure 12.15 Abundance of dominant shell taxa.

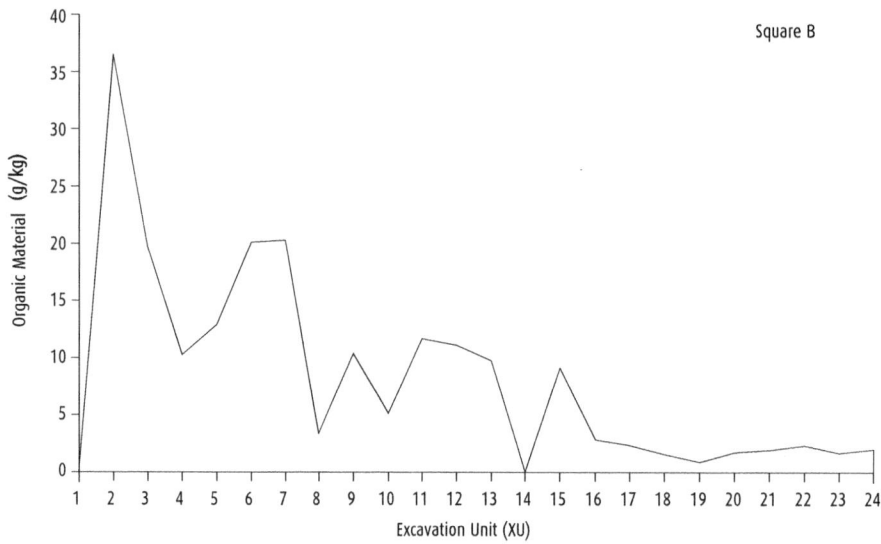

Figure 12.16 Abundance of organic material.

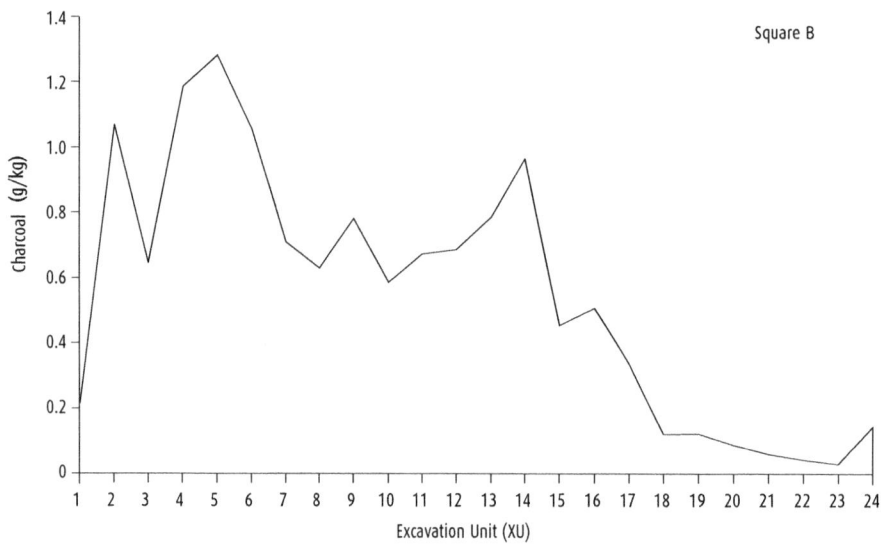

Figure 12.17 Abundance of charcoal.

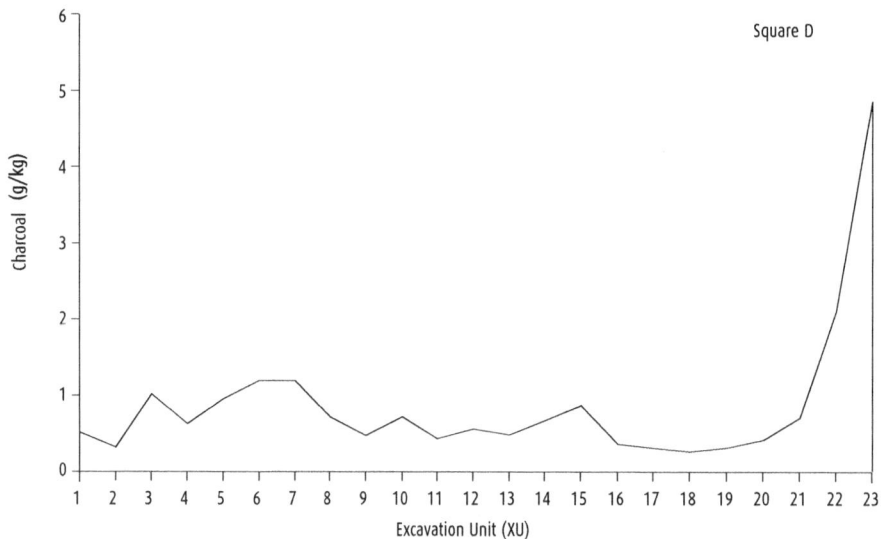

Figure 12.18 Abundance of charcoal.

Table 12.6 Presence/absence of shellfish identified in Eurimbula Site 1, Squares A–D.

FAMILY	SPECIES	SQUARE	1	2	3	4	5	6	7	8	9	10	11	12	13	14	15	16	17	18	19	20	21	22	23	24	25	26	TOTAL (g)
												MARINE BIVALVIA																	
Arcidae	*Anadara trapezia*	A		X	X	X	X	X	X	X	X	X	X	X	X	X	X	X	X			X							
		B		X	X	X	X	X	X	X	X	X	X	X	X	X	X	X	X		X								
		C				X	X	X	X	X	X	X	X	X	X	X	X	X											
		D				X	X	X	X	X	X	X	X		X		X	X		X									1936.7273
Donacidae	*Donax deltoides*	B							X																				0.5711
Mytilidae	*Trichomya hirsutus*	B			X																								
		C				X						X																	0.1631
Ostreidae	*Saccostrea glomerata*	A	X	X	X	X	X	X	X	X	X	X	X	X	X	X	X	X	X	X	X	X	X	X	X		X		
		B	X	X	X	X	X	X	X	X	X	X	X	X	X	X	X	X	X	X	X	X	X	X	X	X	X	X	
		C	X	X	X	X	X	X	X	X	X	X	X	X	X	X	X												
		D	X	X	X	X	X	X	X	X	X	X	X	X	X	X	X	X	X	X	X	X		X					4648.3022
Pteriidae	*Pinctada albina sugillata*	A			X																								
		B		X	X	X	X																						
		C				X	X																						
		D					X	X																					11.8758
Tellinidae	*Tellina* sp.	A								X																			0.9566
								MARINE GASTROPODA																						
Batillariidae	*Pyrazus ebininus*	A				X																							18.2000
Batillariidae	*Velacumantus australis*	C													X														1.2748
Ellobiidae	*Ophicardelus sulcatus*	A	X																										
		B		X	X																								
		C		X	X																								
		D	X																										1.7967
Lottidae	*Acmaeid* sp.	A				X																							
		B				X																							0.1241
Littorinidae	*Bembicium nanum*	A				X	X																						
		B			X	X	X				X		X		X			X											
		C		X	X		X	X				X																	
		D				X	X	X	X			X																	7.5868
Littorinidae	*Littoraria* sp.	C	X																										0.0347
Neritidae	*Nerita balteata*	A			X	X	X			X																			
		B				X	X				X																		
		D		X																									10.7585
Trochidae	*Thalotia* sp.	A	X			X	X	X																					1.8349
								TERRESTRIAL GASTROPODA																						
Camaenidae	*Figuladra* sp.	A					X	X	X	X	X			X		X													
		B			X	X	X	X			X			X	X									X					
		C			X	X			X																				
		D					X	X	X																				24.3205
Camaenidae	*Trachiopsis mucosa*	A	X								X																		
		B	X			X	X	X	X			X		X											X				
		C	X			X		X																					
		D	X					X	X		X																		2.0204
Pupillidae	*Pupoides pacificus*	D	X																										0.0058
Subulinidae	*Eremopeas tuckeri*	A	X	X																									
		B	X																										
		D													X														0.7191

Table 12.7 Metrical data for intact and broken (with umbo) *A. trapezia* valves from Eurimbula Site 1, Squares A–D. Note that excavation units for each square have been collapsed for purposes of analysis. Excavation unit depth and size is approximately equivalent across squares for broad comparison.

XU	MEAN LENGTH			MEAN WIDTH			MEAN HEIGHT			MEAN WEIGHT			MEAN HINGE		
	n	mm	±	n	mm	±	n	mm	±	n	g	±	n	mm	±
4	0	0	0	0	0	0	3	14.9	3.1	3	9.5	5.9	1	20.0	0
5	1	57.7	0	1	51.9	0	3	16.5	4.2	3	15.7	10.9	2	28.0	9.9
6	2	52.7	2.8	1	52.3	0	7	16.7	1.7	8	15.7	5.6	5	30.4	5.0
7	3	45.2	1.9	5	41.0	4.7	7	16.5	3.1	9	19.0	8.8	7	30.1	6.2
8	0	0	0	0	0	0	5	17.8	2.5	5	15.2	4.8	4	31.0	7.1
9	0	0	0	1	47.2	0	6	17.2	3.8	13	12.3	9.6	5	32.6	5.7
10	0	0	0	0	0	0	1	13.7	0	5	11.6	8.9	2	34.5	0.7
11	2	46.2	12.4	2	41.9	10.2	12	13.8	3.2	27	5.9	4.0	6	26.7	5.1
12	1	50.7	0	1	43.3	0	3	16.5	2.6	11	8.8	8.8	1	35.0	0
13	0	0	0	0	0	0	3	12.1	3.0	10	5.7	2.1	0	0	0
14	0	0	0	0	0	0	0	0	0	1	5.4	0	0	0	0

Although a more diverse assemblage of shellfish taxa were identified in Squares A–D than during the initial excavations, the dominance of rock oyster and mud ark was reinforced (Ulm et al. 1999a:112). The bimodal trend in the vertical distribution of these taxa observed previously was also documented in the Squares A–D assemblage (Fig. 12.15). Overall, 86.1% (4,000.4g) of rock oyster by weight was recovered from the top 13.5cm of the deposit, with 88.3% (1,709.5g) of the mud ark from 13.5–60cm. The possible implications of this trend are discussed below.

There is no significant change in the mean size of *A. trapezia* throughout the deposit as measured by height (owing to the high rates of damage to valve margins few valves are amenable to measurement of other attributes) (χ^2=1.9624, df=9, p≤0.05) (Table 12.7). The mean length of *A. trapezia* (as approximated by regression from the height measurements, y=1.8847x+14.491) is c.44mm.

Vertebrate remains

Fish bone is present throughout the cultural deposit, totalling 26.93g and consisting of 1,345 pieces of bone with a NISP of 18. A total MNI of 15 was calculated by summing the MNI for each excavation unit (Table 12.8). Identified taxa in descending order of abundance include Sparidae (bream, tarwhine, snapper) (NISP=8; MNI=8), Sillaginidae (whiting) (NISP=10; MNI=7) and Mugilidae (mullet) (NISP=1; MNI=1). Although fish bone occurs in every unit, it is most abundant in units where shell is also abundant (compare Figs 12.12 and 12.13).

In Square A three Sparidae identifications were made, two from XU6 identified from a vertebra and an otolith and one from XU7 from a dentary. The otolith was identified as bream (*Acanthopagrus australis*). In Square B, XU4, a single vertebra each was identified as belonging to Sillaginidae and Mugilidae and a Sparidae otolith was identified as snapper (*Chrysophrys auratus*). XU5 contained two vertebrae identified as Sparidae and Sillaginidae, and two otoliths identified as Sillaginidae. The centrum diameter of the Sillaginidae vertebrae from the adjacent XUs is similar suggesting that they derive from the same individual. A single vertebra and three otoliths in Square C, XU6, were identified to the Sparidae family, two of which are snapper, and two further otoliths were identified as Sillaginidae. XU9 contained an otolith identified as snapper. Square D, XU4, contained a Sparidae otolith (see Vale 2002, 2004 for further details).

Of the eight vertebral samples subjected to DNA analysis, only one returned a positive fish-like polymerase chain reaction (PCR) product, although this extract did not produce a product when sequenced (Hlinka et al. 2002). Taphonomic factors are thought to be responsible for the low amplification success rate.

Table 12.8 Fish bone abundance, Eurimbula Site 1, Squares A–D.

XU	SQUARE A				SQUARE B				SQUARE C				SQUARE D			
	NUMBER SPECIMENS	TOTAL WEIGHT (g)	NISP	MNI	NUMBER SPECIMENS	TOTAL WEIGHT (g)	NISP	MNI	NUMBER SPECIMENS	TOTAL WEIGHT (g)	NISP	MNI	NUMBER SPECIMENS	TOTAL WEIGHT (g)	NISP	MNI
2	0	0	0	0	2	0.02	0	0	5	0.50	0	0	5	0.01	0	0
3	9	0.06	0	0	32	0.69	0	0	7	0.09	0	0	2	0.01	0	0
4	20	0.25	0	0	187	4.04	3	3	20	0.33	0	0	15	0.31	1	1
5	49	1.54	0	0	28	0.75	4	2	95	1.98	0	0	128	1.86	0	0
6	72	2.05	2	2	50	0.57	0	0	97	3.99	6	5	44	0.87	0	0
7	52	0.82	1	1	36	0.44	0	0	75	0.97	0	0	40	1.41	0	0
8	25	0.36	0	0	3	0.01	0	0	26	0.46	0	0	2	0.01	0	0
9	16	0.24	0	0	4	0.09	0	0	8	0.28	1	1	13	0.19	0	0
10	20	0.44	0	0	25	0.37	0	0	13	0.08	0	0	4	0.04	0	0
11	12	0.13	0	0	8	0.07	0	0	9	0.10	0	0	1	0.19	0	0
12	5	0.08	0	0	5	0.01	0	0	6	0.02	0	0	6	0.06	0	0
13	5	0.08	0	0	4	0.04	0	0	2	0.01	0	0	1	0.03	0	0
14	7	0.11	0	0	4	0.02	0	0	3	0.01	0	0	0	0	0	0
15	6	0.04	0	0	0	0	0	0	0	0	0	0	1	0.01	0	0
16	2	0.01	0	0	4	0.02	0	0	2	0.01	0	0	0	0	0	0
17	0	0	0	0	6	0.06	0	0	0	0	0	0	0	0	0	0
18	0	0	0	0	0	0	0	0	0	0	0	0	0	0	0	0
19	0	0	0	0	5	0.01	0	0	0	0	0	0	0	0	0	0
20	0	0	0	0	2	0.02	0	0	2	0.01	0	0	0	0	0	0
21	1	0.01	0	0	2	0.01	0	0	1	0.03	0	0	0	0	0	0
22	1	0.01	0	0	0	0	0	0	0	0	0	0	0	0	0	0
23	0	0	0	0	0	0	0	0	2	0.01	0	0	0	0	0	0
24	1	0.03	0	0	0	0	0	0	0	0	0	0	NA	NA	NA	NA
25	0	0	0	0	NA	NA	NA	NA	0	0	0	0	NA	NA	NA	NA
26	NA	NA	NA	NA	NA	NA	NA	NA	0	0	0	0	NA	NA	NA	NA
Total	303	6.26	3	3	407	7.24	7	5	373	8.43	7	6	262	5.00	1	1

Seventeen pieces of bone weighing 1.0916g in total could not be assigned to a fish skeletal element. The generally small size of these specimens and the lack of diagnostic attributes prevented identification to taxon.

Stone artefacts

Stone artefacts are distributed throughout the cultural deposit between 1–73cm depth (Fig. 12.11). A total of 70 stone artefacts weighing 166.3g was identified in Squares A–D (Table 12.9). Eight of these were plotted *in situ* between 4.9–55.4cm during excavation with the remainder recovered from the sieve residue. Virtually the entire assemblage is manufactured on rhyolitic tuff (n=45) with the remainder on quartz (n=23), ignimbrite (n=1) and volcanic ash (n=1). Rhyolitic tuff and quartz are available from local headlands, while the source/s of the ignimbrite and volcanic ash are unknown. Most artefacts are extremely small, with an average maximum dimension of 10.9mm and average weight of 2.4g. Two of the largest artefacts recovered from the excavation were located towards the base of the cultural deposit. Both are flaked pieces manufactured on rhyolitic tuff recovered from Square B, XU19, and Square D, XU13, at respective depths of 55.4cm and 38cm below ground surface.

The only identified technological types in Squares A–D are a single flake and broken flake recovered from 45cm and 51–54cm respectively. The two artefacts manufactured on non-local raw materials also appear lower in the sequence at 34cm (ignimbrite) and 45cm (volcanic ash) in deposits dating to before 1,500 years ago. This pattern was also evident in the previous excavations. Few technological categories were represented with just two flakes, one broken flake, a backed artefact and 57 flaked pieces identified. Notably, the backed artefact was manufactured on silcrete and recovered from the basal units of Square E2 (see below). A radiocarbon date from the unit below that containing the backed artefact returned an age of 3,020±70 BP (Wk-3945). A small silcrete flake was recovered from the basal units of Square E1 and a silcrete flaked piece was found in the basal deposits of Square E3. Although the current data do not support earlier findings suggesting a dramatic shift from non-local raw material use to exclusive use of rhyolitic tuff in the recent sequence (Ulm et al. 1999a), several patterns are suggested. Artefacts manufactured on rhyolitic tuff are represented throughout the excavated sequence. However, artefacts manufactured on high quality siliceous stone (i.e. silcrete) tend to be located towards the base of the cultural deposit. These artefacts commonly exhibit retouch and rarely retain cortex suggesting careful manufacture and maintenance. The rhyolitic tuff assemblage, in contrast, is best characterised as the result of an expedient technology. Few artefacts have evidence for retouch or extensive use. In addition, many of the large stone artefacts observed on the erosion bank appear to be manufactured from rhyolitic tuff and to have derived from shallow deposits dating to the last 500 years. The presence of retouched tools, including the backed artefact, manufactured on non-local stone in the lower part of the sequence and the dominance of expedient artefacts on local stone in more recent deposits, including the large stone artefacts on the bank, may indicate a shift in both raw material use and technology.

A small sample of stone artefacts from Squares A–D have been subject to limited residue analysis. Francis (1999) examined six stone artefacts (FS19, 37, 60, 142, 161, 176) from a range of depths and manufactured on a variety of raw materials ranging in weight from 0.8–79.2g. All specimens were examined using an Olympus® metallurgical incident-light microscope under low level (<800x) magnification (see Loy 1994 for a discussion of techniques). All six artefacts exhibited surface features associated with post-depositional processes including rootlets, sand grains, spores and mycelium. All but one (FS60) of the artefacts was found to exhibit archaeological residues, comprising resin, cellulose, starch grains and charcoal. These residues are consistent with use of stone artefacts in a variety of plant processing activities.

Table 12.9 Stone artefacts from Eurimbula Site 1, Squares A–D.

SQUARE	RHYOLITIC TUFF				QUARTZ		IGNIMBRITE		VOLCANIC ASH		TOTAL	
	FLAKE		FLAKED PIECE		FLAKED PIECE		FLAKED PIECE		BROKEN FLAKE			
	#	(g)	#	(g)	#	(g)	#	(g)	#	(g)	#	(g)
A	–	–	15	0.9398	4	0.1851	–	–	1	1.7721	20	2.897
B	1	1.9297	10	22.7418	5	1.2640	–	–	–	–	16	25.9355
C	–	–	10	4.0993	7	0.4709	–	–	–	–	17	4.5702
D	–	–	9	82.0836	7	2.2371	1	48.5000	–	–	17	132.8207
Total	1	1.9297	44	109.8645	23	4.1571	1	48.5000	1	1.7721	70	166.2234

Lamb (2003) examined two large artefacts (FS233/1999; FS1/2001) both exhibiting roughly triangular cross-sections and probable bevelling along at least one margin, surface-collected from the erosion bank bordering Round Hill Creek. In addition to microscopic examination of *in situ* residues, aqueous samples from multiple locations on each artefact were extracted for microscopic examination on slides before staining with Congo Red solution to detect the presence of cooked or otherwise damaged starch in the archaeological residues. Both tools exhibited plant resides, comprising cellulose, parenchyma tissue, sclereids and bordered pits with quantities of both raw and cooked starch grains observed (Lamb 2003). Cooked, partially cooked and otherwise damaged starch grains were abundant and found concentrated along the tool margins. The presence of both undamaged and damaged starch grains indicates the processing of raw as well as cooked plants. Some grains exhibited similar sizes and morphology to reference specimens of cooked swamp fern (*B. indicum*). This finding is not only consistent with other studies linking heavy stone implements of this form to swamp fern processing (Hall et al. 1989; Robertson 1994), but extends the known range of use of this artefact type and associated plant processing technologies north from previously documented occurrences in the Moreton Bay, Cooloola and Fraser Island regions. Both residue and use-wear studies clearly indicate that plant processing was an important activity undertaken at the site.

Other remains

A range of other material was recovered from the site. Small pieces of pumice totalling 8.2g were recovered, concentrated towards the base of the deposit. Charcoal, totalling some 631.4g, is present throughout the deposit (Figs 12.17–12.18). Some 25% (156.2g) of this charcoal was recovered from Square D, XU22–23, and is not associated with cultural material. The presence of charcoal in culturally-sterile sediments below the shell deposit suggests that some of the other small quantities of charcoal in the assemblage may be natural. Large quantities of organic material, totalling 9,507.6g, were recovered, mainly comprising tree roots (Fig. 12.16). Small quantities of non-artefactual stone were also recovered (207.7g), consisting of mainly the native ironstone in dune deposits.

Discussion

Excavations and analyses presented here expand the results available from previous investigations revealing a low density sequence of cultural material distributed over a large area with the most intensive and widespread period of occupation dating to the last 500 years. The distribution of cultural material indicates tightly-focussed settlement strategies in a linear pattern parallel to the creek margin, suggesting a pattern of estuary-focussed settlement. Dates from the basal units of Squares A–D support the isolated date from Square E2, confirming a pre-3,000 BP antiquity for site occupation. Site chronology and stratigraphy indicate that occupation post-dates regional sea-level highstand between c.5,500–3,700 BP (Larcombe et al. 1995). Two lines of evidence support this argument. First, the oldest excavated cultural materials are dated to shortly before 3,000 years ago and occur within 50cm of the modern ground surface. Second, all cultural materials excavated or observed *in situ* in the erosion section are located in sediments stratigraphically above the humicrete or coffee rock unit which probably dates to shortly before the mid-Holocene highstand. This finding has several important implications. If the model of coffee rock formation outlines above is accurate, it indicates that dune landforms supporting barrier swamps were in place prior to the mid-Holocene. However, no cultural materials have been observed in or below the level of the coffee rock, suggesting that people either did not use the area, or only used it ephemerally, prior to the late Holocene despite the presence of rich wetland resources.

The excavations and analyses also demonstrate that although the cultural deposits at Eurimbula Site 1 are of relatively low density, the extent and recent chronology of the majority of the site complex indicates that a large quantity of cultural material was discarded over a relatively short period of time. To put this into perspective, we can extrapolate the figures available from the analysed excavations to the estimated site area for illustration purposes. The deposits cover a minimum area of 100,000m^2. Exposures of shell and stone artefacts are visible discontinuously along the 2km long erosion bank and surface surveys and excavations have demonstrated the presence of cultural material up to at least 50m inland. Erosion at the southern end of the site suggests that at least 30m of bank recession has occurred in some parts of the site, reinforcing the above site area estimate as an absolute minimum. Excavation of Squares E1–E9 and A–D indicates that the average density of shell recovered is 3.35kg/m^2. If this average density figure is taken as a basis for calculations, the entire site area is likely to exhibit a minimum of 335,000kg of shell. These figures become more plausible when it is considered that an unknown quantity of possibly denser deposits once located seaward of the current site has been removed by erosion.

Changes in use of faunal and stone resources are also documented at the site. Oyster and mud ark exhibit a distinctly bimodal distribution. Mud ark dominates earlier deposits and oyster the more recent part of the sequence, indicating a shift to more intensive exploitation of oyster. By far the majority of oyster is dated to the last 500 years. Reid (1997:17) hypothesised that there may have been a recent change in estuary conditions which favoured oysters, replacing the earlier populations of mud arks, although radiocarbon dates indicate that at least some mud arks were available into the historic period. Habitat change may well have expanded and contracted the niches available for oysters and mud arks respectively. Mud arks are found on or just below the surface of muddy and sandy substrates in estuaries and are frequently associated with seagrass, while oysters generally prefer clear water and require a rocky or mangrove root substrate. Rocky substrates do not occur in the estuarine sections of Round Hill Creek, with the nearest intertidal rocky substrates suitable for oyster spat attachment located near the Town of Seventeen Seventy at some distance from the site (Fig. 12.1). This suggests that the majority of oysters entering the site probably grew on mangrove substrates as they do in the estuary today. An increase in the abundance of oyster may therefore have been related to an expansion of mangrove communities in the estuary. An expansion of mangroves would also have reduced the area available for seagrass colonisation and therefore associated mud ark habitats. A recent expansion of mangrove communities is also indicated by the late appearance of the telescope mud whelk (*T. telescopium*), a mangrove gastropod which dominates the modern mollusc biomass but virtually absent from archaeological deposits.

The presence of carefully made and retouched tools, including a backed artefact, manufactured on possibly non-local highly siliceous stone in older deposits and the dominance of expedient artefacts on local raw materials in more recent deposits may indicate a shift in both access to raw material and changes in technology. The small sample sizes do not permit this trend to be examined in more detail at this stage, although the large stone artefacts manufactured on local materials identified in near-surface contexts along the erosion bank support this general pattern.

The distribution of surface and subsurface material suggests that occupation was concentrated along the creek margin, immediately adjacent to the diverse resources it offered. The presence of heavy stone implements exhibiting plant residues indicates that a range of subsistence activities took place at the site. As noted above, these artefacts are morphologically similar to bevelled-edged implements functionally associated with plant food processing in southeast Queensland. McNiven (1992b) found fragments of such implements in deposits at Cooloola dating to c.5,000 BP, although they are thought to only have come into widespread use in southeast Queensland in the late Holocene, when they are associated with restructuring of settlement and subsistence systems including more sedentary camps and increasing localisation of resource use (Ulm and Hall 1996). Until now, the geographical distribution of this implement type was thought to be restricted to the Moreton Bay area north to Cooloola and Fraser Island. The identification of heavy stone artefacts associated with plant processing on the southern Curtis Coast is therefore significant. The concentration of cultural remains along Transect A may thus reflect strategic decisions in site location. This transect is situated close to a variety of vegetation zones, including open forest habitats, extensive estuarine mangrove communities and tidal flats at the southern end of Round Hill Creek and freshwater swamps to the southwest (Olsen 1980a; QDEH 1994). The diversity of resources offered by these environments may have been a factor in the more intensive use of this area of the site. Conversely, evidence for the decrease in cultural material seaward from this transect may simply be related to variability in local resource availability, with a reduction in the area of intertidal flats towards the ocean.

This pattern cannot simply be explained in terms of differential preservation. Archaeological deposits at Eurimbula Site 1 are located towards the top of long, stratified dune sequences formed several thousand years earlier. These findings contrast with those originally presented in Ulm et al. (1999a:111) suggesting that 'the location of the excavations towards the seaward and thus more recently-formed edge of a prograding shoreline … suggest survey and excavation in older deposits to landward may locate material dating to at least the time of sea-level stabilization 6,000–7,000 years ago'. In fact, it appears that the landform and the broad estuarine resource mosaic was in place well before initial human settlement of the lands bordering Round Hill Creek.

As concluded by the previous investigations of the site (Ulm et al. 1999a), results do not suggest any obvious connection between the deposition of cultural remains and the formation of beach ridges. The quantity and location of cultural remains recovered in the excavations, however, strongly suggest that resource availability was a major factor in structuring local settlement patterns and hence deposition of cultural material. Regardless of whether the beach ridges at Eurimbula formed continuously or episodically, evidence suggests that the geomorphological occurrences of the last 3,000 years did not affect subsistence patterns, which were strongly focussed on Round Hill Creek rather than the ocean beach.

Summary

Eurimbula Site 1 was occupied before 3,000 BP, and is the only site in the region with indications of occupation around 2,000 years ago. An initial long period of low intensity occupation focussed at the southern end of the site is followed by widespread evidence for occupation across the entire site complex area over the last 500 years. The increase in evidence for use of the site over the last 1,500 years is accompanied by a general shift in the representation of particular shellfish taxa (from mud ark to oyster) and greater representation of local raw materials in the stone artefact inventory, with a higher degree of curation evident in the earlier assemblage.

13

Tom's Creek Site Complex

Introduction

This chapter reports archaeological investigations at the Tom's Creek Site Complex, located at the confluence of Round Hill Creek and Tom's Creek. It is the only excavated site to the east of Round Hill Creek in this study. Findings are consistent with those from sites to the north, showing repeated deposition of cultural materials from around 1,000 years ago. Like the Ironbark Site Complex, the Tom's Creek Site Complex provides direct evidence for persistence of occupation into the historical period, with flaked bottle glass recovered from the surface of the deposits. Survey and excavation methods are outlined before results of excavations are presented. As the Tom's Creek Site Complex is situated on the same estuary as Eurimbula Site 1, information about general estuary ecology is not repeated here (see Chapter 12).

Site description and setting

The Tom's Creek Site Complex is a large, stratified midden intermittently exposed over low dunes abutting the base of a rhyolitic scree slope at the junction of Round Hill Creek and Tom's Creek (Figs 13.1–13.2). The scree slope is part of a rocky outcrop forming the core of a small peninsula oriented roughly north-south. The site is situated 5km south-southwest of Round Hill Head (Latitude: 24°11'35"S; Longitude: 151°52'21"E). Several low, sandy ridges less than 1m in elevation and exhibiting surface shell are located on extensive mudflats between the mainland and the channel of Round Hill Creek. These stranded residual ridges were almost certainly formed by the movement of the creek channels in the past (Errol Stock, Australian School of Environmental Studies, Griffith University, pers. comm., 2000). The site covers a minimum area of approximately 12,500m², but is thought to be continuous with other deposits recorded to the north (see below). The densest exposures of cultural material occur closest to the coastal fringe.

Shell material is dominated by mud ark (*Anadara trapezia*) and oyster (*Saccostrea glomerata*) with occasional pearl oyster (*Pinctada albina sugillata*). Shell material was also observed on the intertidal flats adjacent to the terrestrial deposits, indicating recent erosion of the site's margins. Stone artefacts manufactured on rhyolitic tuff are commonly associated with the shell deposits as are occasional water-rounded microgranite cobbles, some exhibiting impact-pitting. Close to the base of the scree slope, surveys located a scatter of dark green bottle glass covering an area of c.10m^2, including several thick base sherds, two retouched artefacts and a large retouched artefact in the fork of a nearby tree.

A limited geomorphological coring program revealed that the low sand ridges abutting the rhyolitic core of the peninsula and adjacent sand ridge residuals on the intertidal flats are underlain by bluish-grey muds and clays, suggesting deposition in an estuarine environment. Preliminary examination indicates that these sediments are rich in organic material and dominated by mangrove pollen and are therefore interpreted as mangrove facies (Maria Cotter, School of Human and Environmental Studies, University of New England, pers. comm., 2000). A working model of local landscape evolution suggests that an expanded mangrove-vegetated intertidal zone extended to the base of the scree slope at a time of higher local sea-level around the mid-Holocene. Falling sea-levels after 4,500 BP mobilised sediments to form the transgressive dunes running parallel to the orientation of the peninsula. Some of these dune ridges were subsequently eroded by changes in the orientation of the channel of Round Hill Creek and the mouth of Tom's Creek.

The entire site area is surrounded by the dense mangroves which fringe Round Hill Creek and Tom's Creek and the minor inlets originating from these major waterways (Fig. 13.1). These mangroves largely protect the site from the direct impact of wave and wind action. Vegetation is dominated by spotted mangroves (*Rhizophora stylosa*) backed by yellow mangroves (*Ceriops tagal*) (Olsen 1980a). The narrow swale between the frontal and secondary dune on the mainland is dominated by teatrees (*Melaleuca* sp.) and tall (>15m) weeping cabbage palms (*Livistona decipiens*), while the area from the secondary dune inland to the base of the scree slope has an open canopy of large eucalypts interspersed with clumps of tall cabbage palms. The understorey is composed of common lantana (*Lantana camara*), immature cabbage palms, ground vines and shrubs. A stand of large (>5m high) cycads (*Cycas megacarpa*) occurs on a rocky slope at the south of the site. Residual landforms on the intertidal flats exhibit tall vegetation rising above the surrounding mangrove canopy, consisting mainly of teatrees and cabbage palms.

Although the general area in which the site is contained has been subject to significant European modification, the actual site area appears to have been minimally impacted by these activities. The main terraforming activities are a gravel quarry across the isthmus of the peninsula to the north of the site and a gravel road terminating at a boat ramp into Tom's Creek, which has been constructed along the eastern margin of the peninsula (Fig. 13.1). Surveys also located a small marijuana (*Cannabis sativa*) plantation concealed in dense lantana on the site itself. Use of plastic planting containers has prevented major impact of these cultivation activities on the archaeological deposits. The Tom's Creek Site Complex area is situated on a Recreation Reserve controlled by the Miriam Vale Shire Council. Tom's Creek is promoted to tourists as an estuarine fishing destination and several local brochures show the 4WD track from Captain Cook Drive to the Tom's Creek boat ramp (e.g. Desert Ridge 1998). Observed shore-based recreational fishing activities, however, are concentrated in the area of the boat ramp and are associated with recent human defecation sites, alcohol bottles and camp fires — well away from the main area of cultural deposits located to the northwest. Occasional flotsam is washed onto the intertidal flats through the mangroves and is sometimes blown onto the archaeological deposits.

Low density Aboriginal cultural materials occur throughout the area enclosed by the Town of Seventeen Seventy in the north, Tom's Creek in the south, Captain Cook Drive in the east and Round Hill Creek in the west (see Figs 12.1 and 13.1). Evidence for major landscape modification is

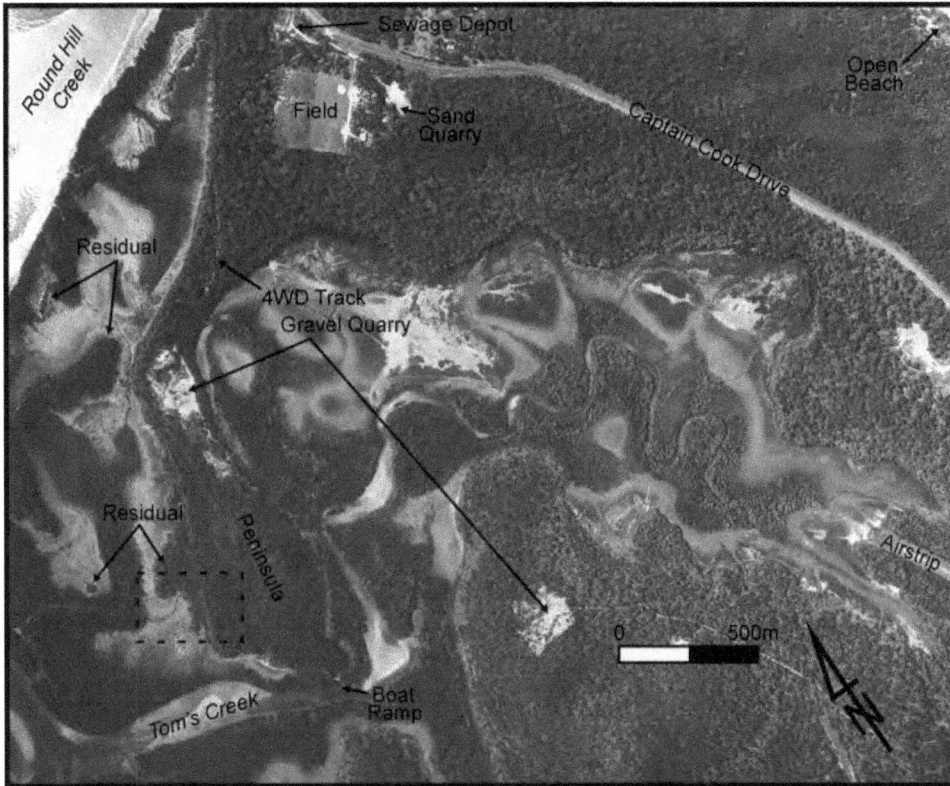

Figure 13.1 Aerial view of the Tom's Creek Site Complex area (after BPA Run 10B/57, 30 July 1996). The dashed box encloses the approximate area shown in Fig. 13.2. Based on data provided by the Department of Natural Resources and Mines, Queensland 2006, which gives no warranty in relation to the data (including accuracy, reliability, completeness or suitability) and accepts no liability (including without limitation, liability in negligence) for any loss, damage or costs (including consequential damage) relating to any use of the data.

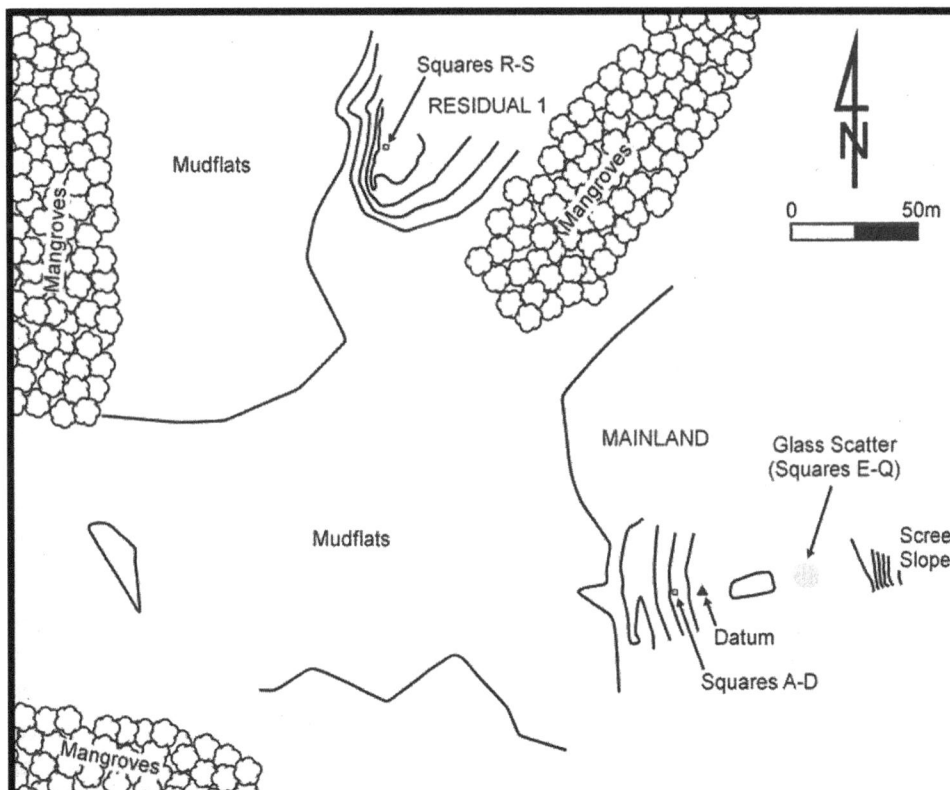

Figure 13.2 Site plan of area of Squares A–S. Contours are in 0.5m intervals.

evident in the northern two-thirds of this area, including quarrying, road-building and vegetation clearing. Numerous disused 4WD tracks also occur in this area. These indicators point to a diminution of the integrity of Aboriginal cultural deposits in this area. Examination of the 1959 aerial photographs show no evidence for quarrying or road-building on the Tom's Creek peninsula. Both the 1964 and 1988 photographs show the same small amount of quarrying at the narrowest part of the peninsula, suggesting that the majority of these activities took place between 1959 and 1964. By 1993 the aerial photograph shows extensive quarrying almost across the width of the peninsula and penetrating to the south. A road from the quarry area to Captain Cook Drive is also clearly visible with its construction probably associated with the more intensive gravel extraction activities occurring between 1964 and 1993.

The site appears to be an extension of the deposits recorded by Burke (1993) as KE:A33 on the Queensland Environmental Protection Agency's (EPA) Indigenous Sites Database. Burke observed very sparse shell scatters on and around 4WD tracks in open woodland, suggesting the area of the gravel boat ramp at Tom's Creek. Rowland (1987) recorded an extensive shell scatter bordering Round Hill Creek as KE:A11 just to the north of the isthmus of the Tom's Creek peninsula. Some of this material was subsequently disturbed by the construction of a large field for a sewage irrigation program that was later abandoned. Stone artefacts manufactured on a range of raw materials and scattered shell is visible across the field, particularly along its northwest margin. Based on findings elsewhere, such diverse lithologies may point to an early chronology for archaeological deposits in this area (see Chapter 14). Burke (1993) subsequently recorded two more middens to the northeast of this area: KE:A62 — a linear stratified midden exposed in a pit behind the sewage treatment depot; and KE:A63 — a sparse midden visible in 4WD tracks in this area (see Figs 12.1 and 13.1).

Intensive pedestrian surveys conducted as part of the Gooreng Gooreng Cultural Heritage Project (GGCHP) demonstrated that all of these deposits are likely to be continuous, with no major disjunctions identified between the sites recorded as KE:A33, KE:A11, KE:A62 and KE:A63 (Fig. 12.1). For the purposes of this site report, however, the Tom's Creek Site Complex is defined as the roughly triangular area bordered by Round Hill Creek and Tom's Creek to the west and south respectively and to the east by the scree slope and ridge that forms the core of the peninsula. To the north the area is truncated by a gravel quarry across the narrowest part of the peninsula. The site definition also includes four residuals exhibiting cultural material which occur on the intertidal flats to the immediate west (Fig. 13.1). The site is recorded as Gooreng Gooreng Cultural Heritage Project Site CC57 and Queensland Museum Scientific Collection Number S230.

Excavation methods

A detailed surface examination of the entire coastal fringe between the junction of Tom's Creek and Round Hill Creek north to the Town of Seventeen Seventy was undertaken before final selection of the areas for excavation (see Fig. 12.1). Surface visibility was usually limited by leaf litter and understorey vegetation away from the creek margin, although residual landforms on the intertidal flats generally exhibited high surface visibility. The three areas selected for excavation and/or systematic surface collection coincided with: (1) the densest exposure of shell material observed on the mainland, located on the frontal slope of the secondary dune (Squares A–D); (2) the densest shell exposure on the residual landforms stranded on the adjacent tidal flats (Squares R–S); and, (3) an area where flaked bottle glass was located (Squares E–Q) (Fig. 13.2). Detailed mapping was undertaken in the immediate vicinity of areas designated for excavation and collection. The large area over which cultural material is distributed prohibited detailed mapping of the entire site area. Excavations were conducted between 6–22 April 1999.

Squares A–D were located in an open, level area exhibiting surface shell and stone artefacts on a secondary dune c.15m east of the edge of the mudflat fringe and c.80m west of the base of the scree slope (Latitude: 24°11′35″S; Longitude: 151°52′21″E) (Figs 13.2–13.3). A site datum was established 384.5cm east of the middle of the east wall of Square B. Excavation proceeded in shallow, arbitrary excavation units averaging 3.8cm in depth and 12.6kg in weight. Excavation ceased at a maximum depth of 90.5cm below ground surface after several units of culturally-sterile sediments had been removed (Fig. 13.6). A total of 94 XUs was removed, distributed as follows: Square A (23 XUs), Square B (24 XUs), Square C (24 XUs), Square D (23 XUs). A total of 1,185.7kg of sediment was excavated. Excavated sediments were gently dry-sieved through 3mm screens onto a plastic tarpaulin located 5m north of the excavation. Stone (n=76), charcoal (n=23), shell (n=8) and bone (n=2) specimens encountered *in situ* during excavation were plotted three-dimensionally. The excavation was backfilled with a layer of plastic sample bags, followed by 50l of mangrove mud from the adjacent fringe and then the material that had passed through the 3mm sieve. A further 50l of mangrove mud was placed at the top (see Chapter 3 for a detailed discussion of the standard excavation methods employed at all sites).

Squares E–Q were situated in a level area c.50m east of Squares A–D and c.20m west of the base of the scree slope (Latitude: 24°11′34″S; Longitude: 151°52′20″E) (Figs 13.2–13.3). A grid was established over an area exhibiting a surface scatter of dark green bottle glass fragments, shell and stone artefacts manufactured on rhyolitic tuff (Fig. 13.4). Surface leaf litter and undergrowth were systematically removed from a c.40m² area around the glass scatter. A large fragment of glass was identified during this clearing operation in the fork of a quinine tree (*Petalostigma pubescens*) c.145cm above the ground (Fig. 13.5). A 2m × 1.5m grid comprising 12 × 50cm × 50cm squares (E–P) was established over the main concentration of bottle glass, with a further single isolated square (Q) established 4.35m to the northeast over two more glass fragments (Fig. 13.3).

The surface of each square was systematically mapped, photographed and described to show the presence, location and orientation of surface materials, including glass, stone artefacts and shell material. All the surface glass fragments were then plotted in three dimensions using a local datum before being removed

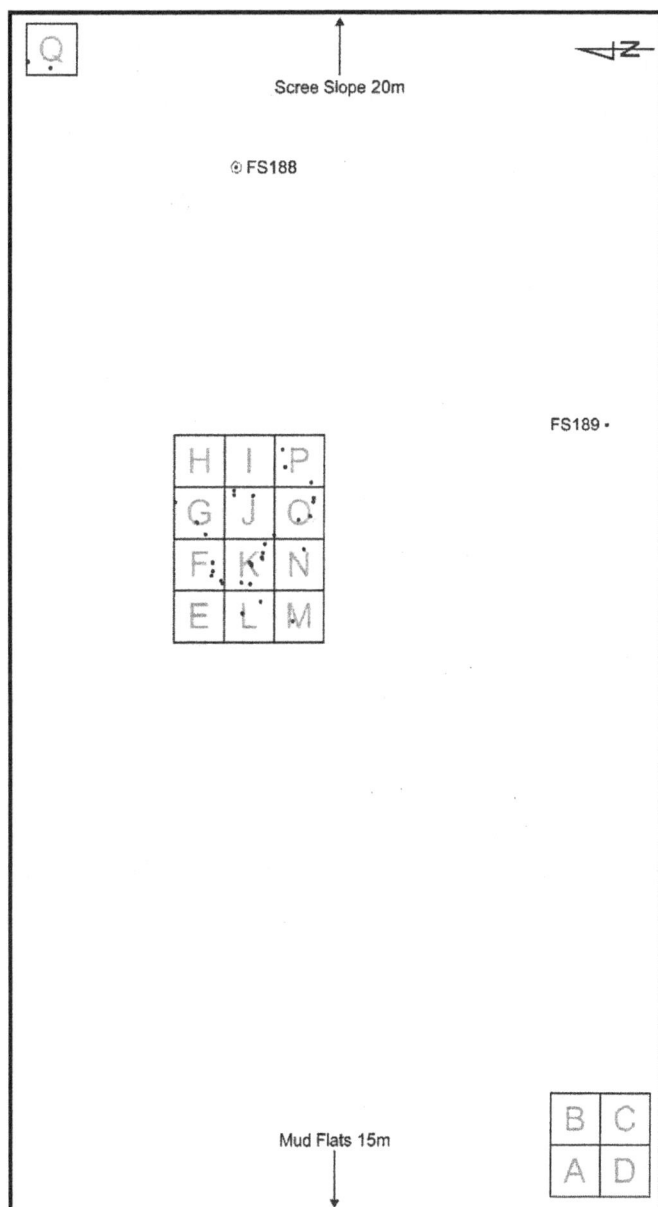

Figure 13.3 Schematic diagram of the layout of Squares A–D and E–Q, showing distribution of collected glass artefacts as solid dots (not to scale).

Figure 13.4 General view of area of glass scatter after removal of leaf litter. Flags indicate the position of glass artefacts. Facing northeast.

Figure 13.5 Glass artefact (FS188) cached in quinine tree. Facing north.

Figure 13.6 General view of completed excavation, Squares A–D. Facing northwest.

with tweezers wrapped in cling wrap, which was changed between samples to avoid cross-contamination of any residues between objects, placed into individual plastic press-seal bags and wrapped in bubblewrap for transport. In total, 36 pieces of glass, two valves of mud ark (*A. trapezia*) and a rhyolitic tuff flaked piece were collected from Squares E–Q. The glass artefact in the fork of the quinine tree was also plotted, photographed and removed. Sediment samples were collected from the surface of Squares K and Q to examine background starch and cellulose.

Squares R–S were situated on the first residual identified (Residual 1) which is a low sandy ridge (<1m), c.80m long and c.30m wide, and isolated from the mainland by a 50m wide section of mudflats densely vegetated with mangroves (Latitude: 24°11′29″S; Longitude: 151°52′18″E) (Figs 13.1–13.2). The residual is oriented roughly north-south, paralleling the orientation of Round Hill Creek, and decreases in elevation from west to east. The feature is roughly oval in shape and has a low erosion bank along its entire western and southwestern margins (Fig. 13.7). Abundant shell dominated by mud ark (>200 valves/m²) is exposed along the entire length of the erosion bank, with a smaller contribution from oyster and whelk. Flaked stone artefacts manufactured on rhyolitic tuff are also visible along the erosion bank (>3/m²), with no other raw materials observed.

Squares R–S were excavated as conjoining 50cm × 50cm pits to form a 1m × 50cm trench oriented at right angles to the frontal erosion bank, where the densest concentration of shell material is exposed 3.7m to the east. The trench was excavated on the first flat area west of the erosion bank. A local excavation datum was established 500cm north of the centre of the trench. Excavation proceeded in shallow, arbitrary excavation units averaging 3.6cm in depth and 11.4kg in weight. Excavation ceased at a maximum depth of 67.2cm below ground surface after several units of culturally-sterile sediments had been removed (Fig. 13.8). A total of 37 XUs was removed, distributed as follows: Square R (18 XUs), Square S (19 XUs). A total of 421.7kg of sediment was excavated. Excavated sediments were gently dry-sieved through 3mm screens onto a plastic tarpaulin located 3m northeast of the excavation. Stone (n=17), charcoal (n=5) and shell (n=1) specimens encountered *in situ* during excavation were plotted three-dimensionally. A sediment core was extracted from Square S below the base of the excavation (see below) (Fig. 13.9).

Cultural deposit and stratigraphy

Excavation revealed a shallow archaeological deposit containing a range of shellfish remains and stone artefacts overlying culturally-sterile sands and muds (Table 13.1). As both Squares A–D and R–S were situated on sandy ridges, bedrock was not reached, although excavation continued below the last cultural material encountered *in situ* to aid in identification of the base of the archaeological deposit. Squares A–D yielded quantities of shellfish remains, dominated by oyster (*S. glomerata*) and mud ark (*A. trapezia*), concentrated in the top 40cm of the deposit. Numbers of stone artefacts were also recovered from the upper deposit. Degraded shell and small shell fragments were recovered to the base of the excavation. Fish bone was recovered throughout the excavation. Artefacts manufactured on non-local raw materials, including chert and volcanic ash, were recovered towards the base of excavation. This is in contrast to the stone artefacts recovered from upper units, which are almost exclusively manufactured on rhyolitic tuff. Basal sediments contained quantities of pumice and large pieces of highly degraded rhyolitic tuff. The deposit can be divided into three major stratigraphic units (SUs) on the basis of sediment colour and texture (Table 13.2, Fig. 13.10) grading from dark greyish brown in the top half of the profile to light grey at its base. The base of the excavation appears to be culturally-sterile. Acidity (pH) values increase with depth from 6.0 to 8.0.

Squares R–S revealed a shallow cultural deposit largely confined to a layer of shell which extended across the trench at a depth of 20–30cm. The layer is dominated by oyster and mud ark and is associated with rhyolitic tuff artefacts and scattered charcoal. The location and

Figure 13.7 General view of shell material concentrated on erosion bank of Residual 1. Facing south.

Figure 13.8 General view of completed excavation, Squares R–S. Note position of shell material in the upper deposit. Facing north.

Figure 13.9 Core taken from below the limits of excavation, Square S. Note distinctive break (at c.75cm below ground surface) between coarse light yellow sands and dark organic muds.

composition of the layer is consistent with the material visible on the adjacent erosion bank. Very little cultural material was recovered below the shell layer. Occasional small shell fragments were recovered to the base of excavations, although these minute fragments may have been displaced by crab burrowing (see below). All sediments comprise brown sands and have been divided into three major stratigraphic units (SUs) (Table 13.3, Fig. 13.11). The majority of the basal unit (SUIII) appears to be culturally-sterile. The pH values range from slightly acidic (6.0) to neutral (7.0), with the higher values towards the base of the deposit.

Table 13.1 Tom's Creek Site Complex, Squares A–D, R–S: summary excavation data and dominant materials.

SQUARE	XUs (#)	DEPTH (cm)	WEIGHT (kg)	SHELL (g)	BONE (g)	CHARCOAL (g)	ARTEFACTS (g)	STONE (g)	ORGANIC (g)
A	23	89.5	280.1	3276.5	13.1	122.7	44.4	456.4	1362.2
B	24	90.5	296.2	2944.1	9.9	120.6	60.6	659.4	1334.9
C	24	89.7	319.3	2658.8	10.9	151.6	30.5	405.8	1406.6
D	23	88.4	290.1	2426.5	9.6	149.1	69.6	200.7	1805.7
R	18	66.8	213.4	2311.5	0.5	86.7	21.6	181.7	1635.1
S	19	67.2	208.3	2475.2	0.4	93.2	16.2	401.8	1900.0
Total	131	–	1607.4	16092.6	44.4	723.9	242.9	2305.8	9444.5

Table 13.2 Stratigraphic Unit descriptions, Tom's Creek Site Complex, Squares A–D.

SU	DESCRIPTION
I	Extends across the entire square with an average depth of c.10cm and a maximum depth of 17cm below ground surface. The unit comprises loosely consolidated dark grey (10YR-4/1) to dark greyish brown (10YR-4/2) fine to medium subrounded and poorly-sorted humic sands. The unit contains numerous small, fibrous roots with occasional larger roots present. Cultural materials include oyster, mud ark and charcoal. pH values are slightly acidic (6.0-6.5).
II	Extends across the entire square with a maximum thickness of 46cm and a maximum depth of 56cm. Comprises greyish brown (10YR-5/2) to dark greyish brown (10YR-4/2) poorly-sorted, fine and subrounded sediments. Although small roots are common, there is a general decrease in the abundance of roots with depth. This unit contains the majority of cultural material recovered from the excavation, including shell, stone artefacts and charcoal. Shell is more abundant in the top half of this unit, decreasing with depth. pH values are slightly acidic (6.5) to alkaline (8.0).
III	Extends across the entire base of the excavation with a minimum thickness of 40cm and a maximum depth of at least 90cm below the surface. The base of this unit was not reached. Sediments comprise loosely consolidated greyish brown (10YR-5/2) to light brownish grey (10YR-6/2) sediments. The unit is defined by the almost complete absence of shell and a dramatic reduction in the presence of roots. The top half of this unit includes artefacts made on non-local stone as well as occasional rhyolitic tuff artefacts. The basal units of the excavation appear to be culturally-sterile with occasional non-artefactual stone and pumice. pH values are slightly alkaline (7.5) to alkaline (8.0).

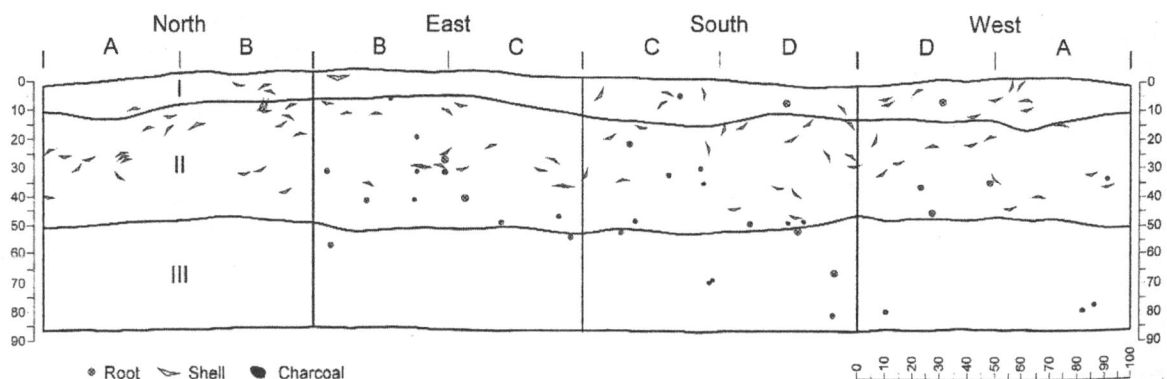

Figure 13.10 Stratigraphic section, Tom's Creek Site Complex, Squares A–D.

Table 13.3 Stratigraphic Unit descriptions, Tom's Creek Site Complex, Squares R–S.

SU	DESCRIPTION
I	Extends across the entire trench with an average depth of 5cm and a maximum depth of 7cm below the surface. Sediments are loosely consolidated pale brown (10YR-6/3) to greyish brown (10YR-5/2) and medium, poorly-sorted and subrounded. Few roots were encountered. Small quantities of shell and stone fragments occur. pH values are slightly acidic (6.0–6.5).
II	Extends across the entire trench with a maximum thickness of 27cm and a maximum depth of 33cm. It comprises well-consolidated greyish brown (10YR-5/2) sediments with numerous roots ranging in size from fine to large. Sediments are poorly-sorted, fine to medium and subrounded. This unit is defined by an abundance of shell material. Oyster dominates at the top, grading to a dense layer of shell dominated by mud ark across the base of the unit. Stone artefacts were found throughout the shell layer. Scattered blocky charcoal and small pieces of pumice are also present. pH values are slightly acidic (6.5) to neutral (7.0).
III	Extends across the entire base of the trench with a minimum thickness of 40cm and a maximum depth of at least 69cm below the surface. The base of this unit was not reached. Subsequent coring of the sediments below the limits of excavation showed that mangrove muds and clays underlie this unit. Sediments are loosely consolidated light grey (10YR-7/1) to pale brown (10YR-6/3) and are dominated by well-sorted, fine and subrounded grains. Occasional roots present, though decreasing with depth. This SU immediately underlies the base of the dense shell layer at the base of SUII. Small quantities of fragmented shell occur at the top of this unit but appear to be absent at the base. The majority of this unit therefore appears to be culturally-sterile. Scattered charcoal present. Pumice nodules increase in abundance and size with depth. pH values are neutral (7.0) to slightly alkaline (7.5).

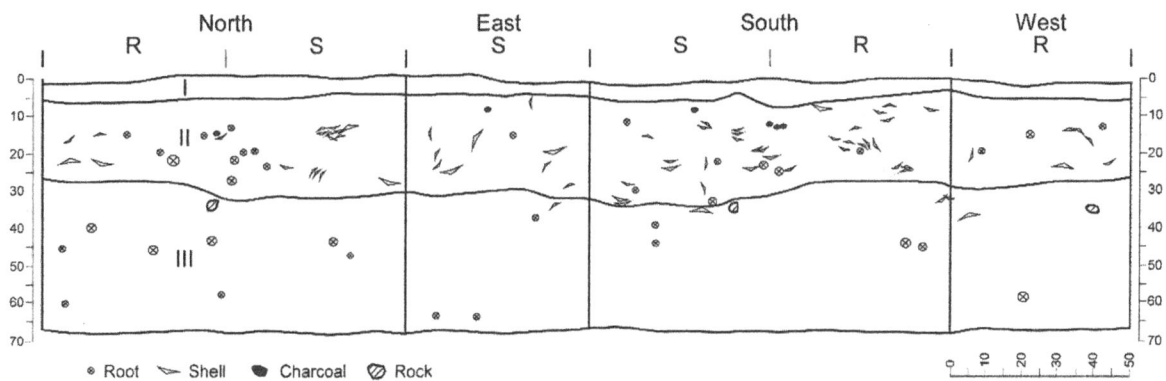

Figure 13.11 Stratigraphic section, Tom's Creek Site Complex, Squares R–S.

Radiocarbon dating and chronology

Ten radiocarbon dates are available for the site (Table 13.4). Six conventional radiocarbon dates were obtained on charcoal and three on mud ark (*A. trapezia*). In addition, a single accelerator mass spectrometry (AMS) date was obtained on organic material from the sediment core recovered below the base of excavation of Square S. The available dates include two shell/charcoal paired samples (Wk-7838/Wk-7686 and Wk-7682/Wk-7681) obtained to investigate local estuarine reservoir conditions. The first pair (Wk-7838/Wk-7686) was from the dense shell layer encountered in Square S, returning an apparent age difference of 90 ^{14}C years with ΔR= –305±61. Unfortunately, no absolute result was obtained from the second pair (Wk-7682/Wk-7681) from close to the surface of Square D owing to the very recent age of the charcoal sample. Calibration calculations for dates obtained on marine shell samples in Table 13.4 therefore employ a ΔR value of –305±61 as a provisional estimate until more data are available to resolve this value further in the Round Hill Creek estuary (see Chapter 4 for details).

The bottle glass assemblage provides a useful adjunct to the radiocarbon assays in resolving the site chronology. The assemblage consists of so-called 'black' glass, which is actually dark green. 'Black' glass was manufactured throughout the period between AD 1850s–AD 1920s for the storage and transport of beer and wine (Errol Beutel, Queensland Museum, pers. comm., 1999). Although a more precise assignment of age is not possible in the absence of manufacturer's marks or other diagnostic features, the glass in the assemblage dates to the late nineteenth/early twentieth centuries.

The six radiocarbon dates available from Square D are generally in sequence and indicate deposition began in this area around 950 years ago and ended in the historical period. Although the charcoal determination Wk-7681 could not be assigned an absolute age, the sample clearly dates to the last 200 years (%Modern= 99.5±0.6). The ^{14}C ages of Wk-7682 and Wk-10966 appear to be inverted. The calibrated ages of these determinations overlap at two standard deviations owing to the large error reported for Wk-10966 combined with the estimated error in the ΔR value applied in the calibration of Wk-7682. Although it is therefore possible that the two determinations are actually of a similar age, the values point to possible post-depositional disturbance of this section of the deposit. In the absence of stratigraphic evidence indicating a change in depositional history in this part of the site, a simple age-depth relationship suggests that these sediments are likely to be in the order of 500 years old. The lowest three dates are broadly in sequence, with some uncertainty introduced by the shell date Wk-7683, again owing to a lack of confidence in the ΔR value for this estuary (see Chapter 4 for discussion). Despite this uncertainty, the sequence clearly indicates first occupation of this part of the site only in the last 1,000 years, with the modern determination at the top and the glass artefacts found on the surface of the site nearby supporting an early twentieth century *terminus post quem* for Aboriginal use of the site.

Table 13.4 Radiocarbon dates from the Tom's Creek Site Complex (see Appendix 1 for full radiometric data for each determination).

SQUARE	XU	DEPTH (cm)	LAB. NO.	SAMPLE	δ^{13}C (‰)	^{14}C AGE	CALIBRATED AGE/S
D	3	3.9	Wk-7681	charcoal	-27.2±0.2	modern	modern (see text)
D	3	3.3	Wk-7682	A. trapezia	-1.2±0.2	620±50	646(521)425
D	8	22.2-25.5	Wk-10966	charcoal	-25.7±0.2	269±125	517(290)0
D	15	50	Wk-7683	A. trapezia	-1.2±0.2	940±50	939(776)647
D	17	55.7-59.5	Wk-7684	charcoal	-26.8±0.2	880±70	926(734)658
D	18	59.5-64	Wk-7685	charcoal	-27.5±0.2	1110±70	1170(966)794
S	8	20.5-24	Wk-7686	charcoal	-25.3±0.2	540±50	625(524)474
S	8	20.5-24	Wk-7838	A. trapezia	-0.9±0.2	630±50	650(527)432
S	11	31.7-35	Wk-10965	charcoal	-26.4±0.2	1070±115	1227(946,943,933)693
S	Core	c.75	NZA-13385	organics	-26.2±0.2	1956±57	1992(1870)1711

The three dates available for Squares S derive from the shell layer located between c.20–30cm. Wk-10965 dates the transition between SUII and SUIII with the later deposits thought to be largely culturally-sterile. The resulting calibrated date of c.943 cal BP is similar to that obtained for the earliest deposits on the mainland at Squares A–D. The other two dates from within the shell layer itself indicate that the vast majority of cultural deposition at this site occurred between 1,000–500 cal BP. The shallow and limited distribution of cultural materials also support the idea that material in the shell layer was discarded over a relatively short time-span. The date from the organic mangrove facies underlying the archaeological deposit at Squares R–S points to a very recent chronology for the formation of the low dunes abutting the Tom's Creek peninsula. The date indicates that the dunes containing cultural materials in the vicinity of Squares A–D and R–S probably only date to a phase of recent dune-building in the last 2,000 years.

Stratigraphic integrity and disturbance

Several lines of evidence suggest that while deposits at Squares R–S exhibit good integrity, Squares A–D have zones of relatively poor integrity. Root penetration appears to be a major source of disturbance in the area of Squares A–D. Numerous roots were encountered in the top 30cm of this deposit, almost entirely composed of small diameter (<3mm) red roots which are generally oriented vertically in the deposit and most likely derive from the large weeping cabbage palms fringing the swale to the southwest of the excavation. This pattern of root growth would tend to push material up through the deposit. Therefore, cultural material is more likely to be displaced upwards rather than downwards through the deposit over the long-term, providing that the local vegetation composition has remained stable.

Conjoin analysis of the *A. trapezia* assemblage suggests that despite exhibiting an overall low density of cultural materials, both excavated deposits retain reasonable integrity. Out of a total dataset of 228 measured intact and broken valves, 122 were discarded owing to an absence of hinge length or valve length and width indicating the presence of marginal damage. This left 106 relatively intact valves for consideration in the conjoin analysis, distributed as follows: Squares A–D (45 valves); Squares R–S (61 valves). Methods proceeded as described in Chapter 5. A total minimum number of five *A. trapezia* conjoins was identified in Squares A–D and nine in Squares R–S (Tables 13.5–13.6). In Squares A–D most pairs were separated by less than 16cm. A single conjoin exhibited a maximum separation of up to 20cm, although it should be noted that this valve has a minimum separation of 11.54cm. These results suggest that material in the lower part of the cultural deposit in particular (i.e. below 30cm) may have been displaced up to 10–20cm from their original point of deposition. The nine conjoins in Squares R–S are all separated by less than 8cm with most separated by less than 5cm. All of the identified conjoins from Squares R–S derive from the shell layer encountered between c.20–30cm below ground surface, indicating good integrity for this feature. Overall the *A. trapezia* assemblage exhibits a high ratio of broken to intact valves (average 11:1), which is similar between the excavation squares.

Table 13.5 Identified *A. trapezia* conjoin sets, Tom's Creek Site Complex, Squares A–D.

CONJOIN SET	SQUARE/XU		MEAN DEPTH (cm)	MIN. SEPARATION (cm)	MAX. SEPARATION (cm)
	L	R			
Set 1	B/2	C/3	4.37	2.92	8.5
Set 2	B/8	A/8	24.24	0	6.60
Set 3	B/9	B/8	24.55	0	7.22
Set 4	B/9	D/13	34.12	11.54	20.00
Set 5	A/12	C/9	35.37	9.62	15.9

Table 13.6 Identified *A. trapezia* conjoin sets, Tom's Creek Site Complex, Squares R–S.

CONJOIN SET	SQUARE/XU		MEAN DEPTH (cm)	MIN. SEPARATION (cm)	MAX. SEPARATION (cm)
	L	R			
Set 1	S/8	S/8	22.24	0	3.44
Set 2	S/8	R/8	22.45	0	3.02
Set 3	R/8	S/8	22.45	0	3.02
Set 4	S/8	R/9	23.71	0	6.38
Set 5	S/9	S/8	24.30	0	7.56
Set 6	S/9	S/8	24.30	0	7.56
Set 7	R/9	S/9	25.97	0	4.22
Set 8	S/9	S/9	26.02	0	4.12
Set 9	S/9	S/9	26.02	0	4.12

Laboratory methods

Owing to the relatively low density of cultural material recovered from the Tom's Creek Site Complex, all squares were analysed to maximise the available sample (see Chapter 3 for a detailed discussion of the standard laboratory methods employed at all sites). Results from all squares are summarised below, although only selected data from Squares A and R are illustrated in Figures 13.12–13.19. Further summary results for all excavated squares are available in Appendix 4. Use-wear and residue analyses were conducted on the bottle glass and stone artefact assemblages in the Archaeological Sciences Laboratory, University of Queensland.

Cultural materials

Invertebrate remains

Twenty-two taxa of shellfish weighing 16,092.6g were recovered from Squares A–S, consisting of eight marine bivalves, 10 marine gastropods and four terrestrial gastropods (Table 13.7). The shell assemblage is dominated by rock oyster (*S. glomerata*), comprising 64.8% of the shell assemblage by weight (Figs 13.12–13.13), followed by mud ark (*A. trapezia*) (32.6%) and scallop (*P. sugillata*) (1.2%). The remaining 19 taxa are relatively rare, each contributing less than 1% of the shell assemblage by weight, comprising taxa common to mangrove environments such as the common nerite (*Nerita balteata*) and periwinkle (*Bembicium nanum*). The assemblage exhibits low diversity with a calculated Shannon-Weaver Function (H') of 0.942 and Simpson's Index of Diversity (1–D) of 0.41. A bimodal trend in the vertical distribution of oyster and mud ark is evident in the shell layer at Squares R–S, with mud ark more common at the base of the shell layer (Fig. 13.13). This pattern does not occur in deposits in Squares A–D (Fig. 13.12) which overlap in time with those at Squares R–S, indicating that the bimodal pattern in this case may be related to taxa-focussed foraging strategies rather than local environmental availability. A noteworthy inclusion in the shellfish assemblage is the open beach dwelling pipi (*Donax deltoides*) in every excavated square. Although only represented in small quantities, the presence of this taxon suggests the possibility of transport from nearby open beaches 3.5km to the east (Fig. 13.1). The weight of shell recovered from each square is remarkably consistent across the site, with an average of 2,682g/square (range 2,312–3,277g) (Table 13.1). The range of shellfish taxa recovered indicate gathering activities focussed on the intertidal zone and creek margin habitats adjacent to the site.

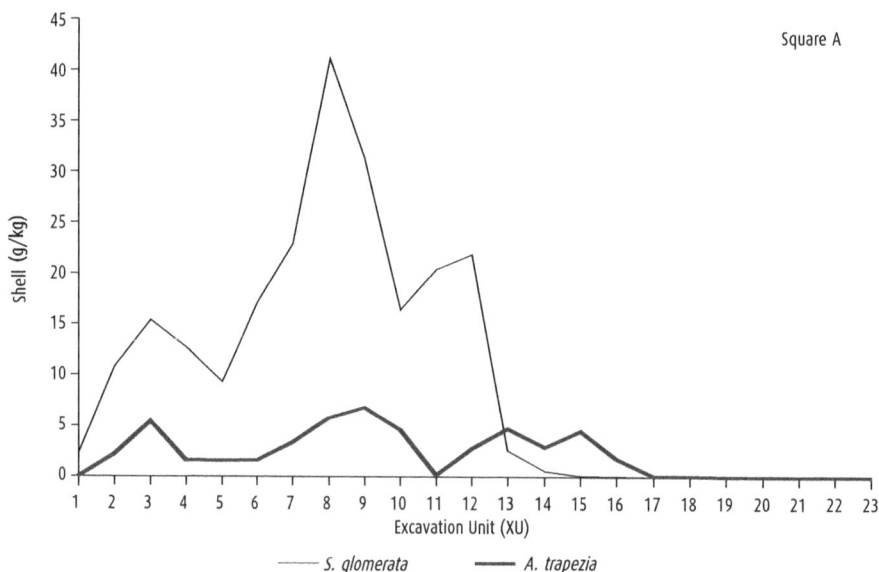

Figure 13.12 Abundance of dominant shell taxa.

Table 13.7 Presence/absence of shellfish identified in the Tom's Creek Site Complex, Squares A–S.

FAMILY	SPECIES	SQUARE	1	2	3	4	5	6	7	8	9	10	11	12	13	14	15	16	17	18	19	20	21	22	23	24	TOTAL (g)
										MARINE BIVALVIA																	
Arcidae	*Anadara trapezia*	A		X	X	X	X	X	X	X	X	X	X	X	X	X	X	X	X	X							
		B	X	X	X	X	X	X	X	X	X	X	X	X	X	X	X	X	X	X							
		C	X	X	X	X	X	X	X	X	X	X	X	X	X	X	X	X	X	X							
		D		X	X	X	X	X	X	X	X	X	X	X	X	X	X	X	X	X							
		R		X	X	X	X	X	X	X	X	X	X	X	X	X											
		S		X	X	X	X	X	X	X	X	X	X	X	X	X	X										5242.6788
Corbulidae	*Corbula crassa*	B			X																						0.0973
Donacidae	*Donax deltoides*	A				X		X		X	X																
		B	X										X														
		C		X					X						X												
		D				X																					
		R					X	X				X	X	X													
		S		X	X		X		X																		14.7604
Mytilidae	*Trichomya hirsutus*	A										X	X														
		B											X														
		C			X			X		X	X	X	X														
		D		X							X	X															
		R						X	X																		
		S			X	X	X				X	X															9.3398
Ostreidae	*Saccostrea glomerata*	A	X	X	X	X	X	X	X	X	X	X	X	X	X	X	X	X	X	X	X	X	X	X	X		
		B	X	X	X	X	X	X	X	X	X	X	X	X	X	X	X	X	X	X		X	X	X	X	X	
		C	X	X	X	X	X	X	X	X	X	X	X	X	X	X	X	X	X	X	X	X	X		X		
		D	X	X	X	X	X	X	X	X	X	X	X	X	X	X	X	X	X	X		X	X	X	X		
		R		X	X	X	X	X	X	X	X	X	X	X	X	X	X	X	X		X						
		S		X	X	X	X	X	X	X	X	X	X	X	X		X										10430.9731
Pteriidae	*Pinctada albina sugillata*	A		X	X	X		X	X	X	X	X	X														
		B		X		X	X	X	X	X	X	X	X						X								
		C			X	X	X	X	X	X		X	X		X												
		D		X	X	X		X		X	X	X	X														
		R		X	X	X	X	X	X	X	X	X															
		S		X	X	X	X	X		X	X	X															188.0454
Veneridae	*Dosinia tumida*	S								X																	1.1078
Veneridae	*Irus* sp.	S								X																	0.2563
										MARINE GASTROPODA																	
Batillariidae	*Pyrazus ebininus*	C							X																		
		D								X																	3.3332
Batillariidae	*Velacumantus australis*	D							X																		0.1444
Ellobiidae	*Ophicardelus sulcatus*	B															X										
		R	X	X	X																						
		S	X	X																							0.8799
Lottidae	*Acmaeid* sp.	S						X																			0.0625
Littorinidae	*Bembicium nanum*	A				X			X			X		X													
		B							X																		
		C		X																							
		D							X	X			X		X												
		R				X		X																			
		S						X		X	X																4.9315
Muricidae	*Bedeva paivae*	A										X	X														
		B			X																						
		D					X																				
		R					X	X																			
		S								X																	1.6023

continued over

Table 13.7 continued

FAMILY	SPECIES	SQUARE	1	2	3	4	5	6	7	8	9	10	11	12	13	14	15	16	17	18	19	20	21	22	23	24	TOTAL (g)
									MARINE GASTROPODA continued																	
Neritidae	*Nerita balteata*	A			X	X			X	X	X	X		X	X													
		B		X	X	X	X	X	X	X	X	X	X	X	X													
		C		X		X	X	X	X	X				X														
		D			X		X	X	X	X	X					X						X						
		R			X	X	X	X	X		X																	
		S		X	X	X	X	X		X	X																	69.9560
Neritidae	*Nerita squamulata*	D								X																		0.2042
Planaxidae	*Planaxis sulcatus*	A						X																				0.1681
Trochidae	*Thalotia* sp.	A											X															
		D				X																						
		R						X		X																		0.9177
									TERRESTRIAL GASTROPODA																			
Camaenidae	*Figuladra* sp.	A		X	X	X	X	X	X	X	X	X	X	X	X													
		B		X	X	X	X	X	X	X	X					X	X											
		C		X	X	X	X	X	X	X	X					X	X											
		D		X	X	X		X	X	X	X	X				X	X			X								
		R		X	X			X		X		X																
		S								X																		108.2116
Camaenidae	*Trachiopsis mucosa*	A	X	X	X	X	X	X	X	X	X	X	X	X	X								X					
		B		X	X	X	X	X	X			X	X		X	X								X				
		C		X		X	X			X			X															
		D		X	X	X	X	X	X	X	X	X	X	X	X		X			X								
		R				X		X																				
		S			X																							9.9973
Pupillidae	*Pupoides pacificus*	B								X																		
		R		X																								0.0748
Subulinidae	*Eremopeas tuckeri*	A		X	X																							
		B		X	X	X				X																		
		C		X	X		X																					
		D		X	X	X																						
		R		X																								4.8858

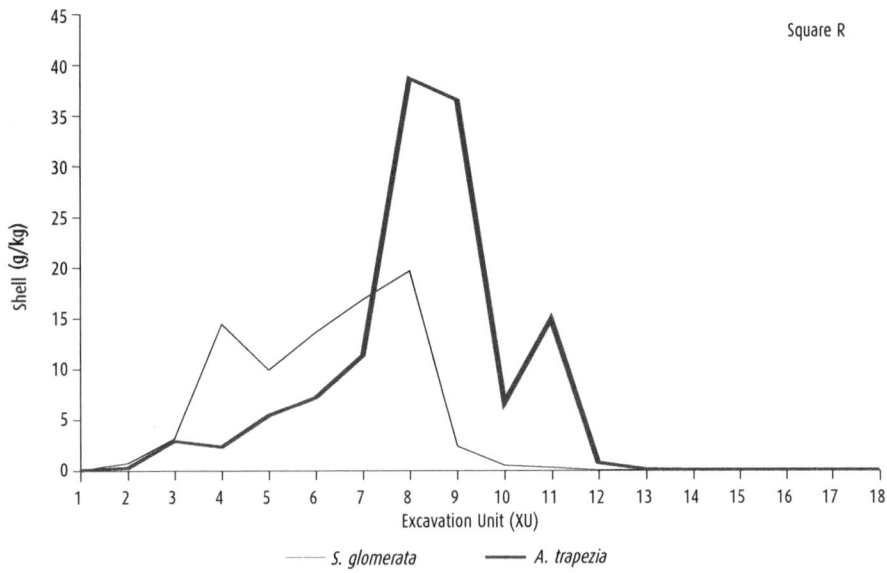

Figure 13.13 Abundance of dominant shell taxa.

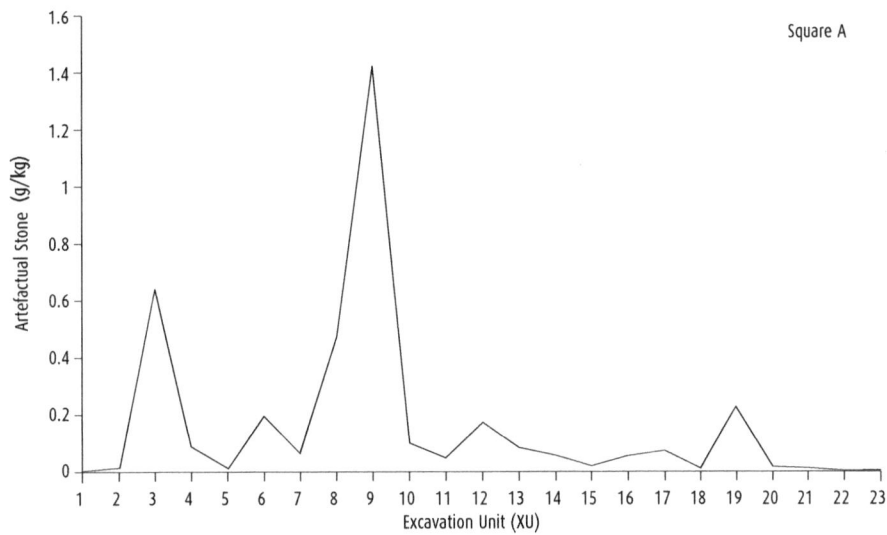

Figure 13.14 Abundance of artefactual stone.

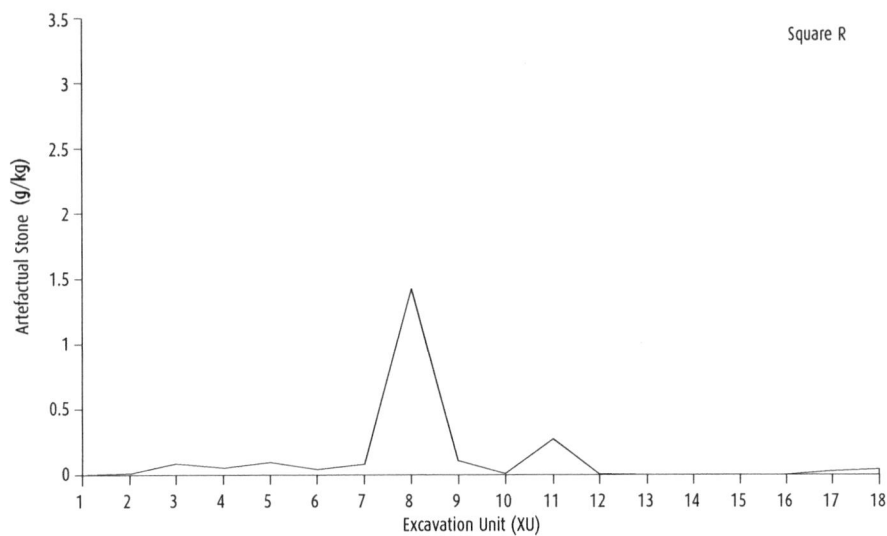

Figure 13.15 Abundance of artefactual stone.

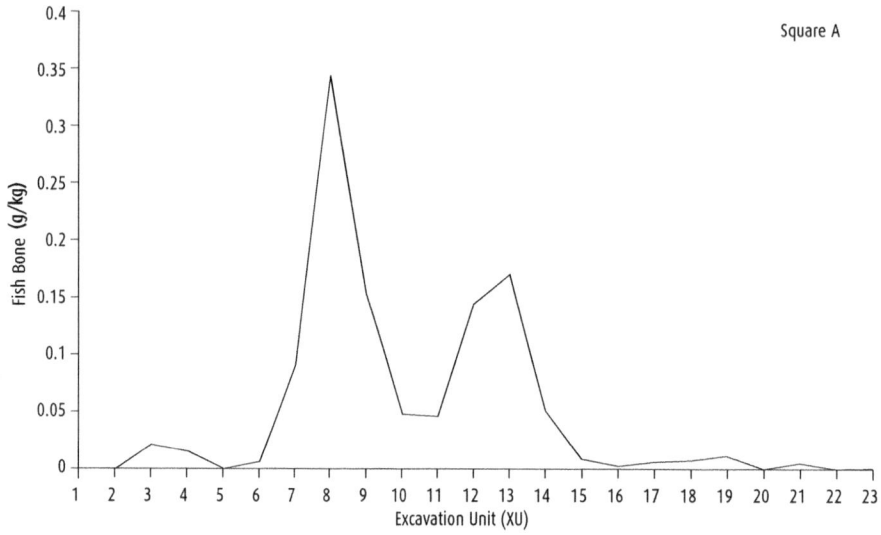

Figure 13.16 Abundance of fish bone.

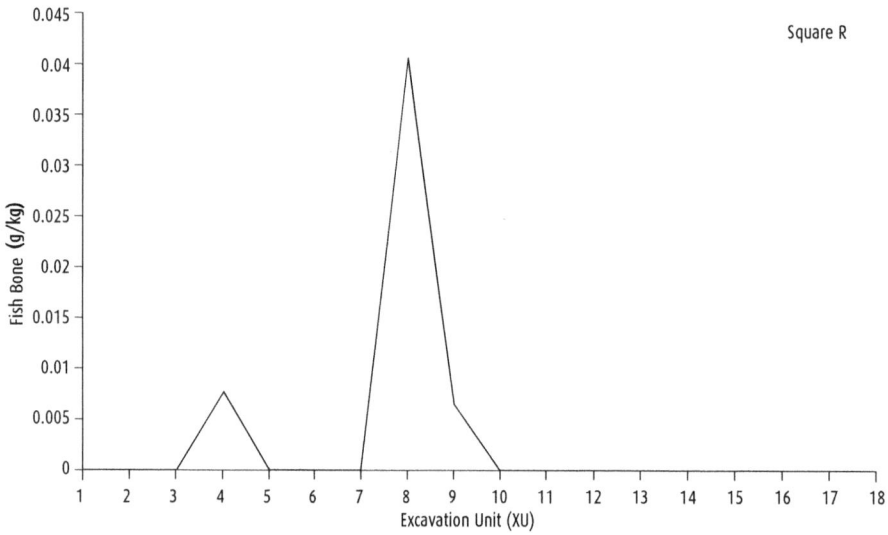

Figure 13.17 Abundance of fish bone.

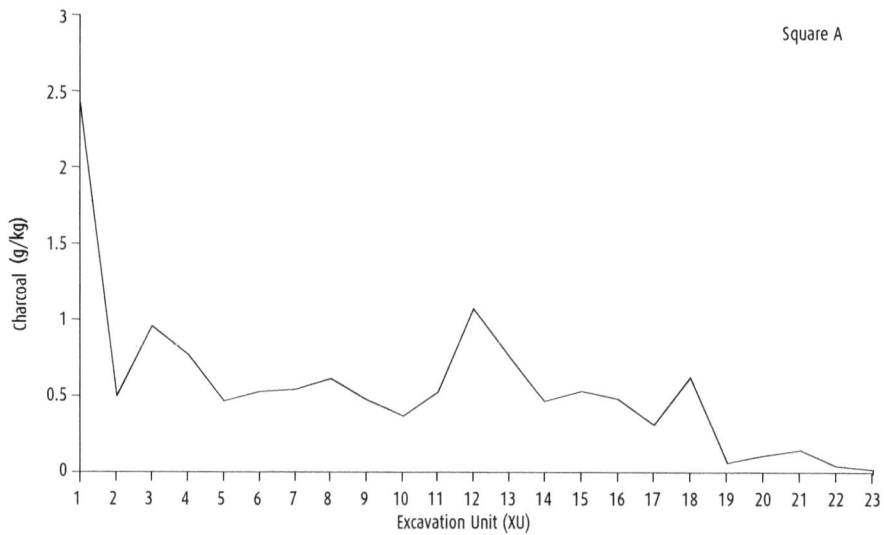

Figure 13.18 Abundance of charcoal.

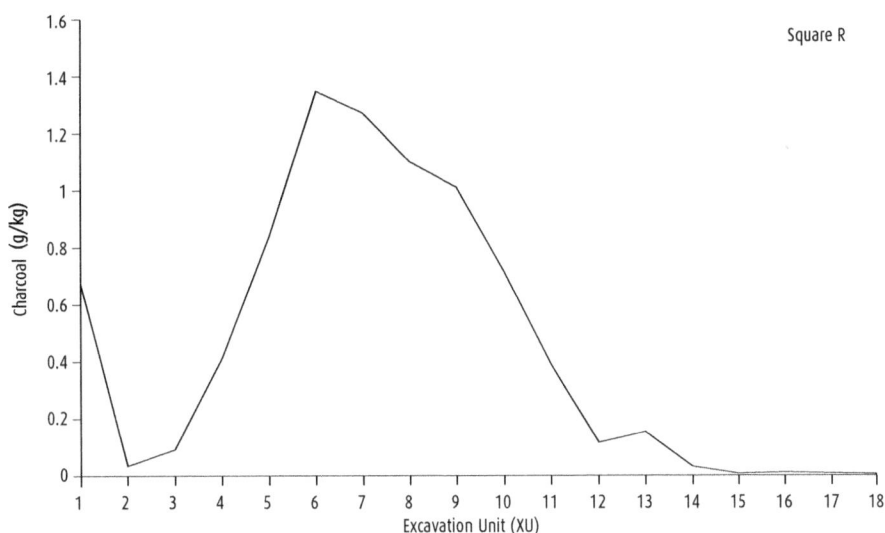

Figure 13.19 Abundance of charcoal

Table 13.8 Metrical data for intact and broken (with umbo) *A. trapezia* valves from the Tom's Creek Site Complex, Squares A–D. Note that excavation units for each square have been collapsed for purposes of analysis. Excavation unit depth and size is approximately equivalent across squares for broad comparison.

XU	MEAN LENGTH			MEAN WIDTH			MEAN HEIGHT			MEAN WEIGHT			MEAN HINGE		
	n	mm	±	n	mm	±	n	mm	±	n	g	±	n	mm	±
2	0	0	0	1	37.8	0	6	14.7	2.0	8	9.2	2.5	2	25.5	6.4
3	3	39.0	4.6	6	26.5	5.6	8	15.0	2.1	9	15.5	6.0	6	29.0	4.6
4	2	20.6	0.4	2	18.4	0.6	5	16.0	0.9	5	11.5	4.1	2	30.0	4.2
5	1	47.1	0	5	30.8	6.5	8	17.6	2.0	8	14.3	5.4	5	28.6	4.0
6	0	0	0	0	0	0	1	17.0	0	1	13.3	0	0	0	0
7	0	0	0	1	38.4	0	2	16.5	1.1	2	12.0	3.4	1	30.0	0
8	3	48.4	6.5	4	43.3	3.3	4	18.4	1.4	4	18.7	3.0	4	31.3	3.6
9	2	54.0	0.4	2	48.3	3.3	6	19.4	1.3	8	20.5	5.9	5	33.6	3.5
10	0	0	0	2	39.7	11.0	5	18.0	3.5	5	18.1	7.0	3	29.7	4.9
11	0	0	0	1	34.7	0	2	15.2	0.6	2	11.3	0	1	28.0	0
12	0	0	0	0	0	0	5	17.6	3.1	6	13.4	5.6	3	25.0	7.2
13	1	59.0	0	2	28.0	6.4	9	19.7	3.2	13	19.3	11.2	7	35.3	5.3
14	0	0	0	1	36.7	0	8	16.1	2.0	9	12.6	5.2	6	30.3	3.4
15	1	55.0	0	1	48.0	0	4	16.3	2.1	5	17.7	7.8	0	0	0
16	0	0	0	0	0	0	1	19.3	0	1	16.8	0	0	0	0

There is no significant change in the mean size of *A. trapezia* throughout either the Squares A–D deposit as measured by height (χ^2=2.1765, df=14, p≤0.05) or hinge length (χ^2=3.1568, df=11, p≤0.05) (Table 13.8) or the Squares R–S deposit as measured by height (χ^2=2.48, df=9, p≤0.05) or hinge length (χ^2=2.3738, df=7, p≤0.05) (Table 13.9). The mean length of *A. trapezia* (as approximated by regression from the height measurements, y=1.8847x+14.491) for Squares A–D is c.47mm and for Squares R–S is c.45mm.

Table 13.9 Metrical data for intact and broken (with umbo) *A. trapezia* valves from the Tom's Creek Site Complex, Squares R–S. Note that excavation units for each square have been collapsed for purposes of analysis. Excavation unit depth and size is approximately equivalent across squares for broad comparison.

XU	MEAN LENGTH			MEAN WIDTH			MEAN HEIGHT			MEAN WEIGHT			MEAN HINGE		
	n	mm	±	n	mm	±	n	mm	±	n	g	±	n	mm	±
2	1	42.2	0	1	40.2	0	1	16.2	0	1	16.4	0	1	28.0	0
3	0	0	0	2	37.6	0.5	3	15.1	2.5	3	10.6	2.7	1	27.0	0
4	0	0	0	0	0	0	3	13.2	2.1	4	8.9	3.9	0	0	0
5	1	30.2	0	2	35.8	6.6	3	14.4	2.0	5	12.4	3.6	1	23.0	0
6	3	18.8	4.0	3	17.0	3.7	13	16.7	2.3	13	15.8	7.5	8	29.9	3.7
7	0	0	0	0	0	0	3	18.7	1.9	5	17.6	3.4	1	32.0	0
8	10	16.6	5.5	16	22.7	5.2	31	18.2	3.4	35	19.7	5.7	21	31.5	39.0
9	2	49.8	0.4	6	43.9	2.7	35	18.5	2.2	39	19.8	7.5	25	33.7	3.9
10	0	0	0	1	44.3	0	2	16.7	2.0	11	6.1	3.3	0	0	0
11	0	0	0	0	0	0	14	13.0	3.4	24	7.4	4.1	3	29.7	3.5
12	0	0	0	0	0	0	0	0	0	2	1.5	1.5	0	0	0

Vertebrate remains

Fish bone is represented in almost every excavation unit in Squares A–D but is rare in Squares R–S (Tables 13.10–13.11). A total of 43.42g of fish bone was recovered from Squares A–D, consisting of 2,703 pieces of bone and a NISP of 30. A total MNI of 27 was calculated by summing the MNI for each excavation unit (Table 13.10). Identified taxa in descending order of abundance include Sillaginidae (NISP=19; MNI=16), Sparidae (NISP=10; MNI=10) and Platycephalidae (NISP=1; MNI=1). Fish bone is most abundant in units where shell is also abundant (compare Figs 13.12 and 13.16).

In Square A six taxonomic identifications were made. An otolith from XU3 was identified as bream (*Acanthopagrus australis*), an abdominal vertebra from XU6 was found to be Sillaginidae (whiting), a supraoccipital from a Sparidae (bream, tarwhine, snapper) was recovered from XU7, a Platycephalidae (flathead) from XU8 was identified from an atlas vertebra, and XU12 had a maxilla and otolith identified to Platycephalidae (flathead) and bream respectively. Five taxonomic identifications were made for Square B. Otoliths identified as Sillaginidae were identified in XU3, 11, 12 and 13 and a supraoccipital assigned to the taxon Sparidae was recovered from XU9. In Square C, 14 taxonomic identifications were made. A single Platycephalidae was identified from a dentary in XU3 and a single Mugilidae abdominal vertebra was found in XU12. Two Sparidae were identified in XU4, 12 and 13 on the basis of molars, an atlas vertebrae and an otolith. It is possible that these Sparidae elements in adjacent excavation units belong to the same fish. Seven Sillaginidae identifications were made in XU12, 13, 14 and 15 on the basis of otoliths and one vertebrae. Five taxonomic identifications were made for Square D. Sparidae were identified in XU2 and eight using an otolith and premaxilla. A Platycephalidae prevomer was found in XU13. Two Sillaginidae verebrae were identified in XU13 (see Vale 2004 for further details).

Only 0.85g of fish bone was recovered from Squares R–S, consisting of eight pieces of bone, a NISP of three and an MNI of three. All bone is associated with the level of the shell layer (compare Figs 13.13 and 13.17). Three taxonomic identifications were made on this material. In Square R, XU8, otoliths were identified as bream (*Acanthopagrus australis*) and Sillaginidae. In Square S, XU4, a Sparidae otolith was recovered (see Vale 2004 for further details).

Table 13.10 Fish bone abundance, Tom's Creek Site Complex, Squares A–D.

XU	SQUARE A				SQUARE B				SQUARE C				SQUARE D			
	NUMBER SPECIMENS	TOTAL WEIGHT (g)	NISP	MNI	NUMBER SPECIMENS	TOTAL WEIGHT (g)	NISP	MNI	NUMBER SPECIMENS	TOTAL WEIGHT (g)	NISP	MNI	NUMBER SPECIMENS	TOTAL WEIGHT (g)	NISP	MNI
1	0	0	0	0	0	0	0	0	5	0.0528	0	0	0	0	0	0
2	0	0	0	0	82	0.5665	0	0	81	1.0401	0	0	7	0.2552	1	1
3	7	0.2526	1	1	39	0.5470	1	1	15	0.2598	2	2	10	0.1290	0	0
4	10	0.1404	0	0	36	0.7016	0	0	62	1.3253	1	1	8	0.0896	0	0
5	0	0	0	0	32	0.4267	0	0	58	0.9688	0	0	0	0	0	0
6	5	0.0606	1	1	43	0.3925	0	0	27	0.3866	0	0	22	0.5261	0	0
7	116	1.1328	1	1	10	0.4233	0	0	48	0.5788	0	0	28	0.7199	0	0
8	108	3.9539	1	1	68	0.6261	0	0	41	0.6590	0	0	72	1.1362	1	1
9	156	1.9140	0	0	41	0.6435	1	1	19	0.1721	0	0	31	0.5145	0	0
10	52	0.5751	0	0	42	0.4773	0	0	44	0.5837	0	0	32	0.3138	0	0
11	19	0.5520	0	0	121	1.0598	1	1	58	0.8494	0	0	33	0.7830	0	0
12	116	1.6424	2	2	100	1.4936	1	1	54	1.1036	6	4	25	0.3966	0	0
13	73	1.7695	0	0	111	1.7689	1	1	81	1.4127	3	2	123	2.1896	2	2
14	36	0.6000	0	0	30	0.4008	0	0	37	0.5668	1	1	52	1.6064	0	0
15	13	0.0973	0	0	14	0.0786	0	0	40	0.7342	1	1	18	0.7915	1	1
16	4	0.0248	0	0	8	0.1666	0	0	5	0.0801	0	0	9	0.0814	0	0
17	4	0.0728	0	0	3	0.0244	0	0	7	0.0603	0	0	0	0	0	0
18	4	0.0887	0	0	2	0.013	0	0	4	0.0200	0	0	0	0	0	0
19	13	0.1438	0	0	3	0.0188	0	0	3	0.0130	0	0	2	0.0127	0	0
20	0	0	0	0	1	0.0074	0	0	0	0	0	0	4	0.0511	0	0
21	11	0.0529	0	0	0	0	0	0	2	0.0245	0	0	0	0	0	0
22	0	0	0	0	0	0	0	0	0	0	0	0	0	0	0	0
23	1	0.0073	0	0	2	0.0154	0	0	0	0	0	0	0	0	0	0
Total	748	13.0809	6	6	788	9.8518	5	5	691	10.8916	14	11	476	9.5966	5	5

Table 13.11 Fish bone abundance, Tom's Creek Site Complex, Squares R–S.

XU	SQUARE R				SQUARE S			
	NUMBER SPECIMENS	TOTAL WEIGHT (g)	NISP	MNI	NUMBER SPECIMENS	TOTAL WEIGHT (g)	NISP	MNI
1	0	0	0	0	0	0	0	0
2	0	0	0	0	0	0	0	0
3	0	0	0	0	0	0	0	0
4	2	0.0766	0	0	1	0.3689	1	1
5	0	0	0	0	0	0	0	0
6	0	0	0	0	1	0.0031	0	0
7	0	0	0	0	0	0	0	0
8	1	0.3168	1	1	0	0	0	0
9	1	0.0783	1	1	0	0	0	0
10	0	0	0	0	1	0.0032	0	0
11	0	0	0	0	1	0.0051	0	0
12	0	0	0	0	0	0	0	0
13	0	0	0	0	0	0	0	0
14	0	0	0	0	0	0	0	0
15	0	0	0	0	0	0	0	0
16	0	0	0	0	0	0	0	0
17	0	0	0	0	0	0	0	0
18	0	0	0	0	0	0	0	0
19	0	0	0	0	0	0	0	0
Total	4	0.4717	2	2	4	0.3803	1	1

Stone artefacts

Stone artefacts were recovered from all excavated squares (Table 13.12). Artefacts are more densely concentrated at this site than in all other excavated deposits in the study except the Ironbark Site Complex, where a major quarry is located (see Chapter 9). A total of 514 stone artefacts, weighing 205g, was identified in Squares A–D, 37 of which were plotted *in situ* between 0–65.4cm during excavation with the remainder recovered from the sieve residue. In Squares R–S, a total of 137 artefacts weighing 37.9g was recovered, including eight plotted *in situ* between 20.4–28.3cm. The assemblage is dominated by rhyolitic tuff, comprising 94% of the stone artefact assemblage by weight, followed by volcanic ash (4.5%), chert (0.9%), microgranite (0.5%), quartz (0.2%) and banded rhyolite (<0.1%). Rhyolitic tuff and quartz are present in local outcrops, including the scree slope demarcating the eastern boundary of the site, while fragments of microgranite recovered from excavations and water-rounded microgranite hammerstones observed on the surface must have been transported to the site, with Bustard Head, 21km to the north, the nearest source of this raw material. The precise sources of the chert, volcanic ash and banded rhyolite are currently unknown. Undiagnostic flaked pieces dominate the assemblage, with only very occasional flakes and cores. As noted, the large proportion of flaked pieces appears to be directly related to the mechanical flaking properties of rhyolitic tuff. Most artefacts are extremely small, with an average weight of 0.4g.

There is marked patterning in the vertical distribution of cultural materials in the deposit. Artefacts manufactured on non-rhyolitic tuff tended to be recovered from stratigraphic contexts below the vast majority of shell material. This pattern is generally consistent with observations made at Eurimbula Site 1 (Chapter 12). In Squares A–D, locally available materials are represented throughout the cultural deposit: rhyolitic tuff (0–75.6cm), quartz (6.5–70.9cm) and microgranite (1.6–13.6cm). In contrast, more siliceous materials that are not available in the immediate site area tend to occur towards the base of the cultural deposit: volcanic ash (55.7–71cm), banded rhyolite

Table 13.12 Stone artefacts from the Tom's Creek Site Complex, Squares A–D, R–S.

| SQUARE | RHYOLITIC TUFF | | | | VOLCANIC ASH | | | | | | | CHERT | | | | BANDED RHYOLITE | | MICROGRANITE | | QUARTZ | | TOTAL | |
| | FLAKE | | FLAKED PIECE | | FLAKE | | FLAKED PIECE | | CORE | | | FLAKE | | FLAKED PIECE | | FLAKED PIECE | | FLAKED PIECE | | FLAKED PIECE | | | |
	#	(g)	#	(g)	#	(g)	#	(g)	#	(g)		#	(g)	#	(g)	#	(g)	#	(g)	#	(g)	#	(g)
A	–	–	128	43.7763	–	–	–	–	–	–		1	0.0217	6	0.3405	1	0.0320	–	–	1	0.0814	137	44.2519
B	2	16.2791	141	44.1316	–	–	–	–	–	–		–	–	1	0.1838	–	–	–	–	1	0.0555	145	60.6500
C	1	4.3243	123	25.7646	–	–	–	–	–	–		–	–	1	0.4457	–	–	–	–	–	–	125	30.5346
D	1	0.9699	94	57.1006	2	0.992	5	1.0072	1	8.2138		–	–	–	–	–	–	2	1.2570	2	0.0599	107	69.6004
R	–	–	75	20.4683	–	–	7	0.5636	–	–		–	–	5	0.4620	–	–	–	–	3	0.1259	90	21.6198
S	–	–	40	15.1678	–	–	1	0.0571	–	–		–	–	4	0.7614	–	–	–	–	2	0.2545	47	16.2408
Total	4	21.5733	601	206.4092	2	0.992	13	1.6279	1	8.2138		1	0.0217	17	2.1934	1	0.0320	2	1.2570	9	0.5772	651	242.8975

(67.8cm). Although a single piece of chert was recovered around 23cm, the remaining eight pieces of chert were recovered between 50.2–89.5cm. These deposits date to before about 700 years ago. This pattern is not as clear cut in Squares R–S, although c.74% (by weight) of the artefacts made on volcanic ash/chert were recovered from below 57cm and all are extremely small in size.

The abundance of rhyolitic tuff in the artefact assemblage is not surprising given the presence of exposures of the raw material on the Tom's Creek Site Complex. However, preliminary examination of the rhyolitic tuff artefacts recovered from the excavations and those observed on the surface indicate that many of them are made on a different form of rhyolitic tuff than that available on the adjacent scree slope (Stephen Cotter, Cooperative Research Centre for Landscape Evolution and Mineral Exploration, University of Canberra, pers. comm., 1999). Rhyolitic tuff exposed on the ridge and scree slope forming the core of the peninsula is distinctive, with numerous fracture plans not common in other observed outcrops of this material in the region. Despite close examination, no unequivocal evidence for stone extraction and reduction was found along the base of the scree slope or on other parts of the peninsula. One possibility is that variation in raw material form within the Tom's Creek peninsula outcrop has been selectively removed by Aboriginal and/or European quarrying activities. Alternative nearby outcrops of rhyolitic tuff include Round Hill Head and the Ironbark Site Complex to the north on Middle Creek.

A small sample of stone artefacts from Squares A–S has been subject to limited residue analysis. Francis (1999) examined eight stone artefacts (FS9, 122, 138, 206, 220, 269, 283, 386) from a range of depths and manufactured on several different raw materials ranging in weight from 0.6–8.2g. All specimens were examined using an Olympus® metallurgical incident-light microscope under low level (<800×) magnification (see Loy 1994 for a discussion of techniques). All artefacts exhibited surface features associated with post-depositional processes including rootlets, sand grains, spores and mycelium. All of the artefacts were found to exhibit a similar suite of archaeological residues, comprising resin, cellulose, starch grains, parenchymal tissue and charcoal. Two artefacts (FS9 and 386) exhibit ridged parenchymal tissue, probably derived from the xylem tissue of a vascular plant

(Francis 1999; Raven et al. 1999), indicating probable use of the artefact in plant processing. These residue elements are consistent with use of stone artefacts in a variety of plant processing activities.

Glass artefacts

Thirty-six bottle glass artefacts, weighing 214.4g, were recovered from surface locations between Squares A-D and the base of the scree slope (Table 13.13, Figs 13.2–13.3). Most (90%) of the glass was scattered over a 3m^2 area (Squares E–P) with two retouched artefacts located c.4m to the northeast (Square Q). Two more artefacts were recovered: a large retouched artefact apparently cached in the fork of a quinine tree c.3m north of the main scatter (Fig. 13.5) and an isolated artefact c.3m south of the main collection grid. The assemblage consists of two bottle base sherds and 34 body sherds. No rim fragments were identified, although neck fragments may be represented in the body sherd count owing to the small size of many sherds (average weight 6g), limiting confidence in ascription of body part. All of the glass is probably from the same bottle. The absence of rim sherds raises the possibility of deliberate breakage of the bottle prior to transport (see Allen and Jones 1980; Freeman 1993). Three of the artefacts (FS186, 187, 188) exhibit marginal retouch and were located away from the main scatter. The spatial separation of the glass scatter from the glass artefacts exhibiting retouch is noteworthy. It is possible that the denser concentration of glass fragments represents a manufacturing area and the two glass flakes in Square Q and the apparently cached artefact in the quinine tree represent artefacts actually used or used more intensively. The retouched artefact in the tree (FS188) was found to conjoin with FS166 from Square K to form part of a bottle base, confirming the suspected association between the glass scatter and cached artefact. The presence of glass at a long-term Aboriginal occupation site some distance from known early European population centres supports the inference that it was discarded by Aboriginal people.

Table 13.13 Glass artefacts from the Tom's Creek Site Complex. * indicates glass artefacts recovered beyond the mapping grid (see Fig. 13.3).

SQUARE	#	FS#	WEIGHT (g)	COMMENTS
F	5	152-156	20.2	FS152 & 153 subject to detailed analysis
G	3	157-158, 163	5.4	
J	4	159-162	10.3	
K	9	164-172	79.9	Includes base fragment
L	2	173, 175	3.6	
M	1	174	1.0	
N	1	176	1.7	
O	4	179-182	41.8	
P	3	183-185	2.9	
Q	2	186-187	9.0	Both retouched & subject to detailed analysis
*	2	188-189	38.6	FS188 retouched artefact in tree; subject to detailed analysis
Total	36	–	214.4	

Vernon (1999) conducted a use-wear and residue analysis on the glass assemblage. All 36 glass fragments were subject to preliminary microscopic examination using an Ibelix® transmitted-light microscope before selection of five artefacts for further study (FS152, 153, 186, 187, 188), including the three retouched artefacts. These five artefacts were examined for use-wear and residues using a Olympus® metallurgical incident-light microscope under low level (<800x) magnification using bright-field and dark field settings and cross-polarised light (see Loy 1994 for a discussion of techniques). Areas of shallow, parallel striations were found on most of the artefacts, which Vernon (1999) speculated to result from sand grains being trapped between the artefact and another object during use. Striations are present at 90°, 60° or 45° to an obtuse edge and are associated with edge-rounding. The two retouched artefacts recovered from Square Q (FS186 and FS187) appear to be well-worked, with deep scratches, internal shatters, Walner lines and Hertzian cones observed. Vernon associates these use-wear characteristics with woodworking activities (after Hardy and Garufi 1998).

All five artefacts were also found to be covered in copious plant residues characterised by quantities of macerated cellulose, minute bark fragments and orange, resinous globules with starch grains associated with a translucent milky film, frequently interspersed with fungal micro-hyphae. Identified structural elements include seeds, a plant scale and abundant raphides (including a bundle of five silicate raphides on FS153). Three sections of xylem (secondary parenchymal wall thickening) were also found, common to all angiosperms (flowering plants). These elements are consistent with use of the artefacts on cycad seeds, tubers or mangrove plants (David Doley, Department of Botany, University of Queensland, pers. comm., 1999). The use-wear and residue elements combined indicate that the glass artefacts were used for a variety of activities including woodworking and probably tuber processing.

Other remains

Charcoal was recovered from all squares, totalling 723.9g, with occasional occurrences of large pieces of blocky charcoal (Figs 13.18–13.19). Pumice, totalling 321.9g, occurs throughout the deposits but is most abundant in Squares R–S on the residual landform. Quantities of organic material were recovered from every square, totalling 9,444.5g, comprising leaf litter and roots. Non-artefactual stone was also recovered, totalling 2,305.8g, dominated by highly weathered material recovered at the base of Squares A–D.

Discussion

Excavation revealed a low density cultural deposit distributed over a large area and dating to the last 1,000 years. The lowest cultural deposits are sparse, with the most intensive period of site use occurring in the last 700 years. Fishing, shellfishing, plant food processing and stone artefact manufacturing are all archaeologically documented. In keeping with the pattern observed at other sites, the shellfish assemblage is dominated by oyster and mud ark, with other taxa making only a minor contribution. The assemblage represents resource procurement strategies focussed on intertidal and near-shore environments. In addition, the presence of small quantities of *D. deltoides* indicates that foraging strategies included all the major coastal environments in the area. The Tom's Creek Site Complex is the closest estuarine site to the open beach sites recorded along the coast adjacent to Agnes Water (see Chapter 2). The deposition of *D. deltoides* at the Tom's Creek Site Complex is coeval with dated specimens of this taxon at the Agnes Beach Midden, suggesting at least occasional movement of people and resources between estuarine and open beach environments.

The Tom's Creek Site Complex is another example of a site only established in the last 1,500 years and remaining in use into the time of European occupation. The association of bottle glass artefacts with the pre-contact Aboriginal occupation of the site demonstrates continuities in site use and knowledge about the site into the recent past. Like the Ironbark Site Complex, post-contact use of the Tom's Creek Site Complex suggests a pattern of regional continuity in land-use from around 1,000 years ago into the historical period. The presence of retouch on several glass artefacts also indicates a persistence of technological strategies from the pre-European repertoire, with such modification/maintenance techniques usually reserved for valued siliceous raw materials (e.g. chert, volcanic ash). These data point towards a pattern of regional occupational continuity from shortly before 1,000 years ago into the historical period.

Summary

Data from the Tom's Creek Site Complex display a pattern common to other sites in the region, with first occupation and significant discard of cultural materials occurring only in the last 1,000 years. Glass artefacts found on the surface of the deposits indicate continuity of use of the site into the early twentieth century, suggesting persistence of knowledge about the location as a campsite into the period of European occupation.

14

Synthesis of results: towards an archaeology of the southern Curtis Coast

Introduction

This chapter synthesises the data detailed in the preceding eight site report chapters and also draws on survey data discussed earlier. It focusses on major indices used to identify continuity and change in the regional archaeological record. These data have quantitative and qualitative dimensions. First, the distribution of radiocarbon dates is used to describe and evaluate regional chronology, with comparisons drawn from the broader southeast Queensland area. Second, site contents are examined with an emphasis on shell, fish bone, charcoal and stone artefacts. Shellfish remains inform on diet breadth, intensity of occupation and habitat preferences. Fish bone and other marine vertebrate remains contribute to an understanding of the antiquity of marine fishing in the wider region and the role of fishing in the local subsistence economy. Charcoal abundance is associated with the intensity of human land-use. Stone artefacts are assessed for raw material procurement patterns, technological strategies and potential exchange relationships. Third, regional variation in site structure is examined to elucidate patterns in the size and form of sites through time and space. Covariation in quantitative and qualitative attributes of these data categories is employed to explore patterns in the regional archaeological record. Finally, a descriptive model is proposed linking the identified patterns to broader frameworks of interpretation with a view to building an understanding of Aboriginal lifeways on the southern Curtis Coast throughout the late Holocene.

This synthesis employs broad chronological units (500 year intervals), commensurate with the general degree of confidence available for individual sites in terms of both site integrity and temporal resolution. This approach balances the need for maximum detail against limitations in data resolution. Although site integrity was generally high — based on the consideration of stratigraphy, age-depth relationships and bivalve conjoin analyses — detailed analyses of integrity were not undertaken for deposits lacking bivalve remains suitable for conjoin analysis (see Chapters 6–13 for details). Similarly, temporal resolution for some assemblages is constrained by variation in regional ΔR signatures related to estuary-specific patterns of water circulation, sedimentation and hinterland geology (see Chapter 4). While these broad time-blocks will mask some variability in individual assemblages, the method is a robust one for indicating temporal trends across the region and broadly characterising regional patterns in assemblage variability.

Regional site chronology

The radiocarbon chronology of Aboriginal occupation of the southern Curtis Coast is informed by studies of local estuarine and marine reservoir effects undertaken as part of this research (Chapter 4) and the site-specific evaluations of the precision and accuracy of individual radiocarbon determinations presented in site report chapters (Chapters 6–13; see also Chapter 2). The comparability and reliability of the radiocarbon dataset are further enhanced by the restriction of sample preparation to a single laboratory (University of Waikato Radiocarbon Dating Laboratory) and actual dating to two laboratories: the University of Waikato Radiocarbon Dating Laboratory for conventional radiocarbon dates and the Rafter Radiocarbon Laboratory for accelerator mass spectrometry dates. This strategy minimises the effects of interlaboratory variability in sample preparation and counting procedures and as a result a high level of confidence can be placed in the validity of the individual assays which underpin the general patterns revealed by analysis of the radiocarbon date assemblage.

Several techniques were used to examine regional occupational chronology based on the distribution of available radiocarbon dates (Figs 14.1–14.6). Despite the limitations inherent in all analytical methods employed (see below), they are useful for characterising broad, long-term patterns in the distribution of radiocarbon dates. If a detailed understanding of the structure of the radiocarbon date record is available, these techniques should provide a general view of the structure of regional occupation trajectories. In addition to the technical issues noted above, the validity of trends identified in the radiocarbon record is dependent on a number of basic assumptions. Chief among these are those concerning sample adequacy: are cultural deposits adequately dated and are there biases in the strategy adopted in the selection of samples that might skew the regional radiocarbon record towards certain periods or site types? A large suite of dates from a wide range of site types, contexts and time periods was obtained in an attempt to ensure sample adequacy. Particular attention was paid to dating termination deposits critical to identifying patterns of abandonment (David and Wilson 1999).

In total, 66 radiocarbon dates from 12 sites on the southern Curtis Coast are considered in this analysis (Appendix 1). After Pettitt et al. (2003:1690), outliers or 'rogue' determinations are defined as 'dates that are statistically distinct from the main group/sequence at 2σ'. Individual determinations were assessed on a site-by-site basis to identify and remove rogue dates. Various chronometric hygiene procedures for large suites of radiocarbon dates are available (e.g. Meltzer and Mead 1985; Spriggs and Anderson 1993), but to avoid introducing further biases through selective exclusion of dates, dates were only rejected where a clear inversion or disjunction could be identified on the basis of comparing individual determinations to a sequence of dates (Ulm and Hall 1996). Following this protocol, eight dates from five sites were excluded from the analysis owing to problems of association or non-cultural status (Table 14.1).

Figure 14.1 shows the minimum and maximum 1σ calibrated age-ranges for the youngest and oldest determinations respectively available for each site on the southern Curtis Coast. The 1σ age-range was selected to reduce the potential for overestimating the occupation span of sites and attempt to avoid techniques which emphasise long-term continuity of occupation. Caution should be exercised in interpreting the span presented for the Round Hill Creek Mound as it is based on a single determination of unknown provenance in the overall sequence of occupation of this site (see Chapter 2). The distribution of occupation spans suggests that the region has been continuously

Table 14.1 Radiocarbon dates from the southern Curtis Coast excluded from chronological analyses.

SITE	LAB. NO.	COMMENT
Eurimbula Site 1	Wk-5215	Anomalous. Out of sequence.
Ironbark Site Complex	OZD-756	Non-cultural. Sediment core.
Mort Creek Site Complex	Wk-3942	Non-cultural. Chenier.
Mort Creek Site Complex	Wk-3943	Non-cultural. Chenier.
Mort Creek Site Complex	Wk-3938	Non-cultural. Chenier.
Mort Creek Site Complex	Wk-3940	Non-cultural. Chenier.
Seven Mile Creek Mound	NZA-12272	Anomalous. Out of sequence.
Tom's Creek Site Complex	NZA-13385	Non-cultural. Sediment core.

occupied from at least 4,000 years ago to the present, with increasing numbers of sites occupied through time. Abandonment of at least two sites — Seven Mile Creek Mound and the Mort Creek Site Complex — is clear, although less confidence can be placed in the termination dates available for the Pancake Creek Site Complex and Middle Island Sandblow Site owing to the very large size of these sites. In the early period there appears to be little, if any, overlap in the occupation of Seven Mile Creek Mound and Mort Creek Site Complex. Only Eurimbula Site 1 is coeval with occupation at Mort Creek Site Complex.

Figure 14.2 provides an estimate of the number of sites occupied in 500 year periods based on the mid-points of calibrated age-ranges and linear interpolations between radiocarbon dates. Other studies have shown that this measure tends to inflate the number of recent sites occupied because in the absence of termination dates, sites are assumed to be occupied to the present if there are no major stratigraphic disconformities to suggest otherwise (e.g. Smith and Sharp 1993; Ulm and Hall 1996). For the southern Curtis Coast data, however, considerable confidence can be placed in this measure as dating strategies specifically targetted both initiation and termination deposits. Figure 14.2 shows initial low numbers of occupied sites with major expansion across the region only taking place in the last 1,000 years after a period of reduced site occupation between 1,000 and 1,500 years ago. These data indicate that more individual places in the landscape were incorporated into settlement-subsistence strategies through time.

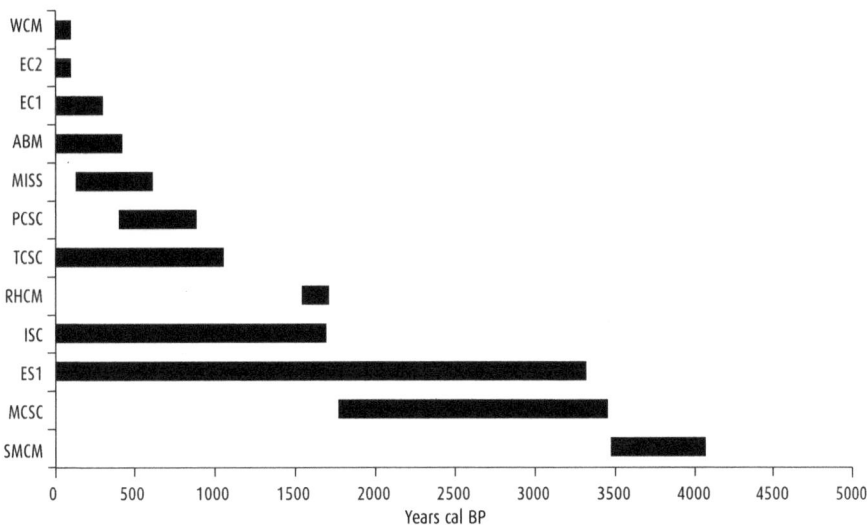

Figure 14.1 Occupation spans of dated sites on the southern Curtis Coast, based on 1σ calibrated age-ranges. Note that a span of 100 years is estimated for the modern dates reported for WCM and EC2.

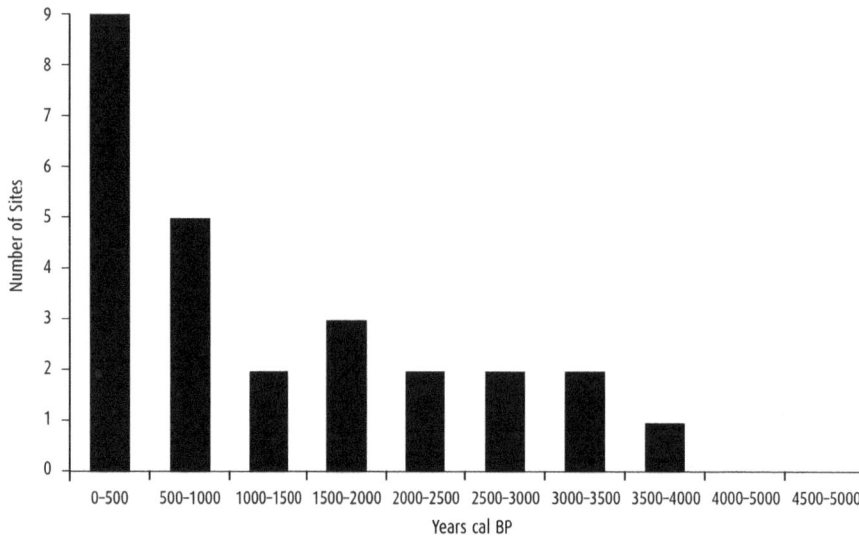

Figure 14.2 Estimated number of dated sites occupied on the southern Curtis Coast in each 500 year period, based on the mid-points of calibrated age-ranges.

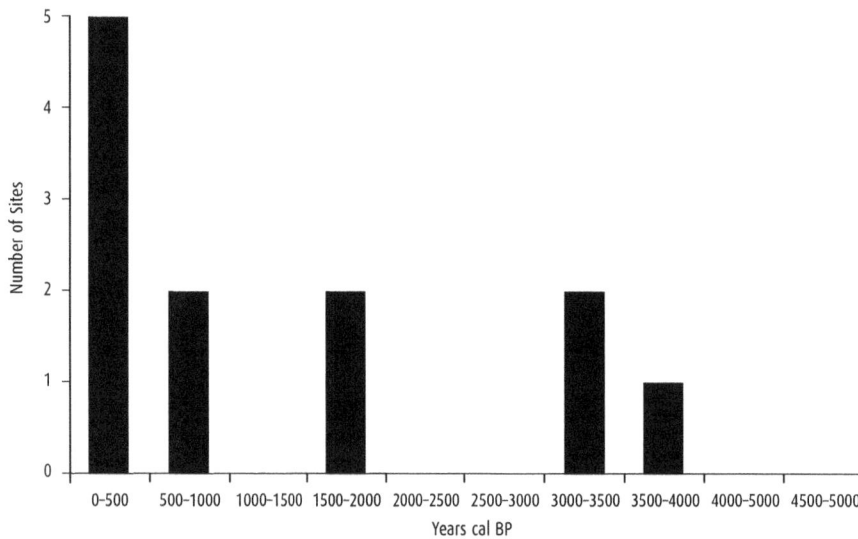

Figure 14.3 Estimated number of new sites established on the southern Curtis Coast in each 500 year period. Note that the mid-point of the calibrated age-range of the oldest date available for each site is assumed to be the basal age.

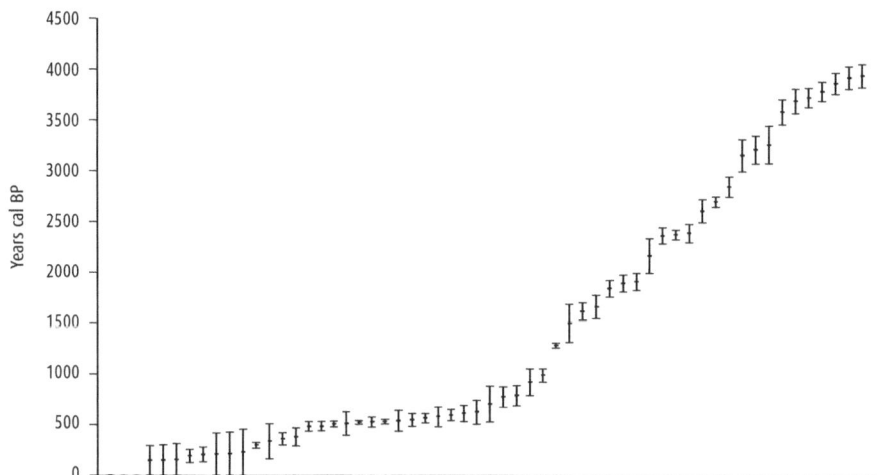

Figure 14.4 Calibrated radiocarbon ages from the southern Curtis Coast (n=58) arranged in order of increasing age. Error bars show the 1σ calibrated age-range. Note the apparent gap between 1,050–1,250 years ago.

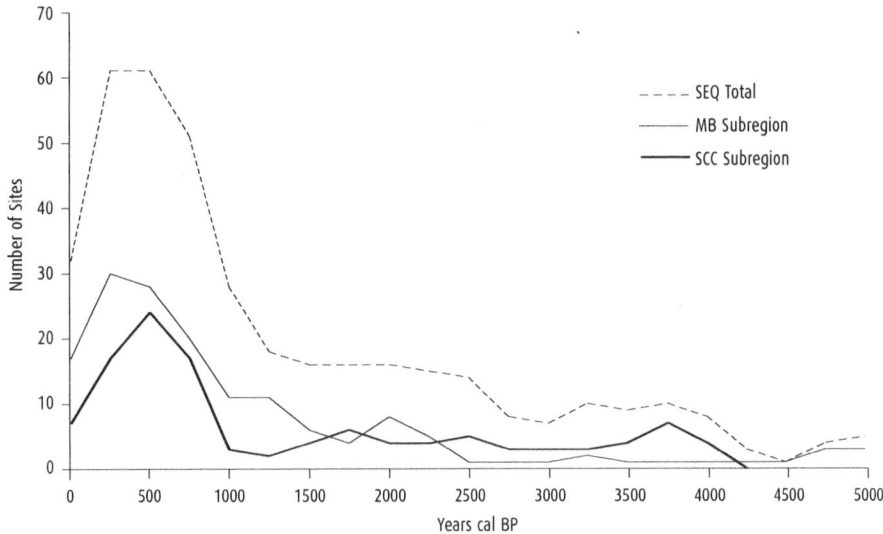

Figure 14.5 Number of sites on the southern Curtis Coast with central calibrated radiocarbon dates falling in each 500 year period, measured at 250 year intervals. For comparison, the same data are shown for all of southeast Queensland and the Moreton Bay region.

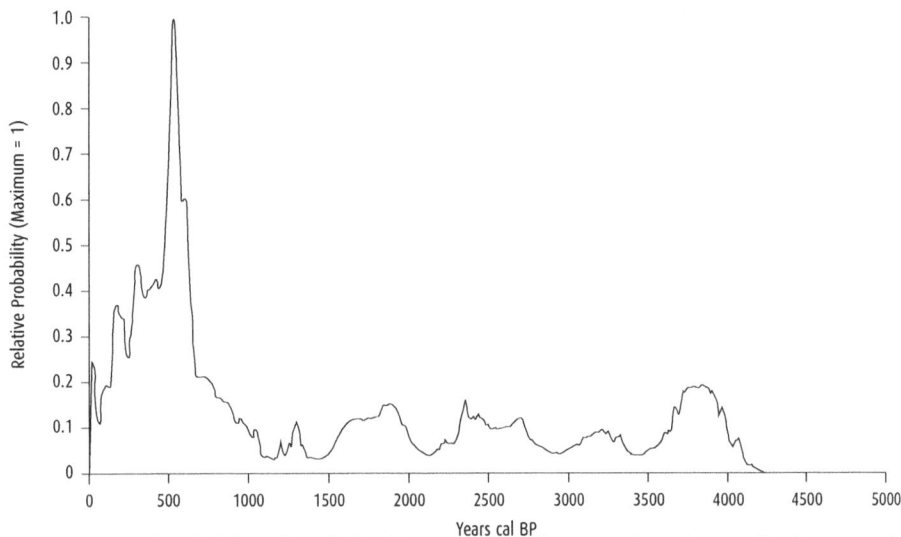

Figure 14.6 Summed probability plot of all calibrated radiocarbon ages (n=56) normalised to a maximum of one. Note that the two modern dates reported for WCM and EC2 are excluded.

Figure 14.3 uses the oldest date available for each site to estimate the rate of establishment of new sites through time. This technique highlights patterns of site creation rather than occupation and use of sites. Bird and Frankel (1991a:4) have criticised this method for, among other things, 'assuming continuity of occupation after first use, and lumping together dates from sequences and short-term occupations', thereby denying 'the possibility of demonstrating discontinuity or the reuse of sites after a significant gap in occupation'. These criticisms do not apply here, however, as the method does not assume continuity of site use and examines site establishment, requiring the use of initial dates for site occupation (and, in a variation of this technique, reoccupation after long breaks in occupation; see Smith and Sharp 1993). Two clearly separated periods of site establishment are evident: sites established before 3,000 BP and those established after 2,000 BP. A comparison of Figures 14.1 and 14.3 shows that in the earlier group, two sites (Seven Mile Creek Mound and Mort Creek Site Complex) were established and then abandoned before the more recent period of site establishment over the last 2,000 years. The only site where occupation spans these two periods (Eurimbula Site 1) exhibits quantitatively and qualitatively different patterns of use between the earlier and later periods (see below). There is also a gap in site establishment between 1,000 and 1,500 cal BP.

Figure 14.4 shows the calibrated radiocarbon ages from the southern Curtis Coast arranged in order of increasing age. The error bars denoting the 1σ calibrated age-range of each date overlap one another for almost the entire sequence, except for an apparent gap between 1,050 and 1,250 years ago. This gap coincides with the gap in site creation (Fig. 14.3) and a reduction in the estimated number of occupied sites (Fig. 14.2) noted above. For reasons discussed below, it may be significant that the first two dates older than this gap are both from the quarry at the Ironbark Site Complex.

Figure 14.5 deploys a modified version of Rick's (1987) method to examine broad patterns in the frequency of radiocarbon dates on the southern Curtis Coast and comparative datasets drawn from southeast Queensland. Rick's method is based on the premise 'that the number of dates is *related* to the magnitude of occupation' (Rick 1987:55, original emphasis) and that 'all things being equal, more occupation produces more carbon dates' (Rick 1987:56). Implicitly, this method assumes that radiocarbon samples are selected at random from an unbiased archaeological record. Rick (1987) and Holdaway and Porch (1995) have noted several limitations of this approach, particularly those relating to the non-random nature of the availability of charcoal for dating structured by taphonomic factors affecting the representation both of sites and of charcoal within deposits, as well as sampling and research biases. However, as Lourandos and David (1998) have noted, some of these problems are minimised by the preferential dating of older deposits by archaeologists in Australia, which should create a bias towards the representation of older dates in the moving average, rather than more recent occupation. Therefore, any log-decline in preservation of materials with increasing age is offset by a log-increase in representation (i.e. sampling) of older material for radiocarbon dating. A moving average was calculated of the number of calibrated radiocarbon dates in sliding 500 year intervals measured every 250 years. Therefore, the measured interval at 1,500 years includes all of the dates between 1,250 and 1,750 BP and the interval at 1,750 years all the dates between 1,500 and 2,000 cal BP. The moving average method broadens the temporal influence of each date and results in a smoothed curve. This method has been adopted in several Australian studies using large regional suites of radiocarbon dates such as Bird and Frankel (1991a) for western Victoria and southeast South Australia, Holdaway and Porch (1995) for southwest Tasmania, David and Lourandos (1997, 1999; see also David 2002; Lourandos and David 1998) for southeast Cape York Peninsula, the semi-arid zone and arid zone, and Ulm and Hall (1996) for southeast Queensland.

For comparison, Figure 14.5 also incorporates an updated and expanded radiocarbon dataset for all of southeast Queensland which includes a further 15 sites and 96 dates since the compilation by Ulm and Hall (1996). These additional data for southeast Queensland are extracted from Ulm and Reid (2000) and subsequent updates (Ulm and Reid 2004). The southern Curtis Coast radiocarbon dataset contributes 71% (n=66) of the new dates available and 80% (n=12) of the new sites. Other new data include dates from recently excavated sites (Alfredson and Kombumerri 1999; McNiven et al. 2002; Ross and Duffy 2000) and redating of previously investigated deposits (Alfredson 2002; Gowlett et al. 1987; Mackenzie 2002) (see Ulm and Reid 2004 for details). The current southeast Queensland radiocarbon dataset contains 245 dates from 73 sites. Since the focus of this study is on the mid-to-late Holocene, the 21 dates from five sites in this dataset with calibrated age mid-points older than 5,000 cal BP were excluded as were the 26 dates from 12 sites exhibiting problematic or non-cultural associations (see Table 14.1; Ulm and Hall 1996:48; Ulm and Reid 2000, 2004). In total, the southern Curtis Coast radiocarbon dataset comprises 27% (n=66) of the radiocarbon dates available for archaeological sites in southeast Queensland and 16% (n=12) of the dated sites. The 66 radiocarbon dates available for the southern Curtis Coast represent the largest group of dates available for any subregion on the Queensland coast after Moreton Bay (n=81) and by far the highest rate of dates/site of 5.4 (compare with 2.6 dates/site for Moreton Bay). Subregional data for Moreton Bay shown in Figure 14.5 are drawn from this expanded southeast Queensland dataset. The Moreton Bay dataset only contains dates from sites fringing, or on the islands of, Moreton Bay. There is marked synchronicity in the timing of changes in the Moreton Bay and southern Curtis Coast datasets, despite Moreton Bay starting from a higher base. These data support the findings presented in an earlier study by Ulm and Hall (1996) which identified significant increases in the number of occupied sites and the rate of site establishment after 1,200 cal BP. However, this technique allows higher resolution insights than those available for previous techniques using 500 year intervals, showing marked increases around 700 years ago.

Figure 14.6 shows the summed calibrated probability of all calibrated radiocarbon ages by year, normalised to a maximum of one and excluding the two modern determinations (Wk-7689; Wk-7681). The summed probability distribution 'represents the probability that independent events A OR B occured at a particular time' (Stuiver et al. 2002), and can therefore be used as a proxy for the probability of occupation in a particular period. Periods of low regional site use are common prior to 1,000 years ago. The high number of dates obtained for the Seven Mile Creek Mound are responsible for the platykurtic curve between 3,500 and 4,000 cal BP (see also Fig. 14.5). Increasing amplitude of relative probability peaks through time indicate a decrease in the spacing of dated occupation events across the southern Curtis Coast subregion after 1,000 BP. Periods of reduced regional occupation occur at 3,400, 2,900, 2,150, 1,400 and 1,150 BP as indicated by troughs of low probability in Figure 14.6. Only the periods around 1,400 and 1,150 cal BP, however, are coincident with clear quantitative and qualitative changes in the archaeological record of the region (see below). These are also the periods with the lowest probability of the occurrence of calibrated radiocarbon dates (Table 14.2). Again, within the last 1,000 years marked increases are evident around 700 years ago.

Table 14.2 Summed probability distribution of all calibrated radiocarbon dates available from cultural contexts on the southern Curtis Coast (n=56). Excludes the two modern determinations (Wk-7681; Wk-7689). 0* indicates a modern age.

% AREA ENCLOSED	CAL BP AGE-RANGES	RELATIVE AREA UNDER PROBABILITY DISTRIBUTION
95.4	4078–3479	0.130
	3359–2980	0.049
	2855–2156	0.114
	2057–1508	0.106
	1346–1232	0.015
	1207–1184	0.003
	1065–0*	0.584

Discussion

The archaeological dating program on the southern Curtis Coast reveals a near-continuous record of occupation from 4,000 cal BP to the present. From this time the estimated number of dated sites occupied in each 500 year period increases, with a major increase in the rate of site creation and representation of radiocarbon dates in the last 1,000 years, notably from around 700 cal BP. However, dating of surface cultural materials highlights marked differences in the age of archaeological materials exposed at the surface. Large suites of dates clearly demonstrate that both the Seven Mile Creek Mound and the Mort Creek Site Complex were abandoned in antiquity.

There is a clear disjunction in the pattern of regional occupation between sites created before 3,000 BP and those created after 2,000 BP. Although the region may not have been abandoned, significant decreases in the intensity of regional land-use are apparent. Eurimbula Site 1 is the only site which remains in use over this transitional period and even here only a single radiocarbon date (Wk-8553) spans the critical period between abandonment of the Mort Creek Site Complex shortly before 1,900 years ago and first occupation of the Ironbark Site Complex at around 1,500 BP. The date of 1,790±60 BP (Wk-8553) on a sample of *A. trapezia* from Square E3, XU7, is associated with small quantities of cultural material (see Ulm et al. 1999a:120). Throughout the entire region only one other radiocarbon date associated with occupation dates to this time interval. An isolated date of 1,910±42 BP (Wk-10090) is available for a sample of *A. trapezia* obtained from a disturbed section at the Round Hill Creek Mound on the opposite bank of Round Hill Creek to Eurimbula Site 1. Further details about this site are currently unavailable, so the cultural context of this determination cannot be assessed (see Chapter 2). As both of these age determinations are on estuarine shell material, however, the accuracy of these two calibration calculations is dependent on the accuracy of the local estuarine reservoir correction factor calculated for Round Hill Creek. As noted in Chapter 4, although three shell/charcoal date pairs are available for the Round Hill Creek estuary, consideration of major discrepancies between the paired results left a single result (ΔR= –305±61) considered free of obvious interpretation problems. This value was adopted as a first approximation, although less confidence was placed in it than those for other estuaries where more than one result is available. Table 14.3 shows that if these two radiocarbon dates are calibrated using the local open ocean result rather than the single estuary-specific ΔR value, both calculations place the dated material into the last 1,500 years, suggesting the possibility of a short hiatus of occupation at Eurimbula Site 1. Although the results of the marine reservoir study of Round Hill Creek are problematic, they highlight an otherwise unconsidered source of error in evaluating the chronology of occupation.

Table 14.3 Radiocarbon ages from the southern Curtis Coast dating to the 1,500–2,000 cal BP interval, calibrated using various ΔR values.

SITE	LAB. NO.	^{14}C	CALIBRATED AGE/S ΔR= –305±61 (CAL BP)	CALIBRATED AGE/S ΔR= +10±7 (CAL BP)
Eurimbula Site 1	Wk-8553	1790±60	1869(1683)1479	1447(1307)1228
Round Hill Creek Mound	Wk-10090	1910±42	1980(1816)1623	1534(1439)1338

The major trend evident in all analyses is the dramatic increase in the number of sites created, sites occupied and radiocarbon dates represented over the last 1,000 years, and particularly the last 700 years. The trend is not gradual, but rather implies a disjunction in the regional trajectory of occupation. This pattern is quite distinct from that evident in the preceding 3,000 years of occupation and implies a dramatic reordering of land-use patterns, with no precedent in the history of occupation of the region.

Regional site contents

Shellfish remains

Regional shellfish gathering focussed on a small proportion of the total available taxa, with collection strategies concentrated on shallow estuarine environments. Although 53 taxa (21 marine bivalves; 1 freshwater bivalve; 27 marine gastropods; 4 terrestrial gastropods) were recovered from archaeological deposits, three taxa — rock oyster (*Saccostrea glomerata*), mud ark (*Anadara trapezia*) and hairy mussel (*Trichomya hirsutus*) — comprise over 95% of the shell by weight and over 72% by MNI (Table 14.4). Rock oyster alone accounts for over 70% of the regional shell assemblage by weight.

Table 14.4 Top 10 shellfish taxa from all excavated deposits ranked by weight and minimum number of individuals. Note that the four taxa of terrestrial gastropod have been excluded.

SPECIES	COMMON NAME	% WEIGHT	RANK	SPECIES	COMMON NAME	% MNI
S. glomerata	rock oyster	73.38	1	S. glomerata	rock oyster	57.08
A. trapezia	mud ark	15.77	2	T. hirsutus	hairy mussel	9.57
T. hirsutus	hairy mussel	6.74	3	A. trapezia	mud ark	6.28
P. sugillata	pearl oyster	1.34	4	P. sugillata	pearl oyster	2.37
P. ebininus	hercules club whelk	1.28	5	N. balteata	common nerite	2.05
N. balteata	common nerite	0.91	6	B. paivae	oyster drill	1.38
V. australis	Australian mud whelk	0.10	7	V. australis	Australian mud whelk	1.26
C. fibula	spiny oyster	0.08	8	Latirus sp.	–	1.00
B. paivae	oyster drill	0.04	9	P. ebininus	hercules club whelk	0.98
B. nanum	periwinkle	0.02	10	B. nanum	periwinkle	0.92
Total		100		Total		100

Shellfish remains are dominated by the large contribution of the Seven Mile Creek Mound which comprises the entire shell assemblage in the 3,500–4,000 cal BP interval (weighing over 100kg) (Fig. 14.7). If the mound assemblage is removed from consideration, comparatively little shell is deposited over the last 3,500 years. Several other patterns are also evident. There is a clear reduction in the quantity of shell deposited across the region between 1,000–1,500 years ago (less than 1kg) coincident with the period of reduced regional occupation identified in the radiocarbon dataset, with increases in the last 1,000 years mirrored in other indicators for increased regional occupation.

Bimodality in the distribution of key shellfish taxa through time was noted in several assemblages (see Figs 12.15, 13.13). Covariation of mudflat-associated bivalves (*A. trapezia*) versus mangrove-associated bivalves (*S. glomerata*) and gastropods (*T. telescopium, N. balteata*) appears to be directional, with an overall trend towards a decline in the representation of *A. trapezia* through time. Although rock and shell debris beds suitable for oyster colonisation occur in estuaries at the north of the study region (Seven Mile Creek, Mort Creek, Worthington Creek), mangroves provide the major oyster substrate across southern estuaries such as Round Hill Creek, Eurimbula Creek, Middle Creek and Pancake Creek. The presence of oyster is therefore directly related to the availability of mangrove substrates in these estuaries. The timing of the bimodal trend varies between estuaries across the region, reflecting localised patterns of mangrove colonisation and sedimentation conditions. The pattern is clearest at Eurimbula Site 1 and is complemented by geomorphological and palynological data available for Round Hill Creek. Although mangrove pollen is present in cores by 3,000 BP, geomorphological data indicate an expansion of mangrove communities in the last 1,000 years coincident with a recent phase of sedimentation. Similarity coefficients calculated on the shell assemblage of Eurimbula Site 1 show a 70.3% similarity between the 0–500 year and 500–1,000 year assemblages, but only a 40.3% similarity between the 0–1,000 and 1,000–2,000 year assemblages. These values indicate a significant shift in the representation of taxa in the deposit, lending support to an argument for changes in local resource availability linked to mangrove expansion.

Another pattern is discernable in the diversity of shell assemblages through time, as calculated by the Shannon-Weaver Function (H'). If the Seven Mile Creek Mound is excluded (the 3,500–4,000 year cal BP column in Fig. 14.8), there is a weak overall trend indicating greater diversity after 1,000–1,500 BP (see Fig. 14.8). This pattern may indicate a broadening of the suite of taxa targetted by gathering strategies.

Fish remains

Fish are likely to have provided most of the protein for people on the southern Curtis Coast despite the relatively low representation of this material compared with shellfish (but see Erlandson 1988, 1991). Identified fish remains indicate targetting of a range of shallow water estuarine species, including whiting (Sillaginidae), flathead (Platycephalidae), bream, tarwhine and snapper (Sparidae), mullet (Mugilidae) and catfish (Ariidae). The pattern of fish bone distribution closely follows that of shellfish, with a decrease between 1,000–1,500 years ago and major increases in the last 1,000 years (Fig. 14.9). The shell and fish bone datasets should not be considered entirely independent, however, as it is likely that fish bone survival in the archaeological record is closely linked to shell abundance because the shell matrix provides more alkaline conditions (McNiven 1991a).

Two major patterns in the distribution of fish bone stand out. First, the ratio of fish bone to shell is dramatically different between early and later periods of occupation. In the period 3,500–4,000 BP there is 1g of fish bone for every 3,942g of shell whereas over the last 500 years the ratio is an order of magnitude lower at 1:435. A number of factors could contribute to this pattern. For example, a smaller proportion of fish bone may have been deposited in the earlier period, related specifically to behaviours involved in the construction of the Seven Mile Creek Mound. Alternatively, different taphonomic agents may have been operating in the earlier time interval to selectively remove fish bone from the discard assemblage. This seems unlikely, however, as the Seven Mile Creek Mound pre-dates the probable arrival of the dog in southeast Australia, identified as a major contributor to the removal of small vertebrate remains (Walters 1984, 1985). An alternative explanation might be found in shifting subsistence and settlement strategies towards an increased emphasis on a broad range of coastal resources (see below).

The second major pattern is the unexpected antiquity of marine fishing in the region. Fish remains are a rare component in southeast Queensland faunal assemblages pre-dating 1,000 years ago (Ulm 2002a; Walters 1992a). Three sites on the southern Curtis Coast — Seven Mile Creek Mound, Mort Creek Site Complex and Eurimbula Site 1 — contain fish bone before 2,000 BP, effectively doubling the number of sites in southeast Queensland with fish bone assemblages dating to this period.

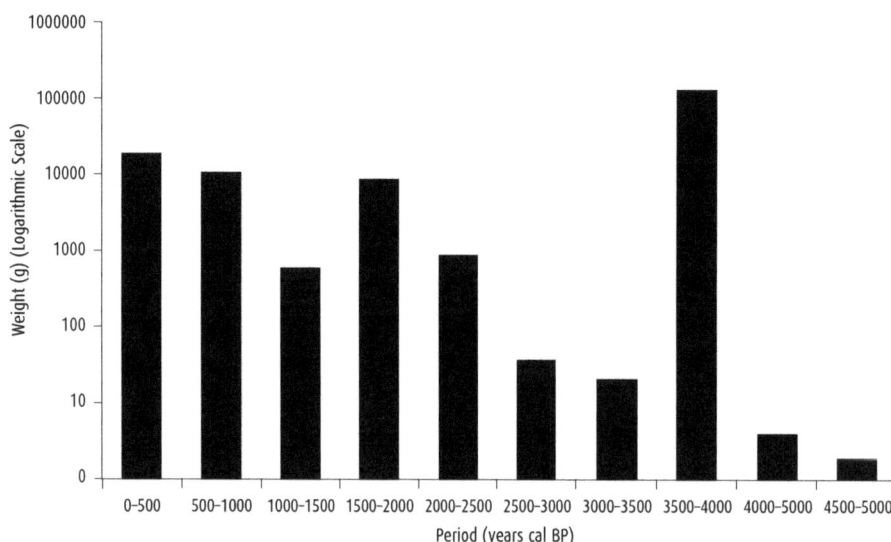

Figure 14.7 Total weight of shell recovered from all excavated sites per 500 year interval. Note logarithmic scale.

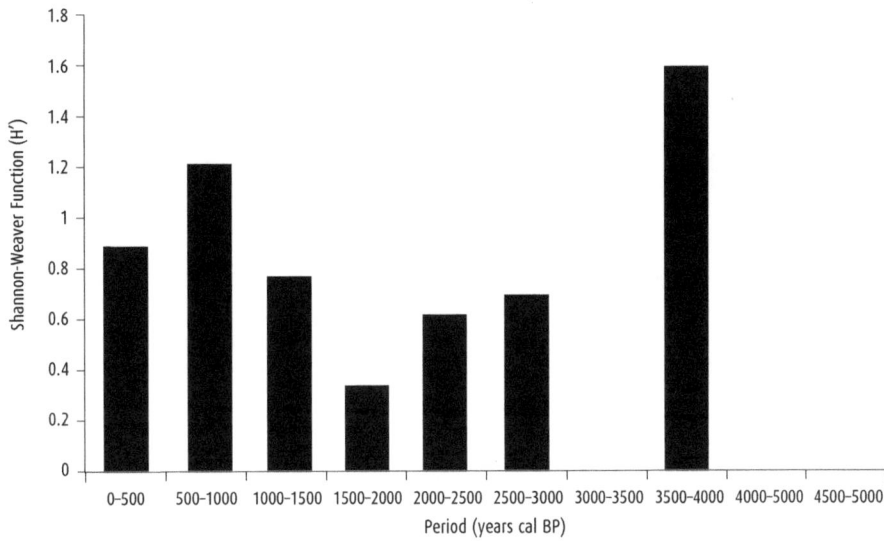

Figure 14.8 Shellfish diversity calculated using the Shannon-Weaver Function (H') per 500 year interval.

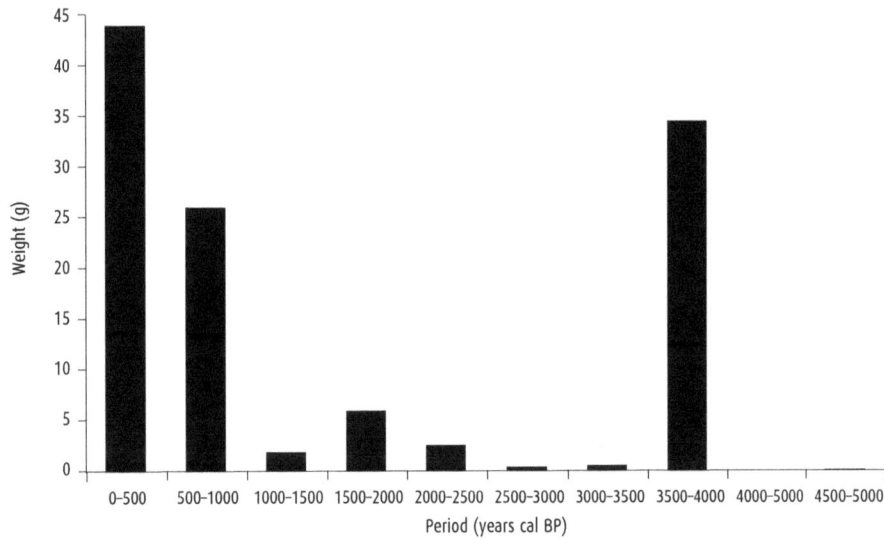

Figure 14.9 Total weight of fish bone recovered from all excavated sites per 500 year interval.

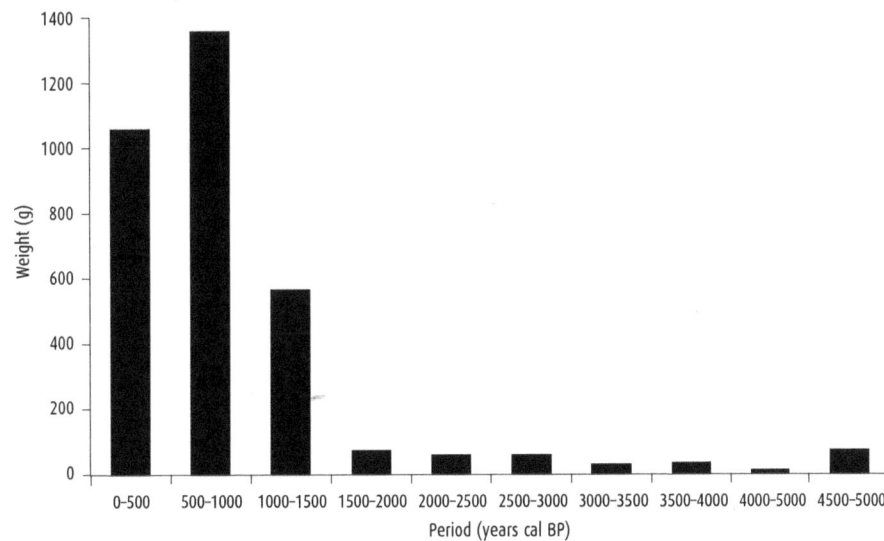

Figure 14.10 Total weight of charcoal recovered from all excavated sites per 500 year interval.

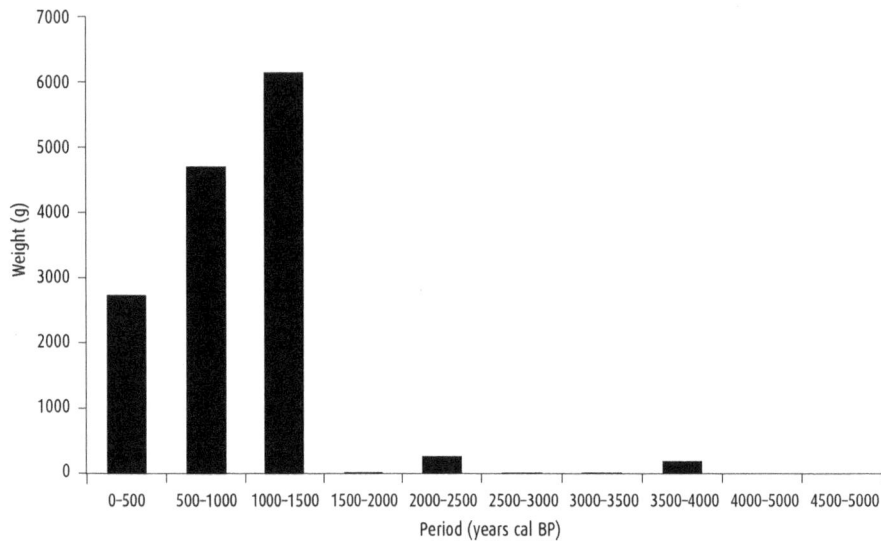

Figure 14.11 Total weight of stone artefacts recovered from all excavated sites per 500 year interval.

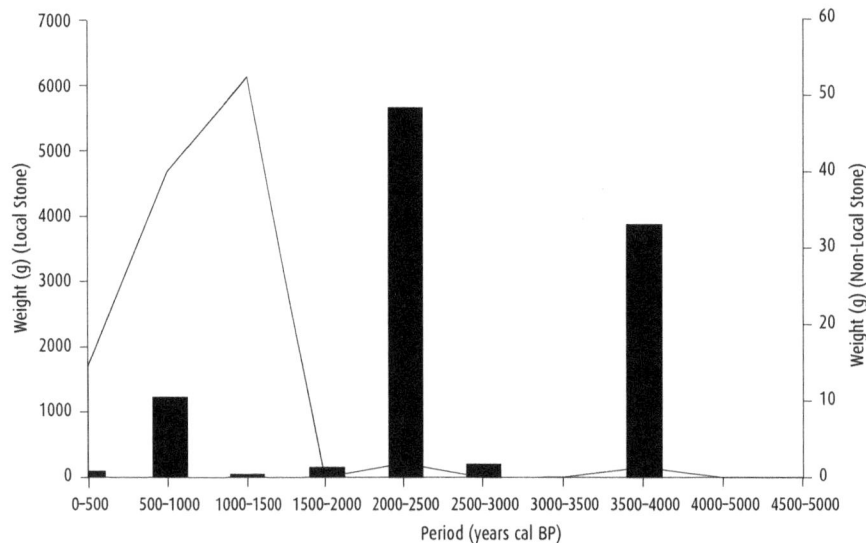

Figure 14.12 Abundance of local (line) versus non-local (black columns) stone artefact raw materials per 500 year interval.

Charcoal

As noted in several of the site reports, low quantities of charcoal are represented in culturally-sterile deposits, indicating continuous deposition of charcoal throughout the coastal landscape. Charcoal representation dramatically increases in the last 1,500 years and coincides with increased rates of regional site occupation and increased rates of shellfish and fish discard (Fig. 14.10).

Stone artefacts

Significant changes in stone artefact technologies and patterns of raw material procurement are evident over the last 4,000 years (Figs 14.11–14.12). Before c.1,500 years ago, stone artefact assemblages are characterised by a larger proportion of high quality siliceous stone (including volcanic ash, chert and silcrete), much of which is thought to have a non-local origin, and which was curated for maximum use-life. From around 1,500 years ago there is a shift towards the almost exclusive use of local stone resources (especially rhyolitic tuff, but also microgranite and quartz) (Fig. 14.12). This localisation in the sourcing of lithic raw materials is accompanied by an alteration in stone reduction strategies towards informal or expedient tool manufacture, use and discard for utilitarian artefacts (i.e. 'minimally altered and are not noticeably specialised' (Mulvaney and Kamminga 1999:221)). Although a similar late Holocene trend identified in southwestern Victoria has been linked to a widespread decline in small tool technology (Fresløv and Frankel 1999; Zobel et al. 1984), the pattern clearly denotes major changes in stone procurement and use strategies on the southern Curtis Coast. McNiven and Hiscock (1988) have linked raw material transport and reduction strategies which maximise reduction potential to the general absence of good quality siliceous stone close to the coast in southeast Queensland. Several indirect lines of evidence indicate that it was during the last 1,500 years that manufacture of edge-ground hatchets on local raw materials commenced, forming the major curated component of recent stone artefact assemblages (see Ulm et al. 2005).

Although the general pattern of stone artefact abundance corresponds with that observed for other cultural materials, there is a clear and dramatic peak between 1,000–1,500 BP, at a time when other indicators suggest an overall reduction in regional occupation (Fig. 14.11). This peak largely comprises material recovered from bank deposits at the Ironbark Site Complex. Although there is indirect evidence for use of the quarry prior to this time in the form of artefacts manufactured on rhyolitic tuff recovered from c.3,000 year old deposits at Eurimbula Site 1, it is not until around 1,500 BP that the first evidence for systematic and large-scale extraction of materials from the quarry occurs. Despite the presence of numerous outcrops of rhyolitic tuff throughout the region, the ubiquitous stone artefacts manufactured on this material noted on the surface of sites and recovered from excavated deposits appear to largely, if not exclusively, derive from the Ironbark Site Complex quarry (Chapter 9; Ulm et al. 2005).

The other major technological change in stone artefact technologies in the last 1,500 years is the appearance of large implements associated with plant processing. Although not recovered from excavated contexts, numerous large stone artefacts, some exhibiting bevelling, were observed eroding from bank deposits dating to the last 500 years at both Eurimbula Site 1 and Ironbark Site Complex, and on the surface of the Middle Island Sandblow Site. Most of these artefacts are manufactured on local rhyolitic tuffs and ignimbrites. Smith's (2003) study of bevelled artefacts from Bribie Island also revealed a high proportion of these raw material types. Residue and use-wear analyses of a small sample of the southern Curtis Coast implements by Lamb (2003) demonstrated the presence of starchy residues associated with processing the rhizome of starch-rich plants, such as fern root (*Blechnum indicum*).

Excavated stone artefact assemblages dating to the last 1,500 years and surface observations indicate that all stages of stone artefact manufacture are represented at major sites in the region (i.e. Tom's Creek Site Complex, Eurimbula Site 1, Ironbark Site Complex, Pancake Creek Site Complex). The high levels of expediency and redundancy and the small quantities of formal tools evident in these assemblages support an interpretation of relatively low regional mobility at these sites, particularly over the last 1,000 years (Parry and Kelly 1987). In general, there is a decrease in the flaking quality of stone used to manufacture artefacts in the region over the last 4,000 years. The dominance of rhyolitic tuff in assemblages dating to the last 1,500 years reflects raw material distribution which is embedded in local settlement-subsistence systems rather than deriving from non-local transport or exchange. The increase in the quantity of local raw materials in sites across the region is related to broader patterns of restructuring of land-use. The low numbers of artefacts coupled with the presence of non-local materials and curated forms in earlier sites provides further indications of more mobile settlement patterns before 1,500 BP.

The flaked stone artefact assemblage is similar to those described elsewhere in coastal southeast Queensland, such as Fraser Island (McNiven and Hiscock 1988; McNiven et al. 2002), Cooloola (McNiven 1990a, 1992a, 1992b) and Bribie Island (Smith 1992, 2003), with few formal artefact types, a dominance of flaked pieces, and generally small artefact size. In fact, McNiven et al. (2002) reported that the entire stone artefact assemblage recovered from Waddy Point 1

Rockshelter on Fraser Island had an average weight of less than 0.8g. Smith (2003:171) has associated this general pattern with 'opportunistic (or expedient) techniques of manufacture and reduction, typical of exploitation of low risk resources and a reliable supply of raw materials'.

Regional patterns in site structure

Site structure describes the properties of size and density of contents. A consideration of regional patterning in site structure is critical to understanding the archaeological record of the southern Curtis Coast, where there is a variety of different site types of various dimensions. While area estimates are relatively straightforward for small or discrete sites (e.g. Seven Mile Creek Mound, Eurimbula Creek 1 and 2), they are less precise for very large sites (e.g. Eurimbula Site 1, Pancake Creek Site Complex). Although dense modern vegetation and restricted subsurface sampling limit confidence in estimates of site area and structure, a number of basic observations can be made.

Table 14.5 reveals a generally inverse relationship between site area and shell density (Figs 14.13–14.16). That is, when we exclude the very small, probably single occupation sites, Eurimbula Creek 1 and 2 from consideration (see Chapters 10 and 11), very large sites tend to have low shell densities and vice versa. For illustrative purposes, Table 14.5 shows a total estimated maximum shell content based on the shell density of shell-bearing excavated deposits and estimated site area. For example, although the Ironbark Site Complex has one of the lowest shell densities of any site (in terms of both g/kg and g/m^2), it probably contains more shell than the Seven Mile Creek Mound, which exhibits the highest excavated shell density. These figures may overestimate or underestimate total shell content owing to sampling and a variety of other considerations. However, in general terms, site size can significantly distort consideration of the regional archaeological record because excavation sample size was not scaled to site area.

By extrapolation from excavated contexts to the whole site area, the vast majority of shell extant in the contemporary landscape can be demonstrated to date to the last 1,000 years. All of the shell at Tom's Creek Site Complex, Ironbark Site Complex, Pancake Creek Site Complex and Eurimbula Creek 1 and 2 dates to the last millennium and the majority of shell at Eurimbula Site 1 dates to this period as well (Fig. 14.16). Clearly, the volume of cultural material deposited on the landscape is related to a change in site morphology from relatively small, discrete sites with high density remains in the early period, represented by the Seven Mile Creek Mound and Mort Creek Site Complex, to extensive sites characterised by low density cultural remains after 1,500 BP.

Table 14.5 Shell density characteristics of excavated sites arranged in descending order of total estimated shell content.

SITE	SITE AREA (m^2)	AREA EXCAVATED (m^2)	SHELL RECOVERED (g)	SHELL DENSITY (g/m^2)	TOTAL ESTIMATED SHELL (kg)
ES1	100000	1.00	6667.27	6667.27	666727.24
TCSC	50000	1.50	16092.63	10728.42	536420.94
ISC	150000	1.00	903.46	903.46	135518.66
SMCM	200	0.25	135794.85	543179.39	108635.88
MCSC	2500	0.25	8748.79	34995.18	87487.95
PCSC	22500	2.00	6616.87	3308.43	74439.74
EC1	10	1.00	1330.22	1330.22	13.30
EC2	10	0.25	201.29	805.16	8.05

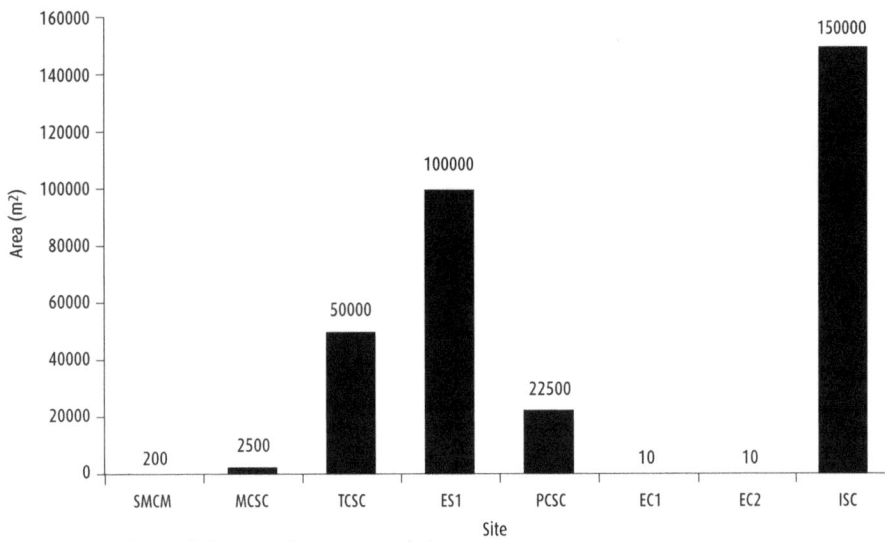

Figure 14.13 Estimated site area for excavated sites.

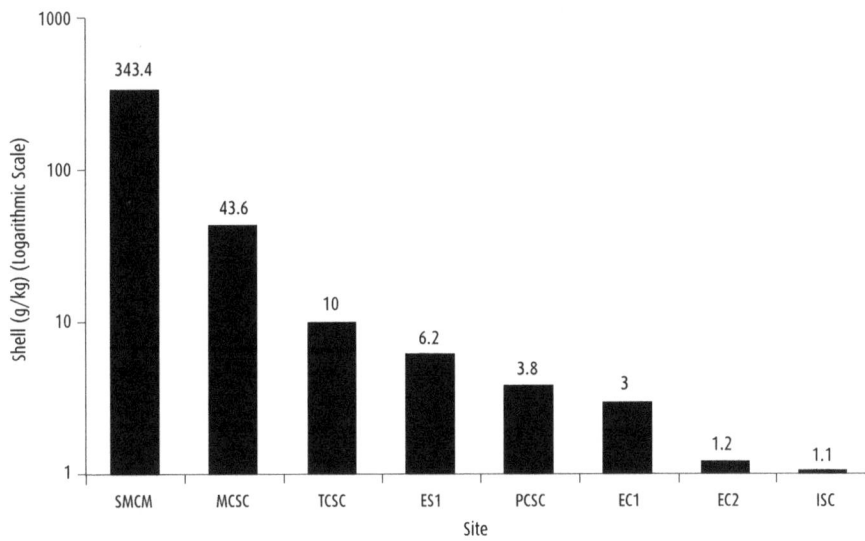

Figure 14.14 Shell density as a unit of weight of excavated deposit for excavated sites. Only shell-bearing squares are included in calculations. Note logarithmic scale.

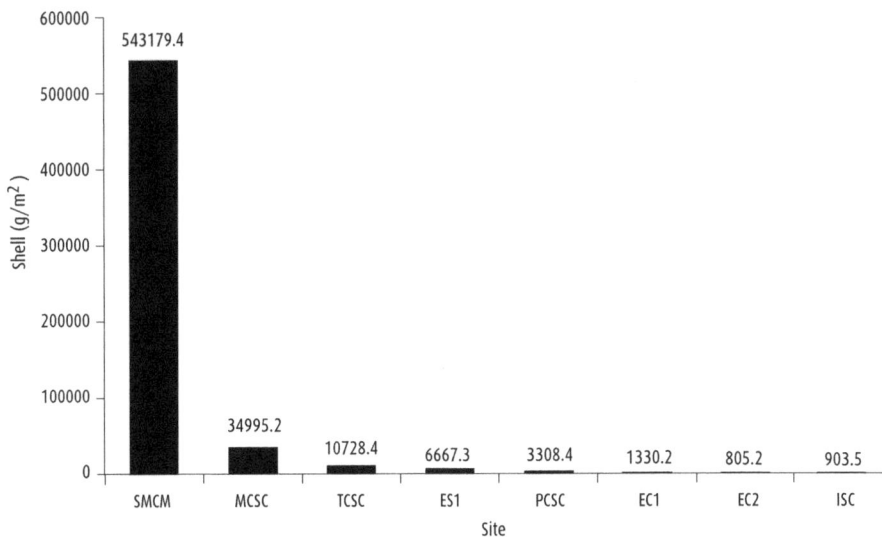

Figure 14.15 Shell density as a unit of area of excavated deposit.

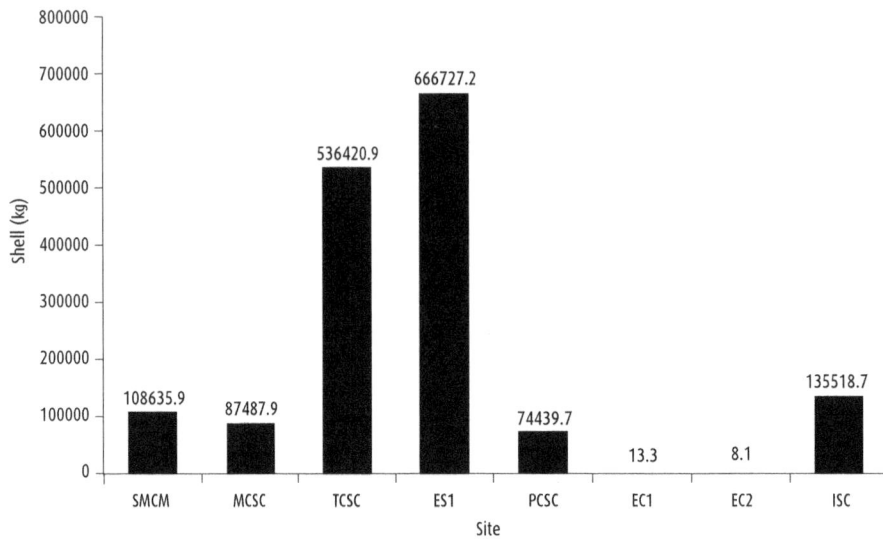

Figure 14.16 Extrapolated total site shell content based on site area and shell density.

Discussion

Clear changes are documented in the archaeological record of the last 4,000 years on the southern Curtis Coast. Two of the three sites occupied before 1,500 BP were abandoned by around 2,000 BP. Deposition of cultural material across the region slowed dramatically shortly after 2,000 years ago and remained low until around 1,000 years ago. A period of regional abandonment is possible at this time, or, at the very least, a general period of reduced use of the coastal zone. It is only after 1,500 years ago that there is evidence for the beginnings of a land-use system that has obvious parallels to that documented in the recent archaeological record, ethnography and Aboriginal oral histories. This most recent phase is characterised by occupation at multiple locations across the coastal region and increasing discard of cultural materials (shell, fish bone, charcoal, stone artefacts) over the last 1,000 years. A distinction can thus be drawn between more persistent occupation post-dating 1,500 BP versus more intermittent use of the coastal zone before 1,500 years ago.

These general patterns cannot be explained in terms of differential site preservation. Archaeological deposits at Eurimbula Site 1, Pancake Creek Site Complex and Eurimbula Creek 2 are located towards the top of long stratified dune sequences which pre-date Aboriginal occupation by millennia. Mort Creek Site Complex and Seven Mile Creek Mound are located on stable landforms and were abandoned in antiquity with no evidence for erosion selectively removing more recent deposits. The situation is less clear at other sites. At Ironbark Site Complex, Tom's Creek Site Complex and Eurimbula Creek 1 at least some of the low dunes containing cultural material probably only date to a more recent phase of dune-building in the last 2,000 years. Elsewhere in Australia the preferential location of middens on or close to the present shoreline has been linked to poor survival potential and differential preservation of this site type (Fresløv and Frankel 1999). However, as argued elsewhere (see Chapter 2), the open coast does not appear to have been a focus of occupation on the southern Curtis Coast, with all major sites located on the protected shores of estuaries. While local landscape depositional and erosional regimes might have impacted on some site-specific assemblages (e.g. storm-surge events), the overall pattern is robust and cannot be explained simply in terms of differential preservation.

Modelling regional settlement histories

The data presented above provide the basis of a descriptive model of regional settlement histories. At least three distinct patterns of land-use can be inferred from structural discontinuities in the regional archaeological record. These patterns can be usefully considered in terms of three phases of occupation:

–   Phase I (pre-4,000 BP–1,500 BP)
–   Phase II (c.1,500 BP–c.AD 1850s)
–   Phase III (c.AD 1850s–c.AD 1920s)

Phase I (pre-4,000 BP–1,500 BP)

Before 1,500 BP occupation of the southern Curtis Coast is geographically tightly-focussed, short-term and discontinuous. Only three sites in the region date to this period: Seven Mile Creek Mound, Mort Creek Site Complex and Eurimbula Site 1. Occupation of only one of these sites, Eurimbula Site 1, continues throughout the last 3,000 years, but even here the changes in site form, discard rates, stone artefact technologies and raw material representation point to changes in the use of the site over this interval.

Use of local raw materials is limited in Phase I and stone artefact assemblages are characterised by the use of highly siliceous stone, such as volcanic ash, silcrete and chert, which appears to have a non-local origin as no major extraction sites for these materials have been located in the study area. Artefacts manufactured on siliceous materials are heavily curated and there is little evidence for onsite reduction, lending support to the idea that most artefacts were manufactured elsewhere, transported into the region and curated for maximum use-life. This pattern is highlighted at Eurimbula Site 1, where deposits dating to before 1,500 BP contain curated artefacts manufactured on silicous materials whereas the post-1,500 BP assemblage is overwhelmingly dominated by expedient artefacts manufactured on local materials.

Site form and content suggest short-term occupation targetting marine fish, gathering of shellfish (especially oyster and mud ark) and capture of large marine animals (dugong and turtle). Deposits at all three sites dating to this period are limited in extent. The Seven Mile Creek Mound is a discrete mound covering a maximum area of 200m². Although uncertainty remains over the formation history of the Mort Creek Site Complex, cultural materials appear to be restricted largely to an area of less than 2,500m². Although Eurimbula Site 1 has an area in excess of 100,000m², deposits pre-dating 1,500 BP are restricted to a small area at the southern end of the site covering less than 2,500m². Neither occupation nor abandonment of the Seven Mile Creek Mound and Mort Creek Site Complex is synchronous, supporting the impression of very low levels of use of the coast during the third and fourth millennia BP.

In the absence of permanent occupation of the coastal zone in Phase I, repeated occupation of certain sites may have been a deliberate strategy to demarcate social geography. For example, the shell mound at Seven Mile Creek and the stone monolith at Mort Creek are imposing landscape features. In particular, mound construction on the southern Curtis Coast represents patterns of landscape use which are fundamentally different from those adopted during subsequent periods, when no mounds occur. The geographically-focussed nature of isolated mound accumulation and the virtual absence of stone artefacts and charcoal in the deposits point to logistical mobility strategies targetting estuarine resources. This signature is not consistent with a pattern of widespread residence on the coast. Instead, following McNiven (1990a), the pattern appears to be one where coastal sites were firmly embedded into regional settlement systems with a subcoastal focus.

The structure of the Seven Mile Creek Mound is of particular interest. The relatively small size of the mound does not appear to offer any obvious practical advantages to its occupants, such as relief from insects, or elevation in a water-logged area (Bailey 1999). On this basis, it is possible that the motivations for mound formation are linked more to social factors than resource availability. As Morrison (2003:5) noted, 'it is very likely that they [the mounds] had inherent symbolic values to the successive generations of people who built and used them'. In the case of the Seven Mile Creek Mound, it appears that the fish, crab and shellfish represented in the mound were consumed elsewhere and deliberately deposited to form the mound. This repeated set of secondary disposal behaviours occurred systematically over a period of around 300 years — patterning that suggests intergenerational transmission of 'ritualised' knowledge relating to mound-building. McNiven and Feldman (2003:171–2) noted that such 'fixed, marked places add to cultural land- or seascapes the dimension of biography, differential knowledge and power ... marked places and their associated rituals not only anchor people to the past, they also provide beacons for the future'. McNiven and Feldman's (2003:188) comments on dugong mound formation in Torres Strait are relevant:

> Whether or not construction of a mound was the long-term deliberate intention of site-users is difficult to know. A mound could simply be the concomitant result of long-term tethering of multiple, ritual, discard events. A growing mound, however, signifies successful hunts (to supply building material), confidence in its functional efficacy and a commitment to continued use. As a mound gradually increased in size, so too would its ritual gravity, because of increasing spiritual, social and historical capital. Each contribution to the mound not only signifies a hunting ritual, but a successful hunt, a dugong, a community feast, sets of social relations and gendered power relations ... Even after construction ceased, the physicality of mounds would continue to remind observers ... that successful dugong hunting depended on successfully-negotiated spiritual and social relationships.

Similarly, the construction of a stone-walled tidal fishtrap at Mort Creek Site Complex associated with deposits pre-dating 2,000 BP can be seen as a type of place-marking. Although the labour investment associated with construction and maintenance of stone-walled tidal fishtraps has been shown to be much less than generally assumed (e.g. Stockton 1982), construction implies intimate knowledge of local hydrology and patterns of fish movement (O'Sullivan 2003). It also imprints a very tangible physical signature onto the landscape, which could be interpreted as place-marking. The evidence strongly indicates that while occupation during Phase I was ephemeral, it was also highly focussed, with systematic extraction of a range of resources from estuarine landscapes.

In summary, use of Phase I sites can be characterised as a coastal component of a geographically wide-ranging settlement-subsistence pattern. That is, the evidence suggests specialised but ephemeral use of the coast by groups occasionally utilising these resources as part of a diffuse and highly-mobile settlement strategy covering a broad area. Groups transported high quality siliceous stone into the area, rather than using local stone, and curated it by retouching artefacts extensively for maximum use-life. This phase is consistent with Hiscock's risk reduction model, in which the adoption of risk-minimising stone artefact technologies is linked to the need for higher mobility in unfamiliar or little-visited country (e.g. Hiscock 1994; McNiven 1994b). Also relevant is Binford's (1979) model of raw material procurement embedded in general land-use strategies, in this case extending over a wide catchment. These groups may have been based primarily around the predictable riverine resources in major catchments such as the Boyne Valley.

As always, caution needs to be exercised in interpreting data from early periods of occupation owing to reduced 'archaeological visibility' of settlement-subsistence strategies 'with low-level seasonal visitation or occupation of regions with unstable and rapidly eroding land surfaces' (Mulvaney and Kamminga 1999:179). Dortch et al. (1984) have also pointed out that low intensity shellfish gathering is unlikely to be represented in the archaeological record owing to the low probability of small shell scatters being preserved (see also Smith 1999). First evidence for occupation of the region should not, therefore, necessarily be taken as indicating the antiquity of coastal settlement or use of the coast. Following the positions advanced by McNiven (1991a) and Hall and Hiscock (1988) and others, I argue that marine and estuarine resources would have always been a consistent and integral feature of broad-based Aboriginal coastal economies. Although access to coastal resources may have been difficult at periods of maximum sea-level fall, expanded river valleys across the continental shelf would have offered a range of resource zones, perhaps with no modern correlates. After rising sea-levels breeched the continental rise and began to invade the continental shelf in the terminal Pleistocene/early Holocene there is no reason to suspect that coastal environments did not always provide a range of resources, which were exploited by people even if these resources were not identical to those available today.

Phase II (c.1,500 BP–c.AD 1850s)

Phase II comprises (1) a diversification and localisation in resource use, evident in the appearance of extraction sites throughout the coastal zone and increased rates of site establishment and use and the increased use of local stone resources; (2) technological investments in processing technologies of plant resources; and (3) more intensive use of existing resources.

Post-1,500 BP use of the southern Curtis Coast appears to have been structured quite differently from that of Phase I. By 1,000 years ago sites appeared throughout the coastal zone, with very large linear sites on the lower margins of major estuaries. Occupation appears to have been unrelated to that at sites occupied before 1,500 years ago, with more frequent and/or more intensive occupation and sustained increases in deposition rates of cultural materials at most sites throughout Phase II. These later sites point to more systematic use of the coastal landscape. Although localised ecological changes, especially those impacting resource availability, could account for site- and estuary-specific patterns of occupation, the synchroneity of changes across the region suggests there were wider processes involved, implicating restructuring of pre-existing patterns of land-use. Archaeological signatures for a transition to increasingly low mobility and logistically-oriented subsistence systems are manifest in widespread regional occupation, increased deposition rates and changes in stone artefact technologies.

The Ironbark Site Complex quarry was heavily exploited for the first time as a major raw material source during Phase II. Although the quarry was used at least occasionally during Phase I, this later period of intensive stone extraction and reduction indicates a fundamental change in the role of the quarry in the pattern of regional land-use. Significantly, the peak in stone artefact discard identified at the quarry between around 1,500–1,000 BP is coeval with an overall reduction in regional occupation identified earlier in patterns of site creation and occupation, as well as in depressed shell and fish bone discard. No evidence for subsistence activities has been found at the Ironbark Site Complex dating to this early period. It appears that while the quarry was targetted for raw material extraction, there was very little use of surrounding coastal landscapes. This use of the quarry as a base for other activities therefore signals the origins of a new system of regional land-use. This pattern suggests that early stone extraction at the quarry was embedded in wider land-use strategies not centred on the coastal zone. This pattern is soon succeeded by the creation and occupation of multiple sites across the region, all of which contain rhyolitic tuff derived (probably exclusively) from the Ironbark Site Complex. While early intensive extraction dating to 1,500 BP may have been related to broad-based and wide-ranging land-use strategies, by 1,000 BP the presence of artefacts manufactured on rhyolitic tuff throughout the coastal zone indicates that use of the quarry is firmly embedded in more localised patterns of resource use. The increasing use of the quarry for artefact manufacture throughout the region signals a refocussing of land-use strategies on local raw materials which may be related to patterns of reduced mobility.

Localisation in the use of stone resources (rhyolitic tuff, microgranite and quartz) in Phase II is accompanied by a change from highly curated to expedient stone reduction strategies associated with reduced mobility (Parry and Kelly 1987). Edge-ground hatchet manufacture on rhyolitic tuff also probably commences during this period, suggesting the possibility of a trade in high prestige items. The use and trade of the distinctively local rhyolitic tuff would also serve to differentiate the distinct identity of coastal groups under a fissioning model like that proposed by McNiven (1999). Significantly in this connection, artefacts manufactured on rhyolitic tuff have a restricted distribution, with expedient forms limited to the coast and curated edge-ground hatchets found up to 100km away, but still within the historically-documented general Gooreng Gooreng language area.

The large size of many Phase II sites, the similarity of their contents and their access to a variety of resource zones suggests a degree of stability in residential precincts on the lower margins of major estuaries. There are also numerous undated scatters of cultural material (stone artefacts and shell) throughout the region which can be chronologically assigned to this period, as widespread use of rhyolitic tuff only occurs after 1,500 BP in dated contexts. Most investigated sites are considered to be residential bases (after Binford 1980), with the exception of two temporary sites (Eurimbula Creek 1 and 2), which were probably associated with specific extraction activities (Binford's 1980:10 field camps and Meehan's 1988:179 dinner-time camps). The residential bases are frequently large and reflect a wide range of activities from food processing and consumption to stone reduction. The distribution of cultural material at large multiple-component sites dating to the last 1,500 years such as Eurimbula Site 1, Ironbark Site Complex, Pancake Creek Site Complex and Tom's Creek Site Complex indicates tightly-focussed settlement strategies in a linear pattern parallel to creek margins, suggesting a pattern of relatively low mobility settlement. The spatial structure of the regional archaeological record reflects largely logistically-organised mobility strategies where large site complexes on the lower margins of major estuaries acted as nodal points in the landscape from which a variety of activities — including specialised exploitation of local microenvironments as represented by small activity-focussed sites — were undertaken, and resources returned to central base camps.

In sum, the features of Phase II are identified by Fresløv and Frankel (1999) as attributes of higher populations, reduced mobility and relative increases in resource use. By 1,000 BP a coastal economy (*sensu* Gaughwin and Fullagar 1995) appears to be firmly in place with evidence for broad-ranging use of the coastal margin and immediate hinterland, implying permanent and structured sedentary mobility strategies throughout the coastal zone. The manufacture and movement of edge-ground hatchets manufactured on rhyolitic tuff beyond the immediate area during this period also hints at the presence of established alliance networks and structures of regional social integration. The broad long-term trend is towards cumulative increases in use of the coastal landscape.

Phase III (c.AD 1850s–c.AD 1920s)

Phase III is defined by the use and discard of artefacts manufactured on European raw materials at long-term occupation sites, demonstrating historical continuities in Aboriginal use of the landscape in the face of European occupation. Colonial impact, notably in the form of frontier violence and introduced diseases, precipitated demographic collapse of local Aboriginal social groups and virtual abandonment of the near-coastal landscapes by the mid-1850s (see Chapter 2). In the main, late nineteenth century Aboriginal populations in the region coalesced into fringe camps at major European townships such as Miriam Vale in the west and Gladstone in the north, or attached themselves to cattle stations established on their traditional lands (Williams 1981). Aboriginal land-owning groups appear to have aggregated along kin-related lines, with Roth (1898) recording a long-term camp at Miriam Vale which comprised people from different local groups across the region. Although Aboriginal people occasionally visited the area after the 1920s from local Aboriginal population centres such as Berajondo and Gladstone, the entire region was effectively depopulated by the removal of Aboriginal people to reserves and missions under the provisions of the *Aboriginal Protection and Restriction on the Sale of Opium Act 1897* (Williams 1981).

Despite disappearing from the European historical record, Aboriginal people continue to use traditional camping places well into the period of European occupation. Flaked bottle glass at both the Ironbark Site Complex and Tom's Creek Site Complex consists of thick bottle bases dating to AD 1890s–AD 1910s. Masses of starch grains and woody tissues observed on several of the artefacts suggest woodworking and plant processing activities, including toxic plant preparation. Contemporary Aboriginal oral histories provide a general historical context for these finds, with Aboriginal families continuing to visit the area from bases at local cattle stations in the early AD 1900s. The location of both sites is far from European settlement of the era. Transportation and use of glass as a medium for artefact manufacture may indicate the presence of small, highly-mobile groups.

The presence of post-contact use of long-term occupation sites points to the persistence of traditional knowledge and continuity of site use. Flaked glass implements dating to the early twentieth century invoke the resilience and hybridity of traditional stone-working technologies in the face of massive social dislocation. The continuing use of these particular locations on the landscape in Phase III provides another line of evidence to determine the antiquity of the recent structure of land-use. The fact that two sites first occupied around 1,500 years ago are targeted for use in the post-contact period is not simply fortuitous, but rather points to a persistence of knowledge of the location of these places on the landscape.

The evidence for post-contact use of these sites coincides with the post-AD 1897 restrictions on the movement of Aboriginal people in Queensland, including forcible removal, suggesting at least three possibilities: use of the sites as resistance and avoidance of European control; use of the sites to fulfill traditional responsibilities; and deliberate surveillance of European activities. A possible fourth phase may be defined as the period from the 1920s to the present, in which Aboriginal people have sought to re-establish connections to country through a range of activities, including residence, camping, fishing, festivals and conducting cultural heritage impact assessments. However, further consideration of this period is beyond the scope of this study.

Summary

A synthesis of archaeological data concerning Aboriginal occupation of the southern Curtis Coast demonstrates initial occupation around 4,000 BP, a major disjunction between c.2,000–1,500 BP and continuous patterns of regional land-use from 1,500 BP into the early twentieth century. The recent trajectory towards localisation of resource use and permanent settlement of the coast can be related to a long-term trajectory of demographic change and social restructuring. Before the late Holocene populations appear to be highly-mobile and wide-ranging, with evidence for only occasional foraging expeditions from subcoastal base camps. After 1,500 BP the coast assumed an even more important role in regional mobility strategies, culminating in permanent occupation from c.1,500 BP. From this time, excavations and analysis document widespread transitions in the archaeological record of the region occurring over relatively short-term periods encompassing increased rates of site establishment and use, localisation in stone raw material sourcing and discard of cultural materials, which together point to relatively more intensive forms of landscape use than previous periods.

15

Wider implications and conclusions

Introduction

This chapter briefly discusses the project's results in the context of key models for southeast Queensland and adjacent regions which emphasise recent intensification of settlement and subsistence strategies. The chapter concludes by considering directions for future research which will improve our understanding of the archaeology of southeast Queensland as well as coastal archaeology in Australia more generally.

Key Findings

The main findings of this study can be summarised as follows:
- significant estuary-specific radiocarbon reservoir offsets of up to $\Delta R = -305\pm61$;
- generally high integrity of open coastal deposits;
- first occupation of the region by 4,000 BP;
- presence of fish bone in deposits pre-dating 3,000 BP;
- a period of reduced regional occupation or abandonment between 2,000 and 1,000 BP;
- creation and occupation of more sites through time, especially after 1,000 BP;
- a general increase in the deposition of shell, fish bone, charcoal and stone artefacts through time;
- a change from small, focussed sites to large, diffuse sites;
- a change in raw material from high quality apparently imported stone to low quality local stone;
- a change in stone artefact technology from curated to expedient tool manufacture;
- the appearance of large stone tools associated with plant processing in the last 1,000 years; and
- continuity in site use from 1,500 BP into the early twentieth century.

A three-phase cultural chronology was developed for the region which proposed initial occupation before 4,000 years ago and significant changes in settlement and resource use after 1,500 BP. Phase

I (pre-4,000 BP–c.1,500 BP) saw ephemeral coastal occupation by groups which occasionally used coastal resources as part of highly-mobile settlement strategies covering a broad area. Phase II (c.1,500 BP–AD 1850s) is characterised by permanent occupation of the coastal zone with relatively low residential mobility systems. This phase is defined by a localisation in resource use and the establishment of large sites throughout the region on the lower margins of major estuaries. Phase III (c.AD 1850s–AD 1920s) saw the emergence of post-European mobility systems. Despite disappearing from the European historical record, Aboriginal people continued to use traditional camping places well into the period of European settlement.

Regional context and implications

As noted in Chapter 1, a number of large-scale archaeological projects have been undertaken along the coast to the north and south of the study area as well as in inland areas to the west (Fig. 15.1). All have yielded evidence broadly consistent with a model of ephemeral low density coastal occupation from before the mid-Holocene with patterns of dramatic change in the late Holocene towards increased rates of occupation. However, while the patterns described for all of these regions are broadly similar, interpretations vary widely and some basic sampling and analytical problems remain to be resolved. In the following sections, key results from the southern Curtis Coast are discussed in terms of major themes in the archaeology of southeast and central Queensland. As will be shown, the results of this study extend and amplify some previous findings while calling some others into question, pointing to issues requiring further research.

Regional occupation before the late Holocene

Findings in the wider region demonstrate the presence of people in inland areas by at least 21,800±400 BP (OxA-806) at Wallen Wallen Creek on the margin of what is now Moreton Bay (Gowlett et al. 1987), 18,800±480 BP (ANU-345) at Kenniff Cave in the Central Queensland Highlands (Callow et al. 1963) and 9,296±119 BP (Wk-9311) at Grinding Groove Cave at Cania Gorge (Tony Eales, Aboriginal and Torres Strait Islander Studies Unit, University of Queensland, pers. comm., 2004). On the coast, the earliest evidence for use of marine resources in Queensland comes from the Whitsunday Islands with dates of 8,150±80 BP (Beta-27835) at Nara Inlet 1 and 6,440±90 BP (Beta-56976) at Border Island 1 (Barker 1989). Unfortunately, few sites are known to span the transition from pre-coastal (terrestrial) to coastal (marine) resource suites. Wallen Wallen Creek on North Stradbroke Island provides one of the few Australian examples. Although the site dates from the late Pleistocene, faunal remains are restricted to the upper shell midden deposit, which is dated to the last c.4,000 years. Neal and Stock (1986:619) concluded that:

> The inhabitants initially hunted terrestrial and aquatic vertebrate fauna, including dugong (*Dugong dugong*), pademelon (*Thylogale* sp.) and snake (*Python spilates*). This was later replaced by an exclusively coastal economy based on the littoral and marine resources of fish and shellfish, with limited dugong hunting.

From around 5,000 BP, and broadly coincident with the final stages of the last marine transgression, increasing numbers of open coastal sites are known, with dates of 4,830±110 BP (Beta-33342) at New Brisbane Airport and 4,350±220 BP (Beta-20799) at Hope Island in the Moreton Bay Region, 4,780±80 BP (Beta-25512) at Teewah Beach 26 in the Great Sandy Region and 4,274±94 BP (NZA-456) at Mazie Bay in the Keppel Islands (Ulm and Reid 2000). The first evidence for occupation on the southern Curtis Coast at the Seven Mile Creek Mound at 3,780±60 BP (Wk-8327) conforms well with these findings.

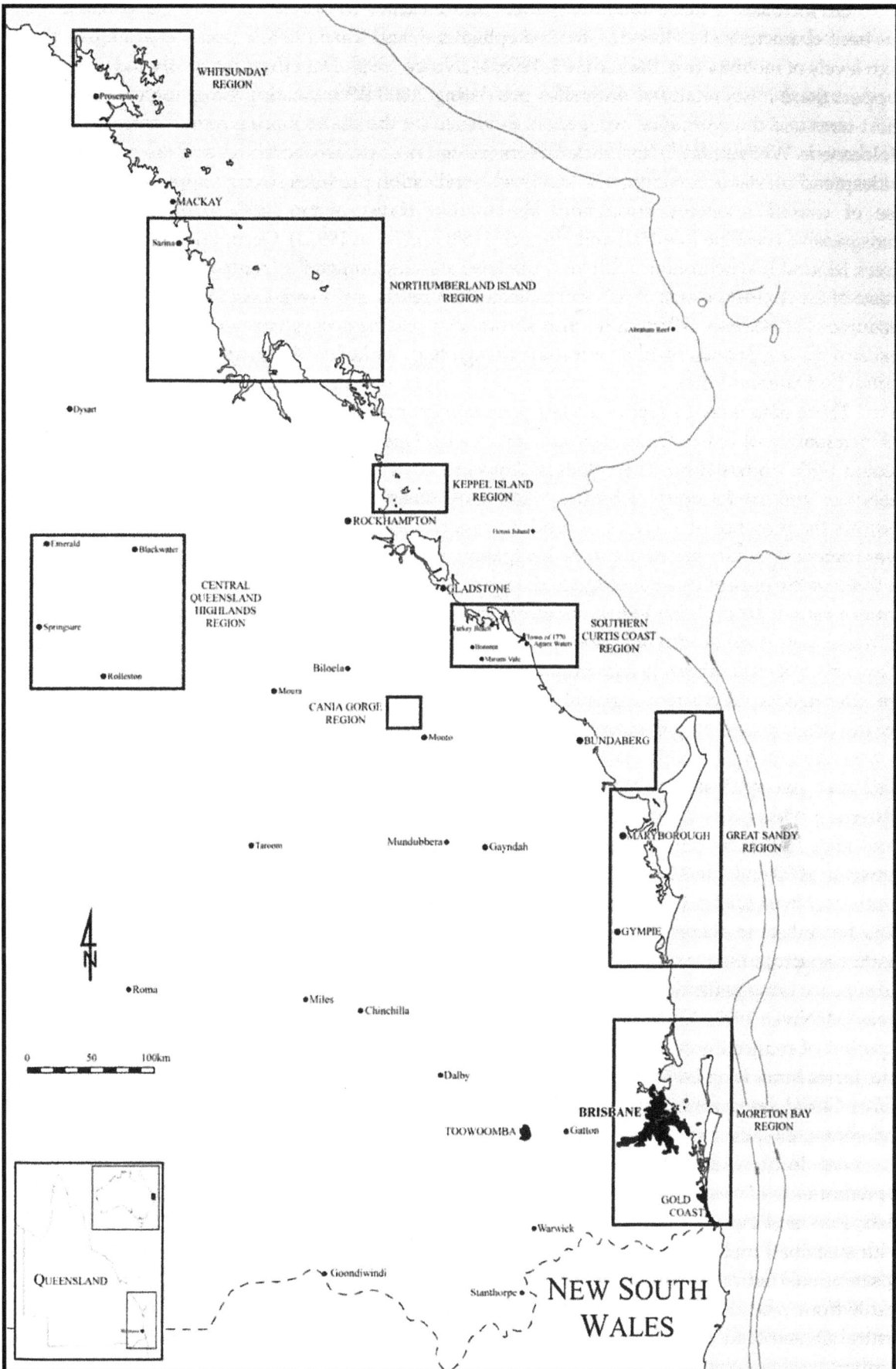

Figure 15.1 Southeast and central Queensland showing the location of major archaeological projects.

On the basis of these data, occupation before c.2,000 BP on the southern Queensland coast has been characterised as low density and ephemeral, and linked to low population densities and high levels of mobility (e.g. Barker 1995, 1996; McNiven 1999). Data from the southern Curtis Coast support these interpretations, with sites pre-dating 2,000 BP indicating geographically-focussed, short-term and discontinuous occupation. Evidence for the use of marine resources from the early Holocene in Whitsunday Island rockshelters located near palaeoshorelines and the coincidence of widespread coastal occupation with sea-level stabilisation provides strong support for continuous use of coastal resources throughout the marine transgression, with people following the transgressive coastline (see Hall and Hiscock 1988; McNiven 1991a). Occupation of the Seven Mile Creek Mound is synchronous with local sea-level stabilisation and its contents (50 taxa of shellfish, 2 taxa of crustaceans, 6 taxa of fish etc) indicate the presence of a well-established suite of estuarine resources at that time. Others have also shown that marine ecosystems are highly adaptable and resilient during periods of environmental change (e.g. Barker 1991; Hutchings and Saenger 1987; Quinn and Beumer 1984).

These data do not support models postulating time-lags between sea-level stabilisation and the availability of coastal resources associated with lags in the timing of coastal settlement (e.g. Beaton 1985; Walters 1986). There does not appear to be any link between the availability of coastal resources and the intensity of human settlement. Palaeoenvironmental and archaeological data indicate the presence of a mosaic of productive estuarine systems by at least 4,500 BP, some 2,500 years before the more recent phase of occupation, even if they were not identical in configuration to those of the present coast. A range of data presented above also suggests that this archaeological pattern cannot be explained in terms of differential preservation or selective sampling (see also Ulm and Hall 1996). As the sample of investigated coastal sites in southeast Queensland increases, the number of sites known to date from the mid-Holocene will increase. Geomorphological studies are now needed to create a regional model of landscape development within which to situate studies of sites pre-dating the late Holocene.

Reduced occupation or abandonment of the coastal zone in the late Holocene

Patterns of ephemeral low intensity occupation on the southern Curtis Coast appear to continue until about 2,000 BP. There is then a period of reduced regional occupation or abandonment between 2,000 and 1,000 BP, which closely matches findings from the Great Sandy Region to the south. McNiven (1992a:12) identified a gap of 1,400 years (2,300–900 BP) in the Cooloola sequence. This he linked to widespread climatically-induced decreases in the availability of rainforest resources across the Great Sandy Region, which prompted restructuring of existing settlement-subsistence arrangements by decreasing use of the coastal region and increasing use of hinterland areas (McNiven 1992a:12). Preliminary results from Cania Gorge complement these findings, with a period of reduced occupation identified between c.2,400 and 1,400 BP (Tony Eales, Aboriginal and Torres Strait Islander Studies Unit, University of Queensland, pers. comm., 2004). Results from Cania Gorge are significant in this context as they are derived from an inland region and from different site types (rockshelters) from those on the coast. This makes it unlikely that the pattern observed in the southern Curtis Coast and Cooloola sequences is related to differential representation of coastal archaeological deposits as has been suggested in other regions (e.g. Head 1983; Rowland 1989). There is no similar pattern discernable in the southeast Queensland data, with sustained increases in site occupation from the mid-Holocene (Ulm and Hall 1996). These observations indicate that the reduced regional occupation on the southern Curtis Coast may result from restructuring of land-use strategies since the mid-Holocene throughout the wider central Queensland region, but not in southeast Queensland, which may have involved at least temporary abandonment (Veth 2003). However, Early Phase occupation on the southern Curtis Coast is not associated with rainforest use, calling into question the significance of reductions in rainforest cited by McNiven (1992a, 1999) as a major factor in the near-abandonment of the Great Sandy Region. Longmore (1997a) has suggested that reductions in rainforest occurred earlier than the late Holocene, further undermining the role of rainforest in the later changes.

These findings run counter to orthodox accounts of late Holocene culture change which emphasise undifferentiated cumulative trajectories towards increased occupation. The identification of major periods of reduced occupation or abandonment challenge us to re-evaluate conventional regional narratives of late Holocene Aboriginal lifeways and more explicitly focus research designs on the identification and definition of variation in trajectories of change.

The antiquity of marine fishing

In southeast Queensland, Walters (1986, 1989, 1992a, 1992b, 1992c, 2001) viewed the commencement of a marine fishery as a necessary precursor to permanent occupation of the southeast Queensland coastal lowlands which he modelled as a marginal terrestrial environment. In fact, he argued that there was no firm evidence for fishing in southeast Queensland before 2,000 years ago (Walters 1992a). Determining the antiquity and nature of marine fishing in the region has therefore been important in various regional studies (e.g. Bowen 1989; Frankland 1990; Hall and Bowen 1989; McNiven 1991a; Ross and Duffy 2000; Walters et al. 1987). However, variability in data recovery techniques and analytical methods combined with poorly developed site chronologies inhibit meaningful integration of these data (Ulm 2002a).

The southern Curtis Coast study not only deployed consistent data recovery and quantification protocols across the region, but is located in the same bioregion as Walters' sites. A major finding of the current study is the antiquity and abundance of fish remains in the region. Fish bone recovered from three sites pre-dates 2,000 BP. The Seven Mile Creek Mound provides the earliest unequivocal evidence for fishing in southeast Queensland at 4,000 BP. The only other site reported to have fish remains pre-dating 3,000 BP is the New Brisbane Airport site where 'fragmentary fish bone' (Hall 1999:174) dating to the mid-Holocene was found encased in the ironstone conglomerate matrix of the lower excavation units. Walters (1992a:35) noted that only a few fragments of fish bone were recovered from this site and argued that these remains have not been demonstrated to be cultural. Stratigraphic and other details published to date do not provide a clear cultural context for the fish remains.

Fish bone from the Mort Creek Site Complex and Eurimbula Site 1 unambiguously supports the antiquity of marine fishing in the southern Curtis Coast region, with assemblages at these sites pre-dating 3,000 and 2,000 BP respectively. These data suggest that fish were always a key resource along the southern Curtis Coast and were not recently incorporated into subsistence production systems to overcome the marginality of the coastal lowlands, as suggested by Walters.

Localisation of resource use

Recent models have highlighted an apparent localisation in the use of animal, plant and stone raw material sources across southeast Queensland in the late Holocene. Morwood (1986, 1987) observed that late Holocene faunal inventories from some rockshelter sites in southeast Queensland exhibited patterns towards representation of smaller-bodied animals, such as possums and koalas, and a more diverse array of species. He argued that this was part of a shift in subsistence strategies from individual encounter-based hunting to more cooperative forms using fire drives and nets with greater production potential.

McNiven (1999) elaborated these ideas in modelling regionalisation in the Great Sandy Region since the mid-Holocene (see David 1991; David and Cole 1990; Hall and Bowen 1989). He defined regionalisation as 'a process whereby social groups segment or fission into smaller social groups with separate and smaller territories. These smaller groups become more localised in their activities tending towards cultural exclusivity' (McNiven 1999:157–8). McNiven defined archaeological correlates for fissioning in terms of increasing localisation of settlement and subsistence patterns and resource use, particularly stone procurement, and the emergence of identity-conscious place-marking strategies such as earthen circles (bora rings) and cemeteries in the last 1,000 years. He argued that access and control of rainforest resources was pivotal to restructuring of local group arrangements. McNiven (2003:339) recently argued that marine stone arrangements in central Queensland were also linked to this recent phase of regionalisation 'whereby newly established residential groups intensify use of local resources and inscribe their social identity into landscapes through place-marking strategies that include formal ritual sites'.

Archaeological investigations in other parts of southeast and central Queensland have revealed similar patterns, in what Lourandos (1997:161) described as 'a more specialised and broad-based coastal emphasis in the economy of the most recent phase' (see also Barker 1995; McNiven 1999; Morwood 1987; Rowland 1982; Ulm 1995; Ulm and Hall 1996; Walters 1989; Westcott et al. 1999a). Overall, these patterns point to a restructuring of land-use over the last 1,500 years towards more systematic and permanent coastal occupation with relatively low levels of residential mobility. Ulm and Hall (1996) dated these changes to around 1,200 BP in southern southeast Queensland and McNiven (1992a, 1999) has argued for a date of 900 BP for the Great Sandy Region. To the north, Barker (1989, 1991, 1995, 1996, 2004) associated increases in diet breadth in the Whitsunday Islands from 600 years ago with the emergence of specialised marine economies akin to those documented in the ethnohistoric record, while Rowland (1982) found that the Keppel Islands were only permanently occupied c.700 years ago.

On the southern Curtis Coast several lines of evidence point to increasingly localised resource use after 1,500 BP. In addition to obvious increases in coastal settlement and use of marine resources, stone raw material sourcing becomes almost exclusively local. The recent exchange of edge-ground hatchets manufactured on rhyolitic tuff also points to the active integration of people into wider networks of social geography in the recent past. In connection with McNiven's (2003) association of place-marking behaviours with recent processes of social fissioning, it appears that mound formation and stone fishtrap construction on the southern Curtis Coast pre-date the appearance of low mobility subsistence-settlement systems associated with localisation by more than 1,000 years, suggesting that caution needs to be exercised in arguments promoting a recent antiquity for place-marking behaviours.

The dramatic increases in the intensity of regional land-use identified on the southern Curtis Coast around 700 BP coincide with similar changes observed in the Whitsunday and Keppel Islands, but post-date changes in southeast Queensland by 500 years. Like the timing of reduced coastal occupation between around 2,000 and 1,000 BP, localisation in central Queensland and northern southeast Queensland are part of the same general system, implying the operation of different factors or at least different timing of these changes from those in southeast Queensland. Further investigation of these findings may shed light on the antiquity of social relationships recorded for the recent past. Intergroup relationships defined on the basis of linguistic and ethnographic data document a disjunction in alliance networks around Rockhampton, with the majority of social relations in the Curtis Coast region focussed to the southeast. The antiquity of these social arrangements is unknown and it could be that this pattern was preceded by an alignment of social affiliations which encompassed central Queensland and the northern half of southeast Queensland.

Wider implications

In the wider context, the recent trajectory towards localisation of resource use and permanent settlement of the coast in southeast and central Queensland can be related to long-term historical trajectories of change. The regional chronology indicates a time-lag between intermittent use of coastal resources, and widespread permanent occupation which occurred much later. People appear to have always used coastal resources, and it is no accident that the first cultural shell deposits on the southern Queensland coast are coeval with the onset of relative local sea-level stability, resulting from people following the transgressive coastline westwards (Hall and Hiscock 1988; McNiven 1991a; Rowland 1999).

It is possible that reduced predictability of coastal resources linked to fluctuations in marine productivity induced by the final stages of marine transgression may have led to a reduction in the use of coastal areas in favour of increased use of subcoastal areas, with only occasional coastal foraging expeditions. Indeed, subcoastal occupation only becomes archaeologically visible in southeast Queensland around the terminal Pleistocene/early Holocene (Hall 1999). This would have resulted in significant depopulation of the coastal zone. Such population redistribution has been suggested as a mechanism to cope with dramatic environmental change elsewhere (e.g. Gamble 1993), necessitating reorganisation of access to land.

In some areas with low offshore gradients, coastal resources may have only been reincorporated as an extension of inland-focussed extractive economies in the late Holocene, after the end of major transgressive fluctuations increased the predictablility of resource abundance and distribution. Smith and Sharp (1993:49) noted the 'presence of sites on or near the Pleistocene coast, wherever the continental shelf is steep enough to allow their preservation'. All known Pleistocene and early Holocene sites in Australia which exhibit use of coastal resources are located in rockshelters situated near rocky and/or precipitous palaeoshorelines (e.g. Barker 1991, 1995; Morse 1988, 1993; Veth 1993). These findings may not simply relate to preservation of sites, but also the relative stability of resource zones in areas where sea-level impacts are primarily vertical rather than horizontal (Beaton 1995). For the Whitsunday Islands, Barker (1995) has argued that the steep rocky coastline supported a similar resource structure throughout the Holocene. However, in southeast Queensland where the continental shelf is relatively wide, has a gentle slope and few near-coastal rock formations, it is perhaps not surprising that archaeological evidence is lacking for coastal occupation prior to mid-Holocene sea-level stablisation. It might be expected that if early coastal use was related to inland-focussed groups, it might be characterised by high mobility and the establishment of activity-specific sites. Hiscock (1994) and Kelly and Todd (1988) have argued that initial incorporation of new territories or use of little-used territories would have focussed on hunting as the detailed knowledge required for successful gathering would not be available. Lithic technologies thought to be indicators of initial colonisation, with curated, reliable forms, are exactly the sorts of stone artefacts recovered from pre-1,500 BP deposits on the southern Curtis Coast.

After 1,500 BP the coast progressively assumed a more important role in regional mobility strategies, culminating in permanent occupation. After this time, excavations reveal rapid and widespread changes in site content, abundance of certain classes of cultural material in deposits and extent of sites. The magnitude and broad synchrony of these sites indicate a reorganisation of demographic structures linked to significant cultural transformation. Occupation in the region in the late Holocene took the form of progressively more nucleated communities. More specifically, these transitions include cessation in the use of non-local stone raw materials, an increasingly diversified subsistence resource base and patterns of increase in site establishment and use. Like McNiven (1990a, 1991a, 1999), I suggest that the pre-1,500 BP occupation phase was characterised by more territorially extensive and open social networks which included both coastal and hinterlands areas. After 1,500 BP regional populations appear more territorially bounded or closed as denoted by marked localisation of resource use and a dramatic increase in the scale of coastal occupation. A recent greater formalisation of social relations may also be evident in the production and exchange of edge-ground hatchets manufactured on rhyolitic tuff. The permanent structured occupation of coastal landscapes provided economic, social and political opportunities for the creation of new identity-conscious groups as part of a general restructuring of the social landscape.

The pattern of change and the emergence of regionalism is not the same everywhere and no single explanation can account for the changes seen in the archaeological record of the late Holocene.

Resolution of the causes of late Holocene modifications in Aboriginal land-use strategies on the southern Queensland coast will require ongoing work with particular attention paid to chronology-building, assessment of site integrity and consistent data recovery strategies.

Methodological implications

Two strands of this research have direct methodological implications for coastal archaeology in Australia as well as archaeology more generally. First is the concern with radiocarbon dating of coastal sites, in particular the dating of marine samples with poorly defined local marine and estuarine reservoir effects and establishing the comparability of these samples vis-à-vis those obtained on terrestrial materials. Second is the development of techniques to assess the stratigraphic integrity of sites. Together these approaches help quantify the temporal limits of the record of human occupation in the region and therefore amplify the interpretative potential of open coastal deposits.

Dating strategies and chronological control

Given the low number of dates available for Holocene open coastal sites in Queensland, a basic objective of this project was to construct a solid chronological framework through a large sample of radiometric determinations obtained on a range of sample materials from all major stratigraphic contexts. Dating of basal and termination deposits in this study contributed key data to defining regional continuities and discontinuities in occupation (David 2002:37). Conventional approaches which assume continuity of occupation from initial site establishment routinely overestimate the number of sites dating to more recent periods and tend to homogenise patterns of regional occupation and apply regionally undifferentiated trajectories of cultural change across vast areas of the continent.

This study revealed major problems in adopting generic open ocean marine reservoir correction estimates in calibrating radiocarbon dates obtained on estuarine shell samples in southeast Queensland. Results obtained from estuarine shell/charcoal pairs demonstrated marked variability in carbon reservoirs between estuaries, with offsets of up to several hundred years different from the recommended open ocean value, suggesting estuary-specific patterns of variation in terrestrial carbon input and exchange with the open ocean. Future studies will need to address this issue on a regional basis to increase confidence in site sequences and to facilitate their interpretation (Ulm 2002d). Large suites of dates including marine samples may be significantly biased by such factors.

Integrity of open sites

Australian archaeology has been dominated by studies of rockshelter deposits and open sites have often been cited as problematic owing to presumed uncertainties in site integrity. As Erlandson and Moss (1999:432) noted 'although many such investigations, especially those focused on large and deeply stratified sites, have helped elucidate general patterns of cultural development in many regions, the lack of comparative data from a broader range of sites in many regions obscures a tremendous amount of variability that helps us to transcend normative cultural reconstructions of the past'. Identification, description, sampling and absolute dating of a wide range of sites is therefore critical to the accurate characterisation of archaeological patterns as coherent over a region and not simply the result of site type and/or assemblage-specific accumulation and representation factors.

Issues of site integrity are basic to our understanding of cultural change. This study has shown that using appropriate data recovery tools at open sites can improve confidence in assessments of site integrity. A novel approach using bivalve conjoin analysis of *Anadara trapezia* yielded data on post-depositional disturbance. These analyses demonstrated that relatively little movement of shell material occurred in most contexts, even in low density deposits without well-defined stratigraphy. The integrity of deposits is also supported by consistent age-depth relationships of dated shell and charcoal samples, as well as the clear disjunctions in the vertical concentration of cultural materials in deposits across the region. Shellfish remains at several sites also form continuous layers with interlocking valves effectively sealing lower deposits.

Bivalve conjoin analyses provide a useful adjunct to the battery of techniques conventionally used to assess post-depositional disturbance and have wide potential application in eastern Australian coastal shell deposits. Allen and O'Connell (2003) recently noted that stone artefact conjoin analyses are needed to help resolve questions about site integrity. Bivalve conjoin analyses provide a new method to apply to coastal sites, which often contain few stone artefacts.

Discussion

Despite the amount of research undertaken over the last three decades, it is still not possible to integrate our understandings of the Holocene on a continental scale, at any but the most abstract level. Few regions have sufficient data to attempt the construction of local archaeological (pre)histories. Major gaps remain in our understanding of the Holocene period owing to an uneven geographic spread of research, biases towards particular site types, and the limitations of conventional methodologies. Large areas of Australia remain virtually unknown archaeologically, especially beyond the southeast corner of the continent. The patchy distribution of studies has encouraged the identification of continuities between widely separated sites and regions, rather than at scales appropriate to the delineation of both continuities and discontinuities at the local level. Although there is a clear disjunction in the archaeological trajectories of many regions between 5,000 and 1,000 years ago, our ability to understand these changes remains hampered by limitations imposed by inadequate sampling and chronological control on the one hand, and the abstract explanatory framework adopted on the other. The continental narrative is important in providing a general heuristic framework but its validity is dependent on data accuracy and the strength of explanatory models. As it is, the continental narrative emerges as a conceptually inadequate framework that actually works to downplay temporal and spatial diversity, diverting attention from the particularity of regionally-specific historical trajectories and archaeological signatures. Obviously, at a continental scale, no historical trajectory is autonomous or exists in isolation, but rather is defined in part by external relationships. This interconnectedness should be the subject of study in its own right and not used to justify homogenising Indigenous Australian histories into a single historical trajectory. An alternative approach emphasising the diversity of regional archaeological and environmental records is much to be preferred. The major task ahead therefore remains a basic one: to construct and compare detailed individual site sequences from a range of site types, at the local and regional level, to establish the existence of trends independent of site-specific taphonomic and/or environmental factors.

Conclusion

Data and interpretations presented in this study form a baseline for future archaeological studies in the southern Curtis Coast region with methodological implications for the conduct of coastal archaeology in southeast Queensland and beyond. Conventional narratives of Australian prehistory emphasise patterns of synchronous change across many areas of the continent in the late Holocene. Although these general patterns have often been presented as robust relationships in the literature, most are confounded by exceptions when subject to detailed scrutiny, underscoring the complexity of cultural diversity, site formation processes and taphonomic histories on both spatial and temporal axes. Techniques adopted in this study attempted to transcend the continental narrative by describing the archaeological record of a specific region and using tools sensitive to change in the archaeological record. Data revealed complex patterns of continuity and disjunction in the record of Aboriginal occupation of the southern Curtis Coast since the mid-Holocene.

Holocene period archaeology in Australia is on the cusp of a major shift in the appreciation of the complexity of temporal and spatial diversity. Accumulating regional archaeological and palaeoenvironmental datasets, coupled with refinements in technical methods, provides the potential for disentangling local and regional variability from amorphous 'long-term, continental narrative and description' (Frankel 1993:31). Although the continental narrative has value, its formative elements need to be rethought, particularly its failure to adequately contextualise sequences within local frameworks, historical trajectories and patterns of research. An exciting challenge in the coming decades will involve the construction of detailed regional archaeological and palaeoenvironmental sequences, and the re-examination of regional datasets, with a view to describing and explicating the complexity and diversity of the archaeological record. Not only will detailed regional studies provide more accurate accounts of the past and more strongly situate people in the context of landscape, but also they will contribute to a more informed, robust and useful continental narrative.

References

Aitken, M. J. 1990. *Science-Based Dating in Archaeology*. London: Longman.

Alfredson, G. 1984. An Archaeological Investigation into the Aboriginal Use of St. Helena Island, Moreton Bay. Unpublished BA(Hons) thesis, Department of Anthropology and Sociology, University of Queensland, Brisbane.

Alfredson, G. 1987. Report on the Archaeological Survey of and Collection from the Proposed Benaraby-Parana Realignment of the North Coast Railway. Unpublished report to McIntyre and Associates.

Alfredson, G. 1989. Report on an Initial Archaeological Survey of the Stuart Oil Shale Project. Unpublished report to Hollingsworth Consultants Pty Ltd, Brisbane.

Alfredson, G. 1990. Report on the Initial Archaeological Survey of the Proposed Monte Cristo Resort at Black Head Curtis Island. Unpublished report to Hollingsworth Consultants Pty Ltd, Brisbane.

Alfredson, G. 1991. Report on an Initial Archaeological Survey for the Gladstone Special Steel Project. Unpublished report to Hollingsworth Consultants Pty Ltd, Brisbane.

Alfredson, G. 1992. Gladstone Industrial Land Project Studies — Archaeology and Heritage — Stage. Unpublished report to Environmental Science Services, Brisbane.

Alfredson, G. 1993. Report on an Archaeological Inspection of a Proposed Residential and Recreational Development on Hummock Hill Island. Unpublished report to AGC Woodward-Clyde Pty Ltd for Hummock Hill Island Pty Ltd.

Alfredson, G. 2002. A Report on an Excavation at Pebble Beach Lot 901 on SP 133223, Parish of Toorbul, Sandstone Point, Caboolture Shire. Unpublished report to Keilor Fox and McGhie.

Alfredson, G. and Kombumerri Aboriginal Corporation for Culture. 1999. Report on a Collection and a Further Cultural Heritage Assessment for a Proposed Golf Course and Residential Development on Lake Coombabah. Unpublished report to Nimmel Partnership, Mudgeeraba, Queensland.

Allen, J. (ed.) 1996. *Report of the Southern Forests Archaeological Project: Site Descriptions, Stratigraphies and Chronologies*. Vol. 1. Bundoora, VIC: La Trobe University Archaeological Publications.

Allen, J. and R. Jones. 1980. Oyster Cove: archaeological traces of the last Tasmanians and notes on the criteria for authentication of flaked glass artefacts. *Papers and Proceedings of the Royal Society of Tasmania* 114: 225–33.

Allen, J. and J. F. O'Connell (eds) 1995. Transitions: Pleistocene to Holocene in Australia and Papua New Guinea. *Antiquity* 69: 649–862.

Allen, J. and J. F. O'Connell. 2003. The long and the short of it: archaeological approaches to determining when humans first colonised Australia and New Guinea. *Australian Archaeology* 57: 5–19.

Anderson, A. J. 1991. The chronology of colonization in New Zealand. *Antiquity* 65: 767–95.

Andrews, J. 1971. *Sea Shells of the Texas Coast*. Austin, TX: University of Texas Press.

Ashmore, P. J. 1999. Radiocarbon dating: avoiding errors by avoiding mixed samples. *Antiquity* 73: 124–30.

Attenbrow, V. 1987. The Upper Mangrove Creek Catchment: A Study of Quantitative Changes in the Archaeological Record. Unpublished PhD thesis, University of Sydney, Sydney.

Attenbrow, V. 1992. Shell bed or shell midden. *Australian Archaeology* 34: 3–21.

Attenbrow, V. 2003. Habitation and land use patterns in the Upper Mangrove Creek catchment, New South Wales central coast, Australia. *Australian Archaeology* 57: 20–31.

Attenbrow, V. 2004. *What's Changing: Population Size or Land-Use Patterns?: The Archaeology of Upper Mangrove Creek, Sydney Basin.* Terra Australis 21. Canberra: Pandanus Books.

Backhouse, J. 1843. *A Narrative of a Visit to the Australian Colonies.* London: Hamilton Adams.

Bailey, G. N. 1983. Problems of site formation and the interpretation of spatial and temporal discontinuities in the distribution of coastal middens. In P. M. Masters and N. C. Flemming (eds), *Quaternary Coastlines and Marine Archaeology*, pp. 559–82. London: Academic Press.

Bailey, G. N. 1993. Shell mounds in 1972 and 1992: reflections on recent controversies at Ballina and Weipa. *Australian Archaeology* 37: 1–17.

Bailey, G. N. 1994. The Weipa shell mounds: natural or cultural. In M. Sullivan, S. Brockwell and A. Webb (eds), *Archaeology in the North: Proceedings of the 1993 Australian Archaeological Association Conference*, pp. 107–29. Darwin: North Australia Research Unit, Australian National University.

Bailey, G. N. 1999. Shell mounds and coastal archaeology in northern Queensland. In J. Hall and I. McNiven (eds), *Australian Coastal Archaeology*, pp. 105–12. Research Papers in Archaeology and Natural History 31. Canberra: Archaeology and Natural History Publications, Research School of Pacific and Asian Studies, Australian National University.

Bailey, G. N., J. Chappell and R. Cribb. 1994. The origin of Anadara shell mounds at Weipa, north Queensland Australia. *Archaeology in Oceania* 29 (2): 69–80.

Ball, L. C. 1915. Letter to Hamlyn-Harris, Queensland Museum, 24 September 1915.

Barker, B. 1989. Nara Inlet 1: a Holocene sequence from the Whitsunday Islands, central Queensland coast. *Queensland Archaeological Research* 6: 53–76.

Barker, B. 1991. Nara Inlet 1: coastal resource use and the Holocene marine transgression in the Whitsunday Islands, central Queensland. *Archaeology in Oceania* 26 (3): 102–9.

Barker, B. 1993. An Archaeological Survey of Eastern Boyne Island, Gladstone, Central Queensland. Unpublished report to Hollingsworth, Dames and Moore.

Barker, B. 1995. The Sea People: Maritime Hunter-Gatherers on the Tropical Coast — A Late Holocene Maritime Specialisation in the Whitsunday Islands, Central Queensland. Unpublished PhD thesis, University of Queensland, Brisbane.

Barker, B. 1996. Maritime hunter-gatherers on the tropical coast: a social model for change. In S. Ulm, I. Lilley and A. Ross (eds), *Australian Archaeology '95: Proceedings of the 1995 Australian Archaeological Association Annual Conference*, pp. 31–43. Tempus 6. St Lucia, QLD: Anthropology Museum, University of Queensland.

Barker, B. 2004. *The Sea People: Late Holocene Maritime Specialisation in the Whitsunday Islands, Central Queensland.* Terra Australis 20. Canberra: Pandanus Books.

Beaglehole, J. C. (ed.) 1963. *The Endeavour Journal of Joseph Banks 1768–1771.* 2nd ed. Sydney: Angus and Robertson.

Beaglehole, J. C. (ed.) 1968. *The Journals of Captain James Cook: The Voyage of the Endeavour, 1768–1771.* Cambridge: Cambridge University Press.

Beaton, J. 1977. Dangerous Harvest: Investigations in the Late Prehistoric Occupation of Upland South-East Central Queensland. Unpublished PhD thesis, Research School of Pacific and Asian Studies, Department of Prehistory, Australian National University, Canberra.

Beaton, J. 1982. Fire and water: aspects of Australian Aboriginal management of cycads. *Archaeology in Oceania* 17 (1): 51–8.

Beaton, J. 1985. Evidence for a coastal occupation time-lag at Princess Charlotte Bay (north Queensland) and implications for coastal colonisation and population growth theories for Aboriginal Australia. *Archaeology in Oceania* 20 (1): 1–20.

Beaton, J. 1990. The importance of past population for prehistory. In B. Meehan and N. White (eds), *Hunter-Gatherer Demography: Past and Present*, pp. 23–40. Oceania Monograph 39. Sydney: Oceania Publications, University of Sydney.

Beaton, J. 1995. The transition on the coastal fringe of Greater Australia. *Antiquity* 69: 798–806.

Beck, W. 1985. Technology, Toxicity and Subsistence: A Study of Australian Aboriginal Plant Food Processing. Unpublished PhD thesis, La Trobe University, Bundoora.

Bedwell, F. P., F. H. S. Bray and E. R. Connor. 1870. *East Coast of Australia, Queensland: Sandy Cape to Keppel Isles* [map]. 1"=1 nautical mile. Sheet XI. London: Royal Navy.

Binford, L. R. 1978. *Nanamiut Ethnoarchaeology.* New York: Academic Press.

Binford, L. R. 1979. Organization and formation processes: looking at curated technologies. *Journal of Anthropological Research* 35 (3): 255–73.

Binford, L. R. 1980. Willow smoke and dogs' tails: hunter-gatherer settlement systems and archaeological site formation. *American Antiquity* 45 (1): 4–20.

Bird, C. F. M. and D. Frankel. 1991a. Chronology and explanation in western Victoria and south-east South Australia. *Archaeology in Oceania* 26 (1): 1–16.

Bird, C. F. M. and D. Frankel. 1991b. Problems in constructing a prehistoric regional sequence: Holocene southeast Australia. *World Archaeology* 23 (2): 179–92.

Bird, D. W. and R. L. Bliege Bird. 1997. Contemporary shellfish gathering strategies among the Meriam of the Torres Strait Islands, Australia: testing predictions of a central place foraging model. *Journal of Archaeological Science* 24: 39–63.

Bird, M. K. 1992. The impact of tropical cyclones on the archaeological record: an Australian example. *Archaeology in Oceania* 27 (2): 75–86.

Bird, M. K. 1995. Coastal morphodynamics and the archaeological record: further evidence from Upstart Bay, north Queensland. *Australian Archaeology* 40: 57–8.

Bladen, F. M. (ed.) 1892 *Historical Records of New South Wales*. Vol. 1. Sydney: Government Printer.

Blake, T. W. 1991. A Dumping Ground: Barambah Aboriginal Settlement 1900–40. Unpublished PhD thesis, Department of History, University of Queensland, Brisbane.

Bollong, C. A. 1994. Analysis of site stratigraphy and formation processes using patterns of pottery sherd dispersion. *Journal of Field Archaeology* 21: 15–28.

Boow, J. 1991. *Early Australian Commercial Glass: Manufacturing Processes*. Sydney: Department of Planning and Heritage Council of New South Wales.

Border, A. 1994. Shoalwater Bay Military Training Area (SWBTA): a review of cultural heritage resources their significance and land use. In *Commonwealth Commission of Inquiry Shoalwater Bay, Capricornia Coast, Queensland, Research Reports*, pp. 173–233. No. 5, Vol. A. Canberra: Australian Government Publishing Service.

Bowdler, S. 1981. Hunters in the highlands: Aboriginal adaptations in the eastern Australian uplands. *Archaeology in Oceania* 16 (2): 99–111.

Bowdler, S. and S. O'Connor. 1991. The dating of the Australian Small Tool Tradition, with new evidence from the Kimberley, WA. *Australian Aboriginal Studies* 1: 53–62.

Bowen, G. 1989. A Model for Moreton Island Prehistory: Colonisation, Settlement and Subsistence. Unpublished BA(Hons) thesis, Department of Anthropology and Sociology, University of Queensland, Brisbane.

Bowen, G. 1998. Towards a generic technique for dating stone fish traps and weirs. *Australian Archaeology* 47: 39–43.

Bowman, G. M. 1985a. Oceanic reservoir correction for marine radiocarbon dates from northwestern Australia. *Australian Archaeology* 20: 58–67.

Bowman, G. M. 1985b Revised radiocarbon oceanic reservoir correction for southern Australia. *Search* 16: 164–65.

Bowman, G. and N. Harvey. 1983. Radiocarbon dating marine shells in South Australia. *Australian Archaeology* 17: 113–23.

Brasch, S. 1975. Gureng Gureng, a Language of the Upper Burnett River, South-East Queensland. Unpublished BA(Hons) thesis, Department of Linguistics, Australian National University, Canberra.

Brian, D. 1994. Shall I Compare Thee to a Fish?: A Comparative Taphonomic Analysis of Vertebrate Remains from Nara Inlet Art Site, Hook Island, Central Queensland Coast. Unpublished BA(Hons) thesis, Department of Anthropology and Sociology, University of Queensland, Brisbane.

Brown, J. A. 1981. Charnel houses and mortuary crypts: disposal of the dead in the Middle Woodland Period. In D. S. Brose and N. Greber (eds), *Hopewell Archaeology: The Chillicothe Conference*, pp. 211–9. Kent: Kent State University Press.

Buchanan, S. 1994. *The Lighthouse Keepers*. Samford, QLD: Coral Coast Publications.

Buchanan, S. 1999. *Lighthouse of Tragedy: The Story of Bustard Head Lighthouse, Queensland's First Coast Light*. Samford, QLD: Coral Coast Publications.

Burke, C. 1993. A Survey of Aboriginal Archaeological Sites on the Curtis Coast, Central Queensland. Unpublished report to Queensland Department of Environment and Heritage, Rockhampton.

Butlin, N. 1983. *Our Original Aggression: Aboriginal Populations of South-Eastern Australia 1788–1850*. London: George Allen and Unwin.

Cadée, G. C. 2002. Floating articulated bivalves, Texel, North Sea. *Palaeogeography, Palaeoclimatology, Palaeoecology* 183: 355–9.

Cahen, D. 1978. New excavations at Gombe (ex Kalinga Point), Kinsasha, Zaire. *Antiquity* 52: 51–6.

Cahen, D., L. H. Keeley and F. L. van Noten. 1979. Stone tools, tool kits, and human behaviour in prehistory. *Current Anthropology* 20: 661–83.

Cahen, D. and J. Moeyersons. 1977. Subsurface movements of stone artefacts and their implications for the prehistory of central Africa. *Nature* 266: 812–5.

Callow, W. J., M. J. Baker and D. H. Pritchard. 1963. National Physical Laboratory radiocarbon measurements I. *Radiocarbon* 5: 34–8.

Campbell, J. 2002. *Invisible Invaders: Smallpox and Other Diseases in Aboriginal Australia 1780–1880*. Melbourne: Melbourne University Press.

Carter, M. n.d. Rodds Peninsula Site Complex: Inventory of Excavations and Summary of Analyses. Unpublished report to Aboriginal and Torres Strait Islander Studies Unit, University of Queensland, Brisbane.

Carter, M. 1997. Chenier and Shell Midden: An Investigation of Cultural and Natural Shell Deposits at Rodds Peninsula, Central Queensland Coast. Unpublished BA(Hons) thesis, Department of Anthropology and Sociology, University of Queensland, Brisbane.

Carter, M. 2002. Recent results of excavations on the Murray Islands, eastern Torres Strait and implications for early links with New Guinea: bridge and barrier revisited. In S. Ulm, C. Westcott, J. Reid, A. Ross, I. Lilley, J. Prangnell and L. Kirkwood (eds), *Barriers, Borders, Boundaries: Proceedings of the 2001 Australian Archaeological Association Annual Conference*, pp. 1–10. Tempus 7. Brisbane: Anthropology Museum, University of Queensland.

Carter, M., I. Lilley, S. Ulm and D. Brian. 1999. Mort Creek Site Complex, Curtis Coast: site report. *Queensland Archaeological Research* 11: 85–104.

Chapman, V. 1999. Drawing the Line: The Rock Paintings of Cania Gorge, South Central Queensland. Unpublished PGDipArts (Anthropology) thesis, Department of Sociology, Anthropology and Archaeology, University of Queensland, Brisbane.

Chapman, V. 2002. Drawing the line: the rock paintings of Cania Gorge, central Queensland. In S. Ulm, C. Westcott, J. Reid, A. Ross, I. Lilley, J. Prangnell and L. Kirkwood (eds), *Barriers, Borders, Boundaries: Proceedings of the 2001 Australian Archaeological Association Annual Conference*, pp. 91–100. Tempus 7. Brisbane: Anthropology Museum, University of Queensland.

Chappell, J. and J. Grindrod 1984. Chenier plain formation in northern Australia. In B. Thom (ed.), *Coastal Geomorphology in Australia*, pp. 197–231. Sydney: Academic Press.

Chivas, A., J. Chappell, H. Polach, B. Pillans and P. Flood. 1986. Radiocarbon evidence for the timing and rate of island development, beach-rock formation and phosphatization at Lady Elliot Island, Queensland, Australia. *Marine Geology* 69: 273–87.

Clarkson, C., M. Williams, I. Lilley and S. Ulm. n.d. Gooreng Gooreng Contemporary Social Landscapes. Unpublished report to Australian Institute of Aboriginal and Torres Strait Islander Studies, Canberra.

Claassen, C. 1986. Temporal patterns in marine shellfish-species use along the Atlantic coast in the south-eastern United States. *Southeastern Archaeology* 5: 120–37.

Claassen, C. 1991. Normative thinking and shell-bearing sites. In M. B. Schiffer (ed.), *Archaeological Method and Theory* 3: 249–98. Tucson: University of Arizona Press.

Claassen, C. 1998. *Shells*. Cambridge: Cambridge University Press.

Coaldrake, J. E. 1961. *The Ecosystems of the Coastal Lowlands ('wallum') of Southern Queensland*. C.S.I.R.O. Bulletin 283. Melbourne: Commonwealth Scientific and Industrial Research Organisation.

Coleman, N. 1981. *What Shell is That?* Sydney: Ure Smith Press.

Collins, M. B. 1991. Rockshelters and the early archaeological record in the Americas. In T. D. Dillehay and D. J. Meltzer (eds), *The First Americans: Search and Research*, pp. 157–82. Boca Raton, FL: CRC Press.

Colliver, F. S. and F. P. Woolston. 1978. Aboriginals in the Brisbane area. In *Brisbane Retrospect: Eight Aspects of Brisbane History: Proceedings of a Seminar Conducted by the John Oxley Library, Centennial Hall State Library of Queensland, 5–6 June, 1976*, pp. 58–88. Brisbane: Library Board of Queensland.

Connah, T. H. 1961. Beach sand heavy mineral deposits of Queensland. *Publications of the Geological Survey of Queensland* 302. Brisbane: Queensland Department of Mines.

Cook, P. J. and W. Mayo. 1977. *Sedimentology and Holocene History of a Tropical Estuary (Broad Sound, Queensland)*. BMR Bulletin 170. Canberra: Australian Department of National Development.

Cook, P. J. and H. A. Polach. 1973. A chenier sequence at Broad Sound, Queensland, and evidence against a Holocene high sea level. *Marine Geology* 14: 253–68.

Cosgrove, R. 1995. *The Illusion of Riches: Scale, Resolution and Explanation in Tasmanian Pleistocene Human Behaviour*. BAR International Series 608. Oxford: Tempus Reparatum Archaeological and Historical Associates Limited.

Cotter, M. 1996. Holocene environmental change in Deception Bay, southeast Queensland: a paleogeographical contribution to MRAP Stage II. In S. Ulm, I. Lilley and A. Ross (eds), *Australian Archaeology '95: Proceedings of the 1995 Australian Archaeological Association Annual Conference*, pp. 193–205. Tempus 6. St Lucia, QLD: Anthropology Museum, University of Queensland.

Courtney, K. and I. J. McNiven. 1998. Clay tobacco pipes from Aboriginal middens on Fraser Island, Queensland. *Australian Archaeology* 47: 44–53.

Cox, J. C. 1888. Notes on two wax figures obtained from an Aboriginal camp at Miriam Vale, near the head of Baffle Creek, Rockhampton. *Proceedings of the Linnean Society of New South Wales* 13: 1223–6.

Cribb, R. 1996. Shell mounds, domiculture and ecosystem manipulation on western Cape York Peninsula. In P. Veth and P. Hiscock (eds), *Archaeology of Northern Australia*, pp. 150–74. Tempus 4. St Lucia, QLD: Anthropology Museum, University of Queensland.

Culbert, N. 1996. The Shell 'Artefact' from A7. Unpublished report submitted for AY269 Independent Study 1. Brisbane: Department of Anthropology and Sociology, University of Queensland.

Curr, E. M. 1887. *The Australian Race*. Melbourne: Government Printer.

Cziesla, E., S. Eickoff, N. Arts and D. Winter (eds) 1990. *The Big Puzzle: International Symposium on Refitting Stone Artifacts*. Studies in Modern Archaeology 1. Bonn: Holos Verlag.

David, B. 1991. Fern Cave, rock art and social formations: rock art regionalisation and demographic changes in southeastern Cape York Peninsula. *Archaeology in Oceania* 26 (2): 41–57.

David, B. 1994. A Space-Time Odyssey: Rock Art and Regionalisation in North Queensland Prehistory. Unpublished PhD thesis, Department of Anthropology and Sociology, University of Queensland, Brisbane.

David, B. 2002. *Landscapes, Rock-Art and the Dreaming: An Archaeology of Preunderstanding*. London: Leicester University Press.

David, B. and D. Chant. 1995. Rock art and regionalisation in north Queensland prehistory. *Memoirs of the Queensland Museum* 37 (2): 357–528.

David, B. and N. Cole. 1990. Rock art and inter-regional interaction in northeastern Australian prehistory. *Antiquity* 64: 788–806.

David, B. and H. Lourandos. 1997. 37,000 years and more in tropical Australia: investigating long-term archaeological trends in Cape York Peninsula. *Proceedings of the Prehistoric Society* 63: 1–23.

David, B. and H. Lourandos. 1999. Landscape as mind: land use, cultural space and change in north Queensland prehistory. *Quaternary International* 59: 107–23.

David, B. and M. Wilson. 1999. Re-reading the landscape: place and identity in NE Australia during the late Holocene. *Cambridge Archaeological Journal* 9 (2): 163–88.

Davie, P. (ed.) 1998. *Wild Guide to Moreton Bay: Wildlife and Habitats of a Beautiful Australian Coast — Noosa to the Tweed*. Brisbane: Queensland Museum.

Davies, S. 1994. An Archaeological Assessment of the Proposed Rail Deviations on the Mainline Upgrade between Bundaberg and Gladstone, Queensland. UQASU Report 237. Brisbane: University of Queensland Archaeological Services Unit.

Desert Ridge. 1998. *The Discovery Centre's Ultimate Survival Guide to Town of 1770 and Agnes Water*. Agnes Water, QLD: Desert Ridge.

Dickson, F. 1981. *Australian Stone Hatchets: A Study in Design and Dynamics*. Sydney: Academic Press.

Dodd, J. R. and R. J. Stanton (Jr). 1981. *Paleoecology: Concepts and Applications*. New York: Wiley Interscience.

Dortch, C. E., G. W. Kendrick and K. Morse. 1984. Aboriginal mollusc exploitation in southwestern Australia. *Archaeology in Oceania* 19 (3): 81–104.

Dowling, R. M. 1980. The mangrove vegetation. In H. F. Olsen, R. M. Dowling and D. Bateman, *Biological Resources Investigation (Estuarine Inventory)*, pp. 45–90. Queensland Fisheries Service Research Bulletin 2. Brisbane: Queensland Fisheries Service.

Druffel, E. R. M. and S. Griffin. 1993. Large variations of surface ocean radiocarbon: evidence of circulation changes in the southwestern Pacific. *Journal of Geophysical Research* 98: 20249–59.

Druffel, E. R. M. and S. Griffin. 1995. Regional variability of surface ocean radiocarbon from southern Great Barrier Reef corals. *Radiocarbon* 37: 517–24.

Druffel, E. R. M. and S. Griffin. 1999. Variability of surface ocean radiocarbon and stable isotopes in the southwestern Pacific. *Journal of Geophysical Research* 104: 23607–13.

Duncum, C. 1991. Archaeological Appraisal of the Southeastern Portion of the Proposed Coral Cove Development, near Bundaberg (Bundaberg KE), Central Queensland, 6 December 1991. Unpublished report to Gutteridge Haskins and Davey.

Dye, T. 1994. Apparent ages of marine shells: implications for archaeological dating in Hawai'i. *Radiocarbon* 36: 51–7.

Eales, T. 1998. Stone Soup: A Residue Analysis of Artefacts from Roof Fall Cave, Cania Gorge, Central Queensland. Unpublished BA(Hons) thesis, Department of Anthropology and Sociology, University of Queensland, Brisbane.

Eales, T., C. Westcott, I. Lilley, S. Ulm, D. Brian and C. Clarkson. 1999. Roof Fall Cave, Cania Gorge: site report. *Queensland Archaeological Research* 11: 29–42.

Elkin, A. P. 1949. The origins and interpretation of petroglyphs in south-eastern Australia. *Oceania* 20: 119–57.

Ellis, P. and W. Whitaker. 1976. *Geology of the Bundaberg 1:250 000 Sheet Area*. Geological Survey of Queensland Report 90. Brisbane: Queensland Department of Mines.

Endean, R., R. Kenny and W. Stephenson. 1956. The ecology and distribution of intertidal organisms on the rocky shores of the Queensland mainland. *Australian Journal of Marine and Freshwater Research* 7 (1): 88–146.

Erlandson, J. M. 1988. The role of shellfish in prehistoric economies: a protein perspective. *American Antiquity* 53 (1): 102–9.

Erlandson, J. M. 1991. Shellfish and seeds as optimal resources: early Holocene subsistence on the Santa Barbara coast. In J. M. Erlandson and R. H. Colten (eds), *Hunter-Gatherers of Early Holocene Coastal California*, pp. 89–100. Perspectives in California Archaeology 1. Los Angeles: Institute of Archaeology, University of California.

Erlandson, J. M. and M. L. Moss. 1999. The systematic use of radiocarbon dating in archaeological surveys in coastal and other erosional environments. *American Antiquity* 64 (1): 431–43.

Evans, R. 1991. *'A Permanent Precedent': Dispossession, Social Control and the Fraser Island Reserve and Mission, 1897–1904*. Ngulaig 5. Brisbane: Aboriginal and Torres Strait Islander Studies Unit, University of Queensland.

Fairholme, J. K. E. 1856. The blacks of Moreton Bay and the porpoises. *Proceedings of the London Zoological Society* 24: 353–4.

Fanning, P. and S. Holdaway. 2001. Temporal limits to the archaeological record in arid western NSW, Australia: lessons from OSL and radiocarbon dating of hearths and sediments. In M. Jones and P. Sheppard (eds), *Australasian Connections and New Directions: Proceedings of the 7th Australasian Archaeometry Conference*, pp. 85–104. Auckland: Department of Anthropology, University of Auckland.

Findlay, J. L. 1979. Results of Research Completed on Lithic Material from Bagara, near Bundaberg, South-East Queensland. Unpublished report to Department of Anthropology and Sociology, University of Queensland, Brisbane.

Flinders, M. 1814. *A Voyage to Terra Australis*. 2 vols. London: G. and W. Nichol.

Flood, J. 1980. *The Moth Hunters*. Canberra: Australian Institute of Aboriginal Studies.

Flood, J. 1999. *Archaeology of the Dreamtime: The Story of Prehistoric Australia and its People*. 3rd ed. Sydney: Angus and Robertson.

Flood, J., B. David, J. Magee and B. English. 1987. Birrigai: a Pleistocene site in the south-eastern highlands. *Archaeology in Oceania* 22 (1): 9–26.

Flood, J. and N. Horsfall. 1986. Excavations of Green Ant and Echidna Shelters, Cape York Peninsula. *Queensland Archaeological Research* 3: 4–64.

Francis, V. 1999. A Residue Analysis of a Sample of Stone Artefacts from the Southern Curtis Coast Region. Unpublished report submitted for ID233 Independent Project in Aboriginal and Torres Strait Islander Studies II. Brisbane: Aboriginal and Torres Strait Islander Studies Unit, University of Queensland.

Frankel, D. 1993. Pleistocene chronological structures and explanations: a challenge. In M. A. Smith, M. Spriggs and B. Fankhauser (eds), *Sahul in Review: Pleistocene Archaeology in Australia, New Guinea and Island Melanesia*, pp. 24–33. Occasional Papers in Prehistory 24. Canberra: Department of Prehistory, Research School of Pacific Studies, Australian National University.

Frankel, D. 1995. The Australian transition: real and perceived boundaries. *Antiquity* 69: 649–55.

Frankland, K. 1990. Booral: A Preliminary Investigation of an Archaeological Site in the Great Sandy Region, Southeast Queensland. Unpublished BA(Hons) thesis, Department of Anthropology and Sociology, University of Queensland, Brisbane.

Freeman, S. 1993. A Preliminary Analysis of the Glass Artefacts Found on the Onkaparinga River Estuary. Unpublished BA(Hons) thesis, Flinders University, Adelaide.

Fresløv, J. and D. Frankel. 1999. Abundant fields?: a review of coastal archaeology in Victoria. In J. Hall and I. McNiven (eds), *Australian Coastal Archaeology*, pp. 239–54. Research Papers in Archaeology and Natural History 31. Canberra: Archaeology and Natural History Publications, Research School of Pacific and Asian Studies, Australian National University.

Friedman, G. M. 1959. Identification of carbonate minerals by staining methods. *Journal of Sedimentary Petrology* 29 (1): 87–97.

Fullagar, R. 1990. A reconstructed obsidian core from the Talasea excavations. *Australian Archaeology* 30: 79–80.

Fullagar, R. and J. Field. 1997. Pleistocene seed-grinding implements from the Australian arid zone. *Antiquity* 71: 300–7.

Fullagar, R., D. M. Price and L. M. Head. 1996. Early human occupation of northern Australia: archaeology and thermoluminescence dating of Jinmium rock-shelter, Northern Territory. *Antiquity* 70: 751–73.

Gamble, C. 1993 People on the move: interpretations of regional variation in Palaeolithic Europe. In J. Chapman and P. Dolukhanov (eds), *Cultural Transformations and Interactions in Eastern Europe*, pp. 37–55. Aldershot: Avebury.

Gaughwin, D. and R. Fullagar. 1995. Victorian offshore islands in a mainland coastal economy. *Australian Archaeology* 40: 38–50.

Genever, M., J. Grindrod and B. Barker. 2003. Holocene palynology of Whitehaven Swamp, Whitsunday Island, Queensland, and implications for the regional archaeological record. *Palaeogeography, Palaeoclimatology, Palaeoecology* 201: 141–56.

Gill, E. 1983. Australian sea levels in the last 15,000 years — Victoria, south-east Australia. In D. Hopley (ed.), *Australian Sea-Levels in the last 15000 Years: A Review*, pp. 59–63. Occasional Papers 3. Townsville: Department of Geography, James Cook University of North Queensland.

Gillespie, R. 1975. The Suitability of Marine Shells for Radiocarbon Dating. Unpublished PhD thesis, University of Sydney, Sydney.

Gillespie, R. 1977. Sydney University natural radiocarbon measurements IV. *Radiocarbon* 19: 101–10.

Gillespie, R. 1982. *Radiocarbon Users Handbook*. North Ryde, NSW: Quaternary Research Unit, Macquarie University.

Gillespie, R. 1991. The Australian marine shell correction factor. In R. Gillespie (ed.), *Quaternary Dating Workshop 1990*, p. 15. Canberra: Department of Biogeography and Geomorphology, Australian National University.

Gillespie, R. and H. A. Polach. 1979. The suitability of marine shells for radiocarbon dating of Australian prehistory. In R. Berger and H. Suess (eds), *Proceedings of the Ninth International Conference on Radiocarbon Dating*, pp. 404–21. Los Angeles: University of California Press.

Gillespie, R. and R. B. Temple. 1977. Radiocarbon dating of shell middens. *Archaeology and Physical Anthropology in Oceania* 12: 26–37.

Gillieson, D. S. and J. Hall. 1982. Bevelling bungwall bashers: a use-wear study from southeast Queensland. *Australian Archaeology* 14: 43–61.

Godfrey, M. C. S. 1989. Shell midden chronology in southwestern Victoria: reflections of change in prehistoric population and subsistence. *Archaeology in Oceania* 24 (2): 65–9.

Godwin, L. 1990. Cultural heritage. In J. McCosker, Eurimbula National Park Draft Management Plan. Unpublished report to Queensland Department of Environment and Heritage, Rockhampton.

Godwin, L. 1997. Little Big Men: alliance and schism in north-eastern New South Wales during the late Holocene. In P. McConvell and N. Evans (eds), *Archaeology and Linguistics: Aboriginal Australia in Global Perspective*, pp. 297–309. Melbourne: Oxford University Press.

Godwin, L. and S. Ulm. 2004. Report on an Assessment of the Cultural Heritage Values Associated with the Proposed Burrumba Village, Buxton, Southeastern Queensland. Unpublished report to Neolido Pty Ltd. Rockhampton, QLD: Central Queensland Cultural Heritage Management Pty Ltd.

Gorecki, P. 1991. Horticulturalists as hunter-gatherers: rock shelter usage in Papua New Guinea. In C. S. Gamble and W. A. Broismier (eds), *Ethnoarchaeological Approaches to Mobile Campsites*, pp. 237–62. Ann Arbor, MI: International Monographs in Prehistory.

Gorecki, P. 1995. The Burnett Shire Council Aboriginal Heritage Study: Stage 1. UQASU Report 268. Brisbane: University of Queensland Archaeological Services Unit.

Gowlett, J. A. J., R. E. M. Hedges, I. A. Law and C. Perry. 1987. Radiocarbon dates from the Oxford AMS system: archaeometry datelist 5. *Archaeometry* 29 (1): 125–55.

Grant, E. 1993. *Grant's Guide to Fishes*. 6th ed. Scarborough, QLD: E. M. Grant.

Growcott, V. and M. Taylor (eds) 1996 *A Short History of Miriam Vale Shire: The Birthplace of Queensland: From the Journals of Arthur Jeffery*. Miriam Vale, QLD: Miriam Vale Historical Society.

Hale, H. and N. B. Tindale. 1930. Notes on some human remains in the Lower Murray Valley, South Australia. *Records of the South Australian Museum* 4: 145–218.

Hall, J. 1980a. An Archaeological Assessment of the Alcan Smelter Site, Gladstone, Queensland. Unpublished report to James B. Croft and Associates, Sydney.

Hall, J. 1980b. Minner Dint: a recent Aboriginal midden on Moreton Island, southeastern Queensland. In P.K. Lauer (ed.), *Occasional Papers in Anthropology* 10: 94–112. St Lucia, QLD: Anthropology Museum, University of Queensland.

Hall, J. 1981. An Archaeological Assessment of the Lend Lease Corporation Coke Plant near Gladstone, Queensland. Unpublished report to Oceanics Australia Pty Ltd.

Hall, J. 1982. Sitting on the crop of the bay: an historical and archaeological sketch of Aboriginal settlement and subsistence in Moreton Bay, southeast Queensland. In S. Bowdler (ed.), *Coastal Archaeology in Eastern Australia: Proceedings of the 1980 Valla Conference on Australian Prehistory*, pp. 79–95. Occasional Papers in Prehistory 11. Canberra: Department of Prehistory, Research School of Pacific Studies, Australian National University.

Hall, J. 1985. An Initial Archaeological Survey of the Proposed Bucca Weir, Kolan River, S.E. Queensland. Unpublished report to Gutteridge Haskins and Davey.

Hall, J. 1999. The impact of sea level rise on the archaeological record of the Moreton region, southeast Queensland. In J. Hall and I. McNiven (eds), *Australian Coastal Archaeology*, pp. 169–84. Research Papers in Archaeology and Natural History 31. Canberra: Archaeology and Natural History Publications, Research School of Pacific and Asian Studies, Australian National University.

Hall, J. 2000. Fishing for fish — no wallabies: an unusual marine intensification strategy for the late Holocene settlement of Moreton Island, southeast Queensland. In A. Anderson and T. Murray (eds), *Australian Archaeologist: Collected Papers in Honour of Jim Allen*, pp. 201–16. Canberra: Coombs Academic Publishing, Australian National University.

Hall, J. and G. Bowen. 1989. An excavation of a midden complex at the Toulkerrie Oysterman's lease, Moreton Island, SE Queensland. *Queensland Archaeological Research* 6: 3–27.

Hall, J., S. Higgins and R. Fullagar. 1989. Plant residues on stone tools: a case for fernroot processing in S.E. Queensland via starch grain analysis. In W. Beck, L. Head and A. Clarke (eds), *Plants in Australian Archaeology*, pp. 136–60. Tempus 1. St Lucia, QLD: Anthropology Museum, University of Queensland.

Hall, J. and P. Hiscock. 1988. The Moreton Regional Archaeological Project (MRAP) — Stage II: an outline of objectives and methods. *Queensland Archaeological Research* 5: 4–24.

Hall, J. and I. Lilley. 1987. Excavation at the New Brisbane Airport site (LB:C69): evidence for early mid-Holocene coastal occupation in Moreton Bay, SE Queensland. *Queensland Archaeological Research* 4: 54–79.

Hall, J. and I. McNiven (eds) 1999. *Australian Coastal Archaeology*. Research Papers in Archaeology and Natural History 31. Canberra: Archaeology and Natural History Publications, Research School of Pacific and Asian Studies, Australian National University.

Hallam, S. 1977. Topographic archaeology and artifactual evidence. In R. V. S. Wright (ed.), *Stone Tools as Cultural Markers: Change, Evolution and Complexity*, pp. 169–77. Canberra: Australian Institute of Aboriginal Studies.

Hamlyn-Harris, R. 1915. Notes on an exhibit of a small Aboriginal 'camp' collection from near Bundaberg. *Proceedings of the Royal Society of Queensland* 27 (2): 103–4.

Hardy, B. and G. T. Garufi. 1998. Identification of woodworking on stone tools through residue and use-wear analyses: experimental results. *Journal of Archaeological Science* 25: 177–84.

Hatte, E. 1992. Archaeological and Anthropological Investigations of the Proposed Route of the Fibre Optic Link: Bundaberg to Ban Ban Springs. Unpublished report to OVE ARUP and Partners.

Hayne, M. and J. Chappell. 2001. Cyclone frequency during the last 5,000 yrs from Curacoa Island, Queensland. *Palaeogeography, Palaeoclimatology, Palaeoecology* 168: 201–19.

Head, J., R. Jones and J. Allen. 1983. Calculation of the 'marine reservoir effect' from the dating of shell-charcoal paired samples from an Aboriginal midden on Great Glennie Island, Bass Strait. *Australian Archaeology* 17: 99–112.

Head, L. 1983. Environment as artefact: a geographic perspective on the Holocene occupation of southwestern Victoria. *Archaeology in Oceania* 18 (2): 73–80.

Head, L. 1986. Palaeoecological contributions to Australian prehistory. *Archaeology in Oceania* 21 (2): 121–9.

Head, L. 1987. The Holocene prehistory of a coastal wetland system: Discovery Bay, southeastern Australia. *Human Ecology* 15 (4): 435–62.

Hedley, C. 1904. Studies on Australian mollusca: part 8. *Proceedings of the Linnean Society of New South Wales* 29: 182–212.

Hedley, C. 1906. The mollusca of Mast Head Reef, Capricorn Group, Queensland. *Proceedings of the Linnean Society of New South Wales* 31: 454–79.

Hedley, C. 1915. Presidential address. *Proceedings of the Royal Society of New South Wales* 49: 1–77.

Heinsohn, G. E. 1991. Dugongs. In R. Strahan (ed.), *The Australian Museum Complete Book of Australian Mammals*, pp. 474–6. North Ryde, NSW: Cornstalk Publishing.

Higgins, S. 1988. Starch Grain Differentiation of Archaeological Residues: A Feasibility Study. Unpublished BA(Hons), Department of Anthropology and Sociology, University of Queensland.

Higham, T. F. G. and A. G. Hogg. 1995. Radiocarbon dating of prehistoric shell from New Zealand and calculation of the ΔR value using fish otoliths. *Radiocarbon* 37: 409–16.

Higham, T. F. G. and A. G. Hogg. 1997. Evidence for late Polynesian colonization of New Zealand: University of Waikato radiocarbon measurements. *Radiocarbon* 39: 149–92.

Hill, I. W. 1978. An Archaeological Report on Boyne Island Sites and Road Corridors. Unpublished report to Archaeology Section, Department of Environment and Heritage, Brisbane.

Hill, K. D. 1992. A preliminary account of *Cycas* (Cycadaceae) in Queensland. *Telopea* 5: 177–205.

Hiscock, P. 1982. An Archaeological Assessment of Site 1, Awoonga Dam, Queensland. Unpublished report to Gladstone Area Water Board, Gladstone.

Hiscock, P. 1986a. Technological change in the Hunter River Valley and its implications for the interpretation of late Holocene change in Australia. *Archaeology in Oceania* 21 (1): 40–50.

Hiscock, P. 1986b. The conjoin sequence diagram: a method of describing conjoin sets. *Queensland Archaeological Research* 3: 159–66.

Hiscock, P. 1988. Developing a relative dating system for the Moreton Region: an assessment of prospects for a technological approach. *Queensland Archaeological Research* 5: 113–32.

Hiscock, P. 1993. Bondaian technology in the Hunter Valley, New South Wales. *Archaeology in Oceania* 28 (2): 65–76.

Hiscock, P. 1994. Technological responses to risk in Holocene Australia. *Journal of World Prehistory* 8: 267–92.

Hiscock, P. 1997. Archaeological evidence for environmental change in Darwin Harbour. In J. R. Hanley, G. Caswell, D. Megirian and H. K. Larson (eds), *Proceedings of the Sixth International Marine Biological Workshop: The Marine Flora and Fauna of Darwin Harbour, Northern Territory, Australia*, pp. 445–9. Darwin: Museums and Art Galleries of the Northern Territory and the Australian Marine Sciences Association.

Hiscock, P. 2001. Late Australian. In *Encyclopedia of Prehistory*, pp. 132–49. New York: Plenum Press.

Hiscock, P. in press. Australian point and core reduction viewed through refitting. In M. de Bie and U. Schurman (eds), *The Big Puzzle Revisited*. British Archaeological Reports.

Hiscock, P. and V. Attenbrow. 1998. Early Holocene backed artefacts from Australia. *Archaeology in Oceania* 33 (2): 49–62.

Hiscock, P. and P. Hughes. 2001. Prehistoric and World War II use of shell mounds in Darwin Harbour. *Australian Archaeology* 52: 41–5.

Hlinka, V., S. Ulm, T. Loy and J. Hall. 2002. The genetic speciation of archaeological fish bone: a feasibility study from southeast Queensland. *Queensland Archaeological Research* 13: 71–8.

Hofman, J. L. 1985. Middle Archaic ritual and shell midden archaeology: considering the significance of cremations. In T. R. Whyte, C. C. Boyd (Jr) and B. H. Riggs (eds), *Exploring Tennessee Prehistory*, pp. 1–27. Report of Investigations 42. Knoxville, TN: Department of Anthropology, University of Tennessee.

Hofman, J. L. 1986. Vertical movements of artifacts in alluvial and stratified deposits. *Current Anthropology* 27: 163–71.

Hofman, J. L. 1992. Putting the pieces together: an introduction to refitting. In J. L. Hofman and J. G. Enloe (eds), *Piecing Together the Past: Applications of Refitting Studies in Archaeology*, pp. 1–20. BAR International Series 578. Oxford: Tempvs Reparatvm Archaeological and Historical Associates Limited.

Hofman, J. L. and J. G. Enloe (eds) 1992. *Piecing Together the Past: Applications of Refitting Studies in Archaeology*. BAR International Series 578. Oxford: Tempvs Reparatvm Archaeological and Historical Associates Limited.

Hogg, A. G., T. F. G. Higham and J. Dahm. 1998. ^{14}C dating of modern marine and estuarine shellfish. *Radiocarbon* 40: 975–84.

Hogg, A. G., F. G. McCormac, T. F. G. Higham, P. J. Reimer, M. G. L. Baillie and J. G. Palmer. 2002. High-precision radiocarbon measurements of contemporaneous tree-ring dated wood from the British Isles and New Zealand: AD 1850–950. *Radiocarbon* 44 (3): 633–40.

Holdaway, S., P. C. Fanning, M. Jones, J. Shiner, D. C. Witter and G. Nicholls. 2002. Variability in the chronology of late Holocene Aboriginal occupation on the arid margin of southeastern Australia. *Journal of Archaeological Science* 29: 351–63.

Holdaway, S. and N. Porch. 1995. Cyclical patterns in the Pleistocene human occupation of southwest Tasmania. *Archaeology in Oceania* 30 (2): 74–82.

Holdaway, S. and N. Porch. 1996. Dates as data: an alternative approach to the construction of chronologies for Pleistocene sites in southwest Tasmania. In J. Allen (ed.), *Report of the Southern Forests Archaeological Project: Site Descriptions, Stratigraphies and Chronologies*, pp. 251–77. Vol. 1. Bundoora, VIC: La Trobe University Archaeological Publications.

Holdaway, S., D. Witter, P. Fanning, R. Musgrave, G. Cochrane, T. Doelman, S. Greenwood, D. Pigdon and J. Reeves. 1998. New approaches to open site spatial archaeology in Sturt National Park, New South Wales, Australia. *Archaeology in Oceania* 33 (1): 1–19.

Hopley, D. 1983. Evidence of 15,000 years of sea level change in tropical Queensland. In D. Hopley (ed.), *Australian Sea-Levels in the Last 15000 Years: A Review*, pp. 93–104. Occasional Paper 3. Townsville: Department of Geography, James Cook University of North Queensland.

Hopley, D. 1985. The Queensland coastline: attributes and issues. In J. H. Holmes (ed.), *Queensland: A Geographical Interpretation*, pp. 73–94. Brisbane: Booralong Publications.

Horsfall, N. 1979. An Analysis of Some Stone Artifacts Collected near Bundaberg. Unpublished report to Department of Anthropology and Sociology, University of Queensland, Brisbane.

Horsfall, N. and J. Findlay. 1979. Eloueras from Southeast Queensland: Museum Collections as an Archaeological Resource. Unpublished report to Queensland Museum, Brisbane.

Horsfall, N. 1987. Living in Rainforest: The Prehistoric Occupation of North Queensland's Humid Tropics. Unpublished PhD thesis, James Cook University, Townsville.

Horton, D. (ed.) 1994. *The Encyclopaedia of Aboriginal Australia: Aboriginal and Torres Strait Islander History, Society and Culture*. Canberra: Aboriginal Studies Press.

Hughes, P. J. and V. Djohadze. 1980. *Radiocarbon Dates from Archaeological Sites on the South Coast of New South Wales and the Use of Depth/Age Curves*. Occasional Papers in Prehistory 1. Canberra: Department of Prehistory, Australian National University.

Hughes, P. J. and R. J. Lampert. 1977. Occupational disturbance and types of archaeological deposit. *Journal of Archaeological Science* 4: 35–40.

Hughes, P. J. and R. J. Lampert. 1982. Prehistoric population change in southern coastal New South Wales. In S. Bowdler (ed.), *Coastal Archaeology in Eastern Australia: Proceedings of the 1980 Valla Conference on Australian Prehistory*, pp. 16–28. Occasional Papers in Prehistory 11. Canberra: Department of Prehistory, Research School of Pacific Studies, Australian National University.

Hutchet, B. M. J. 1991. Conjoins and challenges: a rejoiner to Packard. *Australian Archaeology* 32: 45–7.

Hutchings, P. and P. Saenger. 1987. *Ecology of Mangroves*. St Lucia, QLD: University of Queensland Press.

Inglis, G. J. 1992. Population Ecology and Epibiosis of the Sydney Cockle *Anadara trapezia*. Unpublished PhD thesis, University of Sydney, Sydney.

Ingram, B. L. 1998. Differences in radiocarbon age between shell and charcoal from a Holocene shellmound in northern California. *Quaternary Research* 49: 102–10.

Ingram, B. L. and J. R. Southon. 1996. Reservoir ages in eastern Pacific coastal and estuarine waters. *Radiocarbon* 38: 573–82.

Isaacs, J. 1987. *Bushfood: Aboriginal Food and Herbal Medicine*. Sydney: Ure Smith Press.

Johnson, I. 1979. The Getting of Data: A Case Study from the Recent Industries of Australia. Unpublished PhD thesis, Research School of Pacific and Asian Studies, Department of Prehistory, Australian National University, Canberra.

Jolly, L. 1994. *Gureng Gureng: A Language Program Feasibility Study*. Aboriginal and Torres Strait Islander Studies Unit Research Report Series 1. Brisbane: Aboriginal and Torres Strait Islander Studies Unit, University of Queensland.

Jones, K. 1987. Cunning conjoins: methodological considerations in refitting stone flakes. In G. K. Ward (ed.), *Archaeology at ANZAAS Canberra*, pp. 198–202. Rev. ed. Canberra: Canberra Archaeological Society.

Jones, M. and G. Nicholls. 2001. Reservoir offset models for radiocarbon calibration. *Radiocarbon* 43 (1): 119–24.

Jones, R. 1977. The Tasmanian paradox. In R. V. S. Wright (ed.), *Stone Tools as Cultural Markers: Change, Evolution and Complexity*, pp. 189–204. Canberra: Australian Institute of Aboriginal Studies.

Kelly, M. 1982. *A Practical Reference Source to Radiocarbon Dates Obtained from Archaeological Sites in Queensland*. Cultural Resource Management Monograph Series 4. Brisbane: Archaeology Branch, Department of Community Services.

Kelly, R. and L. Todd. 1988. Coming into the country: early Paleoindian hunting and mobility. *American Antiquity* 53 (2): 231–44.

Kennett, D. J., B. L. Ingram, J. M. Erlandson and P. L. Walker. 1997. Evidence for temporal fluctuations in marine radiocarbon reservoir ages in the Santa Barbara Channel, southern California. *Journal of Archaeological Science* 24: 1051–9.

Kippis, A. 1814. *A Narrative of the Voyages Round the World*. London: Carpenter and Son.

Kirch, P. V. 2001. A radiocarbon chronology for the Mussau Islands. In P. V. Kirch (ed.), *Lapita and its Transformations in Near Oceania: Archaeological Investigations in the Mussau Islands, Papua New Guinea, 1985–88*, pp. 196–236. Berkeley: Archaeological Research Facility, University of California at Berkeley.

Knight, J. 1990. A broken Juin knife from Yandan Creek: some implications. *Archaeology in Oceania* 25 (2): 68–74.

Knox, G. A. 1963. The biogeography and intertidal ecology of the Australasian coasts. *Oceanography and Marine Biology: An Annual Review* 1: 341–404.

Koike, H. 1979. Seasonal dating and the valve-pairing technique in shell-midden analysis. *Journal of Archaeological Science* 6: 63–74.

Krebs, J. R. 1989. *Ecological Methodology*. New York: Harper and Row.

Lamb, J. 2003. The Raw and the Cooked: A Study on the Effects of Cooking on Three Aboriginal Plant Foods Native to Southeast Queensland. Unpublished BA(Hons) thesis, School of Social Science, University of Queensland, Brisbane.

Lambeck, K. and M. Nakada. 1990. Late Pleistocene and Holocene sea-level change along the Australian coast. *Palaeogeography, Palaeoclimatology, Palaeoecology* 89: 143–76.

Lampert, R. J. and P. J. Hughes. 1974. Sea level change and Aboriginal coastal adaptations in southern New South Wales. *Archaeology and Physical Anthropology in Oceania* 9 (3): 226–35.

Lamprell, K. and J. Healy. 1998. *Bivalves of Australia*. Vol. 2. Leiden: Backhuys Publishers.

Lamprell, K. and T. Whitehead. 1992. *Bivalves of Australia*. Vol. 1. Bathurst: Crawford House Press.

Larcombe, P., R. M. Carter, J. Dye, M. K. Gagan and D. P. Johnson. 1995. New evidence for episodic post-glacial sea-level rise, central Great Barrier Reef, Australia. *Marine Geology* 127: 1–44.

Larson, M. L. and E. E. Ingbar. 1992. Perspectives on refitting: critique and a complementary approach. In J. L. Hofman and J. G. Enloe (eds), *Piecing Together the Past: Applications of Refitting Studies in Archaeology*, pp. 151–62. BAR International Series 578. Oxford: Tempvs Reparatvm Archaeological and Historical Associates Limited.

Lauer, P. K. 1977. Report of a preliminary ethnohistorical and archaeological survey of Fraser Island. In P. K. Lauer (ed.), *Occasional Papers in Anthropology* 8: 1–38. St Lucia, QLD: Anthropology Museum, University of Queensland.

Lauer, P. K. 1979. The museum's role in fieldwork: the Fraser Island study. In P. K. Lauer (ed.), *Occasional Papers in Anthropology* 9: 31–72. St Lucia, QLD: Anthropology Museum, University of Queensland.

Leach, H. M. 1984. Jigsaw: reconstructive lithic technology. In J. E. Ericson and B. A. Purdy (eds), *Prehistoric Quarries and Lithic Production*, pp. 107–18. Cambridge: Cambridge University Press.

Leavesley, M. and J. Allen. 1998. Dates, disturbance and artefact distributions: another analysis of Buang Merabak, a Pleistocene site on New Ireland, Papua New Guinea. *Archaeology in Oceania* 33 (2): 63–82.

Lilley, I. 1980. Report on the Archaeological Impact of Proposed Development of the Awoonga Dam, Boyne River, Central Queensland. Unpublished report to Gladstone Area Water Board, Gladstone.

Lilley, I. 1994a. An Archaeological Assessment of Proposed Sand-Mining on Middle Island, Coastal Central Queensland. UQASU Report 244. Brisbane: University of Queensland Archaeological Services Unit.

Lilley, I. 1994b. A Summary Assessment of Aboriginal Cultural Heritage Values in the Proposed Walla Weir and Wallaville Bridge Impact Areas, Bundaberg, Southeast Queensland. Unpublished report to Kinhill Cameron McNamara Pty Ltd.

Lilley, I. 1995a. An Archaeological Assessment of the Cultural Heritage Values of the Swindon Gold Prospect at Mt Rawdon, via Mt Perry, Central Queensland. Unpublished report to Placer Pacific Pty Ltd.

Lilley, I. 1995b. An Archaeological Assessment of the Aboriginal Archaeological Impact of the Proposed Gold Mine at Mt Rawdon, Central Queensland. Unpublished report to Placer Pacific Pty Ltd.

Lilley, I. 1995c. An Archaeological Assessment of the Aboriginal Cultural Heritage Values of the Proposed Mini Excavations Sand Extraction Project, Yarwun, Coastal Central Queensland. Unpublished report to Kinhill Cameron McNamara Pty Ltd.

Lilley, I. 1995d. An Archaeological Assessment of the Cultural Heritage Values of the Proposed Perry River Dam, Mt Rawdon, Central Queensland. Unpublished report to Placer Pacific Pty Ltd.

Lilley, I. 1995e. An Archaeological Assessment of the Cultural Heritage Values of the Proposed Agnes Water Sewage Treatment Plant and Irrigation Area, Agnes Water, Coastal Central Queensland. Unpublished report to Kinhill Cameron McNamara Pty Ltd.

Lilley, I., D. Brian, C. Clarkson and S. Ulm. 1998. Pleistocene Aboriginal occupation at Cania Gorge, central Queensland: preliminary results of fieldwork. *Archaeology in Oceania* 33 (1): 28–31.

Lilley, I., D. Brian and S. Ulm. 1999. The use of foraminifera in the identification and analysis of marine shell middens: a view from Australia. In M-J. Mountain and D. Bowdery (eds), *Taphonomy: The Analysis of Processes from Phytoliths to Megafauna*, pp. 9–16. Research Papers in Archaeology and Natural History 30. Canberra: Archaeology and Natural History Publications, Research School of Pacific and Asian Studies, Australian National University.

Lilley, I. and J. Hall. 1988. An Archaeological Assessment of Northeastern Moreton Island. UQASU Report 126. Brisbane: University of Queensland Archaeological Services Unit.

Lilley, I. and S. Ulm. 1995. The Gooreng Gooreng Cultural Heritage Project: some proposed directions and preliminary results of the archaeological program. *Australian Archaeology* 41: 11–5.

Lilley, I. and S. Ulm. 1999. The Gooreng Gooreng Cultural Heritage Project: preliminary results of archaeological research, 1993–1997. *Queensland Archaeological Research* 11: 1–14.

Lilley, I., S. Ulm and D. Brian. 1996. The Gooreng Gooreng Cultural Heritage Project: first radiocarbon determinations. *Australian Archaeology* 43: 38–40.

Lilley, I., M. Williams and S. Ulm. 1997. The Gooreng Gooreng Cultural Heritage Project: A Report on National Estate Grants Program Research, 1995–1996. 2 vols. Unpublished report to Australian Heritage Commission, Canberra.

Lindauer, O. 1992. Ceramic conjoinability: orphan sherds and reconstructing time. In J. L. Hofman and J. G. Enloe (eds), *Piecing Together the Past: Applications of Refitting Studies in Archaeology*, pp. 210–6. BAR International Series 578. Oxford: Tempvs Reparatvm Archaeological and Historical Associates Limited.

Little, E. A. 1993. Radiocarbon age calibration at archaeological sites of coastal Massachusetts and vicinity. *Journal of Archaeological Science* 20: 457–71.

Longmore, M. E. 1997a. The mid-Holocene 'dry' anomaly on the mid-eastern coast of Australia: calibration of palaeowater depth as a surrogate for effective precipitation using sedimentary loss on ignition in the perched lake sediments of Fraser Island, Queensland. *Palaeoclimates* 4: 1–26.

Longmore, M. E. 1997b. Quaternary palynological records from the perched lake sediments of Fraser Island, Queensland, Australia: rainforest, forest history and climatic control. *Australian Journal of Botany* 45: 507–26.

Longmore, M. E. and H. Heijnis. 1999. Aridity in Australia: Pleistocene records of palaeohydrological and palaeoecological change from the perched lake sediments of Fraser Island, Queensland, Australia. *Quaternary International* 57/58: 35–47.

Lourandos, H. 1980a. Forces of Change: Aboriginal Technology and Population in South-Western Victoria. Unpublished PhD thesis, Department of Anthropology, University of Sydney, Sydney.

Lourandos, H. 1980b. Change or stability?: hydraulics, hunter-gatherers and population in temperate Australia. *World Archaeology* 11 (3): 245–64.

Lourandos, H. 1983. Intensification: a late Pleistocene-Holocene archaeological sequence from southwestern Victoria. *Archaeology in Oceania* 18 (2): 81–94.

Lourandos, H. 1985. Intensification and Australian prehistory. In T. D. Price and J. A. Brown (eds), *Prehistoric Hunter-Gatherers: The Emergence of Cultural Complexity*, pp. 385–423. Orlando: Academic Press.

Lourandos, H. 1988. Palaeopolitics: resource intensification in Aboriginal Australia and Papua New Guinea. In T. Ingold, D. Riches and J. Woodburn (eds), *Hunters and Gatherers 1: History, Evolution and Social Change*, pp. 148–60. Vol. 1. New York: Berg.

Lourandos, H. 1993. Hunter-gatherer cultural dynamics: long- and short-term trends in Australian prehistory. *Journal of Archaeological Research* 1 (1): 67–88.

Lourandos, H. 1996. Change in Australian prehistory: scale, trends and frameworks of interpretation. In S. Ulm, I. Lilley and A. Ross (eds), *Australian Archaeology '95: Proceedings of the 1995 Australian Archaeological Association Annual Conference*, pp. 15–21. Tempus 6. St Lucia, QLD: Anthropology Museum, University of Queensland.

Lourandos, H. 1997. *Continent of Hunter-Gatherers: New Perspectives in Australian Prehistory*. Cambridge: Cambridge University Press.

Lourandos, H. and B. David. 1998. Comparing long-term archaeological and environmental trends: north Queensland, arid and semi-arid Australia. *The Artefact* 21: 105–14.

Lourandos, H. and A. Ross. 1994. The great 'intensification debate': its history and place in Australian archaeology. *Australian Archaeology* 39: 54–63.

Loy, T. H. 1994. Methods in the analysis of starch residues on prehistoric stone tools. In J. G. Hather (ed.), *Tropical Archaeobotany: Applications and New Developments*, pp. 86–114. London: Routledge.

Luebbers, R. A. 1978. Meals and Menus: A Study of Change in Prehistoric Coastal Settlements in South Australia. Unpublished PhD thesis, Research School of Pacific Studies, Australian National University, Canberra.

Lupton, C. J. and M. J. Heidenreich. 1996. *A Fisheries Resource Assessment of the Baffle Creek System in the Wide Bay-Burnett Region of Queensland*. Department of Primary Industries Information Series QI96055(a). Brisbane: Department of Primary Industries.

Lyman, R. L. 1994. *Vertebrate Taphonomy*. Cambridge: Cambridge University Press.

MacGillivray, J. 1852. *Narrative of the Voyage of H.M.S. Rattlesnake*. London: T. and W. Boone.

Mackenzie, E. 2002. Bushranger's Cave: A Technological Analysis in the Moreton Region of Southeast Queensland. Unpublished BA(Hons) thesis, School of Social Science, University of Queensland, Brisbane.

Mangerud, J. 1972. Radiocarbon dating of marine shells, including a discussion of apparent ages of Recent shells from Norway. *Boreas* 1: 143–72.

Mangerud, J. and S. Gulliksen. 1975. Apparent radiocarbon ages of recent marine shells from Norway, Spitsbergen, and arctic Canada. *Quaternary Research* 5: 263–73.

Marks, E. N. 1970. A List of Bora Grounds in South-East Queensland. Unpublished report to Bornong Project.

Marsh, H. and W. K. Saalfeld. 1989. The distribution and abundance of dugong in the southern Great Barrier Reef Marine Park. In H. Marsh (ed.), *Biological Basis for Managing Dugong and Other Large Vertebrates in the Great Barrier Reef Marine Park*. Research Publication 21. Townsville, QLD: Great Barrier Reef Marine Park Authority.

Mathew, J. H. 1914. Note on the Gurang Gurang tribe of Queensland, with vocabulary. *Proceedings of the Australian Association for the Advancement of Science* 14: 433–43.

Mathews, R. H. 1897. Rock carvings and paintings of the Australian Aborigines. *Proceedings of the American Philosophical Society* 36: 466–87.

Mathews, R. H. 1910. Some rock pictures and ceremonial stones of the Australian Aborigines. *Report of the Twelfth Meeting of the Australasian Association for the Advancement of Science Held in Brisbane, 1909* 12: 493–8.

Maynard, L. 1976. An Archaeological Approach to the Study of Australian Rock Art. Unpublished MA thesis, University of Sydney, Sydney.

Maynard, L. 1979. The archaeology of Australian Aboriginal art. In S. M. Mead (ed.), *Exploring the Visual Art of Oceania: Australia, Melanesia, Micronesia and Polynesia*, pp. 83–110. Honolulu: University of Hawaii Press.

McCormac, F. G., A. G. Hogg, T. F. G. Higham, M. G. L. Baillie, J. G. Palmer, L. Xiong, J. R. Pilcher, D. Brown and S. T. Hoper. 1998. Variations of radiocarbon in tree rings: southern hemisphere offset preliminary results. *Radiocarbon* 40 (3): 1153–9.

McCormac, F. G., P. J. Reimer, A. G. Hogg, T. F. G. Higham, M. G. L. Baillie, J. Palmer and M. Stuiver. 2002. Calibration of the radiocarbon time scale for the southern hemisphere: AD 1850–950. *Radiocarbon* 44 (3): 641–51.

McDonald, L. 1988. *Gladstone: City that Waited*. Brisbane: Boolarong Publications.

McNiven, I. 1984. Initiating Archaeological Research in the Cooloola Region, Southeast Queensland. Unpublished BA(Hons) thesis, Department of Anthropology and Sociology, University of Queensland, Brisbane.

McNiven, I. 1985. An archaeological survey of the Cooloola region, S.E. Queensland. *Queensland Archaeological Research* 2: 4–37.

McNiven, I. 1988. Brooyar Rockshelter: a late Holocene seasonal hunting camp from southeast Queensland. *Queensland Archaeological Research* 5: 133–60.

McNiven, I. 1989. Aboriginal shell middens at the mouth of the Maroochy River, southeast Queensland. *Queensland Archaeological Research* 6: 28–52.

McNiven, I. 1990a. Prehistoric Aboriginal Settlement and Subsistence in the Cooloola Region, Coastal Southeast Queensland. Unpublished PhD thesis, Department of Anthropology and Sociology, University of Queensland, Brisbane.

McNiven, I. 1990b. Blowout taphonomy: non-cultural associations between faunal and stone artefact assemblages along the Cooloola coast, southeast Queensland. *Australian Archaeology* 31: 67–74.

McNiven, I. 1991a. Teewah Beach: new evidence for Holocene coastal occupation in coastal southeast Queensland. *Australian Archaeology* 33: 14–27.

McNiven, I. 1991b. Settlement and subsistence activities along Tin Can Bay, southeast Queensland. *Queensland Archaeological Research* 8: 85–107.

McNiven, I. 1992a. Sandblow sites in the Great Sandy Region, coastal southeast Queensland: implications for models of late Holocene rainforest exploitation and settlement restructuring. *Queensland Archaeological Research* 9: 1–16.

McNiven, I. 1992b. Bevel-edged tools from coastal southeast Queensland. *Antiquity* 66: 701–9.

McNiven, I. 1993. Corroboree Beach, Fraser Island: Archaeological Survey and Management Recommendations. 3 vols. Unpublished report to Queensland Department of Environment and Heritage, Maryborough.

McNiven, I. 1994a. *'Relics of a By-Gone Race'?: Managing Aboriginal Sites in the Great Sandy Region*. Ngulaig 12. Brisbane: Aboriginal and Torres Strait Islander Studies Unit, University of Queensland.

McNiven, I. 1994b. Technological organization and settlement in southwest Tasmania after the glacial maximum. *Antiquity* 68: 75–82.

McNiven, I. 1998. Aboriginal archaeology of the Corroboree Beach dune field, Fraser Island: re-survey and re-assessment. *Memoirs of the Queensland Museum, Cultural Heritage Series* 1 (1): 1–22.

McNiven, I. 1999. Fissioning and regionalisation: the social dimensions of changes in Aboriginal use of the Great Sandy Region, southeast Queensland. In J. Hall and I. McNiven (eds), *Australian Coastal Archaeology*, pp. 157–68. Research Papers in Archaeology and Natural History 31. Canberra: Archaeology and Natural History Publications, Research School of Pacific and Asian Studies, Australian National University.

McNiven, I. 2003. Saltwater people: spiritscapes, maritime rituals and the archaeology of Australian indigenous seascapes. *World Archaeology* 35 (3): 329–49.

McNiven, I. and R. Feldman. 2003. Ritually orchestrated seascapes: hunting magic and dugong bone mounds in Torres Strait, NE Australia. *Cambridge Archaeological Journal* 13 (2): 169–94.

McNiven, I. and P. Hiscock. 1988. Small unifacial pebble cores from Fraser Island, southeast Queensland. *Queensland Archaeological Research* 5: 161–5.

McNiven, I., I. Thomas and U. Zoppi. 2002. Fraser Island Archaeological Project (FIAP): background, aims and preliminary results of excavations at Waddy Point 1 Rockshelter. *Queensland Archaeological Research* 13: 1–20.

Meehan, B. 1982. *Shell Bed to Shell Midden*. Canberra: Australian Institute of Aboriginal Studies.

Meehan, B. 1988. The 'dinnertime camp'. In B. Meehan and R. Jones (eds), *Archaeology with Ethnography: An Australian Perspective*, pp. 171–81. Canberra: Department of Prehistory, Research School of Pacific Studies, Australian National University.

Meltzer, D. J. and J. I. Mead. 1985. Dating Late Pleistocene extinctions: theoretical issues, analytical bias, and substantive results. In J. I. Mead and D. J. Meltzer (eds), *Environments and Extinctions: Man in Late Glacial North America*, pp. 145–73. Orono, ME: Centre for the Study of Early Man, University of Maine at Orono.

Minnegal, M. 1982. Dugong Processing as an Archaeological Phenomenon: Evidence from a Small Complex of Sites at Princess Charlotte Bay, North Queensland. Unpublished BA(Hons) thesis, Department of Anthropology and Sociology, University of Queensland, Brisbane.

Mitchell, S. 1993. Shell mound formation in northern Australia: a case study from Croker Island, northwestern Arnhem Land. *The Beagle, Records of the Northern Territory Museum of Arts and Sciences* 10 (1): 179–92.

Morrison, M. 2003. Old boundaries and new horizons: the Weipa shell mounds reconsidered. *Archaeology in Oceania* 38 (1): 1–8.

Morse, K. 1988. Mandu Mandu Creek rockshelter: Pleistocene human coastal occupation of North West Cape, Western Australia. *Archaeology in Oceania* 23 (3): 81–8.

Morse, K. 1993. Who can see the sea?: prehistoric Aboriginal occupation of the Cape Range Peninsula. In W. F. Humphreys (ed.), *The Biogeography of Cape Range, Western Australia*, pp. 227–42. Records of the Western Australian Museum 45. Perth: Western Australian Museum.

Morwood, M. 1979. Art and Stone: Towards a Prehistory of Central Western Queensland. Unpublished PhD thesis, Faculty of Arts, Department of Archaeology and Anthropology, Australian National University, Canberra.

Morwood, M. 1981. Archaeology in the Central Queensland Highlands: the stone component. *Archaeology in Oceania* 16 (1): 1–52.

Morwood, M. 1984. The prehistory of the central Queensland Highlands. *Advances in World Archaeology* 3: 325–79.

Morwood, M. 1986. The archaeology of art: excavations at Gatton and Maidenwell Rockshelters, S.E. Queensland. *Queensland Archaeological Research* 3: 88–132.

Morwood, M. 1987. The archaeology of social complexity in south-east Queensland. *Proceedings of the Prehistoric Society* 53: 337–50.

Morwood, M. and D. Hobbs (eds) 1995. *Quinkan Prehistory: The Archaeology of Aboriginal Art in S.E. Cape York Peninsula, Australia*. Tempus 3. St Lucia, QLD: Anthropology Museum, University of Queensland.

Mowat, F. M. 1995. Variability in Western Arnhem Land Shell Midden Deposits. Unpublished MA thesis, Department of Anthropology, Northern Territory University, Darwin.

Muckle, R. 1985. Archaeological Considerations of Bivalve Shell Taphonomy. Unpublished MA thesis, Department of Anthropology, Simon Fraser University, Vancouver.

Mulvaney, D. 1969. *The Prehistory of Australia*. London: Thames and Hudson.

Mulvaney, D. and J. Golson (eds) 1971. *Aboriginal Man and Environment in Australia*. Canberra: Australian National University Press.

Mulvaney, J. and E. Joyce. 1965. Archaeological and geomorphological investigations on Mt. Moffat Station, Queensland, Australia. *Proceedings of the Prehistoric Society* 31: 147–212.

Mulvaney, J. and J. Kamminga. 1999. *Prehistory of Australia*. St Leonards, NSW: Allen and Unwin.

Murray-Jones, S. 1999. Conservation and conservation biology of the pipi, *Donax deltoides*. Unpublished PhD thesis, School of Biological Sciences, University of Wollongong, Wollongong.

Murray-Wallace, C. V. 1996. Understanding 'deep' time — advances since Archbishop Ussher? *Archaeology in Oceania* 31 (3): 173–7.

Murray-Wallace, C. V., A. G. Beu, G. W. Kendrick, L. J. Brown, A. P. Belperio and J. E. Sherwood. 2000. Palaeoclimatic implications of the occurrence of the arcoid bivalve *Anadara trapezia* (Deshayes) in the Quaternary of Australasia. *Quaternary Science Reviews* 19: 559–90.

Neal, R. 1984. Rescue Archaeology near Blue Lake, North Stradbroke Island. Unpublished report to Associated Minerals Consolidated Ltd and the Archaeology Branch, Department of Community Services.

Neal, R. 1986. Results of the Archaeological Inspection of Proposed Telecom DRCS and Broadband Locations for December 1986. Unpublished report to Telecom Australia.

Neal, R. and E. Stock. 1986. Pleistocene occupation in the south-east Queensland coastal region. *Nature* 323: 618–21.

Nicholson, A. and S. Cane. 1991. Desert camps: analysis of Australian Aboriginal proto-historic campsites. In C. S. Gamble and W. A. Broismier (eds), *Ethnoarchaeological Approaches to Mobile Campsites*, pp. 263–354. Ann Arbor, MI: International Monographs in Prehistory.

Nicholson, A. and S. Cane. 1994. Pre-European coastal settlement and use of the sea. *Australian Archaeology* 39: 108–17.

Nott, J. and M. Hayne. 2001. High frequency of 'super cyclones' along the Great Barrier Reef over the past 5,000 years. *Nature* 413: 508–12.

O'Connor, S. 1999. *30,000 Years of Aboriginal Occupation: Kimberley, North West Australia*. Terra Australis 14. Canberra: Research School of Pacific and Asian Studies, Australian National University.

O'Connor, S. and M. Sullivan. 1994a. Distinguishing middens and cheniers: a case study from the southern Kimberley, Western Australia. *Archaeology in Oceania* 29 (1): 16–28.

O'Connor, S. and M. Sullivan. 1994b. Coastal archaeology in Australia; developments and new directions. *Australian Archaeology* 39: 87–96.

Olsen, H.F. 1980a. Estuarine resource inventory and evaluation for the coastal strip between Round Hill Head and Tannum Sands, Queensland. In H. F. Olsen, R. M. Dowling and D. Bateman, *Biological Resources Investigation (Estuarine Inventory)*, pp. 1–44. Queensland Fisheries Service Research Bulletin 2. Brisbane: Queensland Fisheries Service.

Olsen, H. F. 1980b. Sea-grasses (occurrence and distribution). In H. F. Olsen, R. M. Dowling and D. Bateman, *Biological Resources Investigation (Estuarine Inventory)*, pp. 91–4. Queensland Fisheries Service Research Bulletin 2. Brisbane: Queensland Fisheries Service.

O'Sullivan, A. 2003. Place, memory and identity among estuarine fishing communities: interpreting the archaeology of early medieval fish weirs. *World Archaeology* 35 (3): 449–68.

Oxley, J. 1825. Report of an expedition to survey Port Curtis, Moreton Bay and Port Bowen. In B. Field (ed.), *Geographical Memoirs of New South Wales*, pp. 1–26. London: John Murray.

Parkington, J. and G. Mills. 1991. From space to place: the architecture and social organization of Southern African mobile communities. In C. S. Gamble and W. A. Broismier (eds), *Ethnoarchaeological Approaches to Mobile Campsites*, pp. 355–70. Ann Arbor, MI: International Monographs in Prehistory.

Parkinson, S. 1773. *A Journal of a Voyage to the South Seas*. London: Stanfield Parkinson.

Parry, W. J. and R. L. Kelly. 1987. Expedient core technology and sedentism. In J. K. Johnson and C. A. Morrow (eds), *The Organisation of Core Technology*, pp. 285–304. Boulder, CO: Westview Press.

Peacock, E. 2000. Assessing bias in archaeological shell assemblages. *Journal of Field Archaeology* 27: 185–96.

Peck, L. S. and T. Brey. 1996. Bomb signals in old Antarctic brachiopods. *Nature* 380: 207–8.

Perry, T. M. and D. H. Simpson (eds) 1962. *Drawings by William Westall, Landscape Artist on Board H.M.S. Investigator during the Circum-Navigation of Australia by Captain Matthew Flinders, R.N., in 1801–1803*. London: Royal Commonwealth Society.

Petrie, C. C. 1904. *Tom Petrie's Reminiscences of Early Queensland*. Brisbane: Watson, Ferguson and Co.

Pettitt, P. B., W. Davies, C. S. Gamble and M. B. Richards. 2003. Palaeolithic radiocarbon chronology: quantifying our confidence beyond two half-lives. *Journal of Archaeological Science* 30: 1685–93.

Phelan, M. B. 1999. A ΔR correction value for Samoa from known-age marine shells. *Radiocarbon* 41: 99–101.

Phillips, P. and J. A. Brown. 1978. *Pre-Columbian Shell Engravings from the Craig Mound at Spiro, Oklahoma*. Cambridge, MS: Peabody Museum Press.

Przywolnik, K. 2003. Shell artefacts from northern Cape Range Peninsula, northwest Western Australia. *Australian Archaeology* 56: 12–21.

Pye, K. and E. G. Rhodes. 1985. Holocene development of an episodic transgressive dune barrier, Ramsay Bay, north Queensland, Australia. *Marine Geology* 64: 189–202.

QDEH. 1994. *Curtis Coast Study: Resource Report*. Rockhampton, QLD: Queensland Department of Environment and Heritage.

QDEH. 1997. Vegetation of Rodds Peninsula [map]. 1cm–1km. Rockhampton, QLD: Queensland Department of Environment and Heritage.

QDOT. 1998. *The Official Tide Tables and Boating Safety Guide 1998*. Brisbane: Queensland Department of Transport.

Quinn, R. H. and J. P. Beumer. 1984. Wallum Creek — a study of the regeneration of mangroves. In R. J. Coleman, J. Covacevich and P. Davie (eds), *Focus on Stradbroke: New Information on North Stradbroke Island and Surrounding Areas, 1974–1984*, pp. 238–59. Brisbane: Boolarong Publications.

Quinnell, M. C. 1976. Aboriginal Rock Art in Carnarvon Gorge, South Central Queensland. Unpublished MA thesis, Department of Prehistory and Archaeology, University of New England, Armidale.

Raven, P., R. Evert and S. Eichorn. 1999. *Biology of Plants*. 6th ed. New York: Worth Publishers.

Reid, J. 1997. Results and Analysis of E1: An Investigation of the Archaeological Record of the Eurimbula Shell Midden Complex, Central Queensland Coast. Unpublished report submitted for ID232 Independent Project in Aboriginal and Torres Strait Islander Studies I. Brisbane: Aboriginal and Torres Strait Islander Studies Unit, University of Queensland.

Reid, J. 1998. An Archaeological Approach to Quarry Studies: A Technological Investigation of the Ironbark Site Complex, Southern Curtis Coast, Australia. Unpublished BA(Hons) thesis, Department of Anthropology and Sociology, University of Queensland, Brisbane.

Reid, J., M. Williams and S. Ulm. 2000. An Archaeological Assessment of the Cultural Heritage Values of Lot 20, Captain Cook Drive, Town of Seventeen Seventy, Central Queensland. UQASU Report 331. Brisbane: University of Queensland Archaeological Services Unit.

Reimer, P. J. and R. W. Reimer. 2000. Marine reservoir correction database. Retrieved from http://calib.org/marine.

Reimer, P. J. and R. W. Reimer. 2001. A marine reservoir correction database and on-line interface. *Radiocarbon* 43 (2A): 461–3.

Reitz, E. J. and E. S. Wing. 1999. *Zooarchaeology*. Cambridge: Cambridge University Press.

Rhodes, E. G., H. A. Polach, B. G. Thom and S. R. Wilson. 1980. Age structure of Holocene coastal sediments: Gulf of Carpentaria, Australia. *Radiocarbon* 22: 718–27.

Richardson, N. 1992. Conjoin sets and stratigraphic integrity in a sandstone shelter: Kenniff Cave (Queensland, Australia). *Antiquity* 66: 408–18.

Richardson, N. 1996. Seeing is believing: a graphical illustration of the vertical and horizontal distribution of conjoined artefacts using DesignCAD 3D. In S. Ulm, I. Lilley and A. Ross (eds), *Australian Archaeology '95: Proceedings of the 1995 Australian Archaeological Association Annual Conference*, pp. 81–95. Tempus 6. St Lucia, QLD: Anthropology Museum, University of Queensland.

Richter, J. 1994. A Pound of Bungwall and Other Measures. Unpublished BA(Hons) thesis, Department of Anthropology and Sociology, University of Queensland, Brisbane.

Rick, J. W. 1987. Dates as data: an examination of the Peruvian Preceramic radiocarbon record. *American Antiquity* 52 (1): 55–73.

Ringland, P. 1978. Survey of Facing Island. Unpublished report to Archaeology Section, Department of Environment and Heritage, Brisbane.

Roberts, A. C. 1991. An Analysis of Mound Formation at Milingimbi, N.T. Unpublished MLitt thesis, Department of Archaeology and Palaeoanthropology, University of New England, Armidale.

Robertson, G. R. 1994. An Application of Scanning Electron Microscopy and Image Analysis to the Differentiation of Starch Grains in Archaeological Plant Residues. Unpublished BA(Hons) thesis, Department of Anthropology and Sociology, University of Queensland, Brisbane.

Robins, R. P. 1983. This Widow Land: An Evaluation of Public Archaeology in Queensland using Moreton Island as a Case Study. Unpublished MA thesis, Department of Anthropology and Sociology, University of Queensland, Brisbane.

Robins, R. P. 1984. The excavation of three archaeological sites on Moreton Island: First Ridge and the Little Sandhills. *Queensland Archaeological Research* 1: 51–60.

Robins, R. P. 1999. Clocks for rocks: an archaeological perspective on the Currawinya Lakes. In R.T. Kingsford (ed.), *A Free-Flowing River: The Ecology of the Paroo River*, pp. 150–78. Hurstville, NSW: National Parks and Wildlife Service.

Robins, R. P. and E. C. Stock. 1990. The burning question: a study of molluscan remains from a midden on Moreton Island. In S. Solomon, I. Davidson and D. Watson (eds), *Problem Solving in Taphonomy: Archaeological and Palaeontological Studies from Europe, Africa and Oceania*, pp. 80–100. Tempus 2. St Lucia, QLD: Anthropology Museum, University of Queensland.

Robins, R. P., E. C. Stock and D. S. Trigger. 1998. Saltwater people, saltwater country: geomorphological, anthropological and archaeological investigations of the coastal lands in the southern Gulf country of Queensland. *Memoirs of the Queensland Museum, Cultural Heritage Series* 1 (1): 75–125.

Robinson, S. W. and G. Thompson. 1981. Radiocarbon corrections for marine shell dates with application to southern Pacific Northwest Coast prehistory. *Syesis* 14: 45–57.

Rola-Wojeiechowski, C. 1983. 'A Bit of Bundy': The 'Bundaberg' Engraving Site. Unpublished BA(Hons) thesis, Department of Prehistory and Archaeology, University of New England, Armidale.

Ross, A. 1981. Holocene environments and prehistoric site patterning in the Victorian Mallee. *Archaeology in Oceania* 16 (3): 145–54.

Ross, A., B. Anderson and C. Campbell. 2003. Gunumbah: archaeological and Aboriginal meanings at a quarry site on Moreton Island, southeast Queensland. *Australian Archaeology* 57: 75–81.

Ross, A. and S. Coghill. 2000. Conducting a community-based archaeological project: an archaeologist's and a Koenpul man's perspective. *Australian Aboriginal Studies* 1&2: 76–83.

Ross, A., B. Coghill and S. Coghill. 2000. Results of the Lazaret Midden excavation on Peel Island in a landscape and seascape context. Unpublished paper presented at the Department of Sociology, Anthropology and Archaeology, University of Queensland, Brisbane.

Ross, A. and R. Duffy. 2000. The creation of evidence: sieve size and the evidence for fishing in the Lazaret Midden, Peel Island, Moreton Bay. *Geoarchaeology* 15 (1): 21–41.

Ross, J. 2002. Rocking the boundaries, scratching the surface: an analysis of the relationship between paintings and engravings in the central Australian arid zone. In S. Ulm, C. Westcott, J. Reid, A. Ross, I. Lilley, J. Prangnell and L. Kirkwood (eds), *Barriers, Borders, Boundaries: Proceedings of the 2001 Australian Archaeological Association Annual Conference*, pp. 83–9. Tempus 7. Brisbane: Anthropology Museum, University of Queensland.

Roth, W. E. 1898. The Aborigines of the Rockhampton and Surrounding Districts, A Report to the Commissioner of Police. Unpublished manuscript, Mitchell Library, Sydney.

Roth, W. E. 1904. *Domestic Implements, Arts, and Manufactures*. North Queensland Ethnography Bulletin 7. Brisbane: Department of Public Lands.

Roth, W. E. 1909. *Fighting Weapons*. North Queensland Ethnography Bulletin 13. Brisbane: Department of Lands.

Roughley, T. C. 1928. The Oyster Resources of Queensland. Unpublished report to Queensland Department of Harbours and Marine, Brisbane.

Rowland, M. J. 1981. Radiocarbon dates for a shell fish-hook and disc from Mazie Bay, North Keppel Island. *Australian Archaeology* 12: 63–9.

Rowland, M. J. 1982. Further radiocarbon dates from the Keppel Islands. *Australian Archaeology* 15: 43–8.

Rowland, M. J. 1983. Aborigines and environment in Holocene Australia: changing paradigms. *Australian Aboriginal Studies* 2: 62–77.

Rowland, M. J. 1985. Further radiocarbon dates from Mazie Bay, North Keppel Island. *Australian Archaeology* 21: 113–8.

Rowland, M. J. 1986. The Whitsunday Islands: initial historical and archaeological observations and implications for future work. *Queensland Archaeological Research* 3: 72–87.

Rowland, M. J. 1987. Preliminary Archaeological Survey of Coastal Areas of the Bundaberg 1:250,000 sheet (KE). Unpublished report to Queensland Department of Environment and Heritage, Brisbane.

Rowland, M. J. 1989. Population increase, intensification or a result of preservation?: explaining site distribution patterns on the coast of Queensland. *Australian Aboriginal Studies* 2: 32–41.

Rowland, M. J. 1992. Conservation Plan for Cultural Heritage Sites on the Keppel Island Group, Central Queensland. Unpublished report to Livingstone Shire Council and National Parks and Wildlife Branch, Division of Conservation, Department of Environment and Heritage, Brisbane.

Rowland, M. J. 1994. Size isn't everything: shells in mounds, middens and natural deposits. *Australian Archaeology* 39: 118–24.

Rowland, M. J. 1996. Prehistoric archaeology of the Great Barrier Reef Province — retrospect and prospect. In P. Veth and P. Hiscock (eds), *Archaeology of Northern Australia*, pp. 191–212. Tempus 4. St Lucia, QLD: Anthropology Museum, University of Queensland.

Rowland, M. J. 1999 Holocene environmental variability: Have its impacts been underestimated in Australian prehistory? *The Artefact* 22:11–48.

Rowland, M. J. 2002. 'Crows', swimming logs and auditory exostoses: isolation on the Keppel Islands and broader implications. In S. Ulm, C. Westcott, J. Reid, A. Ross, I. Lilley, J. Prangnell and L. Kirkwood (eds), *Barriers, Borders, Boundaries: Proceedings of the 2001 Australian Archaeological Association Annual Conference*, pp. 61–74. Tempus 7. Brisbane: Anthropology Museum, University of Queensland.

Sattler, P. S. 1999. Introduction. In P. S. Sattler and R. D. Williams (eds), *The Conservation Status of Queensland's Bioregional Ecosystems*, pp. 1–19. Brisbane: Environmental Protection Agency.

Shanco, P and R. Timmins. 1975. Reconnaissance of southern Bustard Bay tidal wetlands. *Operculum* October: 149–54.

Simpson, E. H. 1949. Measurement of diversity. *Nature* 163: 688.

Singer, C. A. 1984. The 63-kilometre fit. In J. E. Ericson and B. A. Purdy (eds), *Prehistoric Quarries and Lithic Production*, pp. 35–48. Cambridge: Cambridge University Press.

Smith, A. D. 1992. An Archaeological Site Location and Subsistence-Settlement Analysis of Bribie Island, Southeast Queensland. Unpublished BA(Hons) thesis, Department of Anthropology and Sociology, University of Queensland, Brisbane.

Smith, A. D. 2003. Archaeological Spatial Variability on Bribie Island, Southeast Queensland. Unpublished MA thesis, School of Social Science, University of Queensland, Brisbane.

Smith, M. 1996. Revisiting Pleistocene Macrozamia. *Australian Archaeology* 42: 52–3.

Smith, M. 1999. Southwest Australian coastal economies: a new review. In J. Hall and I. McNiven (eds), *Australian Coastal Archaeology*, pp. 15–24. Research Papers in Archaeology and Natural History 31. Canberra: Archaeology and Natural History Publications, Research School of Pacific and Asian Studies, Australian National University.

Smith, M. A. 1986. The antiquity of seedgrinding in central Australia. *Archaeology in Oceania* 21 (1): 29–39.

Smith, M. A. 1993. Biogeography, human ecology and prehistory in the sandridge deserts. *Australian Archaeology* 37: 35–50.

Smith, M. A. and N. D. Sharp. 1993. Pleistocene sites in Australia, New Guinea and island Melanesia: geographic and temporal structure of the archaeological record. In M. A. Smith, M. Spriggs and B. Fankhauser (eds), *Sahul in Review: Pleistocene Archaeology in Australia, New Guinea and Island Melanesia*, pp. 37–59. Occasional Papers in Prehistory 24. Canberra: Department of Prehistory, Research School of Pacific Studies, Australian National University.

Smith, P. 1980. Letter to Kate Sutcliffe, Officer in Charge, Archaeology Branch, Department of Aboriginal and Islanders Advancement, 2 September 1980.

Specht, J. 1985. Crabs as disturbance factors in tropical archaeological sites. *Australian Archaeology* 21: 11–8.

Spencer, T. 1995. A Cultural Heritage Survey of Three Mining Lease Application Areas — Upper Burnett River near Monto Southern Queensland. Unpublished report to Lewis Environmental Consultants.

Spennemann, D. H. R. and M. J. Head. 1996. Reservoir modification of radiocarbon signatures in coastal and near-shore waters of eastern Australia: the state of play. *Quaternary Australasia* 14 (1): 32–9.

Spennemann, D. H. R. and M. J. Head. 1998. Togan pottery chronology, ^{14}C dates and the hardwater effect. *Quaternary Geochronology* 17: 1047–56.

Spriggs, M. and A. Anderson. 1993. Late colonization of East Polynesia. *Antiquity* 67: 200–17.

Stein, J. K., J. Deo and L. Phillips. 2003. Big sites-short time: accumulation rates in archaeological sites. *Journal of Archaeological Science* 30: 297–316.

Stern, N. 1980. Taphonomy: Some Observations about its Place in Archaeology. Unpublished BA(Hons) thesis, Department of Prehistoric and Historical Archaeology, University of Sydney, Sydney.

Stevens, N. 1968. *Triassic Volcanic Rocks of Agnes Water, Queensland.* University of Queensland Papers, Department of Geology 6 (6): 147–55.

Stockton, J. 1982. Stone wall fish-traps in Tasmania. *Australian Archaeology* 14: 107–14.

Stone, T. 1989. Origins and environmental significance of shell and earth mounds in northern Australia. *Archaeology in Oceania* 24 (2): 59–64.

Stone, T. 1992. Origins of the Weipa Shell Mounds. Unpublished MSc thesis, Australian National University, Canberra.

Stone, T. 1995. Shell mound formation in coastal northern Australia. *Marine Geology* 129: 77–100.

Stuiver, M. and T. F. Braziunas. 1993. Modeling atmospheric ^{14}C influences and ^{14}C ages of marine samples to 10,000 BC. *Radiocarbon* 35 (1): 137–89.

Stuiver, M., G. W. Pearson and T. Braziunas. 1986. Radiocarbon age calibration of marine samples back to 9000 cal yr BP. *Radiocarbon* 28: 980–1021.

Stuiver, M. and H. A. Polach. 1977. Discussion: reporting of ^{14}C data. *Radiocarbon* 19: 355–63.

Stuiver, M. and P. J. Reimer. 1993. Extended ^{14}C data base and revised CALIB 3.0 ^{14}C age calibration program. *Radiocarbon* 35 (1): 215–30.

Stuiver, M., P. J. Reimer, E. Bard, J. W. Beck, G. S. Burr, K. A. Hughen, B. Kromer, G. McCormac, J. van der Plicht and M. Spurk. 1998a. INTCAL98 Radiocarbon age calibration, 24,000–0 cal BP. *Radiocarbon* 40: 1041–83.

Stuiver, M., P. J. Reimer and T. F. Braziunas. 1998b. High-precision radiocarbon age calibration for terrestrial and marine samples. *Radiocarbon* 40: 1127–51.

Stuiver, M., P. J. Reimer and R. Reimer. 2002. CALIB manual. Retrieved from http://depts.washington.edu /qil/calib/manual/index.html.

Sullivan, A. P., J. M. Skibo and M. Van Buren. 1991. Sherd refitting and the reconstruction of household ceramic technology. *North American Anthropologist* 12 (3): 243–55.

Sullivan, G. E. 1961. Functional morphology, micro-anatomy, and histology of the 'Sydney cockle', *Anadara trapezia* (Deshayes) (Lamellibranchia: Archidae). *Australian Journal of Zoology* 9: 219–57.

Sullivan, M. 1987. The recent prehistoric exploitation of edible mussel in Aboriginal shell middens in southern New South Wales. *Archaeology in Oceania* 22 (2): 97–106.

Sullivan, M. and S. O'Connor. 1993. Middens and cheniers: implications for Australian research. *Antiquity* 67: 776–88.

Sutcliffe, K. A. 1972. Removal of Rock Engravings from the Burnett River, Queensland. Unpublished MA thesis (Draft), Department of Anthropology and Sociology, University of Queensland, Brisbane.

Sutcliffe, K. A. 1974. Removal of Aboriginal Rock Engravings: Burnett River. *Anthropological Society of Queensland Newsletter* 65.

Tanaka, N., M. C. Monaghan and D. M. Rye. 1986. Contribution of metabolic carbon to mollusc and barnacle shell carbonate. *Nature* 320: 520–3.

Taylor, J. C. 1967. Race Relations in Southeast Queensland, 1840–1860. Unpublished BA(Hons) thesis, Department of History, University of Queensland, Brisbane.

Taylor, R. E. 1987. *Radiocarbon Dating: An Archaeological Perspective.* New York: Academic Press.

Taylor, R. E. 1997. Radiocarbon dating. In R. E. Taylor and M. J. Aitken (eds), *Chronometric Dating in Archaeology*, pp. 65–96. New York: Plenum Press.

Thom, B. G., G. M. Bowman, R. Gillespie, R. Temple and M. Barbetti. 1981. *Radiocarbon Dating of Holocene Beach-Ridge Sequences in South-East Australia.* Monograph 11. Duntroon, NSW: Department of Geography, University of New South Wales at Royal Military College.

Thom, B. G. and P. S. Roy. 1983. Sea-level change in New South Wales over the past 15,000 years. In D. Hopley (ed.), *Australian Sea-Levels in the Last 15000 Years: A Review*, pp. 64–84. Occasional Papers 3. Townsville: Department of Geography, James Cook University of North Queensland.

Tindale, N. N. 1974. *Aboriginal Tribes of Australia.* Berkeley: University of California Press.

Todd, L. C. and D. J. Stanford. 1992. Application of conjoined bone data to site structural studies. In J. L. Hofman and J. G. Enloe (eds), *Piecing Together the Past: Applications of Refitting Studies in Archaeology*, pp. 21–35. BAR International Series 578. Oxford: Tempvs Reparatvm Archaeological and Historical Associates Limited.

Ulm, S. 1995. Fishers, Gatherers and Hunters on the Moreton Fringe: Reconsidering the Prehistoric Aboriginal Marine Fishery in Southeast Queensland, Australia. Unpublished BA(Hons) thesis, Department of Anthropology and Sociology, University of Queensland, Brisbane.

Ulm, S. 2000a. A Desktop Archaeological Assessment of the Cultural Heritage Values of the Aldoga Industrial Area, Gladstone State Development Area, Central Queensland. Unpublished report to Kinhill Pty Ltd, Milton.

Ulm, S. 2000b. Evidence for early focussed marine resource exploitation from an open coastal site in central Queensland. *Australian Archaeology* 51: 66–7.

Ulm, S. 2001. An Interim Desktop Assessment of the Cultural Heritage Values of the Proposed Cable and Wireless Optus Low-Impact Telecommunications Facility at Mt Alma, Central Queensland. UQASU Report 347. Brisbane: University of Queensland Archaeological Services Unit.

Ulm, S. 2002a. Reassessing marine fishery intensification in southeast Queensland. *Queensland Archaeological Research* 13: 79–96.

Ulm, S. 2002b. Marine and estuarine reservoir effects in central Queensland, Australia: determination of ΔR values. *Geoarchaeology* 17 (4): 319–48.

Ulm, S. 2002c. The Seven Mile Creek Mound: new evidence for mid-Holocene Aboriginal marine resource exploitation in central Queensland. *Proceedings of the Royal Society of Queensland* 110: 121–6.

Ulm, S. 2002d. Calibrating marine radiocarbon dates: A guide to Australian ΔR values. *AACAI Newsletter* 89:10–4.

Ulm, S. 2004a. Investigations Towards a Late Holocene Archaeology of Aboriginal Lifeways on the Southern Curtis Coast, Australia. Unpublished PhD thesis, School of Social Science, University of Queensland, Brisbane.

Ulm, S. 2004b. Themes in the archaeology of mid-to-late Holocene Australia. In T. Murray (ed.), *Archaeology from Australia*, pp. 187–208. Melbourne: Australian Scholarly Publishing.

Ulm, S., B. Barker, A. Border, J. Hall, I. Lilley, I. McNiven, R. Neal and M. Rowland. 1995. Pre-European coastal settlement and use of the sea: a view from Queensland. *Australian Archaeology* 41: 24–6.

Ulm, S., M. Carter, J. Reid and I. Lilley. 1999a. Eurimbula Site 1, Curtis Coast: site report. *Queensland Archaeological Research* 11: 105–22.

Ulm, S., S. Cotter, M. Cotter, I. Lilley, C. Clarkson and J. Reid. 2005. Edge-ground hatchets on the southern Curtis Coast, central Queensland: a preliminary assessment of technology, chronology and provenance. In I. Macfarlane (ed.), *Many Exchanges: Archaeology, History, Community and the Work of Isabel McBryde*, pp. 323–342. Aboriginal History Monograph II. Canberra: Aboriginal History Inc.

Ulm, S., T. Eales and S. L'Estrange. 1999b. Post-European Aboriginal occupation of the southern Curtis Coast, central Queensland. *Australian Archaeology* 48: 42–3.

Ulm, S. and J. Hall. 1996. Radiocarbon and cultural chronologies in southeast Queensland prehistory. In S. Ulm, I. Lilley and A. Ross (eds), *Australian Archaeology '95: Proceedings of the 1995 Australian Archaeological Association Annual Conference*, pp. 45–62. Tempus 6. St Lucia, QLD: Anthropology Museum, University of Queensland.

Ulm, S. and I. Lilley. 1999. The archaeology of the southern Curtis Coast: an overview. *Queensland Archaeological Research* 11: 59–84.

Ulm, S. and J. Reid. 2000. Index of dates from archaeological sites in Queensland. *Queensland Archaeological Research* 12: 1–129.

Ulm, S. and J. Reid. 2004. Index of dates from archaeological sites in Queensland: Upgrade version 1.4. Retrieved from http://www.atsis.uq.edu.au/resources/index-of-dates/index.html.

Vale, D. 2002. A Report on the Analysis of Archaeological Fishbone Assemblages Retrieved from Shell Midden Sites on the Southern Curtis Coast of Central Queensland. Unpublished report to Aboriginal and Torres Strait Islander Studies Unit, University of Queensland, Brisbane.

Vale, D. 2004. A Report on the Analysis of Archaeological Fishbone Assemblages Retrieved from Shell Midden Sites on the Southern Curtis Coast of Central Queensland: Part II. Unpublished report to Aboriginal and Torres Strait Islander Studies Unit, University of Queensland, Brisbane.

Vale, D. and R. Gargett. 2002. Size matters: 3-mm sieves do not increase richness in a fishbone assemblage from Arrawarra I, an Aboriginal Australian shell midden on the mid-north coast of New South Wales, Australia. *Journal of Archaeological Science* 29: 57–63.

Vanderwal, R. L. 1978. Adaptive technology in southeast Tasmania. *Australian Archaeology* 8: 107–26.

Veitch, B. 1996. Evidence for mid-Holocene change in the Mitchell Plateau, northwest Kimberley, Western Australia. In P. Veth and P. Hiscock (eds), *Archaeology of Northern Australia*, pp. 66–89. Tempus 4. St Lucia, QLD: Anthropology Museum, University of Queensland.

Veitch, B. 1999. Shell middens on the Mitchell Plateau: a reflection of a wider phenomenon? In J. Hall and I. McNiven (eds), *Australian Coastal Archaeology*, pp. 51–64. Research Papers in Archaeology and Natural History 31. Canberra: Archaeology and Natural History Publications, Research School of Pacific and Asian Studies, Australian National University.

Vernon, K. 1999. Post-Contact Cultural Continuity in Central Queensland: Bottle Bases and Retouched Glass Artefacts: A Use-Wear and Residue Analysis of Tom's Creek Glass Assemblage and the Ironbark Ridge Artefacts. Unpublished report submitted for ID233 Independent Project in Aboriginal and Torres Strait Islander Studies II. Brisbane: Aboriginal and Torres Strait Islander Studies Unit, University of Queensland.

Veth, P. 1993. *Islands in the Interior: The Dynamics of Prehistoric Adaptations within the Arid Zone of Australia*. Ann Arbor, MI: International Monographs in Prehistory.

Veth, P. 2003. 'Abandonment' or maintenance of country?: a critical examination of mobility patterns and implications for Native Title. *Land, Rights, Laws: Issues of Native Title* 2 (22): 1–8.

Villa, P. 1982. Conjoinable pieces and site formation processes. *American Antiquity* 47 (2): 276–90.

Walsh, G. L. 1984. *Managing the Archaeological Sites of the Sandstone Belt*. Rockhampton, QLD: Central Queensland Aboriginal Corporation for Cultural Activities and the Queensland National Parks and Wildlife Service.

Walters, I. 1984. Gone to the dogs: a study of bone attrition at a central Australian campsite. *Mankind* 14 (5): 398–400.

Walters, I. 1985. Bone loss: one explicit quantitative guess. *Current Anthropology* 26: 642–3.

Walters, I. 1986. Another Kettle of Fish: The Prehistoric Moreton Bay Fishery. Unpublished PhD thesis, Department of Anthropology and Sociology, University of Queensland, Brisbane.

Walters, I. 1989. Intensified fishery production at Moreton Bay, southeast Queensland, in the late Holocene. *Antiquity* 63: 215–24.

Walters, I. 1992a. Antiquity of marine fishing in south-east Queensland. *Queensland Archaeological Research* 9: 35–7.

Walters, I. 1992b. Farmers and their fires, fishers and their fish: production and productivity in pre-European south-east Queensland. *Dialectical Anthropology* 17: 167–82.

Walters, I. 1992c. Seasonality of fishing in south-east Queensland. *Queensland Archaeological Research* 9: 29–34.

Walters, I. 2001. The Aboriginal Moreton Bay fishery: two things I still know — a response to Ross and Coghill. *Australian Aboriginal Studies* 2: 61–2.

Walters, I., P. Lauer, A. Nolan, G. Dillon and M. Aird. 1987. Hope Island: salvage excavation of a Kombumerri site. *Queensland Archaeological Research* 4: 80–95.

Walthall, J. A. 1998. Rockshelters and hunter-gatherer adaptation to the Pleistocene/Holocene transition. *American Antiquity* 63 (2): 223–38.

Wang, H. and M. van Strydonck. 1997. Chronology of Holocene cheniers and oyster reefs on the coast of Bohai Bay, China. *Quaternary Research* 47: 192–205.

Ward, G. K. 1994. On the use of radiometric determinations to 'date' archaeological events. *Austalian Aboriginal Studies* 2: 106–9.

Ward, G. K. and S. R. Wilson. 1978. Procedures for comparing and combining radiocarbon age determinations: a critique. *Archaeometry* 20: 19–31.

Ward, W. T. and I. P. Little. 2000. Sea-rafted pumice on the Australian east coast: numerical classification and stratigraphy. *Australian Journal of Earth Sciences* 47 (1): 95–109.

Watkins, G. 1891. Notes on the Aboriginals of Stradbroke and Moreton Islands. *Proceedings of the Royal Society of Queensland* 8: 40–51.

Westcott, C. 1997. A Technological Analysis of the Stone Assemblage from Big Foot Art Site, Cania Gorge. Unpublished BA(Hons) thesis, Department of Anthropology and Sociology, University of Queensland, Brisbane.

Westcott, C., I. Lilley and S. Ulm. 1999a. The archaeology of Cania Gorge: an overview. *Queensland Archaeological Research* 11: 15–28.

Westcott, C., I. Lilley, S. Ulm, C. Clarkson and D. Brian. 1999b. Big Foot Art Site, Cania Gorge: site report. *Queensland Archaeological Research* 11: 43–58.

White, J. P. 1971. New Guinea and Australian prehistory: the 'Neolithic Problem'. In D. J. Mulvaney and J. Golson (eds), *Aboriginal Man and Environment in Australia*, pp. 182–95. Canberra: Australian National University Press.

White, J. P. 1994. Australia: the different continent. In G. Burenhult (ed.), *People of the Stone Age: Hunter-Gatherers and Early Farmers*, pp. 206–25. Illustrated History of Humankind 2. St Lucia, QLD: University of Queensland Press.

White, J. P. and J. F. O'Connell. 1982. *A Prehistory of Australia, New Guinea and Sahul*. North Ryde, NSW: Academic Press.

Williams, E. 1988. *Complex Hunter-Gatherers: A Late Holocene Example from Temperate Australia*. BAR International Series 423. Oxford: British Archaeological Reports.

Williams, M. 1981. Traditionally, My Country and its People. Unpublished MPhil (Qual.) thesis, Griffith University, Brisbane.

Wilson, B. R. and K. Gillett. 1979. *A Field Guide to Australian Shells: Prosobranch Gastropods*. Frenchs Forest, NSW: Reed Books.

Woodall, P. F. 1991. Eurimbula National Park excursion, 1989. *Queensland Naturalist* 30 (5–6): 110–1.

Woodall, P. F., L. B. Woodall, K. Lamprell, D. Potter and T. Whitehead. 1991. Molluscs from Eurimbula National Park. *Queensland Naturalist* 30 (5–6): 112–4.

Woodall, P. F., L. B. Woodall and T. Whitehead. 1993. Molluscs from Kinkuna and Deepwater National Parks, central coastal Queensland. *Queensland Naturalist* 32 (3–4): 66–70.

Woodroffe, C. D., J. M. A. Chappell, B. G. Thom and E. Wallensky. 1986. *Geomorphological Dynamics and Evolution of the South Alligator Tidal River and Plains, Northern Territory*. Mangrove Monograph 3. Darwin: North Australian Research Unit, Australian National University.

Woodroffe, C. D. and M. E. Mulrennan. 1993. *Geomorphology of the Lower Mary River Plains Northern Territory.* Darwin: Northern Australian Research Unit, Australian National University and Conservation Commission of the Northern Territory.

Young, P. A. R. and H. A. Dillewaard. 1999. Southeast Queensland. In P. S. Sattler and R. D. Williams (eds), *The Conservation Status of Queensland's Bioregional Ecosystems*, pp. 12/1–12/75. Brisbane: Environmental Protection Agency.

Zobel, D. E., R. L. Vanderwal and D. Frankel. 1984. The Moonlight Head Rockshelter. *Proceedings of the Royal Society of Victoria* 96: 1–24.

APPENDICES

Appendix 1: Radiocarbon dates: technical data

SITE	SQUARE	XU	DEPTH (cm)	LAB. NO.	SAMPLE	WEIGHT (g)	d14C (‰)	δ13C (‰)	D14C (‰)	% MODERN	14C AGE
Agnes Beach Midden	–	–	–	Wk-10969	charcoal	1.6	-36.7±8.5	-27.1±0.2	-32.6±10.4	96.7±1.0	266±87
Agnes Beach Midden	–	–	–	Wk-11280	D. deltoides	35.0	-30.4±4.6	0.8±0.2	-80.4±5.4	92.0±0.5	674±47
Elliott Heads	–	–	–	Wk-6994	D. deltoides	19.6	-0.2±6.1	-0.6±0.2	-49.1±7.1	95.1±0.7	400±60
Eurimbula Creek 1	C	6	14.9-18.3	Wk-7680	charcoal	3.4	-29.9±5.6	-26.1±0.2	-27.7±6.9	97.2±0.7	230±60
Eurimbula Creek 2	A	6	13.1-16.3	Wk-7689	charcoal	2.8	-22.1±6.5	-25.7±0.2	-20.8±8.0	97.9±0.8	modern[a]
Eurimbula Site 1	1	5	9.5	Wk-5601	charcoal	2.5	-30.8±7.6	-27.0±0.2	-26.9±9.3	97.3±0.9	220±80
Eurimbula Site 1	1 (SL)	10	35	Wk-3944	A. trapezia	71.1	-219.8±4.5	-0.8±0.2	-257.6±5.2	74.2±0.5	2390±60
Eurimbula Site 1	1 (SL)	10	35	Wk-5215	charcoal	2.1	-181.3±12.7	-25.3±0.2	-180.8±15.5	81.9±1.5	1600±160
Eurimbula Site 1	2	9	50	Wk-3945	charcoal	10.3	-315.3±4.4	-26.5±0.2	-313.3±5.3	68.7±0.5	3020±70
Eurimbula Site 1	3	7	28.4-34.1	Wk-8553	A. trapezia	20.2	-158.5±5.1	-0.6±0.2	-199.5±6.0	80.1±0.6	1790±60
Eurimbula Site 1	4	4	15-20	Wk-8554	A. trapezia	19.9	-19.6±5.4	-0.9±0.2	-66.9±6.2	93.3±0.6	560±55
Eurimbula Site 1	near 7	surface	0	Wk-3946	A. trapezia	90.7	-17.7±4.8	0.0±0.2	-66.8±5.6	93.3±0.6	560±50
Eurimbula Site 1	7	5	18.8-24	Wk-8555	A. trapezia	21.0	-3.9±5.9	-0.4±0.2	-52.8±6.9	94.7±0.7	440±60
Eurimbula Site 1	A	5	9.7-12.4	Wk-10967	charcoal	1.2	-46.1±11.8	-25.0±0.2	-46.1±14.3	95.4±1.4	379±121
Eurimbula Site 1	A	17	43.7-46.6	Wk-7688	charcoal	4.6	-258.2±5.0	-25.5±0.2	-257.5±6.1	74.2±0.6	2390±70
Eurimbula Site 1	B	12	34.4-38	Wk-10968	charcoal	1.3	-242.8±9.6	-26.0±0.2	-241.3±11.8	75.9±1.2	2218±126
Eurimbula Site 1	D	15	45.4-47.9	Wk-7687	charcoal	2.8	-291.5±7.6	-24.7±0.2	-291.9±9.3	70.8±0.9	2770±110
Gladstone 1	–	–	–	Wk-8456	A. trapezia	11.5	-8.0±4.9	0.3±0.2	-58.2±5.7	94.2±0.6	480±50
Gladstone 2	–	–	–	NZA-12119[b]	A. trapezia	4.6	4.6±6.8	-0.8±0.2	-44.0±6.5	95.6±0.6	360±60
Ironbark Site Complex	M	4	5.4-10.5	Wk-6359	charcoal	4.4	-81.0±5.2	-26.9±0.2	-77.5±6.4	92.3±0.6	650±60
Ironbark Site Complex	M	9	22.9-28.1	Wk-6360	charcoal	4.1	-161.0±5.0	-25.7±0.2	-159.9±6.1	84.0±0.6	1400±60
Ironbark Site Complex	M	17	60-69.3	Wk-6361	charcoal	1.8	-186.2±12.3	-26.2±0.2	-184.3±15.0	81.6±1.5	1640±150
Ironbark Site Complex	O	9a	27.4	Wk-8556	A. trapezia	16.7	-60.7±5.4	-0.5±0.2	-106.7±6.3	89.3±0.6	910±55
Ironbark Site Complex	P	7	16.3	Wk-8557	charcoal	1.0	-26.9±13.8	-26.0±0.2	-25.1±16.8	97.5±1.7	200±140
Ironbark Site Complex	P	7	17.6	Wk-8558	A. trapezia	20.1	-22.8±6.1	-0.3±0.2	-71.1±7.1	92.9±0.7	590±60
Ironbark Site Complex	R	9	17.5-20.4	Wk-10964	charcoal	1.3	-38.9±8.6	-26.8±0.2	-35.5±10.5	96.4±1.1	290±89
Ironbark Site Complex	core	–	25-30	OZD-756[b]	organics	–	–	-25[d]	–	97.4±0.6	215±55
Middle Island Sandblow Site	A	1	0	Wk-7679	D. deltoides	35.0	-66.5±4.8	1.1±0.2	-115.2±5.5	88.5±0.6	980±50
Middle Island Sandblow Site	B	1	0	Wk-10091	D. deltoides	32.3	-37±3.9	0.9±0.2	-86.8±4.5	91.3±0.5	730±39
Middle Island Sandblow Site	C	1	0	Wk-10092	D. deltoides	34.1	-63.4±3.8	1.2±0.2	-112.4±4.5	88.8±0.4	958±40
Middle Island Sandblow Site	D	1	0	Wk-10093	D. deltoides	34.2	-16.2±4.2	0.9±0.2	-67.2±4.9	93.3±0.5	559±42

continued over

Appendix 1: continued

SITE	SQUARE	XU	DEPTH (cm)	LAB. NO.	SAMPLE	WEIGHT (g)	d14C (‰)	δ13C (‰)	D14C (‰)	% MODERN	14C AGE
Mort Creek Site Complex	A7	4	18-20	Wk-5602	A. trapezia	47.3	-264.7±3.7	-0.3±0.2	-301.0±4.3	69.9±0.4	2880±50
Mort Creek Site Complex	A7	6	22.6-26.7	Wk-3937	A. trapezia	75.2	-269.3±4.0	0.1±0.2	-305.9±4.7	69.4±0.5	2930±60
Mort Creek Site Complex	A7	9	32.4-37	Wk-3938	A. trapezia	81.2	-249.3±4.3	0.1±0.2	-286.9±5.0	71.3±0.5	2720±60
Mort Creek Site Complex	Granites	11C	45.5-52.1	Wk-3940	mixed shell[c]	66.7	-296.9±4.4	0.7±0.2	-333.1±5.1	66.7±0.5	3260±70
Mort Creek Site Complex	Granites	11M	45.5-52.1	Wk-3941	A. trapezia	71.3	-246.4±4.5	-0.2±0.2	-283.8±5.3	71.6±0.5	2680±60
Mort Creek Site Complex	WP	4	12.8-18.4	Wk-3942	A. trapezia	79.6	-222.3±5.7	0.6±0.2	-262.2±6.6	73.8±0.7	2440±80
Mort Creek Site Complex	WP	10	37.6-44.8	Wk-3943	A. trapezia	74.8	-235.9±4.4	-0.5±0.2	-273.4±5.1	72.7±0.5	2570±60
Mort Creek Site Complex	C	6	22	Wk-7458	charcoal	2.4	-219.7±6.4	-26.5±0.2	-217.5±7.8	78.3±0.8	1970±80
Mort Creek Site Complex	C	6	22	Wk-7836	A. trapezia	39.2	-213.3±4.1	-1.4±0.2	-250.4±4.8	75.0±0.5	2320±50
Mort Creek Site Complex	C	7	25	Wk-6987	A. trapezia	45.9	-208.2±3.9	-1.5±0.2	-245.3±4.6	75.5±0.5	2260±50
Mort Creek Site Complex	C	18	60	Wk-6988	A. trapezia	8.3	-310.1±6.2	-1.1±0.2	-343.1±7.1	65.7±0.7	3380±90
Mort Creek Site Complex	B	19-20	65	Wk-6986	A. trapezia	6.0	-315.3±9.8	-1.6±0.2	-347.3±11.3	65.3±1.1	3430±140
Pancake Creek Site Complex	A	9	14.3-18.6	Wk-7837	A. trapezia	35.6	-34.3±5.2	-1.1±0.2	-80.5±6.1	92.0±0.6	670±50
Pancake Creek Site Complex	E	7	25	Wk-6989	A. trapezia	5.4	-55.0±12.1	-0.1±0.2	-102.1±14.0	89.8±1.4	870±130
Pancake Creek Site Complex	F	6	25	Wk-6990	A. trapezia	13.9	-27.8±6.3	-0.4±0.2	-75.6±7.3	92.4±0.7	630±70
Pancake Creek Site Complex	G	8	31	Wk-6991	A. trapezia	34.9	-38.9±5.1	0.5±0.2	-87.8±5.9	91.2±0.6	740±60
Pancake Creek Site Complex	H	8	26	Wk-6992	A. trapezia	7.2	-47.2±7.4	-0.3±0.2	-94.3±8.6	90.6±0.9	800±80
Pancake Creek Site Complex	H	8	26	Wk-6993	charcoal	1.2	-86.8±12.2	-26.8±0.2	-83.5±15.0	91.7±1.5	700±140
Port Curtis 1	–	–	–	Wk-8457	V. singaporina	8.8	-5.6±6.1	0.3±0.2	-56.0±7.1	94.4±0.7	460±60
Port Curtis 2	–	–	–	NZA-12120[b]	V. singaporina	1.7	-16.9±6.7	0.9±0.2	-67.9±6.4	93.2±0.6	570±60
Round Hill Creek Mound	–	–	–	Wk-10090	A. trapezia	37.7	-170.6±3.5	-0.3±0.2	-211.6±4.1	78.8±0.4	1910±42
Seven Mile Creek Mound	A	4	6.8-10.4	NZA-12272[b]	charcoal	<0.1	-146.4±8.5	-26.0±0.2	-144.7±8.5	85.5±0.9	1260±80
Seven Mile Creek Mound	A	4	7.14	Wk-8324	A. trapezia	17.5	-323.6±5.5	-0.9±0.2	-356.2±6.4	64.4±0.6	3540±80
Seven Mile Creek Mound	A	13	39-43.6	NZA-12117[b]	charcoal	<0.1	-354.2±4.5	-25.7±0.2	-353.3±4.5	64.7±0.5	3500±60
Seven Mile Creek Mound	A	13	40.4	Wk-8326	A. trapezia	19.5	-329.8±4.9	-0.8±0.2	-363.3±5.7	63.8±0.6	3610±70
Seven Mile Creek Mound	A	20	67.8	Wk-8327	A. trapezia	40.7	-344.2±3.8	-1.2±0.2	-375.4±4.4	62.5±0.4	3780±60
Seven Mile Creek Mound	A	20	67.8-71.5	NZA-12273[b]	charcoal	0.1	-356.6±4.6	-23.4±0.2	-358.7±4.6	64.1±0.5	3570±60
Seven Mile Creek Mound	A	26	88.2	Wk-8328	A. trapezia	33.0	-340.3±3.8	-0.5±0.2	-372.7±4.4	62.7±0.4	3750±60
Seven Mile Creek Mound	A	26	88.7-92.2	NZA-12118[b]	charcoal	0.2	-369.0±4.4	-27.8±0.2	-365.5±4.4	63.4±0.4	3660±60
Tom's Creek Site Complex	D	3	3.3	Wk-7682	A. trapezia	19.7	-28.1±4.9	-1.2±0.2	-74.4±5.7	92.6±0.6	620±50
Tom's Creek Site Complex	D	3	3.9	Wk-7681	charcoal	11.4	-9.6±4.8	-27.2±0.2	-5.1±5.9	99.5±0.6	modern[a]
Tom's Creek Site Complex	D	8	22.2-25.5	Wk-10966	charcoal	1.1	-34.3±12.3	-25.7±0.2	-32.9±14.9	96.7±1.5	269±125

continued over

Appendix 1: continued

SITE	SQUARE	XU	DEPTH (cm)	LAB. NO.	SAMPLE	WEIGHT (g)	d14C (‰)	δ13C (‰)	D14C (‰)	% MODERN	14C AGE
Tom's Creek Site Complex	D	15	50	Wk-7683	A. trapezia	26.7	-66.2±4.5	-1.2±0.2	-110.6±5.2	88.9±0.5	940±50
Tom's Creek Site Complex	D	17	55.7-60	Wk-7684	charcoal	3.2	-106.8±6.6	-26.8±0.2	-103.7±8.1	89.6±0.8	880±70
Tom's Creek Site Complex	D	18	59.5-64	Wk-7685	charcoal	3.3	-133.5±6.4	-27.5±0.2	-129.2±7.8	87.1±0.8	1110±70
Tom's Creek Site Complex	S	8	20.5-24	Wk-7686	charcoal	12.6	-65.1±3.9	-25.3±0.2	-64.5±4.8	93.5±0.5	540±50
Tom's Creek Site Complex	S	8	20.5-24	Wk-7838	A. trapezia	42.9	-29.1±5.3	-0.9±0.2	-75.9±6.1	92.4±0.6	630±50
Tom's Creek Site Complex	S	11	31.7-35	Wk-10965	charcoal	1.2	-127.1±10.2	-26.4±0.2	-124.7±12.4	87.5±1.2	1070±115
Tom's Creek Site Complex	S	-	62.5-67	NZA-13385[b]	organics	30.69	-218±5.5	-26.2±0.2	-216.2±5.5	78.4±0.6	1956±57
Worthington Creek Midden	-	-	5	Wk-10089	S. glomerata	32.4	0.7±4.5	-3.4±0.2	-42.5±5.2	95.7±0.5	349±60

a   The term 'modern' is applied for conventional radiocarbon ages of less than 200 years. Finite ages are problematic in this area of the radiocarbon time-scale owing to high levels of variability in radiocarbon activity in the atmosphere caused by the onset of the industrial revolution and atmospheric testing of thermonuclear devices. ^{14}C ages between 0 and 200 could give ages anywhere from AD 1750 to AD 1950. After 1950, bomb ^{14}C in the atmosphere causes a very rapid increase in sample ^{14}C, peaking around 1965 (Alan Hogg, University of Waikato Radiocarbon Dating Laboratory, pers. comm., 1999).

b   Accelerator Mass Spectrometry (AMS) determination. All other determinations were calculated using Liquid Scintillation Counting (LSC).

c   Mixed shell consisting of Saccostrea, Polynices, Nerita chamaeleon, Placamen calophyllum, Fragum hemicardium, Gafrarium australe, Cymatium sp., Antigona chemnitzii, Trisidos tortuosa, Tapes dorsatus, Meropesta sp., Pinctada sp., Trichomya hirsutus, Bembicium auratum, Calthalotia arruensis and Anadara trapezia.

d   Estimated value.

Appendix 2: Recorded archaeological sites on the southern Curtis Coast

SITE ID	SITE TYPE	RECORDER	LOCATION	DESCRIPTION
JE:A04	Stone Arrangement	G. Alfredson	24°00'43"S 151°28'30"E	Stone arrangement at the summit of Hummock Hill. Alfredson (1993) identified this site 'as a probable surveyor's trig point' although subsequent archival research failed to find any records. References: Alfredson (1993).
JE:A41	Shell Midden/ Artefact Scatter	H. Johnson	23°59'54"S 151°28'20"E	
JE:A42	Shell Midden	H. Johnson	24°02'18"S 151°29'29"E	
JE:A43	Shell Midden/ Artefact Scatter	H. Johnson	24°02'18"S 151°29'22"E	
JE:A60	Shell Midden	C. Burke	24°00'42"S 151°26'37"E	
JE:A61	Artefact Scatter	C. Burke	24°00'51"S 151°29'41"E	
JE:A62	Artefact Scatter	C. Burke	24°00'43"S 151°28'44"E	
JE:A63	Shell Midden	C. Burke	24°01'13"S 151°29'33"E	
JE:A64	Shell Midden	C. Burke	24°01'13"S 151°29'26"E	
JE:A65	Shell Midden	C. Burke	24°02'18"S 151°29'36"E	Shell mound with a depth of up to 40cm. Burke (1993) noted that only a small portion of the site had not been damaged by development activities and water erosion. References: Burke (1993).
JE:A66	Shell Midden	C. Burke	24°00'21"S 151°29'16"E	
JE:A70	Scarred Tree	M. Bird 24.2.1998	24°07'50"S 151°27'44"E	
KE:A05	Stone Quarry	P. Smith 1.11.1980 C. Burke 26.3.1993	24°14'09"S 151°56'10"E	Stone quarry on a silcrete outcrop (c.50m^2) within a granitic headland, adjacent to the coastline at Rocky Point. High density artefact exposure, including a backed blade, scraper, two hammerstones and many small flakes. Burke observed large numbers of artefacts during a 1993 visit, although Reid (1998) failed to identify any unambiguous artefacts during a 1998 visit and called into question the cultural status of the stone exposure. References: Burke (1993); Lilley et al. (1997); Reid (1998).
KE:A06	Axe Grinding Locality	P. Smith 1.11.1980	24°14'09"S 151°56'10"E	Grinding grooves in granite, on a headland adjacent to the coastline at Agnes Water, c.35m from KE:A05 (see above). This site could not be relocated during a field inspection in 1998. References: Reid (1998).
KE:A08	Shell Midden/ Artefact Scatter	R. Neal 25.6.1986	24°04'00"S 151°30'00"E	Sparse shell and stone artefact scatter exposed in sand vehicle tracks on a sloping dune adjacent to a rocky foreshore and mangrove swamp near Seven Mile Creek. Comprises mud ark (60%), whelk (20%) and oyster (20%), and artefacts manufactured on quartz (50%), rhyolite (40%) and black volcanic rock (10%). References: Neal (1986).
KE:A09	Shell Midden	R. Neal 25.6.1986	24°04'00"S 151°31'00"E	Shell midden spotted from the air on a beach ridge adjacent to Seven Mile Creek and bordered by a freshwater swamp. Rowland could not locate this site during a field inspection in 1986 and local informants suggested that it was quartz tailings from quarrying activities rather than a midden deposit. References: Neal (1986); Rowland (1987).
KE:A10	Shell Midden/ Artefact Scatter	M. Rowland A. Border 30.10.1986	24°11'00"S 151°52'00"E	Small, low density surface shell scatter in eroding foredunes and deflated dunes on the ocean beach just north of Agnes Water. Includes pipi, oyster, nerite and occasional stone artefacts. This site is probably the same as KE:A87 (see below). This site is dated on charcoal to 266±87 BP (Wk-10969). References: Lilley et al. (1997); Rowland (1987).

continued over

Appendix 2: continued

SITE ID	SITE TYPE	RECORDER	LOCATION	DESCRIPTION
KE:A11	Shell Midden	M. Rowland 30.10.1986 C. Burke 22.4.1993	24°10'58"S 151°52'47"E	Extensive shell midden complex bordering Round Hill Creek and bounded in the south by Tom's Creek, an eastern tributary of Round Hill Creek. Size not accurately determined owing to heavy vegetation. Material scattered on all 4WD tracks examined in the area. Maximum depth in all locations is 10–20cm. Predominantly mud ark and oyster, with some stone artefacts. Site complex covers a large area and probably subsumes the sites registered separately as KE:A33, KE:A62 and KE:A63 (see below). References: Burke (1993); Rowland (1987).
KE:A12	Stone Quarry/ Shell Midden/ Artefact Scatter	M. Rowland 1.11.1986 C. Burke 29.4.1993	24°09'00"S 151°53'04"E	Site consists of the entire Round Hill Head headland. Isolated stone artefacts and artefact scatters located along exposed walking tracks and ridges. Two large rhyolitic tuff boulders near the navigation beacon at the tip of the headland exhibit a number of negative flake scars. Some oyster shell and flaking debris is scattered in surrounding crevices. Elsewhere, scattered shell fragments and stone artefacts occur, including cobble cores. References: Burke (1993); Lilley et al. (1997); Rowland (1987).
KE:A16	Shell Midden	M. Rowland 24.7.1990 L. Godwin 4.10.1990 C. Burke 30.4.1993	24°12'10"S 151°51'56"E	Multi-component stratified shell mound at least 16m x 10m (c.160m^2) with a depth of more than 50cm, located in open woodland on a low rock terrace c.25m from Round Hill Creek. Extremely high density and spatially discrete shell deposit, dominated by mud ark, but also some oyster, stone artefacts, bone and charcoal. References: Burke (1993); Rowland (1987).
KE:A32	Contact Site/ Story Place	S. Davies 2.2.1994	24°20'20"S 151°34'00"E	Miriam Vale Homestead and Cattle Station built c.1856. Historic and contact site, located just southeast of the modern town of Miriam Vale. The station is the centre of religious and social affiliation to country for many Aboriginal families whose association to the Miriam Vale area spanned the pastoral occupation and into the distant past. This area was the location of several massacres and conflicts between white pastoralists, Native Mounted Police and Aborigines, including a major Aboriginal attack on 12 February 1857. After the establishment of the homestead and until the time of the attacks, local Aborigines had been employed on the station. An Aboriginal camp was situated on the southern bank of House Creek adjacent to the homestead. References: Clarkson et al. (n.d.); Davies (1994).
KE:A33	Shell Midden	C. Burke 22.4.1993	24°11'40"S 151°52'30"E	Large, stratified midden complex (c.100,000m^2) intermittently exposed over low dunes abutting the base of a rhyolitic scree slope on the northern junction of Round Hill and Tom's Creeks. Several low, sandy, residual ridges which exhibit dense midden exposures were also located on the adjacent mudflats. Dominated by mud ark and oyster with occasional other species, stone and flaked glass artefacts. Burke originally recorded part of this site as very sparse oyster and mud ark scatters (c.800m^2) exposed on and around 4WD tracks in open woodland in a gently inclined area 5–20m from mudflats bordering Tom's Creek. This site is dated on charcoal to 1,110±70 BP (Wk-7685). This site is probably part of the more extensive KE:A11 (see above). References: Burke (1993); Ulm (1999).
KE:A34	Shell Midden/ Artefact Scatter	C. Burke 27.1.1993	24°04'25"S 151°45'36"E	Very sparse, surface scatter of shell and stone artefacts (c.2,500m^2) located on a graded survey line on a sand ridge c.1km inland from the central east coast of Middle Island. Oyster (n=20), mud ark (n=10) and stone artefacts (n=3). References: Burke (1993).
KE:A35	Shell Midden	C. Burke 27.1.1993	24°06'49"S 151°46'19"E	Very sparse, surface shell scatter (c.20m^2) on top of a ridge on the southeast of Middle Island. Contains only 5 oyster fragments, no artefacts. This site is probably part of the larger site complex recorded by Lilley, registered as KE:A66 (see below). References: Burke (1993); Lilley (1994).
KE:A36	Shell Midden	C. Burke 27.1.1993	24°06'47"S 151°46'19"E	Very sparse, surface shell scatter (c.1m^2) consisting of only three oyster shells on the southeast of Middle Island. This site is probably part of the larger site complex recorded by Lilley, registered as KE:A66 (see below). References: Burke (1993); Lilley (1994).

continued over

Appendix 2: continued

SITE ID	SITE TYPE	RECORDER	LOCATION	DESCRIPTION
KE:A37	Shell Midden	C. Burke 30.1.1993	24°07'16"S 151°46'41"E	Sparse, surface shell scatter (c.1,500m^2) disturbed by construction activities, located on a dune ridge on the southeastern end of Middle Island. Dominated by oyster, but includes mud ark and mussel. This site is probably part of the larger site complex recorded by Lilley (1994), registered as KE:A66 (see below) and also exposed in the northern erosion bank of Middle Creek. References: Burke (1993); Lilley (1994); Lilley et al. (1997).
KE:A38	Shell Midden	C. Burke 30.1.1993	24°06'08"S 151°45'04"E	Sparse, surface shell scatter (c.2,000m^2) containing one stone artefact on a ridge on the southwest of Middle Island. This site is probably part of the larger site complex recorded by Lilley, registered as KE:A65 (see below). References: Burke (1993); Lilley (1994).
KE:A39	Shell Midden	C. Burke 6.3.1993	24°00'38"S 151°37'26"E	Sparse, surface shell scatter (c.50m^2) in an open area on a bank above the beach fronting Rodds Harbour on the northwestern end of Rodds Peninsula. Dominated by mud ark with some oyster and mussel. References: Burke (1993).
KE:A40	Shell Midden	C. Burke 6.3.1993	24°00'47"S 151°37'37"E	Very sparse, surface shell scatter (c.50m^2), 10m from beach fronting Rodds Harbour and 100m from the sea at low tide c.5m asl, on the northwestern end of Rodds Peninsula. Dominated by oyster with some mud ark. References: Burke (1993).
KE:A41	Shell Midden/ Artefact Scatter	C. Burke 6.3.1993	24°00'51"S 151°37'43"E	Extensive areas of natural shell deposits (cheniers), cultural shell midden deposits and a stone-walled tidal fishtrap located on the western bank of Mort Creek on the western coast of Rodds Peninsula. Shell exposures cover an area in excess of c.6,000m^2. Species include mud ark, oyster and whelk. Stone artefacts and fish bone noted in some excavations. This site is dated on shell to 3,430±140 BP (Wk-6986). References: Burke (1993); Carter (1997); Carter et al. (1999); Lilley et al. (1996); Lilley et al. (1997).
KE:A42	Shell Midden	C. Burke 7.3.1993	24°00'38"S 151°41'18"E	Sparse, stratified shell scatter (c.100m^2) on a ridge near Falls Creek on the central east coast of Rodds Peninsula. Includes oyster and turbo. Site located 500m from sea and rocks. References: Burke (1993).
KE:A43	Shell Midden	C. Burke 7.3.1993	24°03'24"S 151°41'48"E	Group of midden exposures (c.900m^2) located on low residual beach ridges stranded on mudflats at the western extremities of Pancake Creek consisting of a shallow, linear midden exposed in a low erosion bank and three sparse shell scatters. Includes oyster, mud ark, thaid, whelk and mussel. References: Burke (1993).
KE:A44	Shell Midden	C. Burke 9.3.1993	24°02'22"S 151°42'49"E	Large stratified shell midden (c.8,320m^2) on top of a ridge above the beach and mudflats on the northern bank of Pancake Creek, 100m to mudflats and 35m to a small tidal inlet. Dominated by mud ark and includes whelk and oyster to a depth of 5cm determined by auger. This site is dated on charcoal to 700±140 BP (Wk-6993). References: Burke (1993); Lilley et al. (1997); Ulm (1999).
KE:A45	Shell Midden/ Artefact Scatter	C. Burke 9.3.1993	24°02'21"S 151°42'50"E	Stratified linear shell midden (c.5,500m^2) located 50m from the sea and 5-10cm deep, and a shell scatter located 20-30m away from a small tidal inlet and beach flats on the northern bank of Pancake Creek. Dominated by mud ark and includes oyster and whelk, with a single stone artefact noted. This site is dated on charcoal to 700±140 BP (Wk-6993). References: Burke (1993); Lilley et al. (1997); Ulm (1999).
KE:A46	Shell Midden	C. Burke 9.3.1993	24°02'16"S 151°43'03"E	Linear stratified midden (c.7,140m^2) eroding from a creek bank 20m west of a tidal inlet on the northern bank of Pancake Creek. Shell lens is visible in the eroding profile for 238m and continues as a scatter on top of the dune for another 97m. Dominated by oyster and includes mud ark, whelk and charcoal. This site is dated on charcoal to 700±140 BP (Wk-6993). References: Burke (1993); Lilley et al. (1997); Ulm (1999).
KE:A47	Shell Midden	C. Burke 9.3.1993	24°02'05"S 151°43'15"E	Surface shell scatter (c.350m^2) on top of a beach ridge, 30m from mangroves and 15-20m from high water mark on the northern bank of Pancake Creek. Includes oyster, mud ark and thaid. Poor visibility. This site is dated on charcoal to 700±140 BP (Wk-6993). References: Burke (1993); Lilley et al. (1997); Ulm (1999).

continued over

Appendix 2: continued

SITE ID	SITE TYPE	RECORDER	LOCATION	DESCRIPTION
KE:A48	Shell Midden	C. Burke 9.3.1993	24°02'25"S 151°42'50"E	Surface shell scatter (c.1,200m^2) in front of a tidal inlet before mudflats at a shack, 100m from the northern shore of Pancake Creek. Includes oyster, mud ark and whelk. This site is dated on charcoal to 700±140 BP (Wk-6993). References: Burke (1993); Lilley et al. (1997); Ulm (1999).
KE:A49	Shell Midden	C. Burke 25.3.1993	24°11'54"S 151°51'33"E	Burke (1993) recorded three surface shell scatters (c.1,000m^2) located on the eroding western bank of Round Hill Creek. Includes oyster, mud ark and whelk. Ulm et al. (1999) considered this to be part of the extensive Eurimbula Site 1. This site is dated on charcoal to 3,020±70 BP (Wk-3945). References: Burke (1993); Godwin (1990); Lilley et al. (1996); Lilley et al. (1997); Ulm et al. (1999).
KE:A50	Shell Midden	C. Burke 25.3.1993	24°11'44"S 151°51'40"E	Burke (1993) recorded one linear stratified midden (c.100m^2) exposed 5–10cm deep and two surface shell scatters (c.100m^2) located on the eroding western bank of Round Hill Creek. Some shell is eroding out of the bank. Includes oyster and mud ark. Ulm et al. (1999) considered this to be part of the extensive Eurimbula Site 1. This site is dated on charcoal to 3,020±70 BP (Wk-3945). References: Burke (1993); Godwin (1990); Lilley et al. (1996); Lilley et al. (1997); Ulm et al. (1999).
KE:A51	Shell Midden	C. Burke 25.3.1993	24°11'35"S 151°51'42"E	Burke (1993) recorded two shell scatters (c.200m^2) on the western bank of Round Hill Creek exposed on the surface and up to 30cm deep in the erosion section. Dense *in situ* deposit of mud ark and oyster. Ulm et al. (1999) considered this to be part of the extensive Eurimbula Site 1. This site is dated on charcoal to 3,020±70 BP (Wk-3945). References: Burke (1993); Godwin (1990); Lilley et al. (1996); Lilley et al. (1997); Ulm et al. (1999).
KE:A52	Shell Midden	C. Burke 25.3.1993	24°11'28"S 151°51'45"E	Burke (1993) recorded six shell scatters (c.2,000m^2), including linear stratified deposits up to 10cm deep, on the western bank of Round Hill Creek. Dominated by mud ark with some shell eroding out of section, up to 5cm deep. Ulm et al. (1999) considered this to be part of the extensive Eurimbula Site 1. This site is dated on charcoal to 3,020±70 BP (Wk-3945). References: Burke (1993); Godwin (1990); Lilley et al. (1996); Lilley et al. (1997); Ulm et al. (1999).
KE:A53	Shell Midden	C. Burke 26.3.1993	24°11'04"S 151°51'56"E	Burke (1993) recorded three surface shell scatters (c.450m^2) on top of a sand ridge adjacent to Round Hill Creek. Dominated by mud ark with a single large core of granite-like material noted. Ulm et al. (1999) considered this to be part of the extensive Eurimbula Site 1. This site is dated on charcoal to 3,020±70 BP (Wk-3945). References: Burke (1993); Godwin (1990); Lilley et al. (1996); Lilley et al. (1997); Ulm et al. (1999).
KE:A54	Shell Midden	C. Burke 26.3.1993	24°10'56"S 151°51'50"E	Burke (1993) recorded two surface shell scatters on top of a sand ridge and on a tidal flat (c.700m^2), and a linear stratified deposit (c.750m^2) on a sand ridge adjacent to Round Hill Creek. Dominated by mud ark and includes oyster. Ulm et al. (1999) considered this to be part of the extensive Eurimbula Site 1. This site is dated on charcoal to 3,020±70 BP (Wk-3945). References: Burke (1993); Godwin (1990); Lilley et al. (1996); Lilley et al. (1997); Ulm et al. (1999).
KE:A55	Shell Midden/ Artefact Scatter	C. Burke 1.6.1993	24°01'00"S 151°45'46"E	Sparse scatter of oyster shell including 7 stone artefacts (c.400m^2), located on the northern side of Bustard Head. Raw materials may not be local. References: Burke (1993).
KE:A56	Shell Midden	C. Burke 21.4.1993	24°02'31"S 151°33'54"E	Low density surface shell scatter (c.70m^2) located in an open, gently sloping area 10m from the beach on the western side of Innes Head, on the eastern bank of Seven Mile Creek. Dominated by oyster and includes mussel. References: Burke (1993).
KE:A57	Shell Midden/ Artefact Scatter	C. Burke 22.4.1993	24°05'10"S 151°38'52"E	Three small surface shell scatters (c.70m^2) located on and around a graded dirt road c.50m from mangroves, on the eastern edge of an unnamed embayment on the western side of the Turkey Beach peninsula. Dominated by mud ark and includes oyster and a single white quartz flaked piece. References: Burke (1993).
KE:A58	Artefact Scatter	C. Burke 22.4.1993	24°05'44"S 151°38'10"E	Isolated stone artefact manufactured on banded chert located on mudflats on the eastern edge of an unnamed embayment on the western side of the Turkey Beach peninsula, c.50m from mangroves. References: Burke (1993).

continued over

Appendix 2: continued

SITE ID	SITE TYPE	RECORDER	LOCATION	DESCRIPTION
KE:A59	Shell Midden/ Artefact Scatter	C. Burke 1.6.1993	24°01'56"S 151°44'40"E	Very sparse surface shell scatter (c.24,000m^2), including one quartz flaked piece, located on the edge of mudflats on the Jenny Lind Creek side of Bustard Head. Dominated by mud ark and oyster and includes whelk. References: Burke (1993).
KE:A60	Scarred Tree	C. Burke 30.4.1993	24°12'51"S 151°54'16"E	Scarred tree located in the centre of Agnes Water. Scar is located on a large eucalyptus tree (Queensland blue gum or Moreton Bay ash). Scar measures 250cm x 46cm. References: Burke (1993); Lilley et al. (1997).
KE:A61	Shell Midden	C. Burke 17.5.1993	24°01'56"S 151°44'40"E	Fairly dense discrete stratified shell midden with depth of 10cm, located on the eastern bank of Round Hill Creek, c.100m southeast of KE:A16 (see above). Dominated by mud ark. Site damaged by bulldozer activity. References: Burke (1993); Lilley et al. (1997).
KE:A62	Shell Midden/ Artefact Scatter	C. Burke 22.5.1993	24°10'57"S 151°52'53"E	Linear stratified midden (c.4,200m^2) with *in situ* lens of shell c.50cm below ground surface and up to 10cm thick. Subsurface material exposed in a large excavation behind the sewage treatment depot. Dominated by mud ark and includes oyster and stone artefacts manufactured on a variety of raw materials. This site is probably part of the more extensive KE:A11 (see above). References: Burke (1993).
KE:A63	Shell Midden	C. Burke 22.5.1993	24°11'10"S 151°52'33"E	Very sparse surface scatter (c.400m^2) of mud ark and oyster shell located on either side of a 4WD track. This site is probably part of the more extensive KE:A11 (see above). References: Burke (1993).
KE:A64	Shell Midden/ Artefact Scatter	I. Lilley 10.4.1994	24°04'10"S 151°43'35"E	Shell midden complex (c.200,000m^2) up to 15cm deep on the central west coast of Middle Island. Dominated by mud ark and includes oyster and a quartz flake. Located in low swampy melaleuca shrubland adjacent to mudflats on a tidal creek. References: Lilley (1994).
KE:A65	Shell Midden	I. Lilley 10.4.1994	24°05'30"S 151°45'00"E	Shell midden complex (c.800,000m^2) located on high north-south trending dunes extending for c.7km along the central western side of Middle Island. Comprises mud ark, oyster, pipi and whelk. References: Lilley (1994).
KE:A66	Shell Midden/ Artefact Scatter	I. Lilley 10.4.1994	24°06'00"S 151°44'30"E	Shell midden complex (c.800,000m^2) located on high north-south trending dunes extending for c.5km along the central eastern side of Middle Island. Comprises mud ark, oyster and pipi as well as a quartz core. References: Lilley (1994).
KE:A67	Shell Midden	I. Lilley 10.4.1994	24°03'44"S 151°45'56"E	Shell midden complex (c.140,000m^2) dominated by pipi up to 15cm deep on parabolic dunes and sandblows on the northeastern end of Middle Island, bordered in the north and west by Jenny Lind Creek. This site is dated on shell to 980±50 BP (Wk-7679). References: Lilley (1994); Lilley et al. (1997); Ulm (1999).
KE:A87	Shell Midden/ Artefact Scatter	S. Ulm	24°12'15"S 151°54'11"E	Low density scatter of shell and stone artefacts located in a small blowout in the frontal dunes and bordered to the west by a 2m high wire fence. The exposure covers an area of 31m x 12m (372m^2). Maximum densities of 30 shell fragments/m^2, including oyster and mud ark, and 5 stone artefacts/m^2, including quartz, chert and rhyolitic tuff. This site is probably the same as KE:A10 (see above). This site is dated on charcoal to 266±87 BP (Wk-10969). References: Lilley et al. (1997); Rowland (1987).
KE:A88	Shell Midden/ Artefact Scatter	S. Ulm	24°12'21"S 151°54'15"E	Sparse scatter of oyster and pipi fragments eroding out of frontal dunes c.50cm below ground surface. One stone artefact noted, probably manufactured on rhyolitic tuff. References: Lilley et al. (1997).
KE:A89	Shell Midden/ Artefact Scatter	S. Ulm	24°12'22"S 151°54'15"E	Minor scatter of oyster and pipi with one stone artefact located c.20m south of KE:A88 (see above). References: Lilley et al. (1997).
KE:A90	Shell Midden	S. Ulm	24°12'23"S 151°54'15"E	Minor scatter of 12 oyster fragments adjacent to access path to beach, located c.50m south of KE:A89 (see above). References: Lilley et al. (1997).
KE:A91	Shell Midden	S. Ulm	24°10'39"S 151°50'34"E	Sparse shell scatter on southern edge of mangrove fringe of Eurimbula Creek, including whelk, mud ark and pipi. A water-rounded rock was located 20m south of the shell. References: Lilley et al. (1997).

continued over

Appendix 2: continued

SITE ID	SITE TYPE	RECORDER	LOCATION	DESCRIPTION
KE:A92	Shell Midden/ Artefact Scatter	S. Ulm	24°11'00"S 151°49'30"E	Extensive surface scatter of shell and stone artefacts visible on Eurimbula Creek 4WD access track. Bracken fern fringes the track on both sides making it difficult to determine the extent of the scatter owing to lack of visibility. Includes flakes, flaked pieces, cores and manuports manufactured on rhyolitic tuff, quartz and indurated mudstone. Includes mud ark and oyster. References: Lilley et al. (1997).
KE:A93	Shell Midden/ Artefact Scatter	S. Ulm	24°14'24"S 151°56'27"E	Low density shell and stone artefact deposit eroding from subsurface lens. Includes oyster, nerite, mud ark and whelk. Twelve stone artefacts noted including rhyolitic tuff and silcrete. References: Lilley et al. (1997).
KE:A94	Artefact Scatter	S. Ulm	24°14'30"S 151°56'30"E	Stone artefact scatter on 4WD road shoulder on headland. Artefacts manufactured on rhyolitic tuff found eroding out of a nearby road cutting up to 60cm below ground surface. Cores, flakes, flaked pieces, grinding stone made on indurated mudstone, rhyolitic tuff, silcrete, quartz and quartzite. Some retouched artefacts. References: Lilley et al. (1997).
KE:A95	Shell Midden	S. Ulm	24°14'37"S 151°56'36"E	Discrete scatter of oyster lids and bases eroding down orange-yellow dune face covering an area of c.10m^2. Shell densities of up to 28 oyster fragments/m^2. Six unmodified blocks of stone are associated with the shell material. Purple colouration on some oyster valves suggests a recent, perhaps non-Aboriginal, origin. References: Lilley et al. (1997).
KE:A96	Hearth	S. Ulm	24°14'40"S 151°56'34"E	Hearth feature located c.30m south of KE:A94 (see above). Five unmodified blocks of silcrete arranged in a rough circle 46cm x 33cm. No artefactual material is associated with the feature. Possible non-Aboriginal origin. Reference: Lilley et al. (1997).
KE:A97	Artefact Scatter	S. Ulm	24°14'42"S 151°56'37"E	Small artefact scatter comprising 12 stone artefacts manufactured on silcrete exposed over c.10m^2 on a bluff adjacent to low dunes. Large blocks of silcrete embedded in the ground surface may have been modified. Reference: Lilley et al. (1997).
KE:A98	Shell Midden/ Artefact Scatter	S. Ulm	24°15'14"S 151°56'44"E	Sparse scatter of stone artefacts and shell, including oyster, whelk and mussel. Colouration on some shell suggests a recent, perhaps non-Aboriginal, origin. Large silcrete flake collected from adjacent high water mark. Reference: Lilley et al. (1997).
KE:A99	Artefact Scatter	S. Ulm	24°16'00"S 151°56'50"E	Two stone artefacts located on the open coast on the southern side of the Red Rock headland, south of Rocky Point. One broken waterworn pebble manuport with cortex, and one flake on a red igneous rock (possibly silcrete). References: Lilley et al. (1997).
KE:B00	Artefact Scatter	S. Ulm	24°16'00"S 151°56'50"E	Two rhyolitic tuff cores and one andesite flake located on a walking track on the open coast on the southern side of the Red Rock headland, south of Rocky Point. References: Lilley et al. (1997).
KE:B01	Artefact Scatter	S. Ulm	24°14'00"S 151°56'00"E	Low density scatter of rhyolitic tuff flakes and flaked pieces exposed on eroding walking and vehicle tracks across the northern Rocky Point headland. References: Lilley et al. (1997).
KE:B02	Artefact Scatter	S. Ulm	24°14'00"S 151°56'00"E	Scatter of 20 flakes and flaked pieces manufactured on chert, rhyolitic tuff and quartzite exposed in a road cutting on the northern Rocky Point headland. References: Lilley et al. (1997).
KE:B03	Artefact Scatter	S. Ulm	24°14'01"S 151°55'57"E	Six large silcrete flakes and three cores eroding out of secondary orange dune. Located adjacent to two round wooden pillons driven into top of dune. Reference: Lilley et al. (1997).
KE:B04	Shell Midden/ Artefact Scatter	S. Ulm	24°13'47"S 151°55'45"E	Low density scatter of shell and stone material over c.5m^2 area at intersection of 4WD track and beach c.20m west of high water mark. All material may have a non-Aboriginal origin as the site is located in a popular European camping area. Reference: Lilley et al. (1997).
KE:B05	Shell Midden/ Artefact Scatter	S. Ulm	24°13'41"S 151°55'39"E	Stone material eroding down slope of c.3m high frontal dune with material in section c.20cm below ground surface. Located c.50m south of minor headland mid-way between Rocky Point and Agnes Water headlands. Scattered surface shell thought to have modern origin. Test excavation yielded no unambiguously cultural material, although the origin of the stone material in the deposit remains to be explained. Reference: Lilley et al. (1997).

continued over

Appendix 2: continued

SITE ID	SITE TYPE	RECORDER	LOCATION	DESCRIPTION
KE:B06	Shell Midden/ Artefact Scatter	S. Ulm	24°13'28"S 151°55'31"E	Small scatter of shell and stone artefacts in low secondary dune c.50m of minor headland mid-way between Rocky Point and Agnes Water headlands. Includes oyster, mud ark and pipi. References: Lilley et al. (1997).
KE:B07 KE:B08 KE:B09 KE:B10 KE:B22	Shell Midden/ Artefact Scatter/ Stone Quarry	S. Ulm	24°07'00"S 151°46'30"E	Extensive shell midden and quarry site complex (c.140,000m^2) located on the southern bank of Middle Creek close to its mouth. Includes oyster, mud ark, nerite and pipi. Shell material visible in erosion sections up to 25cm deep. Extensive outcrop of modified rhyolitic tuff. Surface artefact densities up to 110/m^2. This site is dated on charcoal to 1,640±150 BP (Wk-6361). References: Lilley et al. (1997); Reid (1998).
KE:B11	Artefact Scatter	S. Ulm	24°09'00"S 151°46'30"E	Stone artefact scatter on salt pan at the southern extremities of Ocean Creek estuary where low mangroves begin at base of creek. Numerous stone artefacts manufactured on rhyolitic tuff spread over c.200m^2 area c.20m west of low casuarina fringe. Several isolated fragments of shell noted along mangrove fringe. References: Lilley et al. (1997).
KE:B12	Artefact Scatter	S. Ulm	24°08'30"S 151°47'00"E	Numerous artefacts manufactured on rhyolitic tuff embedded in a muddy surface on mudflats in the centre of open area on Middle Creek estuary. References: Lilley et al. (1997).
KE:B16	Artefact Scatter	S. Ulm	24°09'30"S 151°48'00"E	Nine rhyolitic tuff artefacts scattered over a 50m^2 area c.30m southeast of bridge on saltpan at the northern extremity of Eurimbula Creek. References: Lilley et al. (1997).
KE:B17	Shell Midden/ Artefact Scatter	S. Ulm	24°09'40"S 151°48'10"E	Low density scatter of oyster shell fragments and two water-worn manuports visible on bank about 10m through mangroves to channel of Eurimbula Creek. References: Lilley et al. (1997).
KE:B18	Shell Midden	S. Ulm	24°09'54"S 151°49'02"E	Scatter of midden shell visible in low (c.30cm high) erosion bank on mangrove fringe of Eurimbula Creek. Some sparse scattered oyster fragments visible on surface. Main scatter c.5m^2 eroding out of bank onto flat mangrove fringe. Shell layer visible in erosion bank c.18cm below surface and c.3cm thick for c.3m along bank. Density is c.108/m^2. Includes oyster and mud ark. This site is dated on charcoal to 230±60 BP (Wk-7680). References: Lilley et al. (1997); Ulm (1999).
KE:B19	Shell Midden	S. Ulm	24°10'04"S 151°49'22"E	Scatter of shell visible on top of a low dune c.20m northeast of mangrove fringe of Eurimbula Creek mainly visible in the burrow of an unknown animal. Scatter spread over an area of c.10m^2. Maximum density is 25/m^2. Predominantly oyster, with some nerite, mud ark, whelk and telescope mud whelk. Located in dry rainforest thicket. Recent excavations in this general site complex have yielded modern radiocarbon dates. References: Lilley et al. (1997); Ulm (1999).
KE:B20	Shell Midden/ Artefact Scatter	S. Ulm	24°10'10"S 151°49'36"E	Very sparse scatter of shell visible on low (c.1m high) erosion bank c.10m north of Eurimbula Creek. Includes mud ark, oyster and whelk as well as several flaked pieces of quartz and rhyolitic tuff and some larger, possibly ground, implements manufactured on the rhyolitic tuff. References: Lilley et al. (1997).
KE:B21	Shell Midden	S. Ulm	24°10'21"S 151°50'17"E	Extensive low density pipi scatter located c.50m from open beach and c.100m west of the mouth of Eurimbula Creek. May be non-cultural. References: Lilley et al. (1997).
KE:B24	Shell Midden	C. Burke	24°01'01"S 151°30'14"E	
KE:B26	Shell Midden	S. Ulm	24°11'50"S 151°52'10"E	Mounded mud ark midden disturbed by brush-turkey nesting located near the eastern bank of Round Hill Creek and the southern bank of Tom's Creek, Agnes Water. References: Lilley et al. (1997).
KE:B28	Scarred Tree	S. Ulm	24°15'30"S 151°53'00"E	Possible scarred tree which has been felled for construction of a power easement on the southern margin of Round Hill National Park. References: Lilley et al. (1997).
KE:B29	Artefact Scatter	S. Ulm	24°15'00"S 151°55'30"E	Low density scatter of stone artefacts located along the northeastern margin of Deepwater Creek, southwest of Rocky Point. References: Lilley et al. (1997).

continued over

Appendix 2: continued

SITE ID	SITE TYPE	RECORDER	LOCATION	DESCRIPTION
KF:A01	Shell Midden	C. Burke 6.3.1993	23°58'52"S 151°36'48"E	Very sparse surface shell scatter dominated by oyster on the eastern side of Richards Point, Rodds Peninsula. Total of 20 shell fragments. References: Burke (1993).
KF:A02	Shell Midden	C. Burke 7.3.1993	23°59'20"S 151°40'06"E	Four low density shell scatters (c.1,200m^2) up to 5cm deep dominated by oyster but also includes chiton, austro, turbo and mud ark, located on the northeastern coast of Rodds Peninsula. Scatters in close proximity to beach, tidal inlet and rock platforms. References: Burke (1993).
KF:A03	Shell Midden	C. Burke 8.3.1993	23°59'23"S 151°39'44"E	Two surface shell scatters (c.1,400m^2) located behind dunes and a tidal inlet 50–60m from ocean and rock platforms, located on the northeastern coast of Rodds Peninsula. Dominated by oyster and includes mud ark, chiton and turbo. References: Burke (1993).
KF:A04	Shell Midden	C. Burke 8.3.1993	23°59'13"S 151°39'29"E	Very sparse surface shell scatter (c.40m^2) of oyster 20m from beach and rock platforms, located on the northeastern coast of Rodds Peninsula, c.10m asl. Site is behind thick scrub. References: Burke (1993).
KF:A05	Shell Midden	C. Burke 8.3.1993	23°59'05"S 151°39'22"E	Two shell scatters (c.1,700m^2) at least 15cm deep situated on a bank near the beach c.20m from the sea and rock platforms, located on the northeastern coast of Rodds Peninsula. Dominated by oyster and includes mud ark, chiton and turbo. References: Burke (1993).
KF:A06	Shell Midden	C. Burke 8.3.1993	23°59'27"S 151°40'14"E	Two sparse oyster scatters (c.150m^2) located 50m from beach and rocks, located in the vicinity of foredunes on the northeastern coast of Rodds Peninsula. Freshwater creeks in close vicinity to deposits. References: Burke (1993).
KF:A07	Shell Midden	C. Burke 9.3.1993	23°58'52"S 151°38'52"E	Sparse oyster deposits (c.200m^2) up to 5cm deep situated c.100m from sea and 5m asl on a bank on top of a ridge, located on the northeastern coast of Rodds Peninsula. Dominated by oyster and includes austro. References: Burke (1993).
KF:A08	Artefact Scatter	C. Burke 9.3.1993	23°58'33"S 151°37'31"E	Isolated stone artefact on a steep rocky slope on top of a headland on the eastern side of Richards Point, Rodds Peninsula, c.20m asl and 20m from rock platforms and ocean. References: Burke (1993).
KF:A09	Shell Midden	C. Burke 9.3.1993	23°58'45"S 151°37'44"E	Surface shell scatter (c.200m^2) on beach c.0.5m asl and c.10m from rock platforms and ocean, c.500m northwest of Richards Point, Rodds Peninsula. Dominated by oyster and includes mud ark and austro. References: Burke (1993).
KF:A10	Shell Midden	C. Burke 9.3.1993	23°58'45"S 151°37'44"E	Surface oyster scatter (c.800m^2) at least 5cm deep situated c.50m from beach in open woodland, located on the northeastern coast of Rodds Peninsula. Ocean and rock platforms c.200m from site. Augering revealed shell to 5cm in depth. References: Burke (1993).
KF:A11	Shell Midden	C. Burke 9.3.1993	23°58'51"S 151°37'18"E	Sparse surface scatters (c.500m^2) of oyster and mud ark, 50–100m from rock platforms and ocean, located on the northeastern coast of Rodds Peninsula. Tidal creek in close vicinity. References: Burke (1993).
KF:A12	Fishtrap	C. Burke 9.3.1993	23°58'40"S 151°37'25"E	Stone-walled fishtrap of unknown dimensions located in a small bay to the immediate west of Richards Point. The trap appears to contain water at both high and low tide. The trap is in the shape of an arc with a formed opening in the centre of it. References: Burke (1993); Lilley et al. (1997).
KF:A13	Shell Midden	C. Burke 8.3.1993	23°59'24"S 151°40'13"E	Very sparse surface shell scatters (c.50m^2) containing mostly oyster 20–50m from rock platforms and ocean. Freshwater creek located 10–50m away. References: Burke (1993).
KF:A14	Shell Midden	C. Burke 8.3.1993	23°59'08"S 151°39'11"E	Sparse surface shell scatters (c.3,650m^2) situated on top of a dune ridge in a clearing behind the beach, located on the northeastern coast of Rodds Peninsula. Dominated by oyster and includes mud ark, chiton and mussel. Tidal creek is located nearby. References: Burke (1993).
KF:A15	Shell Midden	C. Burke 8.3.1993	23°59'08"S 151°39'09"E	Sparse surface shell scatter, containing mostly oyster, situated on top of a dune ridge, located on the northeastern coast of Rodds Peninsula. Tidal creek is located nearby. References: Burke (1993).

continued over

Appendix 2: continued

SITE ID	SITE TYPE	RECORDER	LOCATION	DESCRIPTION
KF:A16	Shell Midden	C. Burke 8.3.1993	23°58'55"S 151°38'50"E	Surface oyster scatter (c.25m^2) situated c.100m from ocean and c.50m from rock platforms, c.5m asl, located on the northeastern coast of Rodds Peninsula. References: Burke (1993).
KF:A23	Shell Midden	S. Ulm	24°59'22"S 151°40'15"E	Low density surface scatter of oyster up to c.150m inland on northeastern Rodds Peninsula associated with large burdekin plum trees. Possible subsurface component. Reference: Lilley et al. (1997).
SCC55	Shell Midden	S. Ulm	24°04'30"S 151°39'00"E	Thin layer of oyster eroding out of low bank c.5cm below ground surface along c.4m of bank at Turkey Beach. Area to the west and south has been levelled for the construction of a small toilet block and BBQ area.
SCC58	Shell Midden	S. Ulm	24°01'40"S 151°44'40"E	Surface scatter of shell on a high dune ridge up to 50m inland on the eastern bank of Pancake Creek immediately behind a navigation beacon opposite Pancake Point. Visible shell appears to be associated with crab burrowing and is probably derived from subsurface deposits. Dominated by oyster but also includes mud ark and whelk. A small silcrete core was also noted. References: Ulm (1999).
SCC64	Shell Midden	S. Ulm	24°07'25"S 151°40'59"E	Extensive linear shell midden exposed in section on the western bank of Worthington Creek. The midden material is located along the top margin of a high (c.4m) creek erosion bank. Sandstone is exposed at the base of the section, overlain by a thick layer of light brown clays and a thin veneer of eroding top soils containing the shell material. Shell is visible along a segment of bank c.350m in length and up to 5cm deep. Includes oyster and scallop. This site is dated on shell to 349±60 BP (Wk-10089). References: Ulm (this volume).

Appendix 3: Site name synonyms for recorded sites on the southern Curtis Coast

EPA REGISTERED SITE NO.	BURKE (1993) FIELD NO.	BURKE (1993) PRE-ALLOCATED SITE NO.	GGCHP SITE ID	OTHER DESIGNATIONS
JE:A41				Hummock Hill Island Site 1
JE:A42				Hummock Hill Island Site 2
JE:A43				Hummock Hill Island Site 3
JE:A60	CC190			
JE:A61	CC192			
JE:A62	CC193			
JE:A63	CC195			
JE:A64	CC196			
JE:A65	CC197			
JE:A66	CC187			
KE:A05	CC132		SCC20	Rocky Point Quarry; Choughs Crossing
KE:A06				Agnes Water Grooves
KE:A08				Boyne Creek I (Neal 1986)
KE:A09			SCC63	Boyne Creek II (Neal 1986); Seven Mile Creek Mound (this volume)
KE:A10			SCC3	MV1 (Rowland 1987)
KE:A11	CC144		SCC65	MV2 (Rowland 1987); Tom's Creek Site Complex (this volume)
KE:A12	CC139		SCC1	MV3 (Rowland 1987)
	CC043	KE:A37		
	CC044	KE:A38		
	CC045	KE:A39		
	CC046	KE:A40		
	CC047	KE:A41		
	CC048	KE:A42		
	CC049	KE:A43		
	CC050	KE:A44		
	CC051	KE:A45		
	CC052	KE:A46		
KE:A16	CC147		SCC53	MV4 (Rowland 1987); Round Hill Creek Mound (this volume)
KE:A32				Miriam Vale Homestead (Davies 1994); BG10 (Davies 1994)
KE:A33	CC141	KE:A31	SCC59	Tom's Creek Site Complex (this volume)
	CC142	KE:A30		
KE:A34	CC005	KE:A32		
KE:A35	CC006	KE:A33		Site Group 4 (Lilley 1994)
KE:A36	CC007	KE:A34		Site Group 4 (Lilley 1994)
KE:A37	CC008	KE:A35	SCC46	Site Group 4 (Lilley 1994)
KE:A38	CC009	KE:A36		
KE:A39	CC065	KE:A47		
KE:A40	CC066	KE:A48		
KE:A41	CC067	KE:A49	SCC42	Rodds Peninsula Site Complex (Carter 1997)
	CC068	KE:A50		Mort Creek Site Complex (this volume)
KE:A42	CC069	KE:A51		
KE:A43	CC090	KE:A52		
	CC091	KE:A53		
	CC092	KE:A54		
	CC093	KE:A55		
KE:A44	CC094	KE:A56	SCC45	Pancake Creek Site Complex (this volume)
KE:A45	CC095	KE:A57	SCC45	Pancake Creek Site Complex (this volume)
	CC096	KE:A58		
KE:A46	CC097	KE:A59	SCC45	Pancake Creek Site Complex (this volume)

continued

Appendix 3: continued

EPA REGISTERED SITE NO.	BURKE (1993) FIELD NO.	BURKE (1993) PRE-ALLOCATED SITE NO.	GGCHP SITE ID	OTHER DESIGNATIONS
KE:A47	CC098	KE:A60	SCC45	Pancake Creek Site Complex (this volume)
KE:A48	CC099	KE:A61	SCC45	Pancake Creek Site Complex (this volume)
KE:A49	CC112A	KE:A62	SCC43	Eurimbula Site 1 (this volume)
	CC113A	KE:A63		
	CC131	KE:A64		
KE:A50	CC114	KE:A65	SCC43	Eurimbula Site 1 (this volume)
	CC115	KE:A66		
	CC116	KE:A67		
KE:A51	CC117	KE:A68	SCC43	Eurimbula Site 1 (this volume)
	CC118	KE:A69		
KE:A52	CC119	KE:A70	SCC43	Eurimbula Site 1 (this volume)
	CC120	KE:A71		
	CC121	KE:A72		
	CC122	KE:A73		
	CC123	KE:A74		
	CC124	KE:A75		
KE:A53	CC125	KE:A76	SCC43	Eurimbula Site 1 (this volume)
	CC126	KE:A77		
	CC127	KE:A78		
KE:A54	CC128	KE:A79	SCC43	Eurimbula Site 1 (this volume)
	CC129	KE:A80		
	CC130	KE:A81		
KE:A55	CC174	KE:A82		
KE:A56	CC133	KE:A83		
KE:A57	CC135	KE:A84		
	CC136	KE:A85		
	CC137	KE:A86		
KE:A58	CC138	KE:A87		
KE:A59	CC173	KE:A88		
KE:A60	CC148	KE:A89	SCC52	
KE:A61	CC169	KE:A90	SCC49	Caravan Midden Scatter (Lilley et al. 1997)
KE:A62	CC140	KE:A91		Tom's Creek Site Complex (this volume)
KE:A63	CC143	KE:A92		Tom's Creek Site Complex (this volume)
KE:A64				Site Group 1 (Lilley 1994)
KE:A65				Site Group 2 (Lilley 1994)
KE:A66				Site Group 4 (Lilley 1994)
KE:A67			SCC47	Site Group 5 (Lilley 1994); Middle Island Sandblow Site (this volume)
KE:A87			SCC3	Agnes Beach Midden (this volume); Agnes Water Shell and Stone Artefact Scatter #2 (Lilley et al. 1997)
KE:A88			SCC4	Agnes Water Shell and Stone Artefact Scatter #3 (Lilley et al. 1997)
KE:A89			SCC5	Agnes Water Shell and Stone Artefact Scatter #4 (Lilley et al. 1997)
KE:A90			SCC6	Agnes Water Shell and Stone Artefact Scatter #5 (Lilley et al. 1997)
KE:A91			SCC7	Eurimbula Creek Shell Scatter (Lilley et al. 1997)
KE:A92			SCC10	Eurimbula Shell and Stone Artefact Scatter (Lilley et al. 1997)
KE:A93			SCC11	Deepwater Shell and Stone Artefact Scatter (+ Lens) (Lilley et al. 1997)
KE:A94			SCC12	Deepwater Stone Artefact Scatter #1 (Lilley et al. 1997)
KE:A95			SCC13	Deepwater Stone Artefact Scatter #2 (Lilley et al. 1997)
KE:A96			SCC14	Deepwater Hearth Features (Lilley et al. 1997)
KE:A97			SCC15	Deepwater Artefact Scatter (Lilley et al. 1997)
KE:A98			SCC16	Deepwater Shell and Stone Artefact Scatter (Lilley et al. 1997)
KE:A99			SCC17	Deepwater Shell and Stone Artefact Scatter (Lilley et al. 1997)
KE:B00			SCC18	Red Rock Stone Artefact Scatter #1 (Lilley et al. 1997)
KE:B01			SCC19	Rocky Point Stone Artefact Scatter (Lilley et al. 1997)

Appendix 3: continued

EPA REGISTERED SITE NO.	BURKE (1993) FIELD NO.	BURKE (1993) PRE-ALLOCATED SITE NO.	GGCHP SITE ID	OTHER DESIGNATIONS
KE:B02			SCC21	Rocky Point Stone Artefact Scatter #1 (Lilley et al. 1997)
KE:B03			SCC22	Rocky Point Stone Artefact Scatter #2 (Lilley et al. 1997)
KE:B04			SCC23	Agnes Water-Rocky Point Stone Artefact Scatter (Lilley et al. 1997)
KE:B05			SCC24	Agnes Water Shell and Stone Artefact Scatter #1 (Lilley et al. 1997)
KE:B06			SCC25	Agnes Water Shell and Stone Artefact Scatter #2 (Lilley et al. 1997)
KE:B07			SCC26-SCC29; SCC41	Ironbark Site Complex (this volume)
KE:B08			SCC26-SCC29; SCC41	Ironbark Site Complex (this volume)
KE:B09			SCC26-SCC29; SCC41	Ironbark Site Complex (this volume)
KE:B10			SCC26-SCC29; SCC41	Ironbark Site Complex (this volume)
KE:B11			SCC30	Middle Creek Stone Artefact Scatter #1
KE:B12			SCC31	Middle Creek Stone Artefact Scatter #2
KE:B16			SCC35	Eurimbula Creek Stone Scatter
KE:B17			SCC36	Middle Creek Shell and Stone Scatter
KE:B18			SCC37	Eurimbula Creek 1 (this volume); Middle Creek Shell Scatter #1
KE:B19			SCC38	Eurimbula Creek 2 (this volume); Middle Creek Shell Scatter #2
KE:B20			SCC39	
KE:B21			SCC40	Eurimbula Creek Shell Scatter (Lilley et al. 1997)
KE:B22			SCC26-SCC29; SCC41	Ironbark Site Complex (this volume)
KE:B24	CC194			
KE:B26			SCC48	Turkey Mound Midden (Lilley et al. 1997)
KE:B28			SCC50	Round Hill National Park Scarred Tree (Lilley et al. 1997)
KE:B29			SCC51	Swamp Artefact Scatter (Lilley et al. 1997)
KF:A01	CC064	KF:A01		
KF:A02	CC071	KF:A02		
	CC072	KF:A03		
	CC073	KF:A04		
	CC074	KF:A05		
KF:A03	CC075	KF:A06		
	CC076	KF:A07		
KF:A04	CC077	KF:A08		
KF:A05	CC078	KF:A09		
	CC080	KF:A10		
KF:A06	CC081	KF:A11		
	CC082	KF:A12		
KF:A07	CC087	KF:A13		
	CC088	KF:A14		
KF:A08	CC101	KF:A15		
KF:A09	CC102	KF:A16		
KF:A10	CC103	KF:A17		
KF:A11	CC104	KF:A18		
	CC105	KF:A19		
KF:A12	CC100	KF:A20	SCC54	
KF:A13	CC070	KF:A21		
	CC083	KF:A22		
KF:A14	CC084	KF:A23		
	CC085	KF:A24		
KF:A15	CC086	KF:A25		
KF:A16	CC089	KF:A26		
KF:A23			SCC44	Plum Tree Site (Lilley et al. 1997)
–			SCC64	Worthington Creek Midden

Appendix 4: Excavation data

Table A4/1 Seven Mile Creek Mound, Square A. *= <0.1g.

XU	MAX. DEPTH (mm)	MEAN SIZE (mm)	WEIGHT (kg)	SHELL (g)	CRUSTACEAN (g)	BONE (g)	CHARCOAL (g)	STONE ARTEFACTS (g)	OTHER STONE (g)
1	3.6	3.6	0.2	1.9	0	0	*	0	0
2	33.4	29.8	9.8	4595.9	7.8	0.7	0.1	0.8	749.8
3	66.2	32.8	11.1	6589.1	29.1	1.3	0.4	1.7	793.8
4	104.0	37.8	12.3	6448.5	55.0	1.6	0.2	0.7	799.2
5	133.8	29.8	11.1	5736.5	59.8	1.2	0.2	2.5	997.2
6	170.4	36.6	12.3	6606.4	20.2	1.3	0.6	0.1	826.4
7	206.2	35.8	10.8	5774.5	42.4	1.7	0.7	8.5	674.2
8	240.2	34.0	10.8	5986.6	22.1	0.2	0.5	16.8	471.8
9	278.6	38.4	11.5	5618.0	19.3	0.8	0.4	2.5	1516.7
10	312.4	33.8	10.2	5437.5	17.7	0.3	0.1	0.5	675.1
11	353.0	40.6	11.6	6114.7	20.5	4.9	0.8	1.4	861.3
12	389.6	36.6	10.1	5441.6	40.6	3.3	0.3	0.7	878.9
13	436.4	46.8	11.1	6000.2	61.8	3.6	0.1	0	979.5
14	466.8	30.4	10.4	5499.1	35.7	5.1	0.2	2.6	1499.1
15	508.0	41.2	11.3	6079.8	17.0	3.9	1.5	31.9	989.6
16	554.4	46.4	12.1	6290.5	5.6	2.0	1.0	16.1	1297.7
17	582.0	27.6	11.7	5572.8	10.8	0.8	1.0	35.8	1176.2
18	643.6	61.6	18.9	4197.7	5.4	0.2	0.2	32.4	1118.8
19	676.2	32.6	13.4	5247.8	8.7	1.5	0.2	2.7	1677.9
20	714.6	38.4	15.3	6478.2	18.6	0.4	0.9	0.9	442.3
21	750.6	36.0	14.0	9679.4	16.1	0.9	1.3	0.7	1358.2
22	785.4	34.8	12.8	4640.6	5.4	0.4	1.4	0	1254.3
23	818.6	33.2	12.7	3921.1	3.8	0.4	1.4	1.0	619.3
24	848.2	29.6	12.2	3368.9	2.6	0.1	1.5	3.7	467.7
25	886.8	38.6	15.1	2496.7	3.7	0.2	1.2	0	394.4
26	921.8	35.0	13.5	1336.9	4.8	1.6	1.0	0	444.5
27	952.6	30.8	12.7	552.9	0.7	*	0.7	4.8	620.7
28	994.6	42.0	12.9	60.6	0	*	0.2	0.3	251.4
29	1030.6	36.0	13.3	16.3	0	*	0.4	0.1	235.1
30	1097.0	66.4	26.1	3.2	0	0	0.1	0	1312.8
31	1163.2	66.2	24.1	0.8	0	0	0.2	0	724.6
Total	–	–	395.4	135794.8	535.2	38.3	18.8	169.1	26108.6

Table A4/2 Mort Creek Site Complex, Square C. *= <0.1g.

XU	MAX. DEPTH (mm)	MEAN SIZE (mm)	WEIGHT (kg)	SHELL (g)	CRUSTACEAN (g)	BONE (g)	CHARCOAL (g)	STONE ARTEFACTS (g)	OTHER STONE (g)
1	0	0	0	0.7	0	0	0	0	0.1
2	47.2	47.2	17.6	723.4	0	1.0	0.2	0	14.1
3	55.8	8.6	9.8	775.0	0	0.9	0.5	0	34.1
4	83.2	27.4	10.0	1661.2	0	2.2	0.5	0	11.8
5	113.4	30.2	9.0	1581.3	0	2.3	0.7	0	7.9
6	158.2	44.8	12.9	2831.6	0	7.9	2.5	0	57.2
7	180.8	22.6	7.4	754.7	0	25.9	0.3	*	34.3
8	222.8	42.0	17.8	291.8	0	9.6	1.1	131.7	252.9
9	255.4	32.6	8.5	70.3	0.1	2.0	0.2	0	9.7
10	291.8	36.4	10.5	13.3	0	1.2	0.3	0	20.1
11	327.2	35.4	11.5	15.0	0	6.8	0.2	0.8	50.2
12	353.4	26.2	8.5	12.7	0	2.2	0.2	0.3	844.1
13	388.4	35.0	10.3	0.4	0	0.6	0.9	0	17.3
14	428.0	39.6	13.6	0.1	0	0.1	5.8	0	32.5
15	457.2	29.2	9.5	0.1	0	0.1	0.1	0	50.2
16	495.6	38.4	9.5	0.5	0	0.1	0.1	0.1	321.0
17	536.4	40.8	14.8	8.3	0	0.3	0.7	2.1	1505.6
18	564.2	27.8	11.5	8.3	0	1.1	1.7	0.8	761.5
19	590.6	26.4	7.8	0.1	0	0.3	0.1	*	1528.7
Total	–	–	200.5	8748.8	0.1	64.5	15.8	135.7	5553.3

Table A4/3 Pancake Creek Site Complex, Square A. *= <0.1g.

XU	MAX. DEPTH (mm)	MEAN SIZE (mm)	WEIGHT (kg)	SHELL (g)	CRUSTACEAN (g)	BONE (g)	CHARCOAL (g)	STONE ARTEFACTS (g)	OTHER STONE (g)
1	7.6	7.6	0.4	0	0	0	0.2	0	*
2	61.4	58.3	16.0	0.5	0	0	3.1	0	4.0
3	87.4	26.0	7.5	0.3	0	0	1.2	0	1.6
4	103.4	16.0	6.2	0	0	0	1.0	0	1.7
5	125.2	21.8	7.5	0	0	0	2.2	0	2.1
6	156.8	31.6	9.5	0.9	0	0	5.7	0	4.5
7	179.8	23.0	8.1	0.5	0	*	13.1	0	4.3
8	204.2	24.4	9.0	4.3	0	0	3.6	0	5.3
9	235.2	31.0	10.5	86.4	0	0	3.8	0	61.3
10	260.4	25.2	9.5	21.1	0	0	12.9	0	9.4
11	294.8	34.4	12.0	2.0	0	0	35.5	*	86.3
12	325.2	30.4	11.3	0.7	0	0	18.8	0	33.7
13	351.0	25.8	10.0	0.1	0	0	6.0	0	30.2
14	378.0	27.0	10.0	0	0	0	4.3	0	31.6
15	434.2	56.2	20.7	0.1	0	0	11.0	0	63.6
16	503.0	68.8	27.0	0	0	0	79.3	0	29.8
17	559.2	56.2	22.4	3.8	0	0	40.7	0	74.3
18	603.2	44.0	16.7	0	0	0	6.0	0	14.3
19	683.2	80.0	33.5	0	0	0	4.6	0	10.8
Total	–	–	247.8	120.6	0	*	252.8	*	468.7

Table A4/4 Pancake Creek Site Complex, Square B. *= <0.1g.

XU	MAX. DEPTH (mm)	MEAN SIZE (mm)	WEIGHT (kg)	SHELL (g)	CRUSTACEAN (g)	BONE (g)	CHARCOAL (g)	STONE ARTEFACTS (g)	OTHER STONE (g)
1	10.8	10.8	0.3	0.1	0	0	0.1	0	0
2	69.0	58.2	16.2	0.4	0	0	2.7	0	3.8
3	87.4	18.4	8.2	0.7	0	0	1.0	0	2.0
4	121.8	34.4	7.7	*	0	0	1.3	0	2.1
5	154.6	32.8	10.4	0.1	0	0	3.6	0	3.3
6	172.6	18.0	9.0	18.0	0	0	2.0	0	3.8
7	206.6	34.0	10.7	18.7	0	0	3.3	0	7.7
8	233.4	26.8	9.0	92.5	0	0	5.1	0	5.5
9	262.4	29.0	9.7	133.9	0	0.5	12.6	0	8.0
10	293.0	30.6	10.3	9.2	0	0	17.0	0	15.4
11	331.2	38.2	11.5	19.2	0	0	11.7	0	16.6
12	358.0	26.8	9.2	1.0	0	0	10.5	0	14.8
13	387.0	29.0	9.1	*	0	0	32.2	0	12.6
14	419.2	32.2	9.7	0	0	0	25.3	0	12.7
15	476.8	57.6	20.5	0	0	0	36.2	0	9.1
16	528.2	51.4	17.9	0	0	0	32.7	0	8.1
17	586.2	58.0	22.9	0	0	0	17.2	*	7.1
18	674.0	87.8	31.1	0	0	0	5.9	0	17.1
Total	–	–	223.3	293.8	0	0.5	220.3	*	149.9

Table A4/5 Pancake Creek Site Complex, Square C. *= <0.1g.

XU	MAX. DEPTH (mm)	MEAN SIZE (mm)	WEIGHT (kg)	SHELL (g)	CRUSTACEAN (g)	BONE (g)	CHARCOAL (g)	STONE ARTEFACTS (g)	OTHER STONE (g)
1	3.6	3.6	0.1	*	0	0	*	0	0
2	80.8	77.2	27.1	44.4	0	0.1	18.0	0	57.5
3	118.0	37.2	11.3	76.7	0	0	1.8	0	1.1
4	151.0	33.0	12.6	83.2	0	0	2.2	0	1.8
5	187.8	36.8	10.8	625.7	0	0	5.6	0	2.2
6	231.8	44.0	12.7	827.1	0	0	7.0	0	12.8
7	270.2	38.4	12.6	157.2	0	0	7.4	0	14.3
8	315.0	44.8	12.3	81.8	0	0	51.2	0	14.6
9	354.6	39.6	11.1	15.7	0	0	38.4	0	0.4
10	393.4	38.8	12.4	6.0	0	0	3.5	0	0.2
11	436.0	42.6	9.2	*	0	0	9.2	0	0.7
12	530.2	94.2	26.4	0.6	0	0	12.4	0	0.1
13	617.0	86.8	31.7	0	0	0	1.5	0	3.9
Total	–	–	190.3	1918.5	0	0.1	158.2	0	109.6

Table A4/6 Pancake Creek Site Complex, Square D. *= <0.1g.

XU	MAX. DEPTH (mm)	MEAN SIZE (mm)	WEIGHT (kg)	SHELL (g)	CRUSTACEAN (g)	BONE (g)	CHARCOAL (g)	STONE ARTEFACTS (g)	OTHER STONE (g)
1	1.2	1.2	0	*	0	0	*	0	*
2	72.0	70.8	24.4	12.9	0	0	11.8	0	1.0
3	140.6	68.6	22.3	47.3	0	0	4.5	0	1.1
4	175.2	34.6	12.6	77.2	0	0	2.5	0	1.9
5	213.4	38.2	12.5	367.6	0	0	3.6	*	4.3
6	246.4	33.0	12.4	100.4	0	0	3.6	0	2.0
7	283.6	32.2	12.4	25.5	0	0	11.5	0	2.4
8	321.6	38.0	13.9	30.5	0	0	27.5	12.8	2.3
9	356.2	34.6	11.8	1.4	0	0	6.1	0	*
10	388.0	31.8	11.6	0.1	0	0	4.6	0	0.2
11	442.2	54.2	17.6	0.1	0	0	8.3	0	0.2
12	525.6	83.4	25.6	0.1	0	0	7.2	0	0.8
13	606.4	80.8	27.6	0.1	0	0	2.5	0	0.5
Total	–	–	204.7	663.1	0	0	93.7	12.8	16.7

Table A4/7 Pancake Creek Site Complex, Square E. *= <0.1g.

XU	MAX. DEPTH (mm)	MEAN SIZE (mm)	WEIGHT (kg)	SHELL (g)	CRUSTACEAN (g)	BONE (g)	CHARCOAL (g)	STONE ARTEFACTS (g)	OTHER STONE (g)
1	1.6	1.6	0	*	0	0	0.5	0	*
2	65.2	63.6	24.3	16.2	0	0	60.9	0	0.7
3	131.0	65.8	12.6	12.5	0	0	35.6	0	1.8
4	165.4	34.4	11.8	45.3	0	0	7.1	0	2.9
5	199.2	33.8	13.4	391.5	0	0	4.5	0	16.9
6	233.0	33.8	12.0	87.9	0	0	5.5	0	29.5
7	269.2	36.2	12.0	33.0	0	0	10.6	0	0.8
8	302.8	33.6	11.8	18.9	0	0	22.0	0	1.1
9	369.8	67.0	23.5	0.1	0	0	8.3	0	0.6
10	448.4	78.6	27.3	3.9	0	0	7.4	0	0.9
11	521.2	72.8	22.4	0.1	0	0	6.5	0	0.5
12	586.4	65.2	25.6	0.4	0	0	2.3	0	7.2
Total	–	–	196.7	609.9	0	0	171.2	0	62.9

Table A4/8 Pancake Creek Site Complex, Square F. *= <0.1g.

XU	MAX. DEPTH (mm)	MEAN SIZE (mm)	WEIGHT (kg)	SHELL (g)	CRUSTACEAN (g)	BONE (g)	CHARCOAL (g)	STONE ARTEFACTS (g)	OTHER STONE (g)
1	1.4	1.4	0	*	0	0	0.1	0	*
2	61.2	59.8	26.2	22.1	0	2.3	57.1	0	0.8
3	119.6	58.4	23.1	48.2	0	0	46.6	0	14.5
4	154.0	34.4	12.9	84.0	0	0	5.3	0	1.9
5	182.6	28.6	11.3	281.2	0	0	6.2	0	8.9
6	218.4	35.8	12.7	503.1	0	0	10.6	0	2.1
7	249.2	30.8	12.4	182.2	0	0	8.4	0.1	2.5
8	282.0	32.8	12.5	56.8	0	0	17.3	*	10.3
9	316.2	34.2	12.6	9.2	0	0	72.5	0	0.7
10	373.6	57.4	24.7	11.4	0	0	18.0	0	0.4
11	458.6	85.0	30.5	0.5	0	0	23.4	0	0.3
12	529.6	71.0	24.0	1.6	0	0	13.9	0	0.1
13	599.2	69.6	23.1	0.5	0	0	2.4	0	2.9
Total	–	–	226.0	1201.0	0	2.3	281.8	0.2	45.5

Table A4/9 Pancake Creek Site Complex, Square G. *= <0.1g.

XU	MAX. DEPTH (mm)	MEAN SIZE (mm)	WEIGHT (kg)	SHELL (g)	CRUSTACEAN (g)	BONE (g)	CHARCOAL (g)	STONE ARTEFACTS (g)	OTHER STONE (g)
1	2.0	2.0	0.2	0.4	0	0	0.1	0	0
2	54.2	52.2	18.7	11.3	0	0	6.9	0	25.6
3	101.0	46.8	15.0	14.0	0	0	3.1	0	2.5
4	158.2	57.2	17.3	50.5	0.2	0	4.6	0	2.2
5	176.0	17.8	7.3	71.5	0	0	3.2	0	2.7
6	208.8	32.8	4.5	337.1	0	0	2.8	0	34.7
7	237.2	28.4	9.3	508.8	0	0	18.4	0	107.6
8	267.4	30.2	10.9	16.9	0	0	44.8	0	3.9
9	311.8	44.4	13.4	29.4	0	0	18.4	0	1.8
10	348.6	36.8	11.4	5.3	0	0	25.1	0	11.3
11	380.0	31.4	10.6	2.0	0	0	21.1	0	0.6
12	403.4	23.4	7.2	163.4	0	0	9.8	0	0.2
13	435.4	32.0	10.8	60.7	0	0	12.1	0	1.6
14	466.6	31.2	11.1	0.2	0	0	9.5	0	0.4
15	496.2	29.6	11.8	1.0	0	0	11.7	0	1.4
16	543.0	46.8	14.6	1.3	0	0	5.4	0	2.9
17	582.2	39.2	13.0	*	0	0	2.8	0	0.6
18	656.4	74.2	25.2	0.2	0	0	2.0	0	17.3
Total	–	–	212.3	1274.1	0.2	0	201.7	0	217.3

Table A4/10 Pancake Creek Site Complex, Square H. *= <0.1g.

XU	MAX. DEPTH (mm)	MEAN SIZE (mm)	WEIGHT (kg)	SHELL (g)	CRUSTACEAN (g)	BONE (g)	CHARCOAL (g)	STONE ARTEFACTS (g)	OTHER STONE (g)
1	2.2	2.2	0.1	0.2	0	0	0.3	0	0.1
2	53.8	51.6	17.8	5.8	0	0	6.7	0.1	1.9
3	97.8	44.0	14.3	7.1	0	0	2.8	0	1.1
4	147.8	50.0	16.8	16.5	0	0	5.3	0	2.1
5	180.6	32.8	11.1	4.4	0	0	4.3	0	1.3
6	210.0	29.4	9.9	41.7	0	0	5.1	*	26.5
7	242.4	32.4	11.1	128.8	0	0	4.4	0	4.5
8	261.4	19.0	7.4	50.0	0	0	13.3	0	2.3
9	297.0	35.6	11.0	131.2	0	0	10.7	0.2	1.8
10	336.0	39.0	11.9	139.5	0	0	13.4	0.1	2.7
11	365.2	29.2	11.2	*	0	*	16.0	0	3.1
12	401.0	35.8	11.8	*	0	0	146.8	0	2.9
13	438.0	37.0	12.2	9.0	0	0	14.0	0	0.7
14	467.2	29.2	10.2	0.7	0	0	8.9	0	0.2
15	530.4	63.2	21.5	0.9	0	0	56.8	0	1.7
16	591.2	60.8	20.6	*	0	0	23.2	0	2.6
17	647.8	56.6	19.8	0.2	0	0	5.2	0	4.0
18	675.6	27.8	9.0	*	0	0	1.5	0	5.5
Total	–	–	227.7	535.9	0	*	338.6	0.4	65.0

Table A4/11 Ironbark Site Complex, Square L. *= <0.1g.

XU	MAX. DEPTH (mm)	MEAN SIZE (mm)	WEIGHT (kg)	SHELL (g)	CRUSTACEAN (g)	BONE (g)	CHARCOAL (g)	STONE ARTEFACTS (g)	OTHER STONE (g)
1	1.0	1.0	0.1	0	0	0	0	0	*
2	51.4	50.4	16.3	3.9	0	0	2.4	68.5	68.4
3	109.4	58.0	17.7	0	0	0	7.4	124.9	275.0
4	136.2	26.8	8.8	0	0	0	6.3	1226.0	230.4
5	200.4	64.2	17.8	0	0	0	6.0	1014.3	756.1
6	263.0	62.6	17.0	0	0	0	2.7	562.0	796.7
7	317.4	54.4	18.0	0	0	0	2.5	3.5	278.1
8	359.2	41.8	8.8	0	0	0	0.7	2.4	32.9
9	393.4	34.2	7.3	0	0	0	0.4	7.7	19.3
10	447.6	54.2	8.4	0	0	0	0.7	45.9	18.2
11	515.4	67.8	9.0	0	0	0	0.4	0.1	11.0
12	608.2	92.8	10.4	0	0	0	1.1	0.9	21.3
13	682.4	74.2	3.5	0	0	0	2.4	2.2	31.3
Total	–	–	143.0	3.9	0	0	33.1	3058.4	2538.9

Table A4/12 Ironbark Site Complex, Square M. *= <0.1g.

XU	MAX. DEPTH (mm)	MEAN SIZE (mm)	WEIGHT (kg)	SHELL (g)	CRUSTACEAN (g)	BONE (g)	CHARCOAL (g)	STONE ARTEFACTS (g)	OTHER STONE (g)
1	0.8	0.8	0	0	0	0	0	0	0
2	26.6	25.8	14.3	0	0	0	1.7	1206.3	53.5
3	53.8	27.2	8.4	0	0	0	2.6	108.2	77.7
4	104.8	51.0	14.3	0	0	0	0.2	1786.5	597.0
5	138.6	33.8	10.7	0	0	0	2.8	1124.2	165.9
6	167.2	28.6	8.7	0	0	0	2.8	220.2	112.6
7	207.2	40.0	19.9	0	0	0	3.6	1878.9	157.4
8	227.8	20.6	8.3	0	0	0	1.4	559.3	107.1
9	281.4	53.6	14.4	0	0	0	0.1	2037.5	193.2
10	326.2	44.8	8.9	0	0	0	2.1	3.7	72.5
11	370.0	43.8	9.0	0	0	0	0.9	1.1	40.8
12	414.0	44.0	8.0	0	0	0	1.3	1.1	15.5
13	452.8	38.8	7.5	0	0	0	0.6	1.1	6.8
14	494.6	41.8	8.0	0	0	0	0.5	0.6	8.0
15	533.8	39.2	8.0	0	0	0	0.3	0.5	11.0
16	600.2	66.4	7.5	0	0	0	1.1	3.9	11.0
17	681.0	80.8	6.5	0	0	0	0	8.0	42.9
Total	–	–	162.3	0	0	0	21.9	8941.1	1672.9

Table A4/13 Ironbark Site Complex, Square N. *= <0.1g.

XU	MAX. DEPTH (mm)	MEAN SIZE (mm)	WEIGHT (kg)	SHELL (g)	CRUSTACEAN (g)	BONE (g)	CHARCOAL (g)	STONE ARTEFACTS (g)	OTHER STONE (g)
1	15.0	15.0	0	5.2	0	0	0.2	0	101.3
2	41.0	26.0	8.2	560.1	0.4	0	2.4	2.8	3835.8
3	69.2	28.2	9.0	338.0	*	0	2.2	2.3	7187.6
4	101.4	32.2	6.1	30.4	0.2	0	1.1	0.1	4074.7
5	130.4	29.0	9.4	6.2	0	*	0.9	0.9	7429.5
6	187.4	57.0	14.0	0.8	0	0	0.8	0	15111.9
7	230.2	42.8	11.3	0.1	0	0	0.5	0	5890.5
8	277.0	46.8	12.6	0.1	0	0	0.7	0	8546.3
Total	–	–	70.6	940.8	0.7	*	8.8	6.0	52177.6

Table A4/14 Ironbark Site Complex, Square O. *= <0.1g.

XU	MAX. DEPTH (mm)	MEAN SIZE (mm)	WEIGHT (kg)	SHELL (g)	CRUSTACEAN (g)	BONE (g)	CHARCOAL (g)	STONE ARTEFACTS (g)	OTHER STONE (g)
1	8.2	8.2	0.8	0.1	0	0	0.2	0.1	2.1
2	39.0	30.8	11.2	10.3	0	0	1.1	*	4.4
3	83.6	44.6	15.4	14.2	2.4	0	2.2	0.1	2.5
4	114.0	30.4	10.8	73.4	19.6	0	2.1	0.1	5.0
5	144.4	30.4	11.0	24.5	0	0	4.0	0.1	2.2
6	180.8	36.4	12.2	29.0	0	0.3	12.3	*	5.9
7	219.4	38.6	12.2	14.5	0	0	8.3	0.2	7.3
8	264.8	45.4	17.1	57.2	0	0.3	15.6	0.2	9.3
9	322.8	58.0	18.3	122.0	0	0	3.4	0.8	24.5
10	356.6	33.8	10.9	0.1	0	0	0.8	0.1	35.7
11	385.6	29.0	10.8	0.2	0	0	1.4	0.5	59.6
12	458.8	73.2	24.0	*	0	0	2.1	0.8	27.6
13	527.2	68.4	24.8	0.3	0	0	1.1	0.3	9.8
Total	–	–	179.5	345.6	22.0	0.6	54.4	3.3	195.9

Table A4/15 Ironbark Site Complex, Square P. *= <0.1g.

XU	MAX. DEPTH (mm)	MEAN SIZE (mm)	WEIGHT (kg)	SHELL (g)	CRUSTACEAN (g)	BONE (g)	CHARCOAL (g)	STONE ARTEFACTS (g)	OTHER STONE (g)
1	6.0	6.0	0.4	3.3	0	0	0.1	0	0.5
2	38.6	32.6	9.1	3.2	0.1	0	1.6	0	0.7
3	74.0	35.4	12.0	9.6	0.5	0	2.6	0	1.9
4	105.0	31.0	10.9	47.6	0	0	1.6	0.1	3.1
5	134.0	29.0	10.1	33.5	*	0	1.9	0.1	1.3
6	159.8	25.8	9.5	15.9	0	0.2	9.1	*	2.9
7	184.0	24.2	8.5	51.0	0	0	9.4	*	4.0
8	215.0	31.0	10.0	15.0	0	0	7.1	*	4.7
9	241.2	26.2	9.0	62.8	0.5	0	17.5	0.6	7.3
10	270.0	28.8	9.7	73.4	0	0	18.2	0.3	6.1
11	302.0	32.0	11.0	18.8	0	0	3.6	0.3	20.9
12	333.2	31.2	9.8	0.2	0	0	1.2	0.4	13.6
13	358.2	25.0	9.0	0.1	0	0	1.0	31.2	19.6
14	382.8	24.6	8.5	0	0	0	0.9	18.0	99.9
15	412.0	29.2	9.7	0	0	0	0.6	0	65.2
16	442.6	30.6	12.0	0	0	0	0.5	0.1	12.1
17	493.8	51.2	17.7	0	0	0	0.3	*	5.5
18	581.8	88.0	32.1	0	0	0	1.0	*	4.7
19	656.6	74.8	35.5	0	0	0	0.6	0	1.8
Total	–	–	234.5	334.4	1.1	0.2	78.9	51.3	275.9

Table A4/16 Ironbark Site Complex, Square Q. *= <0.1g.

XU	MAX. DEPTH (mm)	MEAN SIZE (mm)	WEIGHT (kg)	SHELL (g)	CRUSTACEAN (g)	BONE (g)	CHARCOAL (g)	STONE ARTEFACTS (g)	OTHER STONE (g)
1	6.8	6.8	0	0	0	0	0	0	0
2	36.8	30.0	8.0	0.1	0	0	0.4	0	0.9
3	71.4	34.6	11.0	12.2	0	0	0.6	*	1.0
4	103.2	31.8	9.0	31.6	0	0	0.3	30.4	2.2
5	132.4	29.2	9.4	67.3	0	0	1.0	*	2.5
6	168.0	35.6	10.6	33.1	0	0	1.9	184.3	3.0
7	197.8	29.8	10.6	21.4	0	0	3.0	0.1	5.5
8	252.6	54.8	15.9	6.1	0	0	1.7	0.4	4.4
9	274.8	22.2	7.7	0.6	0	0	1.1	0.6	2.6
10	306.8	32.0	12.3	0.1	0	0	2.1	*	28.2
11	344.2	37.4	11.5	0.1	0	0	1.2	*	1.9
12	376.4	32.2	11.0	0	0	0	3.7	*	4.1
13	421.8	45.4	14.7	0.1	0	0	1.4	0.2	6.1
14	458.6	36.8	11.7	0	0	0	0.4	0	9.0
15	516.0	57.4	19.4	0.2	0	0	0.6	0.1	2.0
16	578.8	62.8	20.4	0	0	0	1.1	0	6.7
17	670.6	91.8	30.3	0.1	0	0	0.7	0	0.3
Total	–	–	213.6	173.0	0	0	21.1	216.2	80.4

Table A4/17 Ironbark Site Complex, Square R. *= <0.1g.

XU	MAX. DEPTH (mm)	MEAN SIZE (mm)	WEIGHT (kg)	SHELL (g)	CRUSTACEAN (g)	BONE (g)	CHARCOAL (g)	STONE ARTEFACTS (g)	OTHER STONE (g)
1	1.2	1.2	0	0.3	0	0	0	0	0
2	17.2	16.0	3.7	0.1	0	0	0.1	0	0.2
3	40.6	23.8	6.4	0.2	0	0	0.4	0	0.5
4	74.4	33.8	11.9	8.2	0	0	0.7	0	1.2
5	102.8	28.4	7.1	17.2	0	0	0.5	0.1	0.7
6	131.0	28.2	9.3	5.1	0	0	1.0	0.1	1.5
7	154.8	23.8	8.8	4.1	0	0	2.7	0.5	2.3
8	177.4	22.6	7.6	0.9	0	0	2.1	0.1	0.8
9	203.6	26.2	8.6	6.3	0	0	7.1	*	1.1
10	237.6	34.0	11.8	0	0	0	3.9	0	1.8
11	268.2	30.6	9.8	0.1	0	0	0.8	0	1.2
12	297.2	29.0	8.8	4.0	0	0	0.7	0.1	1.6
13	337.6	40.4	14.8	0	0	0	1.2	0	2.1
14	367.4	29.8	10.1	0	0	0	0.9	0	4.8
15	394.6	27.2	8.8	0	0	0	0.6	0.2	3.3
16	450.0	55.4	20.2	0	0	0	0.8	0	7.8
17	499.8	49.8	19.0	0	0	0	0	0	4.0
18	548.0	48.2	16.0	0	0	0	0.5	0	2.8
19	621.8	73.8	25.2	0	0	0	0.3	0	1.4
20	662.4	40.6	14.5	0	0	0	0.1	0	0.2
Total	–	–	221.9	46.5	0	0	24.3	1.2	39.3

Table A4/18 Eurimbula Creek 1, Square A. *= <0.1g.

XU	MAX. DEPTH (mm)	MEAN SIZE (mm)	WEIGHT (kg)	SHELL (g)	CRUSTACEAN (g)	BONE (g)	CHARCOAL (g)	STONE ARTEFACTS (g)	OTHER STONE (g)
1	34.0	34.0	10.7	179.1	0.4	*	1.1	0	14.0
2	60.2	26.2	10.0	51.2	0	0	1.5	0	9.1
3	84.6	24.4	11.4	14.8	0	0	1.1	0	14.4
4	110.8	26.2	9.6	0.2	0	0	0.3	0	2.7
5	136.6	25.8	10.4	0.1	0	*	0.2	0	0.7
Total	–	–	52.1	245.3	0.4	*	4.1	0	40.9

Table A4/19 Eurimbula Creek 1, Square B. *= <0.1g.

XU	MAX. DEPTH (mm)	MEAN SIZE (mm)	WEIGHT (kg)	SHELL (g)	CRUSTACEAN (g)	BONE (g)	CHARCOAL (g)	STONE ARTEFACTS (g)	OTHER STONE (g)
1	5.2	5.2	1.0	*	0	0	0	0	0
2	115.0	109.8	8.3	27.9	0	0	2.7	0	26.8
3	144.2	29.2	2.9	65.1	0	0	1.4	0	19.3
4	162.2	18.0	2.2	194.8	0	*	0.7	0	5.4
5	191.4	29.2	3.4	85.9	0	0	1.0	0	4.4
6	251.0	59.6	13.5	120.4	0	0.1	4.4	0	37.5
7	287.2	36.2	10.7	5.3	0	0	1.2	0	43.4
8	328.2	41.0	11.5	0.4	0	0	0.6	0	4.5
9	370.0	41.8	12.3	*	0	0	0.4	0	8.3
10	396.2	26.2	8.3	*	0	0	0.2	0	3.1
11	419.4	23.2	7.4	0	0	0	0.1	0	2.9
Total	–	–	81.5	499.8	0	0.1	12.7	0	155.7

Table A4/20 Eurimbula Creek 1, Square C. *= <0.1g.

XU	MAX. DEPTH (mm)	MEAN SIZE (mm)	WEIGHT (kg)	SHELL (g)	CRUSTACEAN (g)	BONE (g)	CHARCOAL (g)	STONE ARTEFACTS (g)	OTHER STONE (g)
1	1.8	1.8	0.3	0	0	0	0.1	0	0.4
2	30.6	28.8	5.3	1.2	0	0	0.5	0	2.8
3	74.8	44.2	12.9	13.2	0	0	1.6	0	30.4
4	115.6	40.8	12.3	12.7	0	0	2.0	0	35.2
5	149.2	33.6	10.7	41.2	0	0	2.9	0	54.5
6	183.0	33.8	10.1	76.6	0	*	3.4	0	49.9
7	218.4	35.4	10.9	80.2	0	0	5.4	0	28.3
8	248.6	30.2	9.7	45.0	0	0	2.3	0	97.9
9	279.6	31.0	10.4	15.4	0	0	0.8	0	20.0
10	314.2	34.6	10.7	2.2	0	0	1.4	0	21.2
11	337.0	22.8	9.7	0.2	0	0	1.2	0	15.7
12	369.0	32.0	11.2	0.2	0	0	1.2	0	8.4
13	393.2	24.2	10.0	0	0	0	0.4	0	8.2
14	424.0	30.8	11.4	0	0	0	0.5	0	9.1
15	468.2	44.2	16.5	*	0	0	0.2	0	7.1
Total	–	–	152.1	288.2	0	*	23.9	0	389.2

Table A4/21 Eurimbula Creek 1, Square D. *= <0.1g.

XU	MAX. DEPTH (mm)	MEAN SIZE (mm)	WEIGHT (kg)	SHELL (g)	CRUSTACEAN (g)	BONE (g)	CHARCOAL (g)	STONE ARTEFACTS (g)	OTHER STONE (g)
1	4.0	4.0	0.4	0	0	0	0.1	0	0
2	40.8	36.8	4.5	1.8	0	0	0.2	0	1.5
3	88.6	47.8	13.4	9.6	0	0	1.6	0	30.0
4	127.2	38.6	11.6	29.9	0	0	1.8	0	43.0
5	168.6	41.4	12.7	96.6	0	0	3.5	0	63.0
6	201.0	32.4	11.5	67.6	0	0	4.9	0	61.1
7	233.6	32.6	10.2	33.6	0	0	8.1	0	41.3
8	264.0	30.4	10.9	22.3	0	0	3.7	0	47.4
9	286.6	22.6	7.9	6.5	0	*	3.7	0	23.1
10	320.4	33.8	10.7	28.3	0	0	2.7	0	31.2
11	349.8	29.4	10.4	0.6	0	0	1.9	0	30.6
12	375.8	26.0	8.7	0	0	0	0.9	0	14.4
13	404.2	28.4	9.2	0	0	0	0.6	0	8.3
14	437.2	33.0	12.1	0	0	0	0.5	0	18.1
15	463.6	26.4	9.5	0.1	0	0	0.5	0	7.7
16	516.6	53.0	18.8	0	0	0	1.0	0	8.2
Total	–	–	162.5	296.9	0	*	35.8	0	429.0

Table A4/22 Eurimbula Creek 2, Square A. *= <0.1g.

XU	MAX. DEPTH (mm)	MEAN SIZE (mm)	WEIGHT (kg)	SHELL (g)	CRUSTACEAN (g)	BONE (g)	CHARCOAL (g)	STONE ARTEFACTS (g)	OTHER STONE (g)
1	5.2	5.2	0.6	4.7	0	0	0.3	0	0
2	34.4	29.2	9.7	44.7	0	0	1.1	0	0.2
3	64.2	29.8	8.2	12.2	0	0	1.0	0	0
4	88.8	24.6	9.4	28.2	0	0	2.5	0	0.1
5	131.0	42.2	14.5	33.4	0	0	6.5	0	1.1
6	163.2	32.2	10.8	31.5	0	0	2.8	0	0.2
7	194.8	31.6	11.2	3.9	0	0	1.0	0	0.2
8	227.4	32.6	11.9	17.4	0	0	0.5	0	0.1
9	257.0	29.6	11.6	18.0	0	0	0.4	0	2.1
10	276.8	19.8	9.7	1.9	0	0	0.2	0	1.2
11	308.0	31.2	8.9	1.2	0	0	0.1	0	0.1
12	339.8	31.8	13.1	0	0	0	0.1	0	1.1
13	375.2	35.4	14.2	0	0	0	0.1	0	1.3
14	405.2	30.0	10.6	0	0	0	0.1	0	0.3
15	456.8	51.6	22.8	4.3	0	0	0.4	0	2.6
Total	–	–	167.2	201.3	0	0	20.9	0	10.5

Table A4/23 Eurimbula Site 1, Square A. *= <0.1g.

XU	MAX. DEPTH (mm)	MEAN SIZE (mm)	WEIGHT (kg)	SHELL (g)	CRUSTACEAN (g)	BONE (g)	CHARCOAL (g)	STONE ARTEFACTS (g)	OTHER STONE (g)
1	2.0	2.0	0.2	0.4	0	0	0.2	0	0
2	32.2	30.2	7.4	43.9	0	0	2.8	0	0.5
3	59.8	27.6	8.5	63.7	0	0.1	5.8	0	0.5
4	96.8	37.0	12.9	347.9	0	0.3	6.4	0.1	0.7
5	124.4	27.6	10.1	475.1	0	1.6	8.2	0.1	0.8
6	157.0	32.6	10.2	200.7	0	2.2	7.1	0	0.4
7	183.8	26.8	10.4	74.7	0	0.8	6.4	0.2	0.9
8	218.6	34.8	10.2	37.5	0	0.4	6.6	0	0.6
9	247.8	29.2	10.6	29.3	0	0.2	4.8	0	0.7
10	277.6	29.8	9.9	31.1	0	0.3	6.0	*	5.2
11	305.8	28.2	9.9	51.2	0	0.1	5.0	0	0.5
12	335.0	29.2	9.6	22.4	0	0.1	4.7	0	0.5
13	362.6	27.6	11.9	56.8	0	0.1	7.7	0.1	0.5
14	396.0	33.4	11.0	9.3	0	0.1	7.5	0	0.5
15	415.8	19.8	8.6	1.6	0	*	4.2	0.1	0.5
16	436.6	20.8	6.5	6.8	0	0.1	2.6	0.1	0.1
17	466.4	29.8	9.8	3.8	0	0	0.7	1.8	0.4
18	497.0	30.6	10.1	2.0	0	0	2.9	0.2	0.3
19	525.8	28.8	10.6	0.1	0	0	2.3	0.2	0.2
20	555.6	29.8	9.9	0.1	0	0	2.5	0	0.2
21	585.4	29.8	9.8	0.3	0	*	1.7	0	0.8
22	616.2	30.8	9.9	0.4	0	*	1.3	0	1.1
23	648.2	32.0	9.3	0	0	0	1.7	0	0.7
24	711.0	62.8	25.4	0.4	0	*	7.8	0.1	0.6
25	776.0	65.0	21.9	0	0	0	11.6	0	0.8
Total	–	–	264.3	1459.6	0	6.6	118.6	2.9	18.0

Table A4/24 Eurimbula Site 1, Square B. *= <0.1g.

XU	MAX. DEPTH (mm)	MEAN SIZE (mm)	WEIGHT (kg)	SHELL (g)	CRUSTACEAN (g)	BONE (g)	CHARCOAL (g)	STONE ARTEFACTS (g)	OTHER STONE (g)
1	3.4	3.4	0.2	0.1	0	0	*	0	*
2	44.6	41.2	9.9	94.2	0	*	10.6	0.2	0.3
3	75.0	30.4	9.0	587.0	0	0.7	5.8	0	0.5
4	104.0	29.0	9.1	474.4	0	4.0	10.8	0.3	1.4
5	138.6	34.6	9.2	130.6	0	0.8	11.8	0	0.5
6	171.6	33.0	10.7	54.9	0	0.6	11.3	7.4	0.4
7	194.8	23.2	9.9	30.3	0	0.4	7.0	0	0.6
8	237.0	42.2	11.0	54.1	0	*	6.9	0	0.5
9	273.8	36.8	12.8	34.7	0	0.1	10.0	0.8	0.7
10	304.6	30.8	9.8	48.6	0	0.4	5.7	0	0.5
11	344.0	39.4	11.7	252.9	0	0.1	7.9	0.1	5.5
12	380.4	36.4	11.0	289.4	0	*	7.5	0.1	1.0
13	414.8	34.4	11.2	26.6	0	*	8.8	0.4	1.2
14	453.6	38.8	12.0	1.8	0	0.4	11.6	0	0.9
15	461.6	8.0	2.0	3.2	0	0	0.9	0	0.1
16	488.0	26.4	8.3	1.8	0	*	4.2	0	11.3
17	513.0	25.0	9.3	0.4	0	0.1	3.1	0.2	35.8
18	542.4	29.4	11.3	1.0	0	0	1.4	1.9	0.3
19	575.6	33.2	13.3	*	0	*	1.6	14.3	0.7
20	613.0	37.4	13.1	0.3	0	*	1.1	0	0.5
21	642.8	29.8	10.9	*	0	*	0.6	0	0.5
22	673.0	30.2	10.4	0.1	0	0	0.4	0	0.2
23	734.8	61.8	22.3	1.5	0	0	0.6	0.1	1.3
24	804.8	70.0	27.7	8.4	0	0	4.0	0	0.7
Total	–	–	266.1	2096.4	0	7.6	133.8	25.9	65.4

Table A4/25 Eurimbula Site 1, Square C. *= <0.1g.

XU	MAX. DEPTH (mm)	MEAN SIZE (mm)	WEIGHT (kg)	SHELL (g)	CRUSTACEAN (g)	BONE (g)	CHARCOAL (g)	STONE ARTEFACTS (g)	OTHER STONE (g)
1	2.2	2.2	0.3	2.5	0	0	0.3	0	*
2	29.6	27.4	8.9	73.0	0	0.1	3.4	0	0.3
3	53.2	23.6	7.5	92.8	0	0.1	0	1.2	0.6
4	75.8	22.6	7.8	517.8	0	0.3	4.1	0.1	0.7
5	106.8	31.0	9.6	501.5	0	2.0	12.0	0.4	2.6
6	134.0	27.2	9.9	122.0	0	4.0	9.9	0.5	1.6
7	167.2	33.2	10.0	112.5	0	1.2	7.8	0.3	51.0
8	193.8	26.6	10.4	72.9	0	0.5	3.6	0	1.2
9	226.4	32.6	10.2	111.7	0	0.3	3.9	0	0.4
10	255.4	29.0	12.1	87.4	0	0.1	5.3	0	0.7
11	287.0	31.6	10.8	45.9	0	0.1	3.5	0.1	0.7
12	316.6	29.6	9.9	14.1	0	*	3.6	0.1	0.4
13	348.6	32.0	11.7	54.8	0	*	3.6	0	0.8
14	377.0	28.4	11.4	86.3	0	*	3.5	0	0.4
15	417.8	40.8	11.1	20.9	0	0	5.4	0.1	0.3
16	437.4	19.6	11.8	2.1	0	*	5.8	0.1	0.5
17	469.0	31.6	13.0	0.7	0	0	6.5	0.3	0.4
18	489.0	20.0	6.8	0	0	0	2.7	0	0.1
19	518.4	29.4	9.9	0	0	0	3.6	1.2	0.2
20	541.4	23.0	8.7	0	0	*	2.2	0	0.3
21	565.6	24.2	9.3	0	0	*	1.4	0	0.2
22	595.2	29.6	10.7	0	0	0	1.0	0.4	0.6
23	628.2	33.0	11.2	0	0	*	0.7	0	0.5
24	658.2	30.0	11.3	0	0	0	0.5	0	0.6
25	731.8	73.6	25.8	0	0	0	0.7	0	0.6
26	802.6	70.8	22.4	0	0	0	1.0	0	0.5
Total	–	–	282.5	1919.3	0	8.7	96.1	4.6	66.3

Table A4/26 Eurimbula Site 1, Square D. *= <0.1g.

XU	MAX. DEPTH (mm)	MEAN SIZE (mm)	WEIGHT (kg)	SHELL (g)	CRUSTACEAN (g)	BONE (g)	CHARCOAL (g)	STONE ARTEFACTS (g)	OTHER STONE (g)
1	5.6	5.6	0.3	1.0	0	0	0.1	0	0
2	9.8	4.2	8.6	39.3	0	*	2.6	0.1	0.4
3	70.8	61.0	9.3	91.2	0	*	9.3	0	0.6
4	99.0	28.2	8.4	322.3	0	0.3	5.1	0.1	0.9
5	136.2	37.2	10.4	309.5	0	2.0	9.8	0	4.3
6	171.8	35.6	10.2	99.4	0	0.9	12.0	0	0.8
7	203.4	31.6	9.0	160.5	0	1.4	10.6	0	32.1
8	239.8	36.4	11.3	38.0	0	*	7.9	1.1	0.6
9	274.0	34.2	10.8	74.8	0	0.2	4.9	0.8	0.6
10	310.2	36.2	10.9	16.8	0	*	7.6	0	1.0
11	345.2	35.0	11.9	14.3	0	0.2	5.0	0.1	0.2
12	387.0	41.8	13.2	19.4	0	0.1	7.2	51.0	1.4
13	420.0	33.0	11.4	0.5	0	*	5.3	78.4	8.9
14	455.0	35.0	10.6	0.1	0	0	6.9	0	0.6
15	479.2	24.2	7.9	2.2	0	*	6.7	0	0.2
16	509.6	30.4	10.7	2.1	0	0	3.6	0.1	0.4
17	538.6	29.0	11.5	0.2	0	0	3.3	0	1.3
18	567.2	28.6	9.8	0.3	0	0	2.4	0	0.5
19	599.2	32.0	10.9	*	0	0	3.2	1.1	0.9
20	629.0	29.8	10.6	0	0	0	4.2	0	0.3
21	661.8	32.8	12.8	0	0	0	8.8	0	0.3
22	731.2	69.4	24.7	0.1	0	0	51.8	0	1.1
23	797.0	65.8	21.5	0	0	0	104.4	0	0.5
Total	–	–	256.6	1192.0	0	5.1	282.9	132.8	58.0

Table A4/27 Tom's Creek Site Complex, Square A. *= <0.1g.

XU	MAX. DEPTH (mm)	MEAN SIZE (mm)	WEIGHT (kg)	SHELL (g)	CRUSTACEAN (g)	BONE (g)	CHARCOAL (g)	STONE ARTEFACTS (g)	OTHER STONE (g)
1	1.8	1.8	0.1	0.3	0	0	0.2	0	*
2	49.4	47.6	15.3	201.2	0	0	7.7	0.2	127.4
3	91.4	42.0	12.0	264.1	0	0.3	11.6	7.7	5.4
4	120.2	28.8	9.5	139.6	0	0.1	7.4	0.8	42.3
5	156.0	35.8	11.0	120.6	0	0	5.2	0.1	12.8
6	188.8	32.8	9.9	187.8	0	0.1	5.3	1.9	0.9
7	238.4	49.6	12.5	337.1	4.2	1.1	6.9	0.8	12.3
8	275.4	37.0	11.5	551.6	0	4.0	7.2	5.4	99.4
9	317.8	42.4	12.5	483.4	0	1.9	6.0	17.7	2.1
10	355.6	37.8	12.0	256.4	0	0.6	4.5	1.2	0.9
11	399.0	43.4	12.1	253.4	0	0.6	6.4	0.5	14.0
12	433.2	34.2	11.4	285.9	0	1.6	12.3	1.9	14.9
13	468.6	35.4	10.4	77.7	0	1.8	7.9	0.8	0.5
14	501.8	33.2	11.8	41.4	0	0.6	5.6	0.6	100.5
15	536.6	34.8	11.3	52.2	0	0.1	6.1	0.2	0.7
16	564.0	27.4	10.8	19.9	0	*	5.3	0.6	1.2
17	599.6	35.6	12.1	0.8	0	0.1	3.9	0.8	3.4
18	635.6	36.0	11.6	2.5	0	0.1	7.3	0.1	3.0
19	678.4	42.8	12.8	0.1	0	0.1	0.9	2.8	2.1
20	709.4	31.0	11.8	0.1	0	0	1.4	0.2	0.6
21	746.8	37.4	11.0	0.1	0	0.1	1.7	0.1	1.1
22	826.2	79.4	25.7	0.1	0	0	1.2	0	9.8
23	895.0	68.8	21.0	0.1	0	*	0.5	*	1.1
Total	–	–	280.1	3276.5	4.2	13.1	122.7	44.4	456.4

Table A4/28 Tom's Creek Site Complex, Square B. *= <0.1g.

XU	MAX. DEPTH (mm)	MEAN SIZE (mm)	WEIGHT (kg)	SHELL (g)	CRUSTACEAN (g)	BONE (g)	CHARCOAL (g)	STONE ARTEFACTS (g)	OTHER STONE (g)
1	1.2	1.2	0.2	45.5	0	0	0.3	0	0
2	29.4	28.2	9.8	161.3	0	0.6	5.6	10.3	4.7
3	65.0	35.6	10.7	236.6	0.3	0.5	8.9	3.8	2.3
4	99.6	34.6	10.4	286.2	0	0.7	8.1	0.9	2.5
5	137.2	37.6	11.7	258.7	0	0.4	8.1	0.7	35.7
6	179.2	42.0	10.6	198.0	0	0.4	7.0	1.8	4.1
7	209.4	30.2	11.9	248.7	0	0.4	7.2	0.5	0.6
8	241.2	31.8	11.2	333.6	0	0.6	6.6	5.3	9.0
9	281.6	40.4	11.2	445.3	1.1	0.6	5.6	9.2	24.0
10	323.8	42.2	13.5	177.4	0.7	0.5	5.6	0.8	3.1
11	360.0	36.2	12.5	163.9	0.3	1.1	7.3	1.2	3.2
12	397.4	37.4	11.9	214.8	0.5	1.5	17.4	16.5	0.9
13	434.6	37.2	11.8	126.8	0	1.8	10.3	4.1	3.0
14	472.4	37.8	12.0	34.9	0	0.4	7.7	0	22.4
15	515.4	43.0	13.5	1.2	0	0.1	5.1	0.5	0.9
16	546.0	30.6	12.1	2.8	0	0.2	2.2	4.1	0.9
17	592.2	46.2	13.8	3.5	0	*	1.8	0.4	7.5
18	616.8	24.6	9.9	0.3	0	*	1.3	0.1	5.3
19	658.4	41.6	14.5	4.0	0	*	1.5	0.4	2.5
20	691.6	33.2	12.4	0.1	0	*	0.9	*	0.8
21	728.0	36.4	11.4	0.1	0	0	0.6	*	123.6
22	756.4	28.4	11.5	*	0	0	0.4	0.1	142.1
23	847.0	90.6	26.9	0.1	0	*	0.7	0	259.7
24	905.4	58.4	20.8	0.1	0	0	0.4	0	0.6
Total	–	–	296.2	2944.1	2.9	9.9	120.6	60.6	659.4

Table A4/29 Tom's Creek Site Complex, Square C. *= <0.1g.

XU	MAX. DEPTH (mm)	MEAN SIZE (mm)	WEIGHT (kg)	SHELL (g)	CRUSTACEAN (g)	BONE (g)	CHARCOAL (g)	STONE ARTEFACTS (g)	OTHER STONE (g)
1	0.6	0.6	0.1	6.8	0	0.1	0.6	0	0
2	58.6	58.0	19.3	366.9	0	1.0	0	5.2	13.7
3	86.2	27.6	9.1	170.6	0	0.3	9.1	0.7	52.5
4	125.4	39.2	11.9	181.3	0	1.3	10.6	3.2	31.4
5	167.0	41.6	14.8	371.7	0	1.0	8.6	4.8	5.0
6	199.8	32.8	11.4	215.9	0	0.4	5.8	0.4	1.6
7	229.6	29.8	11.8	223.6	0	0.6	6.0	1.0	1.3
8	274.2	44.6	11.7	130.3	0	0.7	5.4	2.4	1.3
9	302.8	28.6	11.0	124.1	0	0.2	4.2	0.2	0.5
10	332.2	29.4	12.0	161.0	0.2	0.6	7.7	0.6	1.3
11	361.0	28.8	9.7	262.5	0.4	0.8	7.6	0.2	4.0
12	392.6	31.6	11.6	143.3	0	1.1	14.1	1.8	0.8
13	426.8	34.2	11.7	217.6	0	1.4	20.9	3.4	2.3
14	454.0	27.2	11.6	36.4	0	0.6	18.2	4.7	2.3
15	494.4	40.4	12.4	29.3	0	0.7	13.7	0.4	153.3
16	523.6	29.2	11.7	12.7	0	0.1	5.1	0.1	0.7
17	557.6	34.0	12.9	4.1	0	0.1	3.8	0.1	0.4
18	612.6	55.0	23.5	0.4	0	*	3.4	1.2	2.1
19	654.2	41.6	15.0	0.2	0	*	2.0	0	2.3
20	687.0	32.8	11.8	*	0	0	1.7	0	1.8
21	718.2	31.2	11.7	*	0	*	1.4	0	36.5
22	757.4	39.2	13.0	0	0	0	0.7	0	89.0
23	834.0	76.6	27.5	*	0	0	0.7	0	1.6
24	897.0	63.0	22.1	0	0	0	0.4	0	0.2
Total	–	–	319.3	2658.8	0.5	10.9	151.6	30.5	405.8

Table A4/30 Tom's Creek Site Complex, Square D. *= <0.1g.

XU	MAX. DEPTH (mm)	MEAN SIZE (mm)	WEIGHT (kg)	SHELL (g)	CRUSTACEAN (g)	BONE (g)	CHARCOAL (g)	STONE ARTEFACTS (g)	OTHER STONE (g)
1	2.8	2.8	0.1	0.1	0	0	*	0	0
2	44.6	41.8	12.3	224.4	0	0.3	14.3	1.9	0.6
3	74.4	29.8	9.8	245.6	0	0.1	20.3	0.1	1.9
4	105.2	30.8	10.2	124.8	0	0.1	3.5	0.9	8.2
5	135.8	30.6	10.2	196.4	0	0	5.5	0.2	3.6
6	184.8	49.0	14.5	325.6	0	0.5	7.7	0.9	2.9
7	222.4	37.6	12.0	230.8	0	0.7	6.0	0.4	0.7
8	255.2	32.8	11.7	210.7	0	1.1	4.9	4.5	1.1
9	293.8	38.6	12.2	156.1	0.1	0.5	4.7	0.3	2.3
10	334.4	40.6	13.2	14.6	0.2	0.3	5.1	0.3	12.5
11	364.0	29.6	11.5	103.5	0.4	0.8	5.8	0.1	27.1
12	397.0	33.0	11.0	123.6	0.1	0.4	9.1	0.4	0.8
13	441.2	44.2	14.3	222.7	0	2.2	15.8	3.2	1.9
14	478.8	37.6	12.0	79.3	0	1.6	15.7	42.8	51.3
15	522.8	44.0	14.5	58.6	0	0.8	11.2	2.0	0.8
16	557.0	34.2	11.9	8.0	0	0.1	4.9	0.7	1.0
17	595.0	38.0	11.1	0	0	0	3.2	9.5	0.7
18	640.2	45.2	16.9	1.6	0	0	3.3	1.2	6.9
19	677.4	37.2	11.3	*	0	*	1.2	*	29.2
20	709.6	32.2	11.7	0.1	0	0.1	3.2	0.2	24.4
21	747.0	37.4	13.6	0.1	0	0	1.6	0	3.2
22	824.0	77.0	24.9	0	0	0	1.4	0	12.0
23	883.8	59.8	19.2	*	0	0	0.6	0	7.4
Total	–	–	290.1	2426.5	0.8	9.6	149.1	69.6	200.7

Table A4/31 Tom's Creek Site Complex, Square R. *= <0.1g.

XU	MAX. DEPTH (mm)	MEAN SIZE (mm)	WEIGHT (kg)	SHELL (g)	CRUSTACEAN (g)	BONE (g)	CHARCOAL (g)	STONE ARTEFACTS (g)	OTHER STONE (g)
1	2.4	2.4	0.1	0.1	0	0	0.1	0	0
2	32.4	30.0	9.9	9.9	0	0	0.3	0.1	0.1
3	67.6	35.2	11.3	74.4	0	0	1.0	1.0	1.5
4	101.0	33.4	10.0	170.1	0	0.1	4.1	0.5	2.7
5	133.6	32.6	11.1	175.3	0	0	9.3	1.0	7.2
6	166.2	32.6	10.1	221.3	0	0	13.6	0.4	2.2
7	209.4	43.2	14.1	421.4	0	0	17.9	1.1	7.6
8	238.6	29.2	7.8	456.2	0	0.3	8.6	11.1	9.0
9	269.0	30.4	12.1	472.2	0	0.1	12.2	1.3	8.9
10	315.0	46.0	13.7	98.0	0	0	9.8	0.1	17.6
11	351.2	36.2	13.2	200.5	0	0	5.2	3.6	60.4
12	406.2	55.0	15.8	11.6	0	0	1.8	0.1	47.8
13	447.6	41.4	13.4	0.3	0	0	2.0	0	5.1
14	488.0	40.4	13.6	*	0	0	0.4	0	9.1
15	532.0	44.0	13.5	*	0	0	0.1	0	1.3
16	568.6	36.6	12.1	0.2	0	0	0.1	0	0.2
17	603.4	34.8	10.8	0	0	0	0.1	0.3	0.1
18	667.8	64.4	20.8	*	0	0	0.1	0.9	0.9
Total	–	–	213.4	2311.5	0	0.5	86.7	21.6	181.7

Table A4/32 Tom's Creek Site Complex, Square S. *= <0.1g.

XU	MAX. DEPTH (mm)	MEAN SIZE (mm)	WEIGHT (kg)	SHELL (g)	CRUSTACEAN (g)	BONE (g)	CHARCOAL (g)	STONE ARTEFACTS (g)	OTHER STONE (g)
1	2.6	2.6	0.1	0	0	0	0	0	0.8
2	39.0	36.4	10.6	50.6	0	0	0.6	0.1	0.5
3	88.8	49.8	13.1	93.5	0	0	10.5	0	0.7
4	124.6	35.8	10.0	179.8	0	0.4	3.3	0	2.5
5	164.0	39.4	10.5	262.9	0	0	11.4	0	3.2
6	202.2	38.2	10.2	331.1	0	*	14.2	0.2	6.7
7	205.2	3.0	1.9	13.2	0	0	0	*	43.6
8	239.6	34.4	9.7	736.1	0	0	13.3	4.1	132.0
9	280.8	41.2	12.1	451.3	0	0	18.1	10.7	10.8
10	317.0	36.2	11.6	44.8	0	*	12.9	0.3	9.8
11	350.2	33.2	11.1	292.9	0	*	6.2	0.1	135.1
12	386.2	36.0	11.8	15.5	0	0	2.0	0	43.9
13	419.2	33.0	11.0	1.6	0	0	0.3	*	1.1
14	459.2	40.0	13.7	1.9	0	0	0.2	0	5.7
15	497.6	38.4	13.1	0	0	0	0.1	0	0.4
16	532.2	34.6	11.3	0	0	0	*	0	0.1
17	572.2	40.0	12.3	0	0	0	0.1	0.1	3.8
18	608.4	36.2	12.4	0	0	0	*	*	0.6
19	671.6	63.2	21.8	0	0	0	0.1	0.7	0.4
Total	–	–	208.3	2475.2	0	0.4	93.2	16.2	401.8

Appendix 5: Shellfish reference collection*

FAMILY	SPECIES	COMMON NAME/S	PREFERRED ENVIRONMENT/SIZE
		MARINE BIVALVIA	
Anomiidae	*Anomia trigonopsis*	(Hutton, 1877) jingle shell	To 10m among shell debris; to 75mm
Arcidae	*Anadara trapezia*	(Deshayes, 1840) Sydney cockle; blood cockle; mud ark	Intertidal mangroves; estuarine tidal flats; seagrass beds; to 70mm
Cardiidae	*Acrosterigma vertebratum*		In muddy sand of intertidal flats
Carditidae	*Venericardia* sp.		In sand in shallow water
Chamidae	*Chama fibula*	(Reeve, 1846) spiny oyster	Attached to shell or coral debris to 10m; to 30mm
Corbulidae	*Corbula (Serracorbula) crassa*	(Reeve, 1843)	Sandy/muddy substrates; to 18mm
Donacidae	*Donax (Plebidonax) deltoides*	(Lamarck, 1818) pipi; eugarie; wong	Littoral sand; to 60mm
Mactridae	*Mactrid* sp.		Littoral sand
Mytilidae	*Trichomya hirsutus*	(Lamarck, 1819) hairy mussel	Tidal estuary; attached to rocks from low tide level to 16m; to 65mm
Noetiidae	*Arcopsis deliciosa*	(Iredale, 1939)	Rocky substrates to 81m; to 10mm
Noetiidae	*Arcopsis symmetrica*	(Reeve, 1844)	Rocky substrates; shallow water; to 16mm
Ostreidae	*Saccostrea glomerata* syn. *S. cuccullata* syn. *S. commercialis*	(Gould, 1850) Sydney rock oyster; rock oyster; commercial oyster	Sheltered rocky shores and mangroves; mid-intertidal; to 100mm
Pteriidae	*Pinctada albina sugillata*	(Reeve, 1857) pearl oyster	Attached to rocks and corals to 22m; to 110mm
Tellinidae	*Tellina* sp.		Littoral sand
Tellinidae	*Tellina (Cyclotellina) remies*	(Linnaeus, 1758)	Littoral sand; to 70mm
Trapeziidae	*Trapezium (Neotrapezium) sublaevigatum*	(Lamarck, 1819)	Littoral shell debris, coral crevices or in oyster clumps; 3–10m; to 65mm
Ungulinidae	*Felaniella (Zemysia) subglobosa* syn. *F. subglobosa*	(E.A. Smith, 1885)	Coral/mud to 13m; to 4.5mm
Veneridae	*Antigona chemnitzii*	(Hanley, 1844)	Littoral sand; to 100mm
Veneridae	*Dosinia tumida*	(Gray, 1838)	Littoral sand; to 58mm
Veneridae	*Gafrarium australe*	(Sowerby, 1850)	Intertidal, muddy sand; to 25mm
Veneridae	*Irus* sp.		Intertidal and subtidal sandy/rocky areas
Veneridae	*Placamen* sp.		Littoral sand
Veneridae	*Venerid* sp.		Littoral sand
		MARINE GASTROPODA	
Batillariidae	*Pyrazus ebininus*	(Bruguière, 1792) hercules club whelk	Mudflats/mangrove swamps; to 110mm
Batillariidae	*Velacumantus australis* syn. *Batillaria australis*	(Quoy & Gaimard, 1834) Australian mud whelk; mud creeper	Sandy estuarine substrates among algae/seagrass/mangroves; to 35mm
Cerithiidae	*Cerithiid* sp.		Sandy intertidal/shallow subtidal
Cerithiidae	*Cerithium* sp.	creeper	Intertidal/shallow subtidal in sandy areas
Cerithiidae	*Clypeomorus bifasciata*	(Sowerby, 1855)	Intertidal/shallow subtidal in sandy areas
Colubrariidae	*Colubraria maculosa*	(Gmelin, 1791) giant false triton	to 90mm
Columbellidae	*Zafra avicennia*	(Hedley, 1914)	On rocks or sand in shallow water; to 5mm
Conidae	*Conus* sp.		In sand in shallow water
Costellariidae	*Vexillum* sp.		Intertidal/subtidal sand/rock/coral

continued over

Appendix 5: continued

FAMILY	SPECIES	COMMON NAME/S	PREFERRED ENVIRONMENT/SIZE
Cypraeidae	*Cypraea* sp.	cowrie	Muddy rocks inshore
Ellobiidae	*Ophicardelus sulcatus*	(H. & A. Adams, 1855)	Interdial and above high tide mark on rocks in mudflat areas/mangrove swamps
Epitoniidae	*Epitonium* sp.		Among rocks on coral; subtidal in sand
Fasciolariidae	*Fasciolariid* sp.		In sand or coral
Fasciolariidae	*Latirus* sp.		Intertidal/subtidal sand to coral
Fissurellidae	*Diodora ticaonica*	(L.A. Reeve, 1850)	Intertidal/shallow subtidal on rocks; to 22mm
Lottiidae	*Acmaeid* sp.		On rocks in intertidal zone
Littorinidae	*Bembicium nanum*	(Lamarck, 1822) periwinkle	Intertidal, rocky reefs; to 12mm
Littorinidae	*Littoraria* sp.	periwinkle	On rocks or mangroves in intertidal zone
Mitridae	*Mitra* sp.		Subtidal in sand or mud
Muricidae	*Bedeva paivae* syn. *B. hanleyi*	(Crosse, 1864) oyster drill	Muddy habitats in lower intertidal/shallow subtidal zone; to 20mm
Muricidae	*Morula marginalba*	(Blainville, 1832) mulberry whelk	Intertidal and subtidal on rocky shores/rocky reefs often on oyster beds; to 30mm
Nassariidae	*Nassarius burchardi*	(Dunker in Philippi, 1849) dog whelk	Intertidal sand and mudflats; to 12mm
Nassariidae	*Nassarius pauperus*	(Gould, 1850)	Intertidal/shallow subtidal, sand flats; to 14mm
Naticidae	*Natica* sp.		In sand or mud
Neritidae	*Nerita balteata* syn. *N. lineata*	(Reeve, 1855) common nerite	On and in logs; on prop roots and on lower trunks of mangroves; to 40mm
Neritidae	*Nerita squamulata*	(Guillou, Le, 1841) variable nerite	Rock platforms; intertidal zone; to 35mm
Planaxidae	*Planaxis sulcatus*	(Born, I. von, 1778)	Intertidal on rocks; 18–35mm
Potamididae	*Telescopium telescopium*	(Linnaeus, 1758) telescope mud whelk	Mudflats/mangrove swamps; to 110mm
Skeneidae	*Pseudoliotia* sp.		Under rocks intertidal/shallow subtidal zones
Triphoridae	*Metaxia* sp.		Shallow to deep water, in sponges
Triphoridae	*Subulphora* sp.		
Trochidae	*Herpetopoma atrata* syn. *Echelus atratus* syn. *Euchelus atratus*	(Gmelin, 1791) beaded top shell	Intertidal rocky reefs; 15–20mm
Trochidae	*Thalotia* sp.		Intertidal, rocky reefs/shores
TERRESTRIAL GASTROPODA			
Camaenidae	*Figuladra* sp.		Coastal vine thicket
Camaenidae	*Trachiopsis mucosa*		
Pupillidae	*Pupoides pacificus*		Coastal vine thicket
Subulinidae	*Eremopeas tuckeri*	(Pfeiffer, 1846)	Coastal vine thicket in leaf litter
FRESHWATER BIVALVIA			
Corbiculidae	*Corbicula (Corbiculina) australis*	(Deshayes, 1830)	Coastal rivers and streams; to 20mm
Mutelidae	*Velesunio ambiguus*	(Philippi, 1847)	
Mutelidae	*Alathyria pertexta*	(Iredale, 1934)	

* Details after Coleman 1981; Lamprell and Healy 1998; Lamprell and Whitehead 1992; Wilson and Gillet 1979.

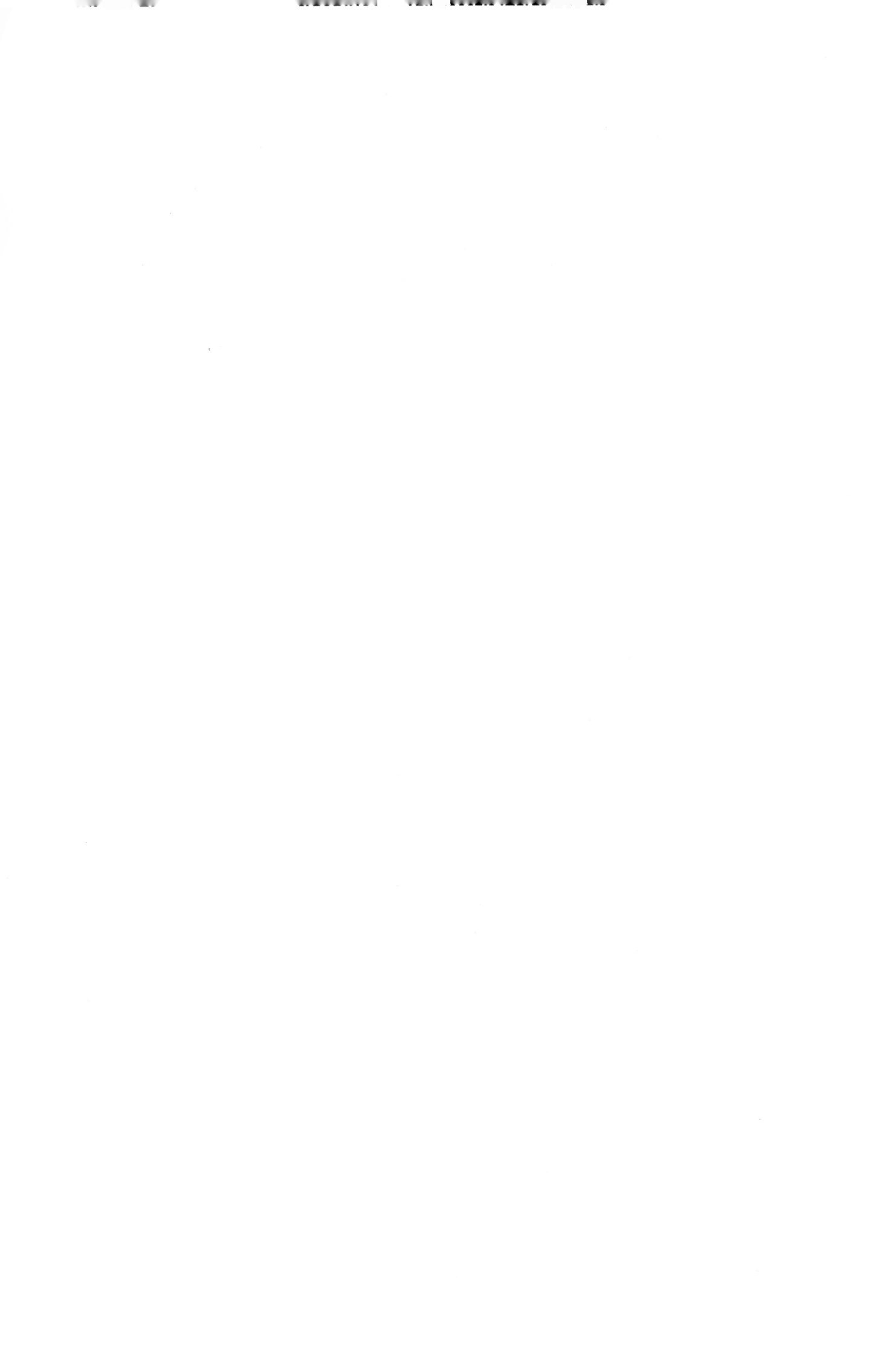

www.ingramcontent.com/pod-product-compliance
Lightning Source LLC
Chambersburg PA
CBHW051310270326
41929CB00034B/3455